E N.S. Dawe.   Leyland Gas
                  Turbines.
                  Meteor Works
                  Solihull,
                  Warwick.

Lincoln 2018

D1739673

Heat Transfer

Heat Transfer

F. J. BAYLEY, J. M. OWEN, and A. B. TURNER

University of Sussex

NELSON

THOMAS NELSON AND SONS LTD
36 Park Street London W1Y 4DE
PO Box 18123 Nairobi Kenya

THOMAS NELSON (AUSTRALIA) LTD
597 Little Collins Street Melbourne 3000

THOMAS NELSON AND SONS (CANADA) LTD
81 Curlew Drive Don Mills Ontario

THOMAS NELSON (NIGERIA) LTD
PO Box 336 Apapa Lagos

THOMAS NELSON AND SONS (SOUTH AFRICA) (PROPRIETARY) LTD
51 Commissioner Street Johannesburg

First published in Great Britain 1972

0 17 761609 1

Made and printed by offset in Great Britain by
William Clowes & Sons, Limited, London, Beccles and Colchester

Preface

The principles of heat transfer have a part to play in the design of components in most branches of modern technology. Consequently they need to appear in curricula designed for the education of engineers and applied scientists whose main interests may range from electrical engineering, with the need to cope with power dissipated as heat, through civil and structural engineering, where thermal stresses have often crucial effects, to thermal power engineering where heat transfer is the *sine qua non*.

The text has been written to present the principles of heat transfer as they appear in the syllabuses of modern engineering education. In each of the areas dealt with the matter is presented from an initial level which presupposes no prior knowledge of the subject. Thus it is intended that the book will prove suitable for the earlier years of an undergraduate course where heat transfer must take its place as a sub-division of thermo-fluid mechanics, just one of the eight or ten basic courses essential at this level to give a wide-ranging foundation for education in the applied sciences. Equally, by developing the principles of heat transfer from these basic beginnings to as near as is practicable in a rapidly developing technology the present frontiers of knowledge, we have made the text suitable for use in the specialist courses of the later undergraduate years, for postgraduates who intend advance study and research in this area, and for use by practising engineers and scientists faced with problems of heat transfer in their professional work.

In Chapter 1 the basic ideas and concepts of the general area of heat transfer are presented at an elementary level; they are then fully developed in the later specialist chapters.

We have chosen to present conduction and convection as separate modes of heat transfer, despite the approach sometimes adopted in recent years of presenting the former as simply a special case of heat transfer in a continuous medium at rest. In practice, the theoretical and experimental techniques adopted in analysing the two modes are quite different, as indeed this text shows. Thus conduction is dealt with at three distinct levels in separate chapters: at the elementary level in Chapter 2, where non-specialists will find treated the commonly occurring problems of everyday technological life; at the more advanced analytical level in Chapter 3, where mathematical procedures are applied to the idealized approximations of real problems for which they are suitable; and then, in Chapter 4 numerical techniques are dealt with in detail and at the level which has made possible, with the aid of

v

high-speed digital computers, the solution of realistic engineering problems involving the principles of heat transfer by conduction.

Convection, as befits a mode of heat transfer at least as common in engineering as conduction, and more often a controlling factor, is equally treated in three separate chapters. Unlike conduction, however, convection is essentially complex, and in the past simplicity of treatment for the non-specialist has often become confused with unsupported (and unsupportable) empiricism. In this text we have elected to move from the general to the particular in considering convection. The basic equations of fluid motion are derived and the significance of the terms considered in Chapter 5. The theoretical solutions of convection problems made possible by the boundary layer approximations to the general equations are presented in Chapter 6. The empirical correlations derived from experiment and theory are presented in Chapter 7 together with techniques of numerical analysis of convection problems which, with digital computers, enable existing knowledge of the nature of convecting fluid flows to be extrapolated to previously unexplored areas. An especial difficulty was faced in deciding upon the content of this chapter, for there exists a plethora of empirical data and approximate methods of analysis, particularly for the technologically important area of turbulent flows. Some selectivity was inevitable therefore in deciding upon the material for Chapter 7 and we have chosen to present experimental data and a technique of numerical analysis which we have found particularly useful in our own work.

Thermal radiation is dealt with in Chapter 8, beginning with the elementary basic ideas where common ground is shared with the particle physicists, but diverging quickly to the approximate and semi-empirical methods inevitable to the consideration of practical problems of engineering. Heat transfer with change of phase is covered in Chapter 9, equally inevitably on a largely empirical basis, for in this complex area even universally accepted correlations remain to be found, quite apart from generally applicable theories. Heat exchangers are considered in Chapter 10, mainly from the point of view of overall thermal design and without dwelling upon the detailed fluid dynamic and convection problems which are covered in the relevant specialist chapters of the book.

The choice of units to use in this book, as in engineering in general, has been difficult. The United Kingdom has decided to change to metric units and, in particular, to adopt the SI (*Système International d'Unités*) which is based on the *metre, kilogramme, second, degree Kelvin*, with the *newton* (N) as the unit of force and the *newton-metre*, or *joule* (J), as the unit of work, energy, and heat. We consider, however, that it is important to retain a facility in previously used systems of units, both because of their existence in literature which will need to be consulted continuously, but also because despite apparent inconveniences and illogicalities many of the earlier units show a remarkable convenience of scale. The numerical examples worked and set in this text therefore include approximately equal proportions in the SI metric system and in the BS or foot, pound mass (lb), pound-force (lbf), second, Fahrenheit degree system, with the British thermal unit (Btu) as the unit of heat, and Joule's equivalent 778 ft lbf/Btu. Occasionally also we have used the Centigrade heat unit (Chu) with Joule's equivalent 1,400 ft lbf/Chu, since this still finds considerable application in the power industry. In the British system, however, we have not included the Newtonian constant, g_c lb ft/lbf s^2, in algebraic equations unless its position demands special comment and those who prefer this system must pay special attention to dimensional consistency (as indeed, should all).

Finally, while the three authors have collaborated to produce this text and share the responsibility for the choice and manner of presentation of its material, they must acknowledge their debt to colleagues and associates, not only in the Mechanical Engineering Laboratories of the University of Sussex, but in the many other university, research, and industrial establishments with whom they have cooperated in studying and applying the principles of heat transfer.

F. J. B.
J. M. O.
A. B. T.

Contents

Notation

A	area; coefficient matrix; coefficient
A_q, A_τ	turbulent transport coefficients
A^+	van Driest constant
a	coefficient
a_∞	speed of sound
$\mathbf{B}$	Biot number (hL/k)
B	coefficient; constant
$\mathbf{b}$	vector
b	coefficient
$\mathbf{C_f}$	skin friction coefficient $(\tau_s/\frac{1}{2}\rho u^2)$
C	specific heat; coefficient; constant
C_0	conductance
c	mass concentration; coefficient; constant
c^*	mass fraction
D	diffusion coefficient; coefficient
d	diameter; coefficient
d_e	equivalent diameter
$\mathbf{E}$	Eckert number $\{[C_p(T_s - T_\infty)]/U_\infty^2\}$
E	error; potential; internal energy
E^*	law-of-the-wall constant
$\mathbf{e}$	unit vector
e	error; specific internal energy
$\mathbf{F}$	Fourier number $(\alpha t/L^2)$
$\mathbf{F_b}$	body force
F	geometric view factor; coefficient; constant; effective temperature difference/log mean temperature difference
F_τ	friction force
$\mathscr{F}$	generalized view factor
f	friction factor $[-(dp/dz)/\frac{1}{2}\rho\overline{U}^2/d]$; velocity parameter
$\mathbf{G}$	Grashof number $(g\beta\Delta Tl^3/v^2)$
G	mass velocity
$\mathbf{g}$	vector gravitational acceleration
g	scalar gravitational acceleration; velocity parameter
g_c	Newtonian constant
h	heat transfer coefficient
$\overline{h}$	mean heat transfer coefficient
$\tilde{h}$	total enthalpy

h_{fg}	latent enthalpy
I	current; integral
I_0, I_1	modified Bessel functions
i, j, k	unit vectors in x, y, and z directions
i, j, k	grid point references
i	intensity of radiation; specific enthalpy
J	flux
J_0, J_1	Bessel functions
K	mass transfer coefficient; coefficient
K_0, K_1	modified Bessel functions
k	thermal conductivity
L	length
l	length; mixing length
l_e	entry length
M	Mach number (u/a_∞)
M	mass transfer parameter
m	mass; fin parameter (hP/kA); exponent
$\dot{m}$	mass flow rate
$\dot{m}'$	mass flux
N	Nusselt number (hL/k_f)
$\overline{\mathbf{N}}$	mean Nusselt number ($\bar{h}L/k_f$)
N	matrix order; number of finite difference points
Ntu	heat exchanger performance parameter ($hA/\dot{m}C$)
n	normal distance; molecular mass flux; finite difference points per row
o	order
P	Prandtl number ($\mu C_p/k_f$)
P	perimeter; pressure parameter
p	pressure; characteristic coefficient
Q	heat
$\dot{Q}$	heat flow rate
q	heat flux
R	Reynolds number ($\rho u L/\mu$)
R	gas constant; source term in diffusion equation; ratio of heat capacities; resistance
$\mathscr{R}$	recovery factor
r	radial coordinate
r_0	duct radius
Sc	Schmidt number ($\mu/\rho D$)
Sh	Sherwood number (KL/D)
St	Stanton number ($h/\rho u C_p$)
S	surface area; source term
s	specific entropy; gap

$\mathbf{T}$	surface force
T	temperature
t	time
U	overall heat transfer coefficient; duct centre line velocity
U_∞	free stream velocity
$\overline{U}$	bulk or mean velocity
u	a representative velocity; velocity component in x direction; transform variable
u^*	friction velocity $[\sqrt{(\tau_s/\rho)}]$
V	volume
v_r	velocity component in radial direction
v_z	velocity component in axial direction
v	a representative velocity; velocity component in y direction
W	work; total emissive power
$\mathbf{w}$	velocity vector
X, Y, Z	linear dimensions
X_{tt}	forced boiling flow parameter
x	cartesian coordinate
y	cartesian coordinate
y_b	fin base thickness
y_1	mixing-length constant
z	cartesian or polar coordinate

α	thermal diffusivity $(k/\rho C_p)$; absorptivity
β	volume expansion coefficient
Γ	transport parameter
γ	ratio of specific heats (C_p/C_v); constant
Δ	a difference
δ	increment; velocity boundary layer thickness
δ_T	temperature boundary layer thickness
ε	emissivity; heat exchanger effectiveness; function
ζ	heat exchanger parameter; ratio of boundary layer thicknesses
η	heat exchanger parameter; dimensionless cross-stream distance
θ	angular coordinate; temperature difference
κ	mixing-length constant
Λ	heat exchanger parameter $(hA/\dot{m}C)$
λ	wavelength; mixing-length constant
μ	absolute or dynamic viscosity
v	kinematic viscosity
ξ	streamwise coordinate
ρ	density; reflectivity
σ	normal stress; dimensionless fluid property; surface tension

τ	shear stress; transmissivity
Φ	dissipation function; dependent variable
ϕ	fin efficiency; angle
χ	vapour dryness fraction
ψ	stream function
ω	successive-over-relaxation acceleration parameter; solid angle

Subscripts

ad	adiabatic
a	average
c	coolant
crit	critical, transition from laminar to turbulent flow
d	diameter
E	exterior point
eff	effective, in turbulent flow
f	fluid; frictional
F	pertaining to body force
g	gas
i, j, k	grid point references
i	initial; internal
I	interior point
j	chemical species
l	pertaining to a length dimension; viscous sublayer; liquid
m	mean
min	minimum
max	maximum
0	total or stagnation value; outer value; wall value
p	constant pressure
r	radial component
s	surface or wall value
sat	saturation condition
t	turbulent component; total value; tip condition
T	constant temperature
v	vapour condition; constant volume
x	x component
xy	x plane, y direction
y	y component
yx	y plane, x direction
z	z component
∞	free stream value

1

Introduction : modes of heat transfer

There are few branches of modern technology in which the rate at which heat can be transferred is not a significant factor at some stage in design. The efficiency and power of prime movers are limited almost entirely by the effectiveness of the associated heat transfer devices; the feasibility of plant which will be able directly to convert intrinsic chemical energy to mechanical or electrical work is largely determined by rates of heat transfer from the inevitably high-temperature plasmas; future developments in the aerospace industries will hinge in the main upon the ease with which structures and engines can be cooled; modern electrical and electronic plant require efficient dissipation of losses converted to thermal energy; the design of chemical engineering plant is usually governed by heat transfer and the analogous mass transfer processes; and even civil engineers must take account of thermal effects in buildings and structures. There is clear justification, therefore, for the concern of modern technologists with the processes of heat transfer.

Of course, from the earliest times, scientists and engineers have made use of the observation that heat, a convenient manifestation of energy, will flow naturally, without mechanical aid, from a hot body to a cold body. But it is only in recent times, with the pressures to operate plant near the temperature limits of available materials, that heat transfer processes have tended to become controlling factors in design. The comparatively recent interest in the governing principles of heat transfer is shown by the very few books on this subject which ante-date the 1939–45 war. A study of published papers on the subject also shows, for example, that in the *Proceedings of the Institution of Mechanical Engineers* between 1937 and 1962,† 70 per cent of papers under the heading 'heat transmission' have appeared since 1950. In contrast, 'steam turbines', which depend for their energy upon heat transferred from the combustion gases to the working fluid, are dealt with in papers spread uniformly throughout the period reviewed. Further, it is only within the last twenty years that heat transfer, as a subject in its own right, has appeared in the curricula of most courses in engineering and scientific education.

The study of fluid dynamics has been said to consist of 'observations which no one could explain and explanations which no one could observe'. For reasons that will become more obvious later in this book, heat transfer is commonly associated with fluid dynamics, and this very soundly based

† Brief subject and author index of papers published by the Institution of Mechanical Engineers, London, 1937–62.

comment could at one time have been applied to either subject. Much of the earlier work on heat transfer, certainly in the provision of data for use in engineering design, was of a wholly empirical nature, with little or no basis in reasoned analysis. Experimental methods, of course, remain of outstanding importance in the study of heat transfer processes, since in many situations of practical interest analysis is impossibly complex.

However, as we shall show in this book—with the ever-increasing sophistication of analytical techniques combined, where necessary, with the enormous potentialities of digital computers—rates of heat transfer in modern scientific and industrial processes can often, with confidence, be predicted theoretically from the fundamental principles of the subject we are about to explore. Although in many practical situations all modes of heat transfer occur together, it is usual to consider conduction, radiation, and convection, separately.

1-1   Conduction

This is the mode by which heat is transferred through a solid body, or through a fluid at rest. As we shall see, heat transfer is essentially dependent upon motion of some conveying agency, and it is the nature and scale of the agency which categorize the different modes of the transport process. Conduction is a consequence of electron migration between the rigidly fixed molecules in the lattice of a solid substance or of direct molecular interaction in liquids and gases. The energy level of the elementary particles is a function of temperature, and thus as these particles move to regions of lower temperature they give up their excess energies. Over a period of time, therefore, the distribution of energy, or temperature, in a given body tends towards uniformity, as articulated by the second law of thermodynamics.

Newton first observed that the rate at which heat is transferred between regions at different temperatures is proportional to this difference, but the basic law of heat conduction was particularized by Fourier, whose statement postulates that 'the rate of heat conduction is proportional to the area measured normal to the direction of heat flow, and to the temperature gradient in the direction of the heat flow'.

The constant of proportionality implicit in the statement of this law is known as *the coefficient of thermal conductivity*, which is usually given the symbol k. Thus, mathematically, Fourier's law of heat conduction becomes

$$\dot{Q}_x = -kA\frac{dT}{dx} \tag{1-1}$$

where $\dot{Q}_x$ is the rate of heat flow in the direction in which x is measured,† and since the second law of thermodynamics requires the temperature to fall

† In addition to the definitions in the text, a list of the principal notation adopted is given after the Preface.

in this direction, giving a negative temperature gradient, the negative sign is required to yield a positive number for the heat flow. In a situation in which temperature increases in the x direction, then the heat flow is in the negative direction, again as predicted by the right-hand side of Eqn (1-1). The area for heat flow, A, is measured normal to the x direction, and the general nomenclature is shown in Fig. 1-1.

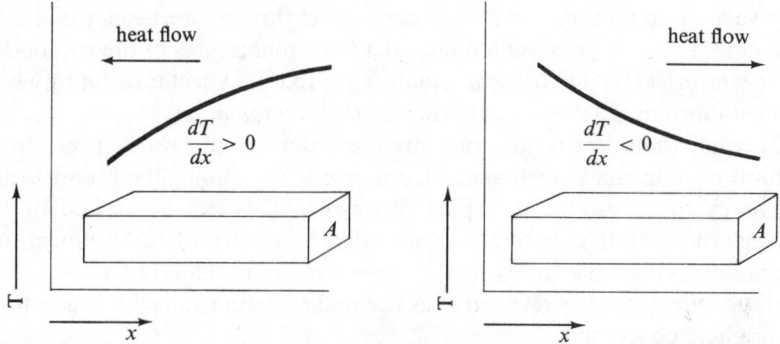

Fig. 1-1 Conduction in one dimension

Some values of thermal conductivities of common materials at normal temperature in the BS and SI systems of units are given in Table 1-1.

Table 1-1 Thermal conductivities of common substances at 60°F (15°C)

Substance	Btu/ft h °R	W/m K
Silver	240·0	415·0
Copper	220·0	380·0
Mild steel	26·0	45·0
Stainless steel	11·0	19·0
Wood	0·10	0·17
Asbestos	0·10	0·17
Water	0·35	0·60
Air	0·015	0·026

In problems of heat transfer by conduction we are often concerned with transient systems in which temperatures, and therefore rates of heat flow, and often thermal conductivities themselves, vary with time as a system tends towards thermodynamic equilibrium. Sometimes, however, we have the situation in which heat is being supplied to a system at a uniform rate and at a given temperature, and removed at the same rate from the system at a lower temperature. In this case the variables concerned are not time-dependent, and solution of the forms of Fourier's equations is correspondingly simpler. In either case closed analytical solutions can sometimes be obtained (as shown in Chapters 2 and 3) but, increasingly commonly, numerical methods of solution provide powerful alternative procedures, as demonstrated in Chapter 4.

1-2 Radiation

If two bodies in an adiabatic enclosure at different temperatures are separated by a vacuum so that there is no mutual contact through the agency of a solid or a fluid, then it is an observed fact that the temperatures of the two bodies will nevertheless tend to become equal. The process of heat transfer by which this equilibrium is achieved is known as *thermal radiation*.

Thermal radiation is just one manifestation of the wide spectrum of natural phenomena known as electromagnetic radiation—the broadcasting of energy by subatomic transport processes. This can be excited by the passage of an electric current—chemically, by electron bombardment and thermally, as a simple consequence of the temperature level of a body. It is the latter effect which is referred to as thermal radiation and with which alone we are here concerned.

By a simple argument based upon principles of thermal equilibrium it will be shown in Chapter 8 that there is a maximum rate at which thermal radiant energy can be emitted, and consequently absorbed, by a body at a given temperature. Bodies and surfaces which interchange radiant energy at these maximum rates are known as *ideal emitters* or, more commonly, *black bodies*. The rate of energy transmission between such bodies is, as might be reasonably inferred, strongly temperature-dependent, and it has been shown both experimentally and theoretically that the *total emissive power* (W_b) of a black body is proportional to the *fourth power of the absolute temperature*. The constant of proportionality, σ, is known as the Stefan–Boltzmann constant after the two Austrian physicists who, towards the end of the nineteenth century, verified the fourth-power relationship. Thus, for a black body the total rate of radiant emission *per unit surface* is given by

$$W_b = \sigma T^4 \tag{1-2}$$

in which $\sigma = 0.171 \times 10^{-8}$ Btu/ft^2 h $^\circ$R^4 in the British system of units, and, in the SI system, $\sigma = 56.7 \times 10^{-12}$ kW/m^2 K^4.

In engineering practice few surfaces behave as black bodies and, by definition, emit and therefore absorb less than the ideal rate of energy flow. To take account of this shortfall the *emissivity*, ε, of a real surface is defined as the fraction of the black body radiant flux emitted, and the absorptivity, α, as the proportion of the incident emission absorbed. *At a given temperature* the principles of thermal equilibrium used in Chapter 8, and already referred to, require that

$$\varepsilon = \alpha \tag{1-3}$$

This equation is of fundamental significance and is used commonly in engineering practice even where the italicized and highly restrictive qualifying phrase, *at a given temperature*, does not apply. To illustrate this point, white hats are often worn as protection from sunlight because, for white surfaces, the absorptivity for energy from a high-temperature source like the sun (say, α_{sun}) is low. Equally, emissivity at body temperatures (ε_{body}) is high. Clearly, Eqn (1-3) does not apply, that is $\varepsilon_{body} > \alpha_{sun}$, for we are dealing with two widely different temperatures in the radiant system. In engineering practice we rarely have to deal with this scale of temperature difference, and surfaces are commonly assumed to behave as if Eqn (1-3) applied, despite the contradictory use of particular surfaces (like that of the white hat in the example above) to disturb the radiation balance. Bodies which behave as if Eqn (1-3) applied without qualification are known as *grey bodies* and the justification for assuming their existence is given in Chapter 8.

1-3 Convection

This is the mode of heat transfer which is the consequence of the motion of a fluid. It is commonly observed that the presence of a heated surface in otherwise stationary atmospheric air sets the air in motion—an effect that is

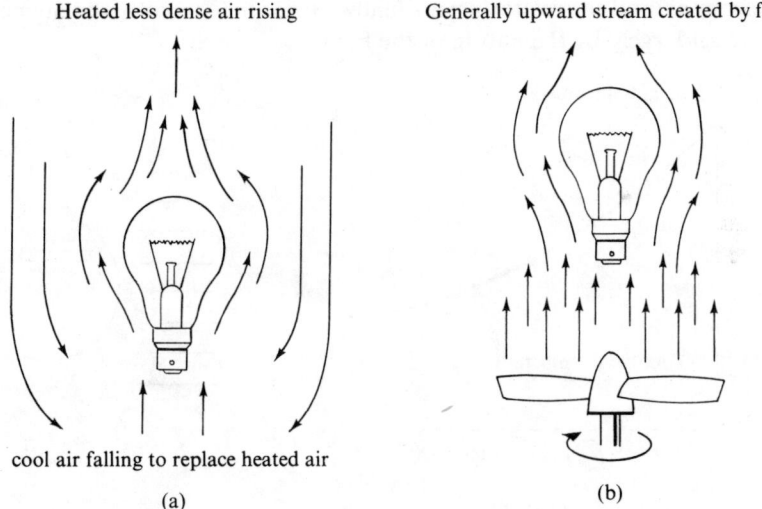

Heated less dense air rising   Generally upward stream created by fan

cool air falling to replace heated air

(a)   (b)

Fig. 1-2 (a) Natural or free convection; (b) forced convection

used, for instance, to produce the 'flicker' in artificial coal fires. The air set in motion by the transfer of heat from an electric bulb is used to rotate a small propeller which, by continually interrupting the light, produces the required effect. Also, light bulbs which are primarily intended to illuminate rather than decorate, and where the air motion is not so manifest, are cooled by the consequent movement of the air initially adjacent to the hot surface which carries the thermal energy not transferred by radiation—in practice by far the greater proportion—to cooler bodies for dissipation. The motion of the air arises from the buoyancy forces in the earth's gravitational field which are due to the differences in density between the heated fluid on the hot surface and the cooler, and therefore denser, surrounding fluid.

When the rate at which heat is transferred by thermally induced buoyancy forces is inadequate, it is a natural further step to enhance the fluid motion mechanically, say, by a fan as in Fig. 1-2(b). The heat transfer rate is then not controlled by the density changes in the ambient fluid but largely by the rate at which it can be forced over the hot surface; that is, in effect, by the power available to drive the fan.

The transfer of heat through the motion induced by the natural volume or density changes associated with temperature differences in a fluid is known as *natural* or *free convection*. It is a very commonly occurring mode of heat transfer, as indeed the name implies. However, the buoyancy forces are a consequence of the physical properties of the fluid concerned, as also, as we shall see, is the resistance to the resultant fluid motion. Thus rates of heat transfer by natural convection are normally outside the control of the engineer, and are set largely by the nature of the fluid.

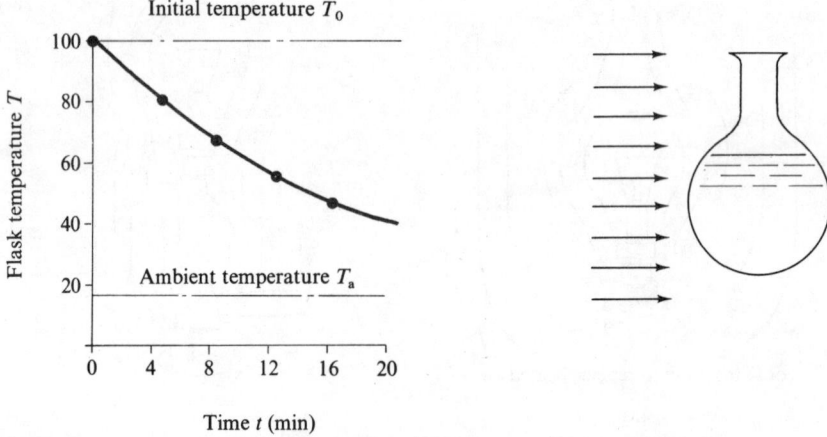

Fig. 1-3   Cooling of a flask of boiling water in an air stream

On the other hand, the rate at which heat is transferred as a result of artificially induced fluid motion, as with the fan of Fig. 1-2(b), is entirely within the control of the engineer. Thus this mode of heat transmission, known as *forced convection*, is of the greatest importance and finds application over the whole technological spectrum, from the cooling of nuclear reactors by circulating gas at the expense of tens of thousands of horsepower, to cooling tea with a hat.

If a heated body is allowed to cool as a result of heat transfer by either forced or natural convection, it is found that the temperature–time history is exponential in form, asymptotically approaching the ambient equilibrium value, as shown in Fig. 1-3.

Such a form is predicted by the assumption that the rate of heat transfer per unit area of the heated body, q, is proportional to the difference in temperature between the hot body, T, and the ambient flow, T_a. If the constant of proportionality is h, then we have

$$\dot{Q} = qA = hA(T - T_a) \qquad (1\text{-}4)$$

and the heat balance for an infinitesimal time, dt, is given by

$$hA(T - T_a)\,dt = -mC\,dT \qquad (1\text{-}5a)$$

where m and C are, respectively, the mass and specific heat of the hot body. The solution to Eqn (1-5a) is the exponential function,

$$\frac{T - T_a}{T_0 - T_a} = \exp\left(-\frac{hAt}{mC}\right) \qquad (1\text{-}5b)$$

in which T_0 is the temperature when $t = 0$.

It was Newton who, as with so many things, first observed the linear dependence of the rate of heat transfer upon temperature difference, but, as with many of his laws, this too is only a good approximation. In natural convection, the temperature differences within the system not only provide the potential driving force for heat transfer according to Newton's law, but also determine the rate of fluid motion which (as we shall see in Chapters 5, 6, and 7) governs the constant of proportionality, h. In this mode we would therefore expect the absolute rate of heat transfer to vary more rapidly than linearly with temperatures, as indeed is shown by exact analysis and experiment.

In the all-important situation of heat transfer by forced convection, however—particularly if differences in temperature within the fluid are not excessive—Newton's law of linear dependence of heat transfer on temperature difference is found to be very close to reality, so that the constant of proportionality, h, in Eqn (1-5a) assumes real significance and is called the *heat transfer coefficient*. The object of most studies of heat transfer by convection, both experimental and analytical, is the determination of the heat transfer coefficient.

The use of the heat transfer coefficient, defined by Eqn (1-4), is traditional in heat transfer, as is the coefficient of thermal conductivity, defined by Eqn (1-1). These parameters are analogous with the generic term *conductance* in electrical practice, which has much in common with heat transfer. Also, in electricity common use is made of the reciprocal of conductance, *resistivity*, particularly in the statement of Ohm's law

$$\text{current} = \frac{\text{potential}}{\text{resistance}} \tag{1-6}$$

Current is analogous with heat flux in thermal problems and potential with the driving temperature difference. Comparing Eqns (1-4) and (1-6) we observe that resistance corresponds to the reciprocal of the heat transfer coefficient. Since, also in thermal problems, we frequently have several thermal barriers in series, it is often convenient to use the concept of thermal resistance, the total of which is then simply the sum of individual resistances, rather than the admittedly more common heat transfer coefficient.

Some typical values of heat transfer coefficients in a number of convective situations are given in Table 1-2 in both the BS and SI systems of units.

Table 1-2

Situation	Btu/ft² h °R	W/m² K
Natural convection in air	2	11
Natural convection in water	100	570
Forced convection in air	30	170
Forced convection in water	200	1,100
Natural convection and boiling in water	1,000	5,700
Forced convection and boiling in water	10,000	57,000
Condensation of water at atmospheric pressure	1,000	5,700

1-4   Heat transfer with change of phase

The last three items in Table 1-2 are concerned with processes in fluids, and therefore associated with convection, but which involve change of phase, particularly from liquid to vapour and vice versa. As we shall see, the processes of convection are complex in themselves but when to the fluid motion, which on a macroscopic scale at least is usually regular and uniform, is added the apparently random formation and collapse of vapour bubbles, it immediately follows that rational analysis of the boiling process in forced or natural convection becomes very difficult indeed, as will be seen in Chapter 9. However, the hydrodynamic complexity of the boiling process associated with the movement of vapour bubbles results in very high heat transfer rates, so that this process is of great technological importance. Since, as with natural convection, the fluid motion is a function of the heat transfer rate, so also are the heat transfer coefficients which rise steeply with temperature difference, but only up to a sharply defined limit beyond which is found one of the few examples of the paradoxical situation in which the rate of heat transfer decreases with driving temperature difference. This situation, demonstrated in Fig. 1-4, is also inherently unstable since further heating causes higher temperatures and still further reduced heat transfer rates, with frequent catastrophic consequences.

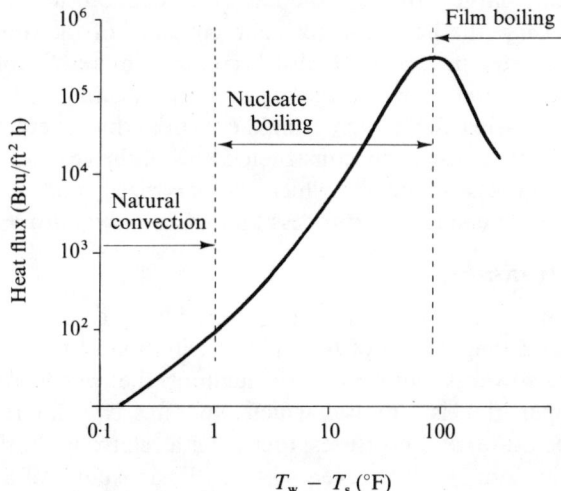

Fig. 1-4 Variation of heat flux with difference between surface temperature, T_w, and liquid saturation temperature, T_s, in boiling heat transfer

The explanation of this startling phenomenon is basically simple and lies in the coalition of the vapour bubbles—which individually add greatly to the heat transfer rate—into a continuous film of low conductivity vapour which insulates the heated surface from the convecting liquid. This regime is thus logically known as *film boiling* and its onset, even more graphically referred to, as *burn-out*. The prediction and avoidance of film boiling are almost the principal concerns of analyses of boiling heat transfer processes.

The reverse process of condensation of a vapour also leads to very high heat transfer coefficients—indeed, the resistance to heat flow in the condensation of a pure vapour can invariably be assumed to be zero. The resistance to heat flow in this mode of heat transfer, so that heat transfer coefficients are not in fact infinite, is found in the film of condensate covering the cooled surface and through which heat has to be transferred by conduction. If this film of condensate can be broken up into drops—the regime of so-called 'dropwise' condensation—very much higher heat transfer coefficients are obtained. Conversely, if the condensate film is whole and the vapour is mixed with a non-condensable gas through which it has to migrate, and which thus represents a resistance to condensation, heat transfer coefficients are substantially reduced.

Rates of heat transfer by both evaporation and condensation, therefore, are liable to wide variations. Thus, while often attractive because of the very high heat transfer coefficients that can be obtained, difficulties of quantitative prediction are severe and gross errors can be incurred through the operation of apparently minor unknown or unappreciated factors.

In recent years, largely through the need for advanced techniques in the aerospace industry, the heat transfer characteristics of the direct solid to vapour phase change process have also become of interest. Ablation, as it is called, is the method by which the aerodynamic heating of high-velocity space vehicles is dissipated on re-entering the earth's atmosphere, through the direct vaporization of a suitably constructed bow of the vehicle. In ablation the controlling process is that by which the vaporized solid is transported away from the parent surface, a process known as *mass transfer*.

1-5   Mass transfer

If a system comprises two or more separately identifiable substances, it is found that over a long period of time the distribution of the separate constituents tends towards uniformity throughout the whole system. For example, if a partition is removed which separates two different gases at identical temperatures and pressures, then after a relatively short period the two gases are uniformly mixed, even without the assistance of any external agency. This mixing is the result of the natural motion of the molecular particles which comprise a gas. Equally, two dissimilar liquids will tend

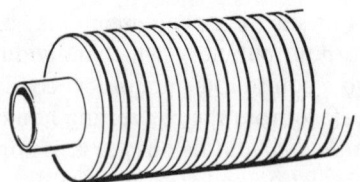

Fig. 1-5   Finned tube from a nuclear power station boiler

to become, by entirely natural means, a uniform solution. Even a continuous solid which initially varies in constituency throughout its mass will tend to a uniform composition, but at a very slow rate which for most practical purposes is negligible since the freedom of molecular motion within a solid is limited by the almost rigid bonds.

Even from this simple qualitative account of the natural tendency for matter to tend towards a uniform distribution in a system, the analogy with heat transfer and the tendency to uniformity of temperature is clear. The analogy applies on a quantitative level also, and if the physical properties appropriate to the two transport processes are known, then knowledge of heat transfer rates in a given situation enables rates of transfer of matter to be predicted, and vice versa. Thus in many common engineering situations exact measurement of heat quantities, particularly on a small local scale, is difficult. For example, it is difficult to determine the variation in heat transfer around the finned nuclear boiler tube of Fig. 1-5 when subject to forced convection cooling in a gas stream, but if the tube is made of a solid which ablates or 'sublimes' into the gas, then the rate at which matter is transferred is indicated by the change in material thickness after a finite period of exposure to the convecting flow. From this the rate of heat transfer in an identical hydrodynamic situation may be deduced. The sublimation of the material of the boiler tube is an example of solid–gas matter transfer, or *mass transfer* as it is more usually known despite the irrelevance of the inertial connotation of 'mass' in this situation. As has been implied above, the rate of mass transfer is exactly analogous with the rate of heat transfer, and both are usually considered as fluxes, that is, transport rates per unit area. The potential, or driving force, in mass transfer that is analogous with temperature in heat transfer is concentration, c, defined as the number of molecules (or, in practice, mass-moles) of one constituent of the system per unit volume of the mixture at a given point. Hence, just as in heat transfer by conduction we had Fourier's law:

heat flux $\propto$ temperature gradient

so, in mass transfer, we have *Fick's law*:

mass flux $\propto$ concentration gradient

Thus, if n is the mass flux in moles/(time $\times$ area)

$$n \propto -dc/dy \qquad (1\text{-}7)$$

$$= -D(dc/dy) \qquad (1\text{-}8)$$

where D is known as the *mass diffusivity*, and has dimensions area/time.

In addition to the solid–gas mass transfer system of the boiler tube example referred to above, other common systems include liquid–solid, liquid–gas, liquid–vapour (liquid and its own vapour), gas–gas, and liquid–liquid mass transfer.

Earlier in this chapter, in extending consideration of heat transfer from conduction in which temperature gradient is controlling to convection in which often only temperature differences over finite distances are known, the concept of *heat transfer coefficient* was introduced. Extending the mass transfer analogy in the same way to convecting systems, a mass transfer coefficient, K, is defined as

$$\frac{\text{rate of mass transfer per unit convecting area}}{\text{potential difference (that is, concentration difference)}}$$

Thus,

$$\dot{n} = K(c - c_a) \tag{1-9}$$

in which c and c_a are the concentrations at the two relevant points in the system just as T and T_a are the appropriate temperatures in Eqn (1-4).

Although the object of this section has been to draw attention to the similarities between the processes of heat and mass transfer, one significant difference in principle should not be overlooked. Although heat transfer takes place only in one direction—that of decreasing temperature—matter transfer can take place simultaneously in either direction. This follows since a point 1 which is of higher concentration of constituent A in constituent B than point 2, is equally of lower concentration of B in A. Thus molecules of A can migrate from 1 to 2, and molecules of B can equally transfer from 2 to 1. The significance of this difference, and of the similarities of mass transfer to heat transfer, will be discussed in more detail in Chapters 5 and 9.

PROBLEMS

1-1 Consider which are the significant modes of heat (or mass) transfer in the following examples: (a) heat loss from a vacuum flask; (b) burning of a finger tip placed on a hot surface; (c) generation of steam in the tubes lining a boiler combustion chamber; (d) a central heating 'radiator'; (e) astronauts' space suits on the moon's surface; (f) re-entry heat shield on a space vehicle; (g) evaporation from a pool of water on the ground; (h) quenching a white hot steel ingot in cold water.

1-2 Lee's disc is an apparatus used to determine the thermal conductivity of poor conductors. A disc of the conductor is clamped between an upper cylinder, usually supplied with steam, to act as a heat source and a lower metal disc, the temperature of which is measured by a thermocouple and which is cooled to act as a heat sink. In an experiment to measure the conductivity of a jointing material a disc 0·3 m in diameter and 5 mm thick was used. In the upper cylinder 0·56 kg/h of steam at 1 atmosphere and 99·6°C (latent heat 2,258 kJ/kg) condensed when the temperature of the lower metal block was maintained at 25°C. Evaluate the thermal conductivity of the jointing material.

Ans: 0·2 W/m K

1-3 What are the likely sources of experimental error in the Lee's disc apparatus of Problem 1-2? Why is it unsuitable for determining the thermal conductivity of 'good' conductors?

1-4 A disc of stainless steel of thermal conductivity 11 Btu/ft h °F is substituted for the jointing material of Problem 1-2. The rate of condensation of steam in the upper cylinder of the apparatus, at 1 atmosphere and 211·3°F (latent heat 971 Btu/lb), is kept as near as possible at the previous value and is observed to be 1·3 lb/h. If the thermocouple recording the temperature of the lower disc reads 1°F low, what is the erroneous value of the thermal conductivity of stainless steel obtained from this experiment? Use the conversion tables in the Appendix as required.

Ans: 6·9 Btu/ft h °F

1-5 A 'radiator' in a domestic central heating system operates at a surface temperature of 131°F (55°C). Determine the rate at which it emits radiant heat per unit area in both the British and SI systems of units if it behaves as a 'black body'.

Ans: 208·5 Btu/ft^2 h; 0·656 kW/m^2

1-6 The temperature at the outer surface of the lagging on a steam pipe is measured as 250°F when the ambient temperature is 70°F. If the heat transfer coefficient due to convection to the surrounding atmosphere is 3 Btu/ft^2 h °F and the emissivity of the lagging surface is 0·4, determine the proportion of the total heat loss due to radiation. Consider carefully the assumptions involved in the calculation.

Ans: 18·1%

1-7 A flask contains 3 kg of water boiling at 100°C and has a surface area of 0·1 m^2. When the source of heat is removed the following temperature–time data are observed on cooling in air at 20°C:

Time (min):	0	2	5	10	15	30	60
Temperature (°C):	100	98	95·3	91	87	76	59

Show that these data are consistent with Newton's law of cooling and correspond to a heat transfer coefficient of about 25 W/m^2 °C. Take the specific heat of water as 4·2 kJ/kg K and neglect the thermal capacity of the flask material.

1-8 To what value will the heat transfer coefficient in Problem 1-7 have to be increased by forced convection for the flask to cool to 50°C in 15 minutes?

Ans: 137 W/m^2 °C

1-9 The temperature and pressure in a pool of water are 16°C and 1 bar (0·1 MN/m^2). Determine the rate at which the surface level will drop through evaporation into still air at the same temperature with a relative humidity (vapour pressure/saturation pressure) of 0·5. It may be assumed that the concentration gradient is linear through an air depth of 0·1 m and that the moisture behaves as a perfect gas with a characteristic constant of 462 J/kg K. Diffusion data from Table A10 in the Appendix apply and the saturation pressure and density of water at 16°C are 0·0182 bar and 1,000 kg/m^3 respectively.

Ans: 6·9 × 10^{-6} m/h

1-10 As a result of air motion over the surface of the pool in Problem 1-9, the water level is found to fall at the rate of 1 mm/h. Determine the corresponding value of the mass transfer coefficient.

Ans: 0·041 m/s

2

Steady-state conduction in simple systems

2-1 Basic concepts of conduction

The results of experiments with different solids and various geometries show that if two parallel planes, δx apart in a solid, are maintained at different temperatures T_1 and T_2 (Fig. 2-1), the quantity of heat, Q, flowing between the planes through the surface A in time δt is given by

$$Q = k \frac{A}{\delta x} (T_1 - T_2) \, \delta t \tag{2-1}$$

where k is a constant for each material and is called its thermal conductivity.

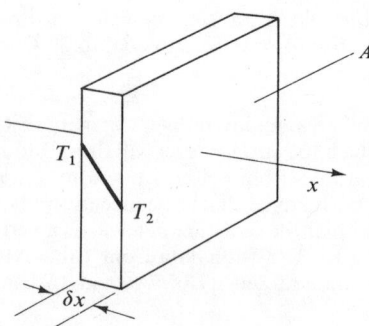

Fig. 2-1 One-dimensional conduction through a plane slab

These experiments suggest, as the fundamental postulate of the conduction of heat, that the rate per unit area at which thermal energy crosses from the inside to the outside of an isothermal surface at a point is given by

$$q \left(= \frac{1}{A} \frac{dQ}{dt} \right) = -k \frac{\partial T}{\partial n} \tag{2-2}$$

where $\partial T / \partial n$ is the temperature gradient along the outward normal to the surface.

The negative sign implies that, by convention, heat flow is positive in the direction of the outward normal. Equation (2-2) expresses what is known as *Fourier's law*, formally stated in Chapter 1.

If the rate of heat flow is time-dependent the problem is said to be one of *transient* heat conduction. In this case the heat flows entering and leaving an element of the solid are not equal and the difference causes a change, dU, in the internal energy† of the element. By definition of the specific heat of a substance C, we can write for the energy change

$$\frac{dU}{dt} = \rho V C \frac{\partial T}{\partial t} \tag{2-3}$$

where ρ is the density of the substance and V the volume.

From these elementary basic concepts we can proceed to the analysis of conducting systems of increasing complexity.

2-2　General heat flow equation

The infinitesimally small rectangular parallelepiped shown in Fig. 2-2 has a volume $V = dx\,dy\,dz$ and is supposed to generate internal heat at a rate $q_g(T)$ per unit volume, in general as a function of local temperature. For simplicity the solid is considered homogeneous (uniform constituency) and isotropic (the properties the same in all directions) and has a thermal conductivity $k(T)$ which is dependent only upon temperature.

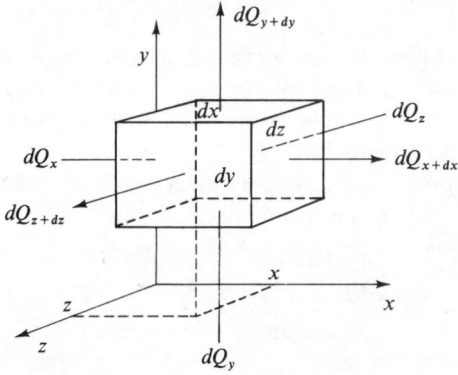

Fig. 2-2　Heat conduction through a differential element

The total quantity of heat entering the face $dydz$ at x in time dt is given by Fourier's law, Eqn (2-2), as

$$dQ_x = -dy\,dz\,k(T)\frac{\partial T}{\partial x}\,dt$$

The heat leaving the element at $x + dx$ may be expressed as

$$dQ_{x+dx} = dQ_x + \frac{\partial}{\partial x}(dQ_x)\,dx$$

† The internal energy of the solid is taken to be a function of temperature only.

which can be rewritten as

$$dQ_{x+dx} = -dy \, dz \left[k(T)\frac{\partial T}{\partial x} + \frac{\partial}{\partial x}\left(k(T)\frac{\partial T}{\partial x} \right) dx \right] dt$$

Similar analyses in the other two coordinate directions give the corresponding expressions.

The internal energy of the volume element is increased in time dt by

$$dU = \rho C \, dx \, dy \, dz \, \frac{\partial T}{\partial t} \, dt$$

and the total internal heat generated in time dt is

$$dQ_g = q_g(T) \, dx \, dy \, dz \, dt$$

Applying the principle of the conservation of energy to the element we obtain

$$dQ_x + dQ_y + dQ_z + dQ_g = dQ_{x+dx} + dQ_{y+dy} + dQ_{z+dz} + dU$$

which finally yields

$$\frac{\partial}{\partial x}\left[k(T)\frac{\partial T}{\partial x} \right] + \frac{\partial}{\partial y}\left[k(T)\frac{\partial T}{\partial y} \right] + \frac{\partial}{\partial z}\left[k(T)\frac{\partial T}{\partial z} \right] + q_g(T) = \rho C \frac{\partial T}{\partial t} \quad (2\text{-}4)$$

This equation can be made more generally applicable by writing the physical properties k, ρ, and C as functions of x, y, z, and T, and q_g as a function of x, y, z, T, and t, although only rarely in engineering is this additional complexity necessary or worthwhile.

If the physical properties can be assumed constant, this general equation reduces to the following important special forms:

Fourier's equation

$$\frac{\partial^2 T}{\partial x^2} + \frac{\partial^2 T}{\partial y^2} + \frac{\partial^2 T}{\partial z^2} = \frac{1}{\alpha}\frac{\partial T}{\partial t} \quad (2\text{-}5)$$

in which the internal heat generation is assumed to be zero. The parameter α is a combination of the physical properties of the material defined as

$$\alpha = k/\rho C$$

This is known as the *thermal diffusivity* of the material and has the dimensions (L^2/t). It occurs constantly in this form in conduction problems.

Poisson's equation

$$\frac{\partial^2 T}{\partial x^2} + \frac{\partial^2 T}{\partial y^2} + \frac{\partial^2 T}{\partial z^2} + \frac{q_g}{k} = 0 \quad (2\text{-}6)$$

which applies to steady heat conduction in three dimensions.

Laplace's equation

$$\frac{\partial^2 T}{\partial x^2} + \frac{\partial^2 T}{\partial y^2} + \frac{\partial^2 T}{\partial z^2} = 0 \qquad (2\text{-}7)$$

This last equation is also known as the potential equation and is probably the most commonly used equation in heat conduction.

Similar equations can be derived for cylindrical and spherical coordinate systems. For instance, by substituting $x = r\cos\theta$ and $y = r\sin\theta$ (where r is the radius, θ the longitude, and z the axial coordinate) Fourier's form of the conduction equation [Eqn (2-5)] can be written in cylindrical coordinates as

$$\frac{\partial^2 T}{\partial r^2} + \frac{1}{r}\frac{\partial T}{\partial r} + \frac{1}{r^2}\frac{\partial^2 T}{\partial \theta^2} + \frac{\partial^2 T}{\partial z^2} = \frac{1}{\alpha}\frac{\partial T}{\partial t} \qquad (2\text{-}8)$$

Poisson's and Laplace's equations [Eqns (2-6) and (2-7)] take the corresponding forms.

2-3 Steady conduction in one dimension

'Steady conduction', as has been seen, is the situation in which the temperature distribution in the system does not vary with time.

Heat flow through a plane wall in which the temperature only varies in one space-coordinate is said to be one-dimensional and is sometimes referred to as conduction through an 'infinite slab', the strictly necessary physical situation. Similarly, conduction along a long thin rod insulated over its curved surface, and conduction through the wall of a long, hollow cylinder which has its curved surfaces maintained at uniform temperatures, are both examples of one-dimensional conduction.

Although the solutions to elementary conduction problems, such as that of steady heat flow through a plane wall with specified surface temperatures, are immediately obvious from Fourier's law [Eqn (2-2)] they are simply solutions to Laplace's equation [Eqn (2-7)], employing the appropriate boundary conditions. Thus, for the 'infinite slab', Laplace's equation applies in the one-dimensional form as

$$\frac{d^2 T}{dx^2} = 0$$

which integrates to

$$T = Bx + C$$

with the boundary conditions $T = T_1$ at $x = x_1$, say, and $T = T_2$ at $x = x_2$ so that

$$B = \frac{T_2 - T_1}{x_2 - x_1}$$

and since

$$q = -k \frac{dT}{dx}$$

$$q = -kB$$

and thus

$$= -k \frac{(T_2 - T_1)}{(x_2 - x_1)} \qquad (2\text{-}9a)$$

2-3-1 Composite walls and slabs

A composite wall constructed of, say, three layers of different materials having thermal conductivities k_{12}, k_{23}, and k_{34} is shown in Fig. 2-3.

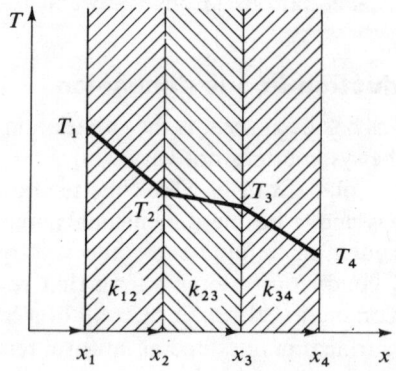

Fig. 2-3 Composite wall with specified boundary temperatures

If the thermal conductivities are assumed constant and the outer boundary surface temperatures T_1 and T_4 are specified, then for steady-state conditions the heat flow per unit area, q, is the same for each layer and we may write

$$q = -k_{12} \frac{T_2 - T_1}{x_2 - x_1} = -k_{23} \frac{T_3 - T_2}{x_3 - x_2} = -k_{34} \frac{T_4 - T_3}{x_4 - x_3}$$

from which

$$T_1 - T_2 = q \frac{x_2 - x_1}{k_{12}}$$

$$T_2 - T_3 = q \frac{x_3 - x_2}{k_{23}}$$

$$T_3 - T_4 = q \frac{x_4 - x_3}{k_{34}}$$

Adding these equations we eliminate the unknown temperatures T_2 and T_3 to give the solution for the heat flow as

$$q = \frac{T_1 - T_4}{(x_2 - x_1)/k_{12} + (x_3 - x_2)/k_{23} + (x_4 - x_3)/k_{34}} \qquad (2\text{-}9b)$$

The distribution of temperature, linear in each different material, then follows. In general, for a composite wall of n layers the heat flow is given by

$$q = \frac{T_1 - T_{n+1}}{\sum_1^n \Delta x/k}$$

2-3-2   Thermal resistance, conductance, and analogies

The thermal conductance is defined as $k/\Delta x$ and the thermal resistance as its reciprocal $\Delta x/k$. There is, therefore, an obvious analogy between heat flow as described by Eqn (2-9b) and the flow of electrical current through a series of resistors, the potential difference being equivalent to the temperature difference in heat conduction. Indeed, the analogy between heat flow and electric current flow is much deeper; the equation of Laplace [Eqn (2-7)] is called the *potential* equation since it describes not only the steady flow of heat and electricity in homogeneous conductors but also the irrotational flow of an incompressible fluid and many other potential problems in electricity, magnetism, and gravitating matter.

Heat flow problems are sometimes solved by constructing resistance network analogies. These, however, can be classed as 'difference schemes'; their applications in conduction problems are discussed in Chapter 4 and their relevance in radiant heat flows is shown in Chapter 8.

2-4   Boundaries with fluids in motion

Problems in which the boundary temperatures are known arise very infrequently in practice. A much more important class of problem is that in

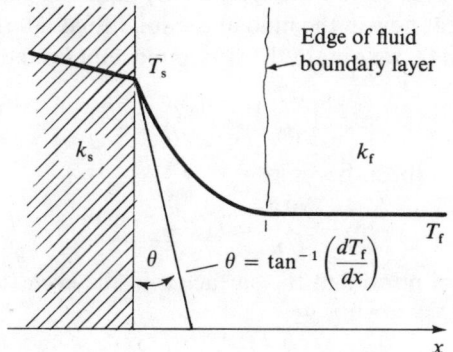

Fig. 2-4   Heat conduction through a solid/fluid interface

which the conducting solid is in contact with a fluid and the surface temperatures are unknown. If heat transfer is taking place the fluid will be in motion either through an external pumping action (forced convection) or through the buoyancy forces created within the fluid by the inevitable temperature differences within it (natural or free convection). In either case the relatively thin velocity and thermal boundary layers (see Chapters 5, 6, and 7) give rise to an appreciable temperature difference between the bulk of the fluid and the solid surface. In Fig. 2-4 the temperature of the surface and of the fluid are denoted by T_s and T_f, respectively, and the solid and fluid thermal conductivities by k_s and k_f, respectively.

The initial heat transfer to the fluid occurs by conduction through the first few molecules adhering to the surface, so that theoretically we could determine the heat flow from Fourier's law, Eqn (2-2),

$$q = -k_f \left(\frac{dT_f}{dx}\right)_s \tag{2-10}$$

However, the temperature gradient in the fluid at the wall $(dT_f/dx)_s$ cannot easily be measured in practice, and for most engineering purposes it is more useful to relate the heat flow rate to a representative temperature difference between the surface and the fluid by an equation of the form

$$q = h(T_s - T_f) \tag{2-11}$$

where h is a *heat transfer coefficient* (sometimes referred to as a *film coefficient* or as a *surface conductance*) as described in Chapter 1.

The heat transfer coefficient h includes the combined effects of conduction and convection in the fluid. In many engineering problems h is difficult to calculate exactly since it is a function of many variables such as the fluid viscosity, density, specific heat, thermal conductivity, and principally the velocity relative to the surface, as shown in Chapters 5, 6, and 7. In addition, the definition of the fluid reference temperature, T_f in Eqn (2-11), is dependent upon the nature of the fluid flow system. By equating the heat flow in the solid wall to the heat flow in the fluid at the solid/fluid interface, by means of Eqns (2-2) and (2-11), we obtain the very important boundary condition

$$-k_s \left(\frac{dT}{dx}\right)_s = h(T_s - T_f)$$

which, in its general form, becomes

$$-k \frac{\partial T}{\partial n} = h(T - T_f) \tag{2-12}$$

N.B. If the outward normal at the surface is in the *negative direction* of the coordinate axis, say x, then

$$\frac{\partial T}{\partial n} = -\frac{\partial T}{\partial x}$$

This ensures that the heat flow is in the opposite direction to that in which T increases algebraically.

Equation (2-11) should strictly be written

$$q = h(T_s - T_{ad,s})$$

where $T_{ad,s}$ is the temperature which the wall would take up in the absence of heat transfer across it (that is, adiabatic conditions). For a large number of flows $T_{ad,s} = T_f$ where, for external boundary layer flows, T_f is the free stream fluid total temperature, and for internal fully developed flows T_f is the average or more precisely the bulk-mean fluid temperature. The heat flow, q, should always go to zero as the representative temperature difference goes to zero.

These aspects will be discussed in more detail in Chapter 6 but for the present it will be assumed that h is a given constant.

2-5 Composite wall with fluid boundaries

A composite three layer wall transferring heat from one fluid to another is shown in Fig. 2-5.

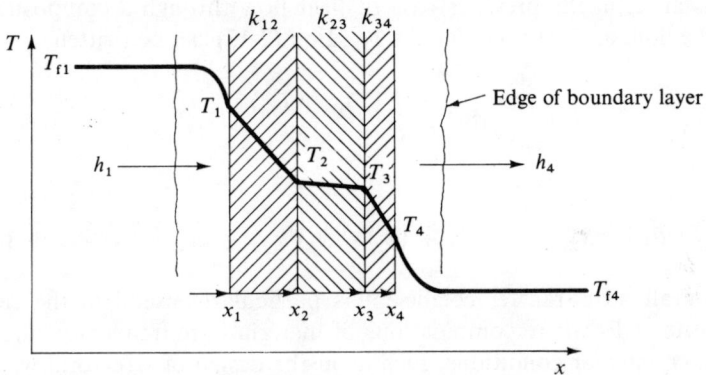

Fig. 2-5 Composite wall with fluid boundaries

The notation used is the same as that for the composite wall with specified boundary temperatures. The bulk fluid temperatures at faces x_1 and x_4 are denoted by T_{f1} and T_{f4} and the heat transfer coefficients at these faces by h_1 and h_4 respectively.

Again the heat flow per unit area through each layer is the same and, using Eqns (2-9a) and (2-11), we can write

$$q = h_1(T_{f1} - T_1) = \frac{T_1 - T_4}{(x_2 - x_1)/k_{12} + (x_3 - x_2)/k_{23} + (x_4 - x_3)/k_{34}}$$

$$= h_4(T_4 - T_{f4})$$

N.B. The second term $h_1(T_{f1} - T_1)$ obtained from Eqn (2-11) is so written because the outward normal at this face is in the negative x direction.

Eliminating the surface temperatures T_1 and T_4 gives the heat flux in terms of the fluid temperatures

$$q = \frac{T_{f1} - T_{f4}}{1/h_1 + (x_2 - x_1)/k_{12} + (x_3 - x_2)/k_{23} + (x_4 - x_3)/k_{34} + 1/h_4} \qquad (2\text{-}13)$$

The reciprocal of the heat transfer coefficients at each surface may therefore be treated as additional thermal resistance to heat flow from the fluid at T_{f1} to the fluid at T_{f4}.

2-6   Overall heat transfer coefficient

In the case of walls and cylinders with fluid boundaries the heat flow rate is sometimes expressed as a product of an overall temperature difference ΔT and an overall heat transfer coefficient U:

$$q = U \, \Delta T \qquad (2\text{-}14)$$

For instance, in the previous case of heat flow through a composite wall from the fluid at T_{f1} to the fluid at T_{f4}, Eqn (2-13) can be written

$$q = U(T_{f1} - T_{f4}) \qquad (2\text{-}15)$$

where

$$U = \frac{1}{1/h_1 + (x_2 - x_1)/k_{12} + (x_3 - x_2)/k_{23} + (x_4 - x_3)/k_{34} + 1/h_4}$$

The overall heat transfer coefficient is particularly useful in the case of composite walls where combinations of materials are frequently employed under very similar conditions, such as in the design of structural walls for centrally heated or air conditioned buildings. In the design of heat exchangers use is also made of an overall heat transfer coefficient to equate the heat flow through the available surface area to the representative temperature difference. This is discussed in more detail in Chapter 10.

2-7   The hollow infinite cylinder

2-7-1   Plain hollow cylinder with specified boundary temperatures

For this geometry, shown in Fig. 2-6, the cylindrical coordinate system is the obvious choice.

If the outer radius, r_2, of an infinitely long hollow cylinder is maintained at temperature T_2, and the inner radius r_1 at T_1, then the heat flow is one-dimensional in the independent coordinate r. Laplace's equation, from the form of Eqn (2-8), therefore reduces to

$$\frac{d^2T}{dr^2} + \frac{1}{r}\frac{dT}{dr} = 0$$

which is equivalent to

$$\frac{d}{dr}\left(r\frac{dT}{dr}\right) = 0 \tag{2-16}$$

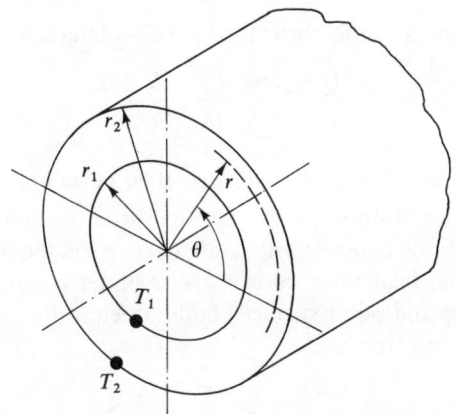

Fig. 2-6 Hollow cylinder with specified boundary temperatures

This equation has the solution

$$T = B\ln r + C \tag{2-17}$$

The boundary conditions

$$T = T_1 \text{ at } r = r_1 \quad \text{and} \quad T = T_2 \text{ at } r = r_2$$

give the arbitrary constants B and C in Eqn (2-17) as:

$$B = \frac{T_2 - T_1}{\ln(r_2/r_1)}; \quad C = T_1 - \frac{\ln r_1(T_2 - T_1)}{\ln(r_2/r_1)}$$

so that the temperature distribution through the cylinder wall is logarithmic and is given by

$$T = T_1 + \ln\left(\frac{r}{r_1}\right)\frac{T_2 - T_1}{\ln(r_2/r_1)}$$

The heat flow per unit length of cylinder, $\dot{Q}/L$, found by applying Fourier's law [Eqn (2-2)] is

$$\frac{\dot{Q}}{L} = -kA\frac{dT}{dr}$$

In this case, for unit length of cylinder of radius r, $A = 2\pi r$. Differentiation of Eqn (2-17), together with the expression for the constant B, gives

$$\frac{dT}{dr} = \frac{1}{r}\frac{T_2 - T_1}{\ln(r_2/r_1)}$$

which gives the expression for heat flow per unit length as

$$\frac{\dot{Q}}{L} = \frac{2\pi k(T_1 - T_2)}{\ln(r_2/r_1)} \tag{2-18}$$

2-7-2   Composite hollow cylinder with specified boundary temperatures

A common practical example of this configuration is a pipe (Fig. 2-7) with an outer coating of low-conductivity insulation. As in the previous example, for one-dimensional heat flow to exist the cylinder is considered infinitely long with the inner and outer surfaces maintained at the constant temperatures T_1 and T_3 respectively.

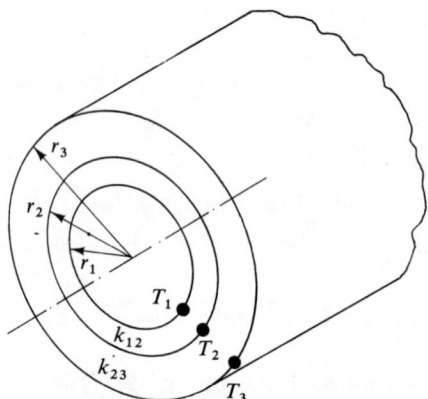

Fig. 2-7   Composite hollow cylinder with specified boundary temperatures

Using the expression for heat flow per unit length for a hollow cylinder [Eqn (2-18)], we can employ the same procedure as was used for a composite slab, that is, the heat flow through each interface is the same; however, in this particular case the areas at the inner and outer radii are different:

$$\frac{\dot{Q}}{L} = \frac{2\pi k_{12}(T_1 - T_2)}{\ln(r_2/r_1)} = \frac{2\pi k_{23}(T_3 - T_2)}{\ln(r_3/r_2)}$$

Eliminating the unknown interface temperature, T_2, gives

$$\frac{\dot{Q}}{L} = \frac{2\pi(T_1 - T_3)}{\dfrac{\ln (r_2/r_1)}{k_{12}} + \dfrac{\ln (r_3/r_2)}{k_{23}}} \tag{2-19}$$

Once $\dot{Q}/L$ is known the intermediate temperature T_2 may be calculated from the expression for heat flow through each layer. Equation (2-19) may be extended to cylinders with any number of different layers.

2-7-3 Composite hollow cylinder with fluid boundary conditions

Using the notation above, a composite double layer tube is shown in Fig. 2-8 separating two fluids at T_{f1} and T_{f3}. The convective heat transfer coefficients

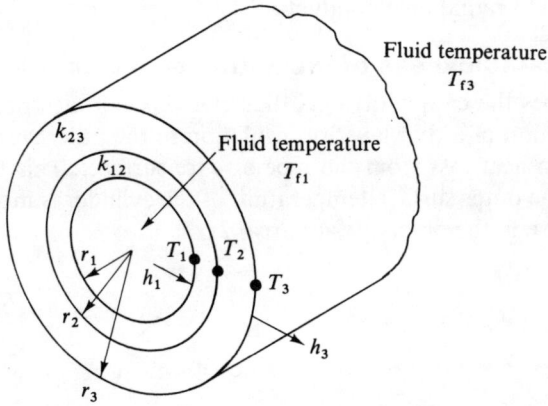

Fig. 2-8 Composite hollow cylinder with fluid boundaries

at the radii r_1 and r_3 are denoted by h_1 and h_3 respectively. As in the previous cases the heat flow through each layer is the same, and by using Eqns (2-11) and (2-19) we may write the heat flow per unit length as

$$\frac{\dot{Q}}{L} = 2\pi r_1 h_1(T_{f1} - T_1) = \frac{2\pi(T_1 - T_3)}{\dfrac{\ln (r_2/r_1)}{k_{12}} + \dfrac{\ln (r_3/r_2)}{k_{23}}} = 2\pi r_3 h_3(T_3 - T_{f3})$$

which, when expressed only in terms of the fluid temperatures, becomes

$$\frac{\dot{Q}}{L} = \frac{2\pi(T_{f1} - T_{f3})}{\dfrac{1}{r_1 h_1} + \dfrac{\ln (r_2/r_1)}{k_{12}} + \dfrac{\ln (r_3/r_2)}{k_{23}} + \dfrac{1}{r_3 h_3}} \tag{2-20a}$$

Thus the convective heat transfer coefficients appear in this equation in their reciprocal form as additional resistances, but coupled with the areas over which they act. Equation (2-20a) could be written as

$$\dot{Q} = \frac{T_{f1} - T_{f3}}{\dfrac{1}{2\pi r_1 L h_1} + \dfrac{1}{2\pi r_3 L h_3} + \dfrac{\ln (r_2/r_1)}{2\pi L k_{12}} + \dfrac{\ln (r_3/r_2)}{2\pi L k_{23}}}$$

$$= \frac{T_{f1} - T_{f3}}{\displaystyle\sum \frac{1}{Ah} + \sum \frac{\ln (r_{n+1}/r_n)}{2\pi L k}} \qquad (2\text{-}20b)$$

in which the terms $1/Ah$ represent the *resistance* (that is, the converse of conductance, Ah) to convective heat transfer, and the term $\ln (r_{n+1}/r_n)/2\pi Lk$ the resistance to radial heat conduction.

2-8 Critical thickness of insulation on a cylinder

It is sometimes the case with small-diameter pipes or electrical conductors that the addition of a thin layer of insulation to the outer surface results in an *increase* in heat loss from the pipe or wire surface. This is because the decrease in the outer surface temperature of the cylinder is more than offset by the increase in the outer surface area, $2\pi rL$.

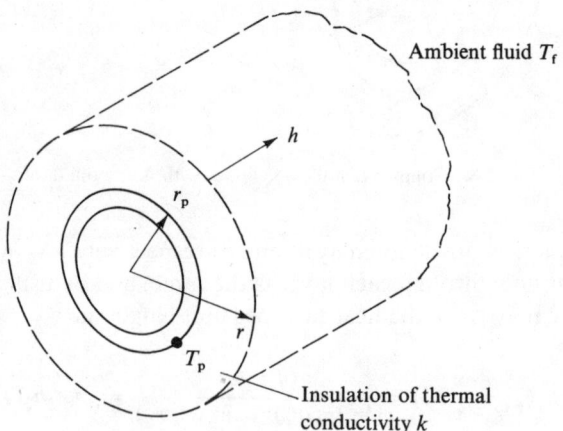

Fig. 2-9 Insulated hollow cylinder

An insulated pipe is shown in Fig. 2-9. For the purposes of analysis the outer surface temperature of the pipe, T_p, is assumed constant as r varies. This assumes an infinite conductance from the pipe fluid to the outer pipe surface. The heat transfer coefficient at the outer surface is also assumed constant as r varies.

With these assumptions Eqn (2-20a) gives the heat flow per unit length of pipe as

$$\frac{\dot{Q}}{L} = \frac{2\pi(T_p - T_f)}{\dfrac{1}{rh} + \dfrac{\ln (r/r_p)}{k}}$$

By differentiating this with respect to r and equating to zero, that is,

$$\frac{d}{dr}\left(\frac{\dot{Q}}{L}\right) = \frac{\left(\dfrac{1}{r^2 h} - \dfrac{1}{rk}\right)}{\left(\dfrac{1}{rh} + \dfrac{\ln (r/r_p)}{k}\right)^2} = 0$$

we see that there is a turning point in the function $(\dot{Q}/L)(r)$ when $r = k/h$. That the heat loss is, in fact, a maximum at this point is verified by the second derivative, $(d^2/dr^2)(\dot{Q}/L)$, at $r = k/h$, being negative.

This value of r is sometimes known as the critical radius, and since k is usually numerically much smaller than h, critical thicknesses of pipe insulation usually only occur with very small diameter tubes and wires.

2-9 Variable thermal conductivity

In all the above cases the thermal conductivity, k, has been assumed constant, and for many engineering problems involving small temperature differences across the materials this is satisfactory. However, most materials (and fluids) have temperature-dependent thermal conductivities. For limited ranges of temperature, k may be taken for most materials to have a linear variation with temperature:

$$k = BT + C \tag{2-21}$$

Fourier's law becomes

$$q = -k\frac{dT}{dx} = -(BT + C)\frac{dT}{dx}$$

In the steady state q is independent of x and we can integrate between x_1 and x_2 in Fig. 2-10 to give

$$q(x_2 - x_1) = -\frac{B}{2}(T_2^2 - T_1^2) - C(T_2 - T_1)$$

$$q = -\left(\frac{B}{2}(T_2 + T_1) + C\right)\frac{(T_2 - T_1)}{(x_2 - x_1)} \tag{2-22}$$

A mean conductivity for the system may be written as

$$k_m = \frac{k(T_1) + k(T_2)}{2} = \frac{(BT_1 + C) + (BT_2 + C)}{2}$$

Hence,

$$k_{\mathrm{m}} = \left(\frac{B}{2} (T_2 + T_1) + C \right) \tag{2-23}$$

By comparing Eqns (2-22) and (2-23) it is clear that the rate of heat transfer, q, may be evaluated using an arithmetic mean of the thermal conductivities at T_1 and T_2 provided a linear variation of k with T exists. However, as shown in Fig. 2-10, the temperature distribution through the wall is now not linear, but quadratic.

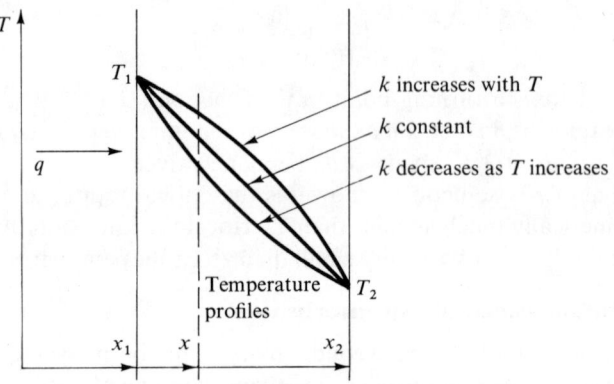

Fig. 2-10   Effect of temperature dependent thermal conductivity

PROBLEMS

2-1   Calculate the rate of heat loss through the vertical walls of a boiler furnace 4 m by 3 m by 3 m high. The walls are constructed from an inner firebrick wall 25 cm thick of thermal conductivity 0·4 W/m K, a layer of ceramic blanket insulation of thermal conductivity 0·2 W/m K, 8 cm thick and a steel protective layer of thermal conductivity 54 W/m K 2 mm thick. The inside temperature of the firebrick layer was measured at 600°C and the temperature of the outside of the insulation at 60°C. What is the temperature drop across the steel layer?

Ans: 22·1 kW, 1·95 × 10⁻² °C

2-2   A high pressure steam pipe 25 cm o.d., 2 cm thick, of thermal conductivity 54 W/m K carries steam at a temperature of 450°C. The pipe is covered by a layer of insulation 12 cm thick of thermal conductivity 0·04 W/m K. If the heat transfer coefficient from the steam to the pipe bore can be considered to be infinitely large and the outer surface temperature of the insulation is found to be 55°C calculate the heat loss from an 8 m length of the pipe.

Ans: 1·18 kW

2-3 A hot gas pipe, 0·3 m o.d., is covered with a layer of insulation A, 0·025 m thick, and a layer of insulation B, 0·04 m thick. The surface temperature of the pipe itself was found to be 400°C when the outer surface of layer B was at 40°C. After an additional layer of insulation, of thickness 0·02 m and thermal conductivity 0·2 W/m K, was added to the outer surface of layer B the pipe temperature was found to be 500°C, the outer surface of layer B 180°C, and the outer surface of the new insulation 30°C. What is the rate of heat loss per metre length of pipe before and after the addition of the new insulation.

Ans: 2·38 kW; 2·12 kW

2-4 A flat-roofed house has the effective dimensions 30 ft by 40 ft by 20 ft high. If the total window area in the house is 290 ft^2 calculate the total heat lost from the house in Btu/h and kW and also the ratio of the convective heat loss through the windows to the total heat loss from the house. For all four walls and the roof the inner and outer heat transfer coefficients are 1·1 and 2·1 Btu/ft^2 h °R and the inner and outer air temperatures are 70 and 35°F respectively. The window glass is $\frac{1}{8}$ inch thick and the 11 inch thick walls are constructed from two layers of brick each $4\frac{1}{2}$ inches thick, separated by a layer of foam insulation 2 inches thick. The roof is constructed from a $\frac{3}{8}$ inch thick layer of fibre insulating board, a 1 inch layer of glass wool and a $\frac{1}{2}$ inch thick layer of tile. The thermal conductivities (Btu/ft h °R) of the building materials may be taken as: glass 0·44, brick 0·26, foam insulation 0·02, fibre insulating board 0·028, glass wool 0·023, tile 0·6.

Ans: 20,956 Btu/h, 6·14 kW, 0·524

2-5 A steel pipe, 10 cm bore 12 cm o.d., carries hot water at 80°C. If the thermal conductivity of the pipe is 54 W/m K, the ambient temperature 15°C and the inner and outer heat transfer coefficients 1 kW/m^2 K and 9 W/m^2 K respectively, calculate the heat loss per metre length of pipe and the surface temperatures. Calculate also the heat loss and the surface temperatures when the pipe is covered with an insulation of thermal conductivity 0·048 W/m K, 4 cm thick with the outer surface heat transfer coefficient reduced to 7 W/m^2 K.

Ans: 217·85 W; 79·31°C; 79·20°C; 33·76 W; 79·89°C; 79·87°C; 22·68°C

2-6 The hot water pipe of Problem 2-5 has an additional 3 cm thick layer of insulation of thermal conductivity 0·03 W/m K added to it. Should it be placed on the outside or the inside of the existing insulation for the best overall insulation effect? What is the ratio of the heat loss for the two alternatives? The inner and outer heat transfer coefficients may be taken as given in Problem 2-5.

Ans: 1 : 0·869

2-7 A boiler tube has an outside diameter of $2\frac{7}{8}$ inches and walls $\frac{1}{4}$ inch thick, and on the inner surface a coating of scale has formed $\frac{1}{10}$ inch thick. In a laboratory test, in which water at 50°F was passed through the tube, a power input of 4 kW applied uniformly to a 1 foot length of tube produced an outer surface temperature of 250°F, compared with 150°F observed on an unscaled tube under identical conditions. In practice the boiler tube operates in gases at 3,000°F, from which the heat transfer coefficient is 30 Btu/ft^2 h °R, and the heat transfer coefficient within the tube is 2,000 Btu/ft^2 h °R to water boiling at 600°F.

Determine the coefficient of thermal conductivity of the scale and compare the maximum temperatures in operation of scaled and clean tubes. (Thermal conductivity of tube material = 26 Btu/ft h °R; 1 kW = 3,413 Btu/h.)

Ans: 2·1 Btu/ft h °R; 1,000°F; 705°F

2-8   A stainless steel hot gas duct 0·3 m square 30 m long has a wall thickness of 2 mm and is insulated with ceramic blanket 38 mm thick. The heat transfer coefficient from gas to inner wall is 0·1 kW/m² K and from outer wall to surroundings 0·01 kW/ m² K. If the mass flow of gas is 1·5 kg/s with an entry temperature of 800°C what is the bulk mean temperature of the gas at exit? Take the thermal conductivity of stainless steel and ceramic blanket to be 25 W/m K and 0·2 W/m K respectively, the specific heat of the gas to be 1·1 kJ/kg K, and the ambient temperature to be 20°C.

Ans: 745·3°C

2-9   An experiment is designed to measure the heat transfer from a gas stream at 400°F to the outer surface of a tube of 1 inch i.d. and 2 inches o.d. Coolant is passed through the bore at a known rate so that the heat transfer coefficient on the inner surface is fixed at 20 Btu/ft² h °R. Under test the bulk mean temperature of the coolant was 60°F when the outer surface mean temperature was 300°F. If the thermal conductivity of the tube material is 10 Btu/ft h °R determine the heat transfer co-efficient from the gas to the tube surface. What is the percentage error in this experi-mental value for an estimated ±5% error in the inner surface heat transfer coefficient and a ±2°F error in the measured surface temperature?

Ans: 22·7 Btu/ft² h °R; ±5%, ±2·9%

2-10   The sensing element of a hot-wire anemometer consists of a tungsten wire 5 μm diameter 1·2 mm long. The tungsten wire dissipates heat into an air stream according to the equation

$$\mathbf{N} = (T_m/T_0)^{0.17}(0.24 + 0.56\,\mathbf{R}^{0.45}).$$

where $\mathbf{N}$ is the Nusselt number defined as hd/k_f (h is the heat transfer coefficient, d is the wire diameter and k_f is the thermal conductivity of the air), $\mathbf{R}$ is the Reynolds number of the flow and (T_m/T_0) is the *absolute* temperature ratio of the mean tem-perature between fluid and wire $[T_m = \frac{1}{2}(T_w + T_0)]$ and T_0 is the ambient fluid temperature. If the Reynolds number has a value of 15 at an ambient temperature of 20°C and the temperature of the wire is 300°C calculate the power dissipated by the wire. Take the thermal conductivity of the air to be 0·043 W/m K and ignore temperature variations in the wire.

Ans: 0·105 W

2-11   An electric heater element is constructed from a nickel wire 2·5 mm diameter insulated with ceramic from a metal outer sheath of inner and outer diameter 6 mm and 7 mm respectively. The heat dissipated in the conducting wire is 2 kW/m; the ambient temperature is 30°C and the effective heat transfer coefficient 100 W/m² K. If the ceramic and metal sheath thermal conductivities are 0·7 W/m K and 13 W/m K respectively, determine the temperature of the outer surfaces of the nickel wire and the metal sheath. If the ambient temperature and the heat transfer coefficient remain constant would the element be physically capable of dissipating 3 kW/metre?

Ans: 1,341°C; 939·3°C; No

2-12   The thermal conductivity of a plane wall varies quadratically as $k = k_0 \times (1 + bT + cT^2)$. If the wall is δx thick and the surface temperatures are T_1 and T_2 show that the heat flux, q, through the wall is given by

$$q = -k_0\frac{(T_2 - T_1)}{\delta x}\left[1 + \frac{b}{2}(T_2 + T_1) + \frac{c}{3}(T_2^2 + T_1T_2 + T_1^2)\right]$$

2-13   An electrical conductor dissipates heat at a constant rate, $\dot{Q}$, per unit length. The conductor is covered by a layer of material of thermal conductivity k. If the heat transfer coefficient between the outer surface of the insulation and the surrounding has a value h, show that the temperature of the conductor surface is a minimum when the outer radius of the insulation has a value k/h.

2-14   An electrical wire 1 mm diameter dissipates 0·5 kW/m suspended in an air stream. If the heat transfer coefficient is 370 W/m^2 K and the ambient temperature is 100°C, determine the temperature of the wire. The temperature variation in the wire may be neglected. An insulation of thermal conductivity 0·277 W/m K is then added to the wire increasing its outer diameter to 1·5 mm. Assuming the heat transfer coefficient remains constant at 370 W/m^2 K, determine the new wire temperature and explain the physical significance of the result.

Ans: 530°C; 503·5°C

2-15   Show that the heat flow rate through a spherical shell of inner and outer radii r_1 and r_2 respectively is given by

$$\dot{Q} = \frac{4\pi k r_1 r_2}{r_2 - r_1}(T_1 - T_2)$$

where k is the thermal conductivity of the shell, and the inner and outer surfaces are maintained at the temperatures T_1 and T_2 respectively. (N.B. the surface area of a sphere is given by $4\pi r^2$.)

2-16   Calculate the heat loss (in Btu/h and kW) from a steel cylindrical steam drum 5 ft internal diameter, 1 inch thick with hemispherical ends. The cylindrical section is 4 ft long and the drum is covered with a 4 inch thick layer of insulation. The inner and outer heat transfer coefficients are 300 and 5 Btu/ft^2 h °R, and the corresponding fluid temperatures are 350°F and 80°F respectively. The thermal conductivities of the steel and the insulation are, respectively, 31 and 0·03 Btu/ft h °R.

Ans: 3,916 Btu/h; 1·15 kW

2-17   A domestic hot-water tank comprises a cylinder 18 inches in diameter and 3 feet long with a hemispherical top and its surface is maintained at 180°F by the fluid inside. Determine the percentage reduction in heat loss if the tank is lagged with a 2 inch thickness of insulation having a thermal conductivity of 0·1 Btu/ft h °R. The external heat transfer coefficient is 1·0 Btu/ft^2 h °R to air at 70°F, and the lower end of the tank may be assumed to be perfectly insulated throughout.

Ans: 56·0%

2-18   A chemical reaction takes place in a spherical pressure vessel. The rate of heat release per unit volume from the reaction in Btu/ft^3 h is

$$q = K\, e^{\theta/1,000}$$

where θ is the absolute temperature in the vessel in °R. The internal diameter of the sphere is 2 feet, its wall, of thermal conductivity 10 Btu/ft h °R, is 6 inches thick and it is covered with a 2 inch thickness of lagging of conductivity 1·0 Btu/ft h °R. If the heat transfer coefficient between the lagging and the atmosphere, which is at 60°F, is 3·0 Btu/ft^2 h °R and the temperature in the vessel at which the reaction proceeds in equilibrium is 740°F, determine the value of the constant, K, in the reaction equation.

Derive from first principles any equation used.

Ans: 2,800

3

Analytical solution of conduction problems

The heat conduction problems considered so far can all be classified under the heading of 'simple conduction'. They have all involved relatively straightforward geometries with a single independent variable in steady equilibrium conditions. Finned or extended surfaces are sometimes treated in this context although many of these problems involve geometries which require a deeper mathematical treatment for their solution. Accordingly, this chapter considers (1) steady, one-dimensional heat conduction in finned or extended surfaces, and (2) heat conduction in two independent variables, which involves partial differential equations.

Analytical solutions are obviously limited in their range of applicability and many practical two- and three-dimensional engineering problems require numerical methods for their solution. However, assumptions can usually be made in most practical problems which reduce them to simple geometrical shapes with straightforward boundary conditions for which many analytical solutions have been tabulated.

The classical works on analytical heat conduction are considered to be those of Fourier[1]† and Carslaw and Jaeger[2]. Fourier, of course, laid the foundations of conduction in solids, while Carslaw and Jaeger later described numerous solutions to many different geometries and situations. This latter work is principally an operational text in that the mathematical methods of solution are only briefly outlined. Carslaw and Jaeger were principally concerned with conduction for its own sake, but other authors such as Arpaci[3] have outlined conduction in an engineering context with the solutions described in more detail.

Although analytical solutions are possible in some cases they require a great deal of analysis and often a great deal of computation to evaluate numerically. However, despite the complexity of some analytical solutions they have obvious advantages in that they are exact solutions and that the majority are relatively easy to evaluate numerically. Also, many standard solutions have great value in the preliminary assessment of numerical algorithms which will eventually be applied to nominally similar problems which have no known analytical solution.

This chapter can in no way attempt a complete presentation of the analytical methods available for the solution of heat conduction problems. We

† Full details of references cited are given at the end of the chapter.

only consider some of the more commonly used and more straightforward procedures employed in engineering practice. Although the emphasis is placed on the 'separation of variables' method and its application to standard geometrical shapes, an outline is given of some alternative methods that can be used particularly where the nature of the geometry considered means that the separation of variables method is inapplicable.

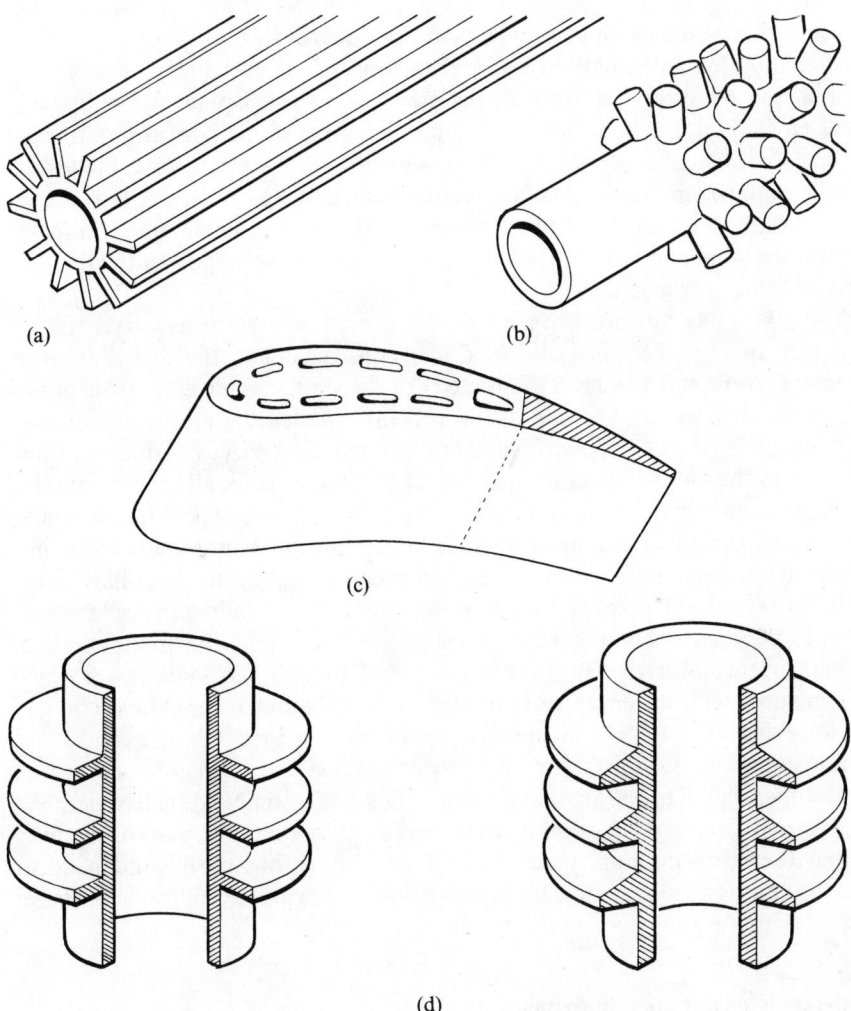

Fig. 3-1 Examples of finned surfaces: (a) straight uniform fins; (b) pin fins or spines; (c) triangular fin; (d) circular or annular fins

3-1   Fins

The rate of convective heat transfer from a primary surface is directly proportional to the surface area presented to the fluid, and it is a common practice in engineering to increase this area by attaching thin strips of metal or *fins* to the primary surface. These heat conducting strips, often referred to as an 'extended surface', exist in a variety of forms, some of which are shown in Fig. 3-1. They have many different practical applications ranging from the fins on the cylinders of air cooled engines and compressors to the pin fins or spines on high-efficiency boiler superheater tubes.

Many physical situations other than finned heat exchanger tubes can be analysed by approximating the geometry as a fin or spine. An example is given in Fig. 3-1 in which the trailing edge of an air cooled gas turbine blade is treated as a truncated triangular fin. Solid gas turbine blades are often approximated as fins conducting heat down their length into a cool disc. In heat transfer instrumentation the error estimation in many temperature measurement installations is made by treating the support or the thermocouple leads as a spine.

Accordingly, fin problems are posed in one of two forms: first, given a fin or spine of certain physical dimensions, calculate the distribution of temperature and the heat dissipated; or, second, design a fin or a finned surface that will give maximum heat transfer efficiency and sometimes, in addition, have adequate strength for minimum cost, size, weight, and resistance to the fluid. The latter problem is, of course, generally more complex than the former since many additional factors have to be taken into account.

In the analytical treatment of steady heat flow in fins it is usual to assume that the thermal conductivity of the fin is constant, that the heat flow to the fin surface at any point is directly proportional to the fluid-surface temperature difference, that the heat transfer coefficient and the fluid reference temperature over the fin are constant, and that the fin thickness is small compared with its length so that, effectively, one-dimensional heat conduction exists. In a review and analysis of fins by Gardner[4] these assumptions are shown to be satisfactory for most practical fins. It is also shown by Gardner that many straight and circular fins can be analysed by first deriving a basic differential equation in which the surface area presented to the fluid and the cross-sectional area of the fin are functions of the independent coordinate and which, in general, has a solution expressed in terms of Bessel functions.

3-1-1   General one-dimensional fin

The general differential equation mentioned above can be derived by considering a one-dimensional fin with an arbitrary geometry as shown in Fig. 3-2. By considering the heat flow to and from an elemental cross-section of this fin δx thick and distance x from the base, we can write a heat balance for the element as

$$\dot{Q}_x = \dot{Q}_{x+\delta x} + \dot{Q}_c \qquad (3\text{-}1)$$

where $\dot{Q}_c$ is the heat flow per unit time convected through the surface of the element δS, considered to be all at temperature T_s, out into the fluid at T_f. As the heat flows $\dot{Q}_x$ and $\dot{Q}_{x+\delta x}$ are only dependent upon the x coordinate, $\dot{Q}_{x+\delta x}$ can be expressed in terms of $\dot{Q}_x$ by means of a Taylor expansion

$$\dot{Q}_{x+\delta x} = \dot{Q}_x + \delta x \frac{d}{dx}(\dot{Q}_x) + \frac{\delta x^2}{2!}\frac{d^2}{dx^2}(\dot{Q}_x) + \cdots$$

which, when substituted into the element heat balance, Eqn (3-1), gives

$$\delta x \frac{d}{dx}(\dot{Q}_x) + \frac{\delta x^2}{2!}\frac{d^2}{dx^2}(\dot{Q}_x) + \cdots + \dot{Q}_c = 0$$

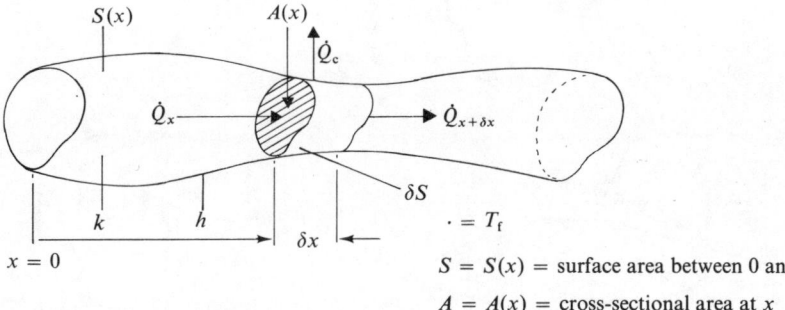

$S = S(x) =$ surface area between 0 and x

$A = A(x) =$ cross-sectional area at x

Fig. 3-2   A general one-dimensional fin

It is usual either to consider the fluid temperature T_f as zero by redefining the variable for temperature, or to express the dependent variable T as the difference $(T_s - T_f)$. This simplification will be used throughout the chapter. We can therefore write

$$\dot{Q}_c = h\,\delta S\,T \quad \text{and} \quad \dot{Q}_x = -kA\frac{dT}{dx}$$

where h is the heat transfer coefficient and k is the thermal conductivity, so that the heat balance becomes

$$\delta x \frac{d}{dx}\left(-kA\frac{dT}{dx}\right) + \frac{\delta x^2}{2!}\frac{d^2}{dx^2}\left(-kA\frac{dT}{dx}\right) + \cdots + h\,\delta S\,T = 0$$

This is true for any position along the fin, so that as $\delta x \to 0$ we obtain

$$\frac{1}{A}\frac{d}{dx}\left(-kA\frac{dT}{dx}\right) + \frac{h}{A}\frac{dS}{dx}T = 0$$

in which the first term can be expanded so that we obtain as the differential equation of a general one-dimensional fin :

$$\frac{d^2T}{dx^2} + \left(\frac{1}{A}\frac{dA}{dx}\right)\frac{dT}{dx} - \frac{h}{k}\left(\frac{1}{A}\frac{dS}{dx}\right)T = 0 \tag{3-2}$$

This equation is applicable to all fins in which the heat flow is dependent upon a single coordinate, as with straight and circular, parabolic, hyperbolic, or triangular profiled fins.

3-1-2   Simple straight fins and spines

In these two simple cases the cross-sectional area A is constant and $S = Px$ so that the general fin equation [Eqn (3-2)] can be simplified to:

$$\frac{d^2T}{dx^2} - \frac{hP}{kA}T = 0 \tag{3-3}$$

or

$$\frac{d^2T}{dx^2} - m^2T = 0$$

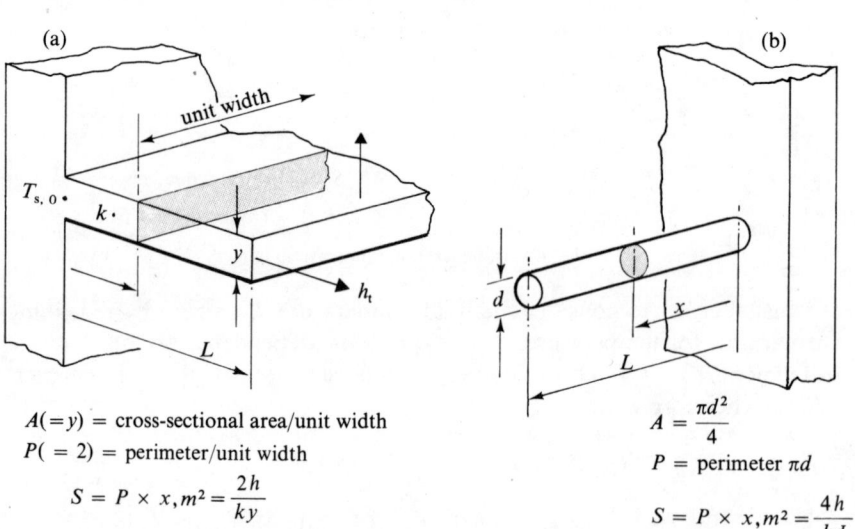

$A(=y)$ = cross-sectional area/unit width

$P(=2)$ = perimeter/unit width

$$S = P \times x, m^2 = \frac{2h}{ky}$$

$$A = \frac{\pi d^2}{4}$$

P = perimeter πd

$$S = P \times x, m^2 = \frac{4h}{kd}$$

Fig. 3-3   Simple straight fin and spine

where $m^2 = hP/kA$, that is, $m^2 = 2h/ky$ for the straight fin (Fig. 3-3(a)). The solution to this differential equation can then be written down simply as

$$T = C_1 e^{-mx} + C_2 e^{mx} \tag{3-4}$$

where C_1 and C_2 are constants which depend upon the boundary conditions imposed on the problem. If, for example, the boundary conditions are those shown in Fig. 3-3, then $T = T_0$ (that is, $T_s = T_{s,0}$) at $x = 0$ and $-k(dT/dx) = h_t T$ [that is, $-k(dT_s/dx) = h_t(T_s - T_f)$] at $x = L$, in which k is the metal conductivity and h_t is the heat transfer coefficient at the tip. Applying these conditions we have:

at $x = 0$: $$T_0 = C_1 + C_2$$

and at $x = L$:

$$m\left[-C_1 e^{-mL} + C_2 e^{mL}\right] = \frac{-h_t}{k}\left[C_1 e^{-mL} + C_2 e^{mL}\right]$$

which yields C_1 and C_2 as

$$C_2 = T_0 - C_1$$

$$C_1 = \frac{T_0(1 + h_t/mk)e^{mL}}{(e^{mL} + e^{-mL}) + (h_t/mk)(e^{mL} - e^{-mL})}$$

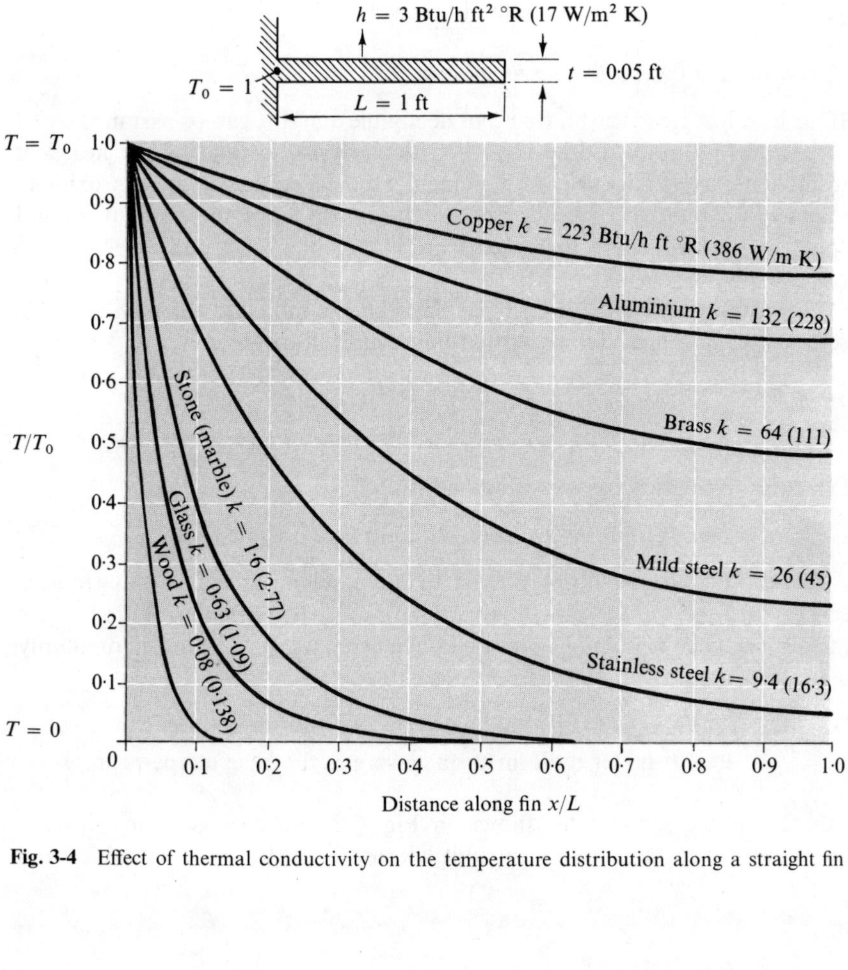

Fig. 3-4 Effect of thermal conductivity on the temperature distribution along a straight fin

which, when substituted into Eqn (3-4) and simplified, gives the temperature distribution along the fin as

$$T = T_0 \left[\frac{\cosh m(L - x) + N \sinh m(L - x)}{\cosh mL + N \sinh mL} \right] \tag{3-5}$$

where $N = h_t/mk$. For many practical fins the heat convected from the fin tip is negligibly small, especially for long fins of low conductivity, and in this case $h_t \approx 0$, that is, $N = 0$ and Eqn (3-5) can be simplified to

$$T = T_0 \frac{\cosh m(L - x)}{\cosh mL} \tag{3-6}$$

To illustrate the effect of thermal conductivity this expression for the temperature distribution in a straight fin has been applied to a fin of unit length in Fig. 3-4 for widely different thermal conductivities. The heat transfer coefficient h was assumed constant at 3 Chu/h ft² K (17·04 W/m² K), a value typical of a slowly moving air stream.

Total heat lost from a fin and fin efficiency

If the heat lost from the fin tip is *not* negligible and if h_t can be assumed equal to h, Eqn (3-6) may still be used—particularly for evaluating the heat lost from a fin—by adding to L an increment which accounts, in an approximate manner, for the extra surface area at the fin tip. For the straight fin and spine in Fig. 3-3 these increments are ($\frac{1}{2}$ × thickness) and ($\frac{1}{2}$ × radius) respectively.

Now any heat lost by a fin must have been conducted through the base so that the *total* heat dissipated per unit time will be given by

$$\dot{Q} = -kA \left(\frac{dT}{dx} \right)_{x=0}$$

Equation (3-6) gives for the simplified case

$$\dot{Q} = mkAT_0 \tanh (mL) \tag{3-7}$$

In heat exchangers fins are used to increase the effective heat transfer surface area presented to the fluid and in practice it is often useful to express the performance of a fin in terms of a fin 'efficiency', ϕ. The most commonly used definition is

$$\phi = \frac{\text{actual heat transfer from fin}}{\text{heat transfer if all fin surface were at the base temperature}}$$

For the simple straight fin shown in Fig. 3-3, the cross-sectional area A and the total surface area S per unit width are, respectively, y and $2L$ and

$m^2 = 2h/ky$, where y is the fin thickness, so that the fin efficiency of a straight, uniform fin is given by

$$\phi = \frac{mkAT_0 \tanh mL}{2LhT_0} = \frac{1}{mL} \tanh mL \qquad (3\text{-}8)$$

The above definition of fin efficiency is satisfactory provided that the effect of the increase in surface area is greater than that of the decrease in the mean surface temperature due to the addition of the fin. By differentiating Eqn (3-5) with respect to x, the heat flow, $\dot{Q}$, through the base of the fin can be shown to be

$$\dot{Q} = kAT_0 m \left[\frac{\tanh mL + N}{1 + N \tanh mL} \right]$$

It can be seen from this expression that for $N = 1$, $\dot{Q}$ is independent of the length of the fin L. If we now assume $h_t = h$ the definitions of N and m lead to the result that the ratio hy/k has an *optimum* value of 2. This implies that when $h < 2k/y$ ($N < 1$) the fin is useful and when $h > 2k/y$ ($N > 1$) the addition of fins will *decrease* the net heat transfer from the primary surface and defeat its original purpose.

This indicates that, in general, fins are very effective with gases, are less effective with liquids, and are disadvantageous with condensation or evaporation vapours. This becomes clear physically if the reciprocal of the heat transfer coefficient h is regarded as a resistance to heat flow, so that if h is low the resistance at the fin surface is high and the conductivity (or resistance) of the fin has little effect on the system; but if h is very high (as in condensing liquids), the surface resistance to heat flow is low and the extra thermal resistance of the fin constitutes an insulation to heat flow.

Fin spacing

One of the original assumptions made in the derivation of the basic differential equation is that the heat transfer coefficient h is constant over the fin length. In practice this is not always the case and it raises the designer's problem of fin spacing since, after fin efficiency, the correct spacing of the fins is of primary importance in the design of secondary surfaces, as fins are sometimes known.

Fins should be placed so that they do not interfere with each other, and theoretically this means that the boundary layers which develop on adjacent fins should be allowed to do so independently. Using the methods outlined in Chapters 6 and 7 it is possible in many situations to calculate the development of the boundary layer, but in practice a closer spacing than that given by boundary layer considerations is satisfactory. For many practical designs experiments have been conducted (see the discussion in reference (4)) in which the fin spacing has been varied, with all other parameters remaining constant, to indicate the spacing which yields the maximum heat dissipation from the primary surface.

3-1-3 The triangular fin

The truncated triangular fin shown in Fig. 3-5 is an important example of a straight fin with a non-uniform cross-section since many fins, for reasons of strength or fabrication, have bases which are thicker than their tips. The

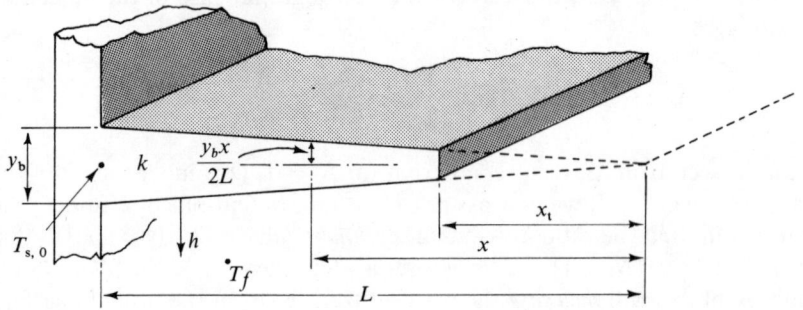

Fig. 3-5 Truncated triangular fin

analysis of the triangular fin is simplified if the origin of the x coordinate is taken at the fin tip as shown in Fig. 3-5. If we consider a unit width of fin, as in the simple fin analysis, the variation of cross-sectional area and surface area with x may be written (using the nomenclature in Fig. 3-5) as

$$A = \frac{y_b x}{L}$$

and

$$S = 2\left\{\sqrt{\left[x^2 + \left(\frac{y_b x}{2L}\right)^2\right]} - \sqrt{\left[x_t^2 + \left(\frac{y_b x_t}{2L}\right)^2\right]}\right\}$$

$$= 2(x - x_t)\sqrt{\left[1 + \left(\frac{y_b}{2L}\right)^2\right]}$$

Usually if the one-dimensional assumption is valid (that is, if the fin thickness is small compared with its length), the term $y_b/2L$ will be small and may be neglected for most practical fins.

The general one-dimensional fin equation [Eqn (3-2)] is applicable to this geometry and by substituting the expressions for cross-sectional and surface area we can obtain the differential equation :

$$\frac{d^2T}{dx^2} + \frac{1}{x}\frac{dT}{dx} - \frac{p^2 T}{x} = 0 \tag{3-9}$$

where $$p^2 = \frac{2hL}{ky_b} \sqrt{\left[1 + \left(\frac{y_b}{2L}\right)^2\right]}$$

This is a disguised form of Bessel's zero order differential equation, and by comparing it with the generalized Bessel equations its general solution is seen to be

$$T = C_1 I_0(2px^{1/2}) + C_2 K_0(2px^{1/2}) \qquad (3\text{-}10)$$

where the constants C_1 and C_2 are determined from the boundary conditions imposed on the problem, and I_0 and K_0 are modified Bessel functions of the first and second kind, respectively.

For a given numerical value of the argument $(2px^{1/2})$ these Bessel functions, I_0 and K_0, also have fixed numerical values, some of which are given in Tables A12 and A13 in the Appendix.

With the temperature specified at the fin base and with convection at the tip we may express the boundary conditions as

$$T = T_0 \quad \text{at } x = L$$

and $$-k\frac{dT}{dx} = h_t T \quad \text{at } x = x_t$$

The first of these conditions gives, from Eqn (3-10),

$$T_0 = C_1 I_0(2pL^{1/2}) + C_2 K_0(2pL^{1/2}) \qquad (3\text{-}11a)$$

and from the second condition, noting that differentiation of Bessel functions yields

$$\frac{d}{dx} I_0(u) = I_1(u)\frac{du}{dx}; \qquad \frac{d}{dx} K_0(u) = -K_1(u)\frac{du}{dx}$$

we obtain the expression

$$-k[C_1 p x_t^{-1/2} I_1(2px_t^{1/2}) - C_2 p x_t^{-1/2} K_1(2px_t^{1/2})]$$
$$= h_t[C_1 I_0(2px_t^{1/2}) + C_2 K_0(2px_t^{1/2})] \quad (3\text{-}12a)$$

Equations (3-11a) and (3-12a) may then be used to evaluate C_1 and C_2.

Example. As an example of the truncated triangular fin we consider the trailing edge region of an air cooled gas turbine blade as shown in Fig. 3-6 in which the temperature of the metal has been measured experimentally at the base and the temperature of the tip region is required. The *mean* heat transfer coefficient, h, over the whole fin is estimated to be 575 Chu/h ft^2 K (3·27 kW/m^2 K) and the thermal conductivity of the metal, k, is taken to be 15·5 Chu/h ft K (26·8 W/m K), a value which is typical of a high-temperature nickel alloy. These values, together with the dimensions given in Fig. 3-6, allow the following parameters to be calculated:

$$p = \sqrt{\left\{\frac{2hL}{ky_b}\left[1 + \left(\frac{y_b}{2L}\right)^2\right]^{1/2}\right\}} = \sqrt{\left(\frac{2 \times 575 \times 0\cdot07}{15\cdot5 \times 0\cdot009}(1\cdot002)\right)} = 24\cdot05$$

$$2px_t^{1/2} = 2 \times 24\cdot05 \times 0\cdot1483 = 7\cdot13$$
$$2pL^{1/2} = 2 \times 24\cdot05 \times 0\cdot265 = 12\cdot7$$

Substituting these parameters into Eqns (3-11a) and (3-12a) gives

$$T_0 = C_1 I_0(12\cdot7) + C_2 K_0(12\cdot7) \tag{3-11b}$$

$$-\frac{15\cdot5 \times 24\cdot05}{575 \times 0\cdot1483}[C_1 I_1(7\cdot13) - C_2 K_1(7\cdot13)] = C_1 I_0(7\cdot13) + C_2 K_0(7\cdot13)$$
$$\tag{3-12b}$$

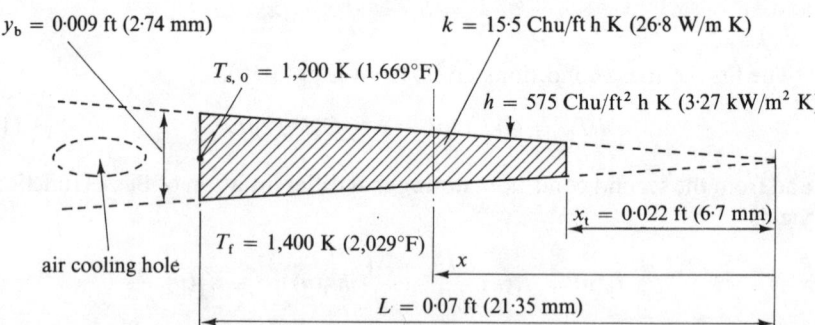

Fig. 3-6   A fin approximation to a gas turbine blade trailing edge

From tables of Bessel functions we obtain the values

$$I_0(7\cdot13) = 189\cdot2 \qquad\qquad K_0(7\cdot13) = 0\cdot376 \times 10^{-3}$$
$$I_0(12\cdot7) = 0\cdot370 \times 10^5 \qquad K_0(12\cdot7) = 0\cdot106 \times 10^{-5}$$
$$I_1(7\cdot13) = 175\cdot3 \qquad\qquad K_1(7\cdot13) = 0\cdot402 \times 10^{-3}$$

These values can be substituted into Eqns (3-11b) and (3-12b), noting that $T_0 = 1,200 - 1,400 = -200$ K, to give the constants

$$C_1 = -0\cdot00539; \qquad C_2 = -1,459\cdot0$$

The fin temperature at the tip can then be found from Eqn (3-10):

$$T = C_1 I_0(2px_t^{1/2}) + C_2 K_0(2px_t^{1/2})$$
$$T = -0.00539 \times 189.2 - 1,459.0 \times 0.000376 = -1.57 \text{ K}$$

This calculation indicates that only a 1·57 K difference in temperature exists between the gas turbine blade trailing edge and the hot gas stream, and that the air cooling hole shown in Fig. 3-7 has very little cooling effect in this particular design.

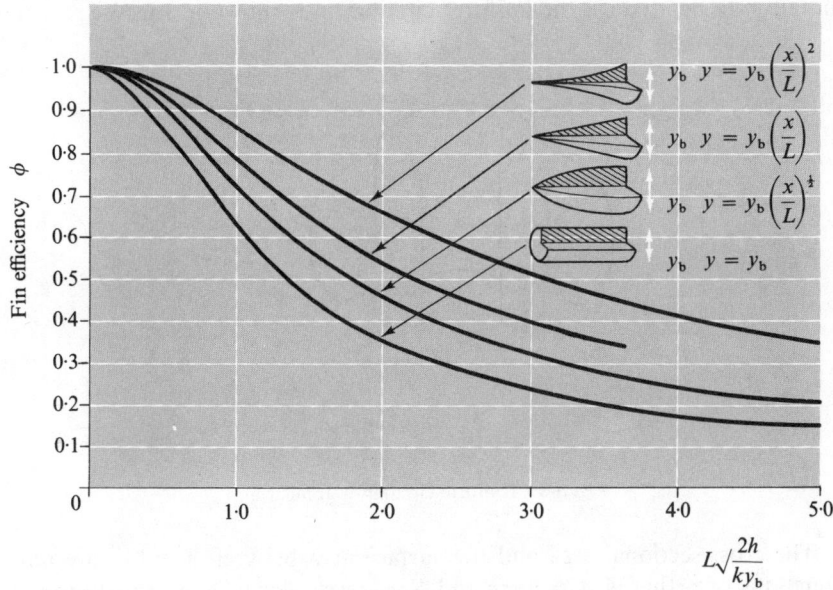

Fig. 3-7 Heat transfer efficiency of four differently profiled spines (after Gardner[4])

Fins and spines with other profiles

Many fins are used in practice with other than uniform or triangular profiles, and the analysis and efficiency of several types is presented by Gardner[4]. To illustrate the effect of the fin profile the efficiency of the straight uniform spine, based on the definition given in Section 3-1-2, is compared with the efficiency of the triangular spine and both types of parabolic profile, that is, thickness $\propto x^2$ and thickness $\propto x^{1/2}$ in Fig. 3-7. In this figure the efficiency is plotted against the square root of a dimensionless group which, for a spine of given physical dimensions and thermal conductivity, may be regarded as a non-dimensional heat transfer coefficient between fin surface and fluid.

The differences between similarly profiled straight fins, as distinct from spines, are obviously not as large but such comparisons are given by Gardner together with efficiency curves for circular fins with uniform and rectangular profiles. The application of fins to the design of heat exchangers is discussed in Chapter 10.

3-1-4 Uniform circular or annular fin

This type of fin is used extensively on radiator tubes and heat exchanger elements of many kinds, and two configurations are shown in Fig. 3-1. The notation to be used for the uniform circular fin is shown in Fig. 3-8.

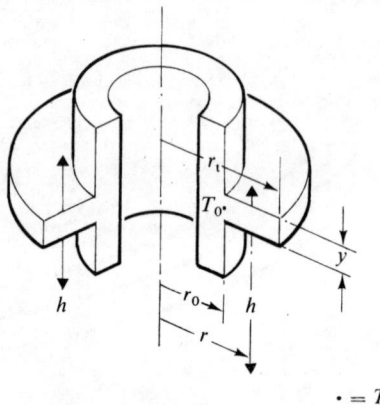

$\bullet = T_f$

Fig. 3-8 Uniform circular or annular fin

The cross-sectional area and the surface area between r and r_0 are now functions of r, that is $A = 2\pi r y$ and $S = 2\pi(r^2 - r_0^2)$. The substitution of these values into the general fin equation [Eqn (3-2)], in which r is now the independent variable, gives

$$\frac{d^2T}{dr^2} + \left(\frac{1}{2\pi r y} 2\pi y\right)\frac{dT}{dr} - \frac{h}{k}\left(\frac{1}{2\pi r y} 4\pi r\right)T = 0$$

$$\frac{d^2T}{dr^2} + \frac{1}{r}\frac{dT}{dr} - \frac{2h}{ky}T = 0 \qquad (3\text{-}13)$$

As in the case of the triangular fin equation, this is a form of Bessel's zero order differential equation and its solution is given in terms of Bessel functions as:

$$T = C_1 I_0(pr) + C_2 K_0(pr) \qquad (3\text{-}14)$$

where $p^2 = 2h/ky$, C_1 and C_2 are constants, and I_0 and K_0 are modified Bessel functions of the first and second kind, respectively. For a given numerical value of (pr), tables of Bessel functions may be consulted to give the corresponding numerical values of I_0 and K_0.

The constants C_1 and C_2 must be determined as before from the boundary conditions on the ends of the fin. The most simple practical boundary conditions we may impose are those of known temperature at the fin base and no heat loss from the fin tip:

$$T_s = T_{s,0} \quad \text{or} \quad T = T_0 \quad \text{at } r = r_0$$

and

$$\frac{dT_s}{dr} = \frac{dT}{dr} = 0 \qquad \text{at } r = r_t$$

$$(3\text{-}15)$$

The first condition in Eqn (3-15) gives

$$T_0 = C_1 I_0(pr_0) + C_2 K_0(pr_0)$$

and by noting that

$$\frac{d}{dr} I_0(ar) = aI_1(ar) \quad \text{and} \quad \frac{d}{dr} K_0(ar) = -aK_1(ar)$$

the second condition in Eqn (3-15) gives

$$pC_1 I_1(pr_t) - pC_2 K_1(pr_t) = 0$$

Thus the constants C_1 and C_2 can be determined from these two expressions to give the particular solution for the variation of temperature with radius:

$$T = T_0 \left[\frac{I_0(pr)K_1(pr_t) + I_1(pr_t)K_0(pr)}{I_0(pr_0)K_1(pr_t) + I_1(pr_t)K_0(pr_0)} \right] \qquad (3\text{-}16)$$

The total heat dissipated by the fin must be that conducted through the fin base at r_0, therefore,

$$\dot{Q} = -kA \left(\frac{dT}{dr} \right)_{r=r_0}$$

$$= -k2\pi r_0 y T_0 p \left[\frac{I_1(pr_0)K_1(pr_t) - I_1(pr_t)K_1(pr_0)}{I_0(pr_0)K_1(pr_t) + I_1(pr_t)K_0(pr_0)} \right] \qquad (3\text{-}17)$$

As in the case of the straight uniform fin, the heat loss from the fin tip may be accounted for, to a first approximation, by using a fictitious outer radius in Eqns (3-16) and (3-17) so that the Bessel functions I_1 and K_1 are evaluated at r_t' where $r_t' = \sqrt{[r_t(r_t + y)]}$.

3-2 Conduction in two dimensions (separation of variables)

This section discusses conduction problems that have to be described in two independent variables. The first part considers steady conduction in two space dimensions, and the second part considers transient heat conduction in one space dimension with time as the second coordinate.

The 'separation of variables' technique is considered to be the classical approach to the solution of two-dimensional boundary-value problems and it represents a straightforward approach which, if an analytical solution is feasible, is usually successful.

It will be shown in Section 3-3 that other mathematical techniques are equally effective and for some problems, such as those which have one or more dimensions extending to infinity, they can yield solutions which cannot be obtained by the separation of variables method. However, for student use the separation of variables technique is the simplest approach. With two independent variables this method reduces the partial differential equation to two ordinary differential equations which, for straightforward geometries such as rectangular plates and circular cylinders, have well-known or easily determined solutions.

3-2-1   Steady-state conduction in rectangular sections and circular cylinders

Rectangular plates and slabs

Figure 3-9 shows a two-dimensional geometry which can be treated using cartesian coordinates. For two-dimensional heat transfer to occur, conduction in the direction normal to the section, the z direction in Fig. 3-9, must be zero. This means that the section illustrated can apply equally to a thin plate or to a section through a slab, provided the above conditions hold.

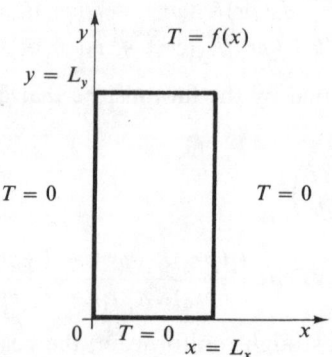

Fig. 3-9   Rectangular plate with temperature specified boundary conditions

For steady equilibrium conditions the equation describing the temperature distribution in the section is Laplace's equation, derived in Chapter 2:

$$\frac{\partial^2 T}{\partial x^2} + \frac{\partial^2 T}{\partial y^2} = 0 \qquad (3\text{-}18)$$

This equation is linear and the separation of variables method assumes that the solution to the equation may be expressed as a product of two functions which are each only functions of one independent variable. The solution may therefore be written as

$$T = X(x)Y(y) \tag{3-19}$$

where X is a function of x only and Y is a function of y only. By substituting this solution into Eqn (3-18) we obtain

$$\frac{1}{X}\frac{d^2X}{dx^2} = -\frac{1}{Y}\frac{d^2Y}{dy^2}$$

Since the variables in this equation are separated, each side is a constant. Taking this constant to be $-p^2$† we obtain two separate equations:

$$\frac{1}{X}\frac{d^2X}{dx^2} = -p^2; \qquad \frac{1}{Y}\frac{d^2Y}{dy^2} = p^2$$

These are equivalent to the following ordinary differential equations which, with their solutions, can be written

$$\frac{d^2X}{dx^2} + p^2X = 0; \qquad X = C_1 \sin(px) + C_2 \cos(px)$$

$$\frac{d^2Y}{dy^2} - p^2Y = 0; \qquad Y = C_3 \sinh(py) + C_4 \cosh(py)$$

These solutions can be verified by simply substituting them into the corresponding differential equation. The product of these solutions for X and Y gives the general solution to Eqn (3-18):

$$T = [C_1 \sin(px) + C_2 \cos(px)][C_3 \sinh(py) + C_4 \cosh(py)] \tag{3-20}$$

where the constants C_1, C_2, C_3, C_4, and p are determined from the boundary conditions imposed on the problem.

Consider as an example the boundary conditions shown in Fig. 3-9 in which three edges of the rectangular slab are held at a constant temperature T_1 while the fourth edge is maintained at a temperature which is a function of the x coordinate $T = f(x)$. In this situation, as with the fin problems discussed in Section 3-1, it is convenient mathematically to redefine the dependent variable so that the edges other than $y = L_y$ can be considered at zero temperature. The boundary conditions can then be written:

(i) $T = 0$ along $x = 0$ (iii) $T = 0$ along $y = 0$
(ii) $T = 0$ along $x = L_x$ (iv) $T = f(x)$ along $y = L_y$

† The sign of this constant is determined by the boundary conditions; in this case the exponential part of the solution is in the y-coordinate.

These conditions allow the arbitrary constants in Eqn (3-20) to be determined. Conditions (i) and (iii) imply that C_2 and C_4 must both be zero and that Eqn (3-20) simplifies to

$$T = C \sin (px) \sinh (py)$$

in which $C = C_1 . C_3$. If this expression is to satisfy condition (ii) for all y we must have

$$\sin (pL_x) = 0$$

which implies that

$$p = \pi/L_x, 2\pi/L_x, 3\pi/L_x, \ldots, n\pi/L_x \quad (n = 1, 2, 3, \ldots)$$

There is accordingly a different solution for each consecutive integer n and a constant C associated with each solution. The complete solution is therefore expressed as the sum of all these separate solutions:

$$T = \sum_{n=1}^{\infty} C_n \sin \left(\frac{n\pi x}{L_x}\right) \sinh \left(\frac{n\pi y}{L_x}\right)$$

This expression must also satisfy condition (iv) above, that is, $T = f(x)$ along $y = L_y$, which gives

$$f(x) = \sum_{n=1}^{\infty} C_n \sin \left(\frac{n\pi x}{L_x}\right) \sinh \left(\frac{n\pi L_y}{L_x}\right)$$

Now in this expression $\sinh (n\pi L_y/L_x)$ is a constant and the series may be recognized as a Fourier expansion of the arbitrary function $f(x)$ in an infinite series of sines. It is shown later in Section 3-3-1 that the constants, or amplitudes, C_n, in this Fourier series are given by

$$C_n = \frac{1}{\sinh (n\pi L_y/L_x)} \frac{2}{L_x} \int_0^{L_x} f(x) \sin \left(\frac{n\pi x}{L_x}\right) dx$$

Now, depending on the specific form of $f(x)$, this integral can be evaluated, although some difficult cases may require numerical quadrature (numerical integration), so that the final solution reads

$$T = \frac{2}{L_x} \sum_{n=1}^{\infty} \sin \left(\frac{n\pi x}{L_x}\right) \frac{\sinh (n\pi y/L_x)}{\sinh (n\pi L_y/L_x)} \int_0^{L_x} f(x) \sin \left(\frac{n\pi x}{L_x}\right) dx$$

If boundary temperatures are specified on two or more sides of the rectangle, the fact that Laplace's equation is linear allows the superposition (addition) of two or more solutions to be made to build up the complete solution. Each separate solution would be found with three sides of the rectangle held at zero.[2]

Regular circular cylinders

It is shown in Chapter 2 that circular cylinders and tubes are best treated using cylindrical-polar coordinates (r, z, θ), and these are illustrated in Fig. 3-10 for a regular circular cylinder. Laplace's equation in this coordinate system can be written in the form

$$\frac{\partial^2 T}{\partial r^2} + \frac{1}{r}\frac{\partial T}{\partial r} + \frac{1}{r^2}\frac{\partial^2 T}{\partial \theta^2} + \frac{\partial^2 T}{\partial z^2} = 0 \qquad (3\text{-}21)$$

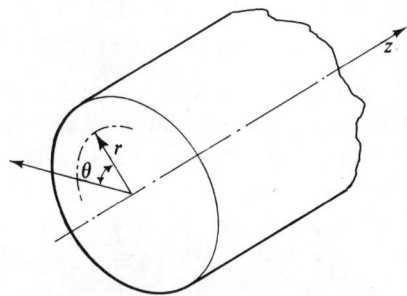

Fig. 3-10 Cylindrical polar coordinates used for a regular circular cylinder

This is a three-dimensional equation but in many practical situations the boundary conditions are such that the problem can be considered in the r, z or the r, θ coordinates only. If the boundary conditions imposed on the problem have no variation in the circumferential direction, θ, then the problem is said to be *axially symmetric* and Laplace's equation reduces to

$$\frac{\partial^2 T}{\partial r^2} + \frac{1}{r}\frac{\partial T}{\partial r} + \frac{\partial^2 T}{\partial z^2} = 0 \qquad (3\text{-}22)$$

The method of separation of variables assumes that the solution can be written as the product of two independent functions, each of which is a function only of a single independent variable:

$$T = R(r)Z(z) \qquad (3\text{-}23)$$

By substituting this into Eqn (3-22) we can obtain the relationship

$$\frac{1}{R}\frac{d^2 R}{dr^2} + \frac{1}{rR}\frac{dR}{dr} = -\frac{1}{Z}\frac{d^2 Z}{dz^2}$$

and, as in the case of the rectangular plate, since each side of this equation is a function of a single independent variable each side must equal a constant. Choosing this to be a positive constant, say p^2,† we can obtain two ordinary

---

† The sign of this constant is, in general, determined by the boundary conditions of the problem.

differential equations, one of which is Bessel's zero order equation and the other of which has already been described:

$$\frac{d^2R}{dr^2} + \frac{1}{r}\frac{dR}{dr} + p^2R = 0, \quad R = C_1 J_0(pr) + C_2 Y_0(pr)$$

$$\frac{d^2Z}{dz^2} - p^2Z = 0, \quad Z = C_3 \sinh(pz) + C_4 \cosh(pz)$$

The solution to Eqn (3-22) can therefore be written as the product of R and Z:

$$T = [C_1 J_0(pr) + C_2 Y_0(pr)][C_3 \sinh(pz) + C_4 \cosh(pz)] \quad (3\text{-}24)$$

where the constants C_1, C_2, C_3, C_4, and p would be determined from the application of the boundary conditions imposed on the problem.

If the boundary conditions vary in the θ direction but not in the z direction, Laplace's equation becomes

$$\frac{\partial^2 T}{\partial r^2} + \frac{1}{r}\frac{\partial T}{\partial r} + \frac{1}{r^2}\frac{\partial^2 T}{\partial \theta^2} = 0 \quad (3\text{-}25)$$

Assuming a solution in the form

$$T = R(r), \Theta(\theta)$$

allows Eqn (3-25) to be separated into the two ordinary differential equations

$$r^2 \frac{d^2R}{dr^2} + r\frac{dR}{dr} - p^2R = 0$$

$$\frac{d^2\Theta}{d\theta^2} + p^2\Theta = 0$$

The solution to this first equation is obtained by setting $R = r^x$, where $x = \pm p$, which gives

$$R = C_1 r^p + C_2 r^{-p}$$

The solution to the second equation has already been given as

$$\Theta = C_3 \sin(p\theta) + C_4 \cos(p\theta)$$

so that the general solution, as the product of these two, becomes

$$T = [C_1 r^p + C_2 r^{-p}][C_3 \sin(p\theta) + C_4 \cos(p\theta)] \quad (3\text{-}26)$$

where again the constants C_1, C_2, C_3, C_4, and p are to be determined from the boundary conditions imposed on each particular problem.

We may observe in the above cases that although simple geometries have been taken, the solutions obtained have been relatively complex. The particular solution to Eqn (3-20) for the rectangular slab with specified boundary temperatures takes the form of a series, and any particular solution to Eqns (3-24) and (3-26) would also take the form of a series. These series do not indicate any qualitative information about the solution; also, they are not necessarily rapidly convergent and an accurate solution could require the evaluation of many terms.

We have only considered boundary conditions of the type in which the function value is specified, and in practice this boundary condition does not occur very frequently. A more common boundary condition is the convective one, Eqn (2-12), which specifies a combination of the function and its first derivative. By examination of the above solutions—Eqns (3-20), (3-24), and (3-26)—it is clear that differentiation of the solution considerably complicates the problem and that the numerical evaluation of the resulting solution would be more involved. However, many solutions with this type of boundary condition have been determined, but only for relatively straightforward geometries such as rectangles, circular cylinders, cones, and spheres.[2] Irregular geometries with two or more space coordinates present considerable analytical difficulties and in many cases, although it may be possible after a great deal of analysis to obtain a solution, modern numerical methods as described in Chapter 4 present easier and quicker alternative approaches.

3-2-2 Transient conduction in one space dimension

These problems are similar to steady conduction in two dimensions in that they involve two independent variables—in this case, one space dimension and time. Mathematically, the problems of using time as a dimension are not as difficult as the problems of an extra space dimension since a variety of geometrical shapes can be described by two space coordinates. In this section we consider not only the case of a solid body in which the boundary temperatures suddenly change but also the more practical example of a solid body that is suddenly exposed to a fluid at a different temperature.

We consider initially the conduction of heat in a rod of small cross-section. The rod is taken to be so thin that the temperature at all points of the section may be considered to be the same. The heat flow is therefore linear and the temperature is specified by time t and distance x measured along the rod. We consider only cases in which there is no heat loss from the curved surfaces of the rod, that is, the rod is insulated as shown in Fig. 3-11. By initially treating the case of a thin rod it is easier to conceive of an initial temperature distribution other than that of constant temperature. This problem is identical to that of transient heat flow in a solid bounded by two parallel planes, sometimes called the 'infinite wall or slab', and we go on to consider the important practical example of an infinite wall with convective boundary conditions on the two faces, one or both of which is suddenly exposed to different fluid conditions.

For both problems the transient temperature distribution is described by Fourier's one-dimensional equation

$$\frac{\partial T}{\partial t} = \alpha \frac{\partial^2 T}{\partial x^2} \tag{3-27}$$

where α is the thermal diffusivity $(k/\rho C)$ previously introduced in Chapter 2.

Using the separation of variables method we seek a solution in the form

$$T = X(x)\tau(t)$$

where $X(x)$ and $\tau(t)$ are functions only of x and t respectively. Substituting this into Eqn (3-27) in a similar manner to the previous cases we obtain the separated equations

$$\frac{1}{X}\frac{d^2 X}{dx^2} = \frac{1}{\alpha\tau}\frac{d\tau}{dt} = -p^2$$

Both parts of this separated form must be constant and are taken equal to a negative constant, $-p^2$; as a consequence the exponential part of the solution decays with time. We therefore obtain two ordinary differential equations which, with their solutions, can be written

$$\frac{d\tau}{dt} + \alpha p^2 \tau = 0; \qquad \tau = A\,e^{-\alpha p^2 t} \tag{3-28}$$

$$\frac{d^2 X}{dx^2} + p^2 X = 0; \qquad X = B\sin(px) + C\cos(px) \tag{3-29}$$

The first solution is simply obtained by putting the differential equation in its 'separate variable' form and integrating. The second solution has been previously dealt with.

The product of these two solutions gives the general solution to Eqn (3-27):

$$T = e^{-\alpha p^2 t}\,[C_1\sin(px) + C_2\cos(px)] \tag{3-30}$$

in which the arbitrary constants C_1, C_2, and p are to be determined from the boundary conditions imposed on each particular problem.

Sudden temperature changes at the ends of a rod

Consider the rod in Fig. 3-11 which is initially heated over its length so that it has the initial temperature distribution $T = f(x)$. This could be arranged experimentally, for instance, by heating the rod with a suitably wound electric coil. If at time $t = 0$ the ends of the rod are reduced to zero and held at that temperature, for example, by holding blocks of ice on the ends, the initial and boundary conditions for the problem can be written as

$$T = f(x) \text{ for } t = 0; \quad T = 0 \text{ at } x = 0 \text{ and } x = L \text{ for } t \geqslant 0$$

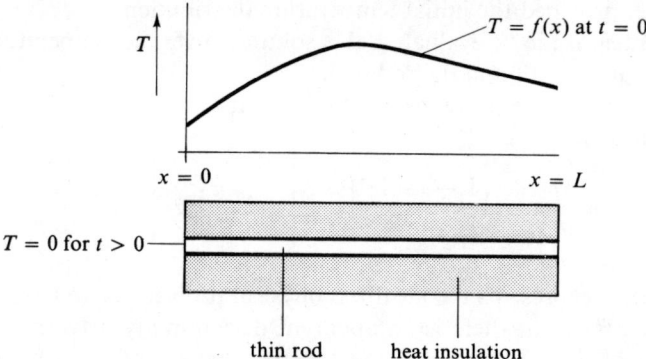

Fig. 3-11 Thin heat conducting rod insulated over its curved surface

To satisfy the condition $T = 0$ at $x = 0$ in Eqn (3-30) we must have $C_2 = 0$, and for $T = 0$ at $x = L$ we must have $\sin(pL) = 0$ which is true for

$$p = \pi/L, \, 2\pi/L, \, 3\pi/L, \ldots, n\pi/L, \quad (n = 1, 2, 3, \ldots)$$

where we do not consider the trivial case when $p = 0$.

Adding all these separate solutions we obtain the complete solution:

$$T = \sum_{n=1}^{\infty} \exp\left[-\left(\frac{n\pi}{L}\right)^2 \alpha t\right] C_n \sin\left(\frac{n\pi x}{L}\right)$$

This equation must satisfy the initial condition $T = f(x)$ at $t = 0$ and therefore we can write

$$f(x) = \sum_{n=1}^{\infty} C_n \sin\left(\frac{n\pi x}{L}\right)$$

and, as in the previous case of heat flow in a rectangle, this expression is a Fourier expansion of $f(x)$ in an infinite series of sines and the constants C_n as shown in Section 3-3-1 are given by

$$C_n = \frac{2}{L} \int_0^L f(x) \sin\left(\frac{n\pi x}{L}\right) dx$$

so that the full solution can be written

$$T = \frac{2}{L} \sum_{n=1}^{\infty} \exp\left[-\left(\frac{n\pi}{L}\right)^2 \alpha t\right] \sin\left(\frac{n\pi x}{L}\right) \int_0^L f(x) \sin\left(\frac{n\pi x}{L}\right) dx \quad (3\text{-}31)$$

Therefore, provided the initial temperature distribution $T = f(x)$ is known and the integral can be evaluated, this solution gives the temperature at any time t for any position in the rod.

If the initial temperature distribution is constant, say $T = T_i$, Eqn (3-31) can be simplified to

$$T = \frac{2T_i}{\pi} \sum_{n=1}^{\infty} \frac{[1 - (-1)^n]}{n} \exp\left[-\left(\frac{n\pi}{L}\right)^2 \alpha t\right] \sin\left(\frac{n\pi x}{L}\right)$$

This solution represents the idealized physical problem as we have stated it, despite the fact that there is an apparent discontinuity between the initial condition of $T = T_i$ at $t = 0$ and the boundary conditions of $T = 0$ at $x = 0$ and L at $t \geqslant 0$ which could not be strictly true in practice.

Transient conduction in an infinite wall with fluid boundaries

As a more important example which occurs frequently in engineering, consider a wall or plate of infinite size and of thickness L which is suddenly exposed to fluids in motion on both its surfaces. An arbitrary initial temperature distribution $T = f(x)$ at $t = 0$ is assumed, and to illustrate the procedure it is also assumed that the fluids on both sides of the wall are at the same temperature. It is shown later in this section how the result can be modified if the fluids are at different temperatures. However, in this case, by redefining the variable T as in the case of the fin, we can consider the wall to be convecting heat into a medium at a temperature of zero, Fig. 3-12.

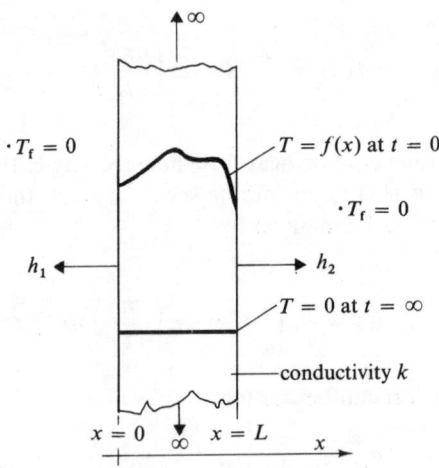

Fig. 3-12   Infinite wall convecting heat into two fluids at zero temperature

The differential equation and the initial and the boundary conditions may therefore be written

$$\frac{\partial T}{\partial t} = \alpha \frac{\partial^2 T}{\partial x^2}, \quad 0 < x < L \tag{3-27}$$

$$T = f(x) \quad \text{at } t = 0 \tag{3-32}$$

$$k\left(\frac{\partial T}{\partial x}\right) = h_1 T \quad \text{at } x = 0 \tag{3-33a}$$

$$-k\left(\frac{\partial T}{\partial x}\right) = h_2 T \quad \text{at } x = L \tag{3-33b}$$

Note that at $x = 0$, with reference to Eqn (2-12), $\partial T/\partial n = -\partial T/\partial x$. By the method of separation of variables we have shown that a solution of Eqn (3-27) is

$$T = e^{-\alpha p^2 t}\left[C_1 \sin (px) + C_2 \cos (px)\right] \tag{3-30}$$

If this expression is to satisfy the boundary conditions (3-33a) and (3-33b) for $0 \leqslant t \leqslant \infty$, we require combinations of T and $\partial T/\partial x$ at the surfaces $x = 0$ and $x = L$.

By differentiating Eqn (3-30) we obtain

$$\frac{\partial T}{\partial x} = e^{-\alpha p^2 t}\, p[C_1 \cos (px) - C_2 \sin (px)]$$

and by setting $x = 0$ and $x = L$, respectively, in this and in Eqn (3-30), Eqns (3-33a) and (3-33b) can be written in the form

$$e^{-\alpha p^2 t}\, pC_1 = \frac{h_1}{k}\, e^{-\alpha p^2 t}\, C_2 \tag{3-34a}$$

$$-e^{-\alpha p^2 t}\, p[C_1 \cos (pL) - C_2 \sin (pL)]$$

$$= \frac{h_2}{k}\, e^{-\alpha p^2 t}\left[C_1 \sin (pL) + C_2 \cos (pL)\right] \tag{3-34b}$$

Equation (3-34a) simplifies to

$$C_2 = C = \frac{pC_1}{h_1/k}$$

which, when inserted into Eqn (3-34b) together with some rearrangement, gives

$$\tan (pL) = \frac{pk(h_1 + h_2)}{p^2 k^2 - h_1 h_2} \tag{3-35}$$

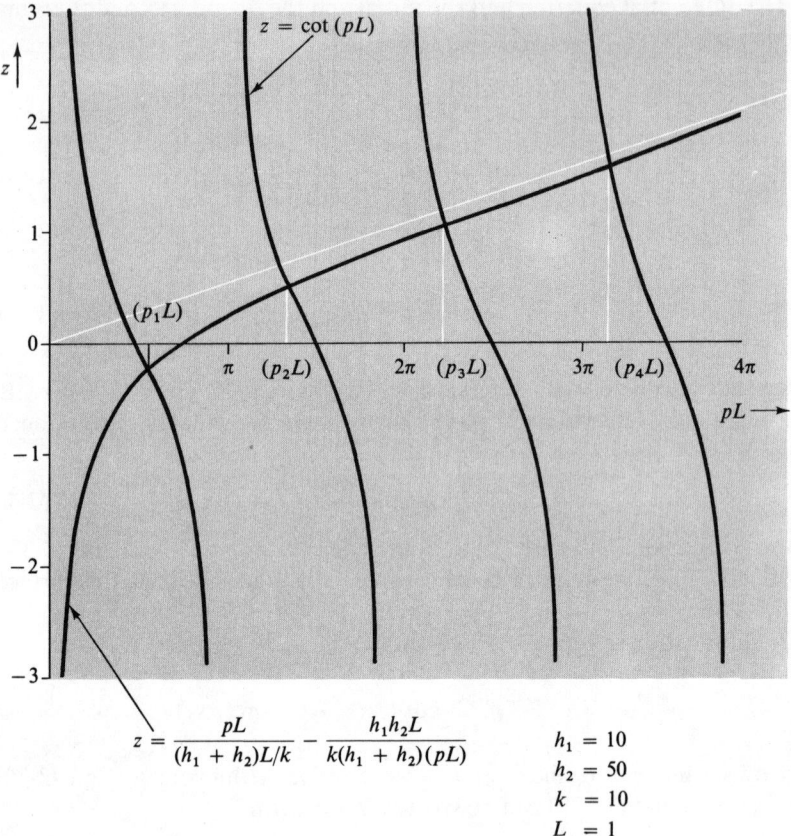

$$z = \frac{pL}{(h_1 + h_2)L/k} - \frac{h_1 h_2 L}{k(h_1 + h_2)(pL)}$$

$$h_1 = 10$$
$$h_2 = 50$$
$$k = 10$$
$$L = 1$$

Fig. 3-13 Graphical illustration of the roots of the equation

$$\cot (pL) = \frac{(pL)}{(h_1 + h_2)L/k} - \frac{h_1 h_2 L}{k(h_1 + h_2)(pL)}$$

At this stage, therefore, we know that the expression

$$T = e^{-\alpha p^2 t} \, C \left[\frac{h_1}{kp} \sin (px) + \cos (px) \right] \qquad (3\text{-}36)$$

satisfies the differential equation (3-27) and the boundary conditions (3-33), provided that p is any root, other than zero, of Eqn (3-35). The roots of Eqn (3-35) can be illustrated graphically by plotting

$$\cot (pL) \quad \text{against} \quad \frac{(pL)k}{(h_1 + h_2)L} - \frac{h_1 h_2 L}{k(h_1 + h_2)(pL)}$$

The second curve is a hyperbola and is shown plotted against the cotangent curve in Fig. 3-13 for a specific example; that is, for given values of h_1, h_2, k, and L. From the intersections of these curves it can be seen that the positive roots lie one in each of the intervals $(0, \pi)$, $(\pi, 2\pi)$, ... The negative roots are equal numerically to the positive roots, and it can be shown that there are no imaginary or complex roots.[2] The values of p can be determined for a given value of L and it is worth noting that, by virtue of the fact that the exponential part of the solution contains p^2, it is unusual for more than six roots to be required. If the numerical evaluation of the solution were to be performed on a computer, an iterative method such as Newton's iteration described in Section 4-2-5 would normally be used.

To determine the constants C in Eqn (3-36) by applying the initial conditions $T = f(x)$ at $t = 0$, we have that $f(x)$ can be expanded in the infinite series

$$f(x) = C_1 X_1 + C_2 X_2 + C_3 X_3 + \cdots \tag{3-37}$$

where, from Eqn (3-36),

$$X_n = \frac{h_1}{kp_n} \sin (p_n x) + \cos (p_n x)$$

in which p_n is the nth root of Eqn (3-35).

It should be noted that (3-37) is not a Fourier series since the p_n are not integers. However, it can be shown[2] that such an expansion is valid, that the series can be integrated term by term, and that the values of the coefficients C_n may be obtained in a similar manner to that in which the coefficients in the Fourier series were determined in the previous example.

This depends on the fact that when the series (3-31) is multiplied by X_n and integrated term by term, we find

$$\int_0^L X_m X_n \, dx = 0 \quad (m \neq n)$$

and
$$\int_0^L X_n^2 \, dx = \frac{1}{p_n^2} \left\{ \left[p_n^2 + \left(\frac{h_1}{k}\right)^2 \right] L + \frac{\left(p_n^2 + \frac{h_1 h_2}{k^2}\right)\frac{h_1 + h_2}{k}}{\left[p_n^2 + \left(\frac{h_2}{k}\right)^2 \right]} \right\} = A_n \tag{3-38}$$

The evaluation of this latter expression takes into consideration, of course, the series at $x = 0$ and at $x = L$ and a proof of this derivation is given, for a similar series, by Carslaw and Jaeger[2].

Therefore, from Eqn (3-37) we have

$$\int_0^L f(x)X_n\,dx = C_n \int_0^L X_n^2\,dx$$

so that the constants C_n are given by

$$C_n = \frac{\int_0^L f(x)X_n\,dx}{A_n} \quad (n = 1, 2, 3, \ldots)$$

where, for each constant C, the denominator A_n is given by Eqn (3-38) and is a constant for given values of h_1, h_2, k, and L—p_n being the roots of Eqn (3-35).

The complete solution can therefore be written as the sum of all the separate solutions arising from Eqn (3-36):

$$T = \sum_{n=1}^{\infty} e^{-\alpha p_n^2 t}\left(\frac{h_1}{kp_n}\sin(p_n x) + \cos(p_n x)\frac{1}{A_n}\right)$$
$$\times \int_0^L f(x)\left(\frac{h_1}{kp_n}\sin(p_n x) + \cos(p_n x)\right)dx \quad (3\text{-}39)$$

where A_n is given by Eqn (3-38).

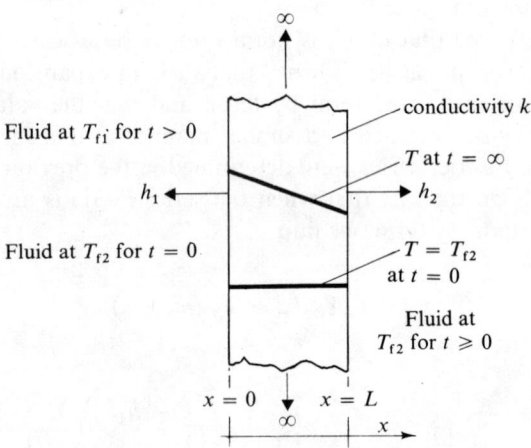

Fig. 3-14   Infinite wall with fluid boundaries with a sudden change in the temperature of one fluid

Sudden changes in conditions on one surface only. The theory above may be extended to many practical problems. A particularly important one occurs in the ignition of flames in which a wall of finite thickness separating two fluids is suddenly exposed to a higher temperature on one face only. Such a situation is shown in Fig. 3-14. Both fluids are assumed to have the same initial temperature T_{f2} with heat transfer coefficients h_1 and h_2 at $x = 0$ and $x = L$ respectively, so that the temperature of the whole wall is initially constant at T_{f2}. If at $t = 0$ the fluid moving over the surface $x = 0$ is suddenly raised to T_{f1}, the temperature of the wall will rise until equilibrium conditions are reached at which point the temperature distribution in the wall will be linear.

This situation can be reduced to the problem considered above [Eqn (3-39)] by putting

$$T = U + V \qquad (3\text{-}40)$$

where U is a function of x *only* and V is a function of x and t. Now, if U is a function of x only it must be the steady-state 'equilibrium' solution for $t = \infty$, and it must therefore satisfy the equations

$$\frac{d^2 U}{dx^2} = 0 \quad (0 < x < L)$$

$$k \frac{dU}{dx} = h_1(U - T_{f1}) \quad \text{when } x = 0$$

$$-k \frac{dU}{dx} = h_2(U - T_{f2}) \quad \text{when } x = L$$

This simple system is discussed in Chapter 2 and the solution is

$$U = B_1 x + B_2$$

where the constants B_1 and B_2 are given by

$$B_1 = \frac{h_1 h_2 (T_{f2} - T_{f1})}{kh_1 + kh_2 + h_1 h_2 L} \qquad (3\text{-}41)$$

$$B_2 = \frac{kh_1 T_{f1} + kh_2 T_{f2} + h_1 h_2 L T_{f1}}{kh_1 + kh_2 + h_1 h_2 L} \qquad (3\text{-}42)$$

The function V in Eqn (3-40) is a function of x and t, and must satisfy the equation

$$\frac{\partial V}{\partial t} = \alpha \frac{\partial^2 V}{\partial x^2} \quad (0 < x < L)$$

the initial condition

$$V = T_{f2} - U \quad \text{when } t = 0$$

and the boundary conditions

$$\frac{\partial V}{\partial x} = h_1 V \quad \text{when } x = 0$$

$$-\frac{\partial V}{\partial x} = h_2 V \quad \text{when } x = L$$

The solution to this latter problem is given by Eqn (3-39) in which $f(x) \equiv T_{f2} - B_1 x - B_2$, where the constants B_1 and B_2 are given by Eqns (3-41) and (3-42). The expression under the integral in Eqn (3-39) can therefore be obtained analytically to give the complete solution, albeit a cumbersome one, to the problem of Fig. 3-14.

Sudden simultaneous change in conditions on both surfaces. This section considers the infinite wall (Fig. 3-12) analysed above but with a symmetrical initial profile $T = f(x)$, as shown in Fig. 3-15. If the temperature of the

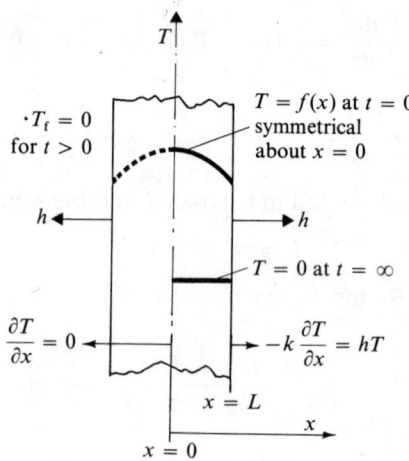

Fig. 3-15 Transient response of a 'quenched' plate

surrounding fluid is suddenly reduced to a lower value we have what is known as the 'quenching' problem, and by redefining the variable T we can take this lower value of fluid temperature to be zero. The heat transfer coefficient on both surfaces of the plate, h, is assumed constant with time.

Now, although the general solution, Eqn (3-39), could be reduced to yield the solution for this particular case the solution would still be a series containing sines and cosines, and we can obtain a simpler solution by considering this problem as a special case.

Since the initial temperature distribution is symmetrical we need only consider one half of the plate thickness by taking the coordinate origin at the centre-line of the plate, as shown in Fig. 3-15. The merit of this will also be demonstrated later when the results for the infinite plate of thickness $2L$ are compared with those for the infinite cylinder of diameter $2R$ in chart form. A solution of the governing differential equation has been shown to be

$$T = e^{-\alpha p^2 t}[C_1 \sin (px) + C_2 \cos (px)] \qquad (3\text{-}30)$$

in which the arbitrary constants are determined by the initial condition

$$T = f(x) \quad \text{at } t = 0 \qquad (3\text{-}43)$$

and the boundary conditions

$$\frac{\partial T}{\partial x} = 0 \quad \text{at } x = 0 \qquad (3\text{-}44a)$$

$$-k \frac{\partial T}{\partial x} = hT \quad \text{at } x = L \qquad (3\text{-}44b)$$

For Eqn (3-30) to satisfy condition (3-44a) we must have

$$\left(\frac{\partial T}{\partial x}\right)_{x=0} = e^{-\alpha p^2 t} p[C_1 \cos (px) - C_2 \sin (px)] = 0$$

so that $C_1 = 0$, which reduces Eqn (3-30) to

$$T = e^{-\alpha p^2 t} C \cos (px) \qquad (3\text{-}45)$$

Condition (3-44b) can therefore be written as

$$-k e^{-\alpha p^2 t} pC[-\sin (pL)] = h e^{-\alpha p^2 t} C \cos (pL)$$

so that we have

$$kp \sin (pL) = h \cos (pL) \quad \text{or} \quad \tan (pL) = \frac{hL}{k(pL)} \qquad (3\text{-}46)$$

This equation, which is rather more simple than Eqn (3-35), is also satisfied for an infinite number of values of (pL) which may be determined graphically by plotting $\cot (pL)$ against $pL/(hL/k)$ or by using an iterative method; see Section 4-2-5. However, the first six roots of this equation are tabulated by Carslaw and Jaeger[2] for various values of hL/k.

The application of the initial condition, $T = f(x)$ at $t = 0$, to the sum of all the separate solutions of Eqn (3-45) gives

$$f(x) = \sum_{n=1}^{\infty} C_n \cos (p_n x)$$

which is an expansion of the function $f(x)$ in an infinite series of cosines when p_n is defined as a root of Eqn (3-46), and it can be shown in a manner similar to that in the previous case that the constants C_n must be given by

$$C_n = \frac{2p_n \int_0^L f(x) \cos (p_n x)\, dx}{(p_n L) + \sin (p_n L) \cos (p_n L)}$$

The final solution is then given by

$$T = 2 \sum_{n=1}^{\infty} e^{-\alpha p_n^2 t}\, \frac{\cos (p_n x)}{[(p_n L) + \sin (p_n L) \cos (p_n L)]} \int_0^L f(x) \cos (p_n x)\, dx \quad (3\text{-}47)$$

A common practical assumption in the 'quenching' problem as defined above is that the plate is initially at constant temperature, $T = T_i$, for which Eqn (3-47) reduces to

$$T = 2T_i \sum_{n=1}^{\infty} e^{-p_n^2 \alpha t}\, \frac{\cos (p_n x) \sin (p_n L)}{[(p_n L) + \sin (p_n L) \cos (p_n L)]} \quad (3\text{-}48)$$

where p_n is the nth root of

$$(p_n L) \tan (p_n L) = \frac{hL}{k}$$

Heat loss from the plate

The heat loss from the surface of the plate $x = L$ to the surrounding fluid per unit area up to time t can be expressed in three separate ways:
  (i) from the rate at which heat is conducted out of the surface

$$\int_0^t - k(\partial T/\partial x)_{x=L}\, dt$$

  (ii) from rate of convection at the surface

$$\int_0^t hT_{x=L}\, dt$$

or (iii) from the heat contained in the half-plate

$$\int_0^L \rho CT\, dx$$

Using method (iii), and noting that the initial heat, Q_i, contained between $x = 0$ and L is $Q_i = \rho LCT_i$, the heat *contained*, Q, in one half of the plate, $0 < x < L$, per unit surface area at time t can be expressed as

$$\frac{Q}{Q_i} = 2 \sum_{n=1}^{\infty} \frac{1}{p_n L}\, e^{-p_n^2 \alpha t}\, \frac{\sin^2 (p_n L)}{p_n L + \sin (p_n L) \cos (p_n L)} \quad (3\text{-}49)$$

Dimensionless groups and charts

The above solutions to the transient conduction problems considered—namely, Eqns (3-31), (3-39), (3-48), and (3-49)—are relatively cumbersome for practical applications, and in engineering practice wide use is made of non-dimensional charts. Consider, for example, Eqn (3-48). The temperature of the plate at any time and any location is a function of hL, αt, and x so that if these factors are converted into dimensionless form the solution can be expressed as

$$T = f\left(\frac{hL}{k}, \frac{\alpha t}{L^2}, \frac{x}{L}\right)$$

The first of these groups is known as the *Biot number*, $\mathbf{B} = hL/k$, and it indicates the ratio of thermal conductance at the surface to the thermal conductivity of the solid. The second group is known as the Fourier number, $\mathbf{F} = \alpha t/L^2$, and this indicates the speed of heating or cooling of a solid body. Low values of the Fourier number imply that a long period of time is required to heat or cool the body, and vice versa.

Analytical solutions such as Eqns (3-48) and (3-49) have been evaluated numerically and plotted non-dimensionally for a wide variety of shapes which include plates, cylinders, and spheres. Although we have not outlined the application of the separation of variables to the problem of transient conduction in a circular cylinder, an examination of the derivation of Eqn (3-24), which is the solution for steady conduction in a cylinder, shows that the solutions of transient conduction problems in cylinders involve Bessel functions.

A number of investigators[5] have plotted these theoretical relations in terms of the dimensionless groups involved, and the results of Gröber[6] and Heisler[7] are particularly useful. Gröber plotted dimensionless values of temperature and heat *content* of the body against Biot number $\mathbf{B}$ for various values of the Fourier number $\mathbf{F}$. Heisler, however, plotted temperature against Fourier number $\mathbf{F}$ for various values of the inverse of the Biot number $1/\mathbf{B}$, and where the Heisler or the Gröber charts can be used, more accurate results can be obtained from the Heisler charts.

As examples, the temperature T_c at the centres of an infinite plate and an infinite cylinder initially at temperature T_i and placed in a fluid at zero temperature are given in the form of Heisler charts in Fig. 3-16. It should be remembered that the plate *half*-thickness L has been used throughout for the infinite plate problem. Although Fig. 3-16 gives only the temperature at the centres of the plate and the cylinder, the temperature at any position in the solid can be found from Heisler's position-correction charts (Fig. 3-17) in which the temperature, as a ratio of the centre temperature, is plotted against the inverse of the Biot number, $1/\mathbf{B}$.

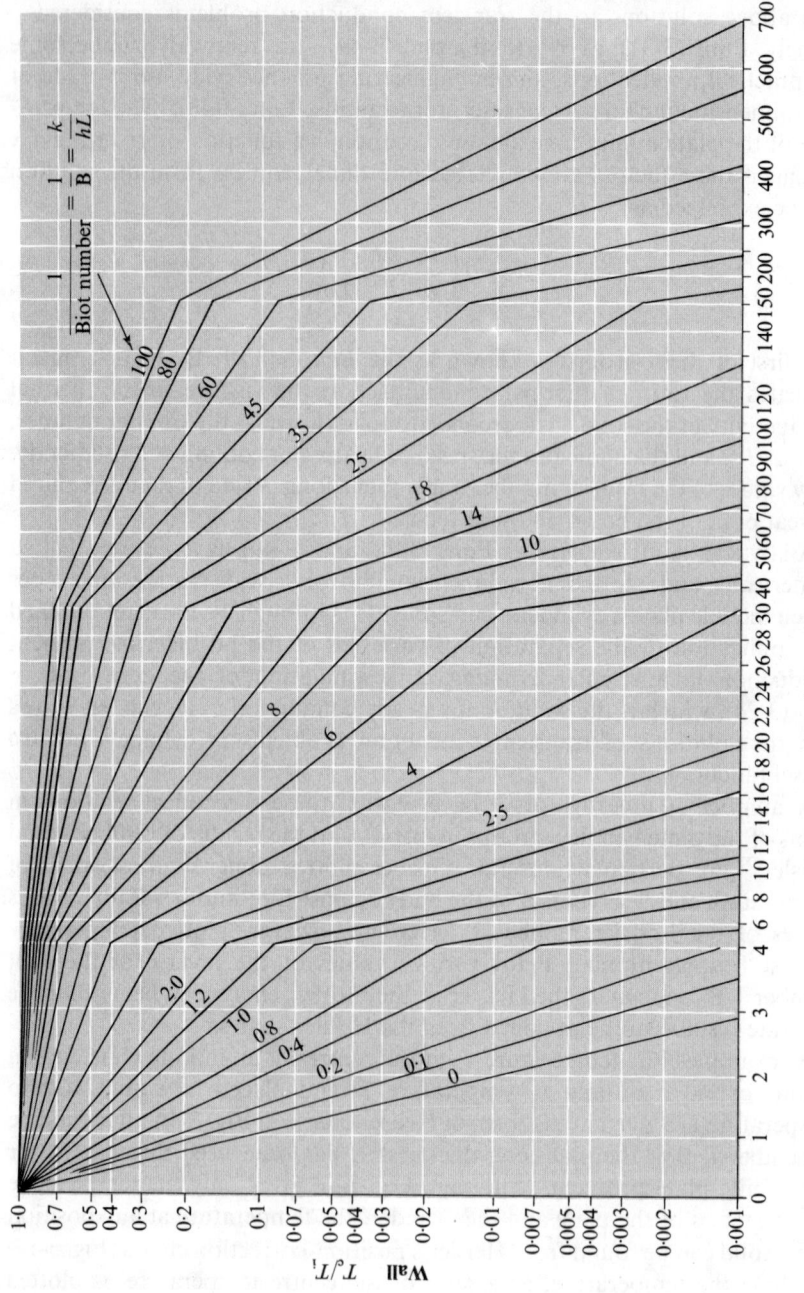

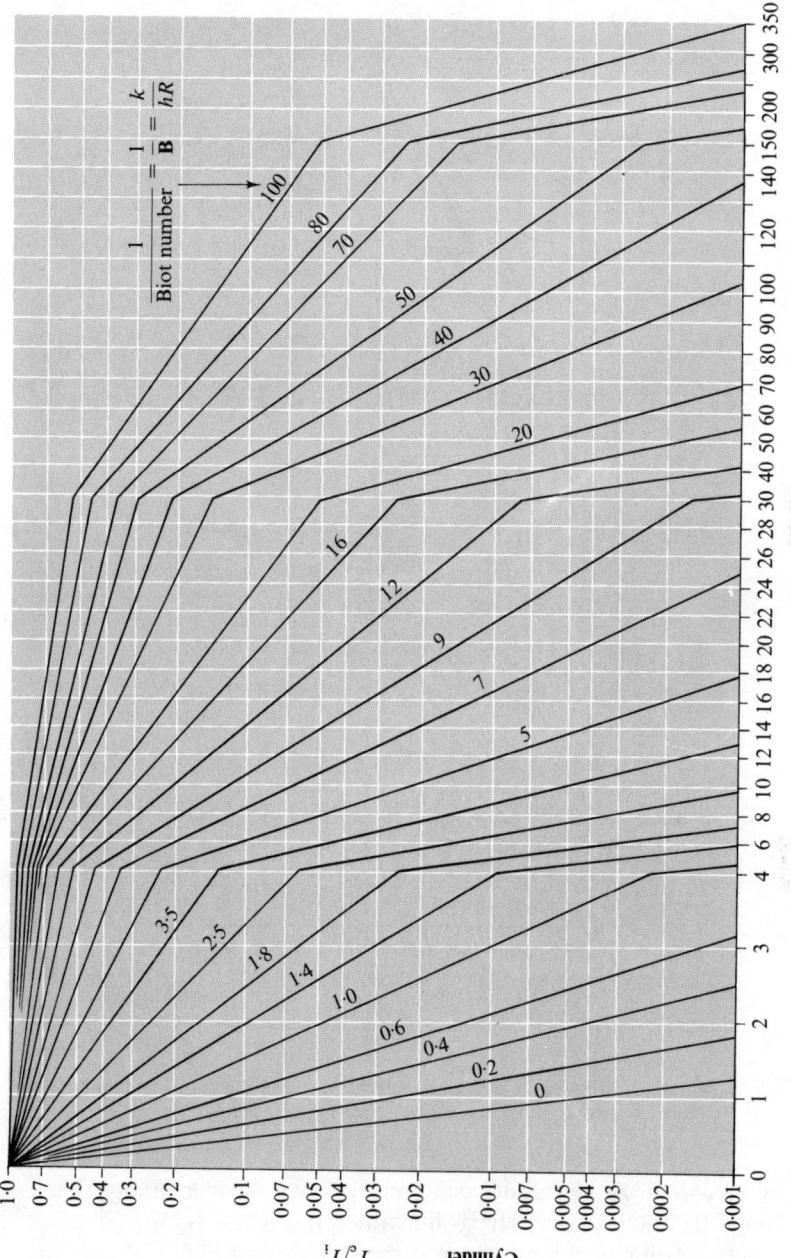

Fig. 3-16 Transient response of the centre of an infinite wall ($L = \frac{1}{2}$ wall thickness) and an infinite cylinder, at T_i initially, in a fluid at zero temperature (after Heisler[7])

Fourier number $\mathbf{F} = \alpha t / R^2$

$$\frac{1}{\mathbf{B}} = \frac{1}{\text{Biot number}} = \frac{k}{hR}$$

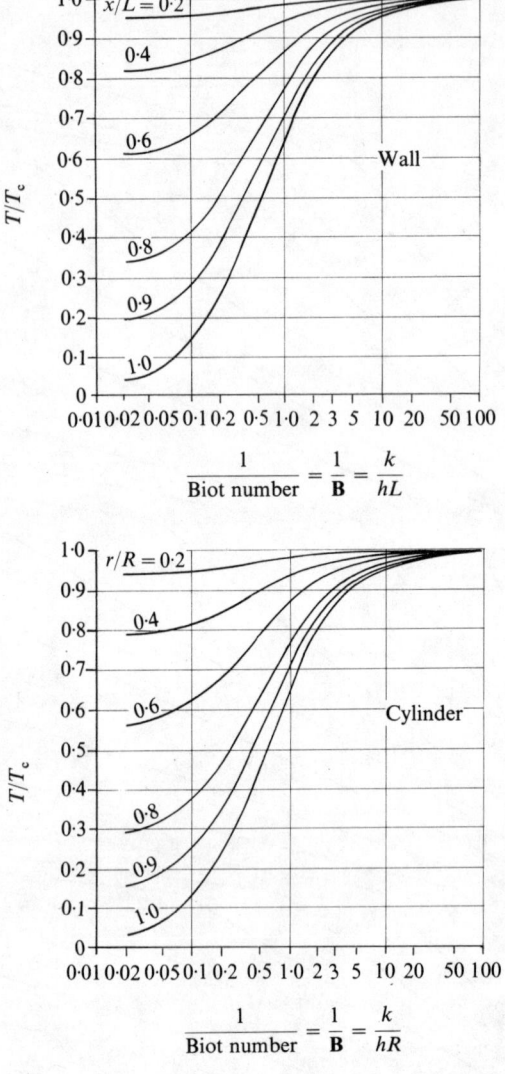

Fig. 3-17 Temperature distribution in an infinite wall ($L = \frac{1}{2}$ wall thickness) and an infinite cylinder relative to the centre temperature T_c (after Heisler[7])

The heat *content* of the infinite plate and infinite cylinder at any time t can be found from the Gröber charts illustrated in Fig. 3-18.

A geometry of practical importance is the sphere, and Fig. 3-19 presents Gröber charts for the temperature of the centre, T_c, the surface temperature, T_s, and the heat contained in the sphere, Q, at any time t.

As the charts in Figs 3-16–3-19 do not give very accurate results for low values of Fourier number, Heisler developed another set of charts for plates, cylinders, and spheres to give more accurate results for values of $\alpha t/L$ between 0 and 0·2.

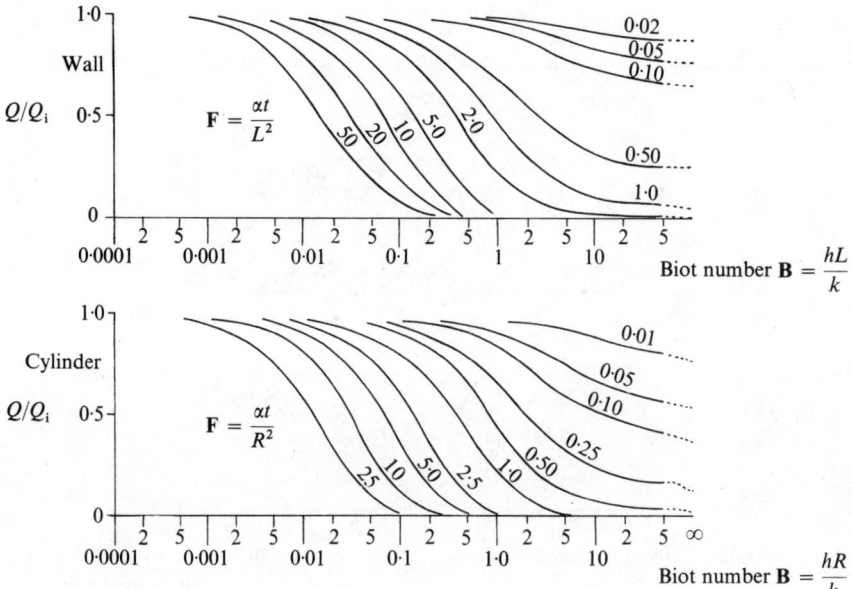

Fig. 3-18   The heat, Q, contained in an infinite wall ($L = \frac{1}{2}$ wall thickness) and an infinite cylinder relative to the initial heat, Q_i (after Gröber[6])

As an illustration of the use of these charts, consider the following two examples.

Example 1. A sheet of steel, 0·25 in thick, density $\rho = 489$ lb/ft³ (7,835 kg/m³), thermal conductivity $k = 17$ Btu/h ft °R (29·42 W/m K), specific heat capacity $C = 0·111$ Btu/lb °R (0·465 kJ/kg K) initially at 2,000°F (1,093°C) is immersed in a bath of water at 60°F (15·6°C). The heat transfer coefficient h is estimated to be 1,000 Btu/h ft² °R (5,680 W/m² K). It is required to calculate the time taken by the centre of the plate to reach 640°F (338°C) and also the surface temperature at this point.

$$\text{Half thickness} = L = \frac{0·25}{2 \times 12} = 0·014 \text{ ft}$$

$$\alpha = \frac{k}{\rho C} = \frac{17}{489 \times 0·111} = 0·313 \text{ ft}^2/\text{h}$$

$$\frac{1}{B} = \frac{k}{hL} = \frac{17}{1,000 \times 0·0104} = 1·635$$

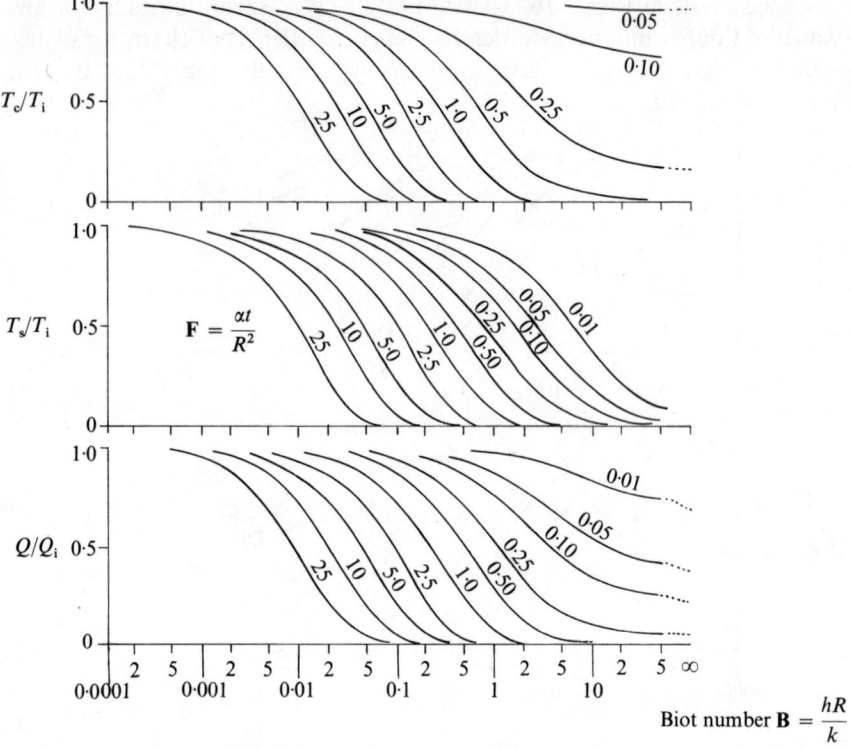

Fig. 3-19 Transient response of a sphere initially at T_i in a fluid at zero temperature. T_c, T_s are the temperatures of the centre and surface respectively (after Gröber[6])

Datum $= 60°F$; therefore

$$T_i = 2,000 - 60 = 1,940°F$$

At $640°F$,

$$T_c = 640 - 60 = 580°F$$

Thus,

$$\frac{T_c}{T_i} = \frac{580}{1,940} = 0·299$$

From Fig. 3-16,

$$\text{Fourier number } F \ (= \alpha t/L^2) \approx 2·5$$

$$\text{Time } (t) \approx \frac{(0·0104)^2 \times 2·5}{0·313} \approx 8·6 \times 10^{-4} \text{ h} \approx \textbf{3·1 s}$$

From Fig. 3-17, the surface temperature at this point is:

$$0.77 \times 580 + 60 = \textbf{507°F}$$

Example 2. A long cylindrical concrete column 0·4 m diameter, density $\rho = 2,000$ kg/m^3 (125 lb/ft^3), thermal conductivity $k = 1.4$ W/m K (0·81 Btu/h ft °R), specific heat capacity $C = 0.879$ kJ/kg K (0·21 Btu/lb °R) initially at 250°C (482°F) is suddenly exposed to air at 20°C (68°F). The heat transfer coefficient h is estimated to be 15 W/m^2 K (2·64 Btu/h ft^2 °R).

It is required to calculate the surface temperature after 10 hours.

$$R = \frac{0.4}{2} = 0.2 \text{ m}$$

$$\alpha = \frac{k}{\rho C} = \frac{1.4}{2,000 \times 0.879 \times 10^3} = 0.797 \times 10^{-6} \text{ m}^2/\text{s}$$

$$\frac{1}{B} = \frac{k}{hR} = \frac{1.4}{15 \times 0.2} = 0.467$$

After 10 hours

$$t = 3.6 \times 10^4 \text{ s}$$

Thus,

$$\text{Fourier number, } \mathbf{F}\left(=\frac{\alpha t}{R^2}\right) = \frac{0.797 \times 10^{-6} \times 3.6 \times 10^4}{0.04} = 0.717$$

From Fig. 3-16,

$$\frac{T_c}{T_i} \approx 0.2$$

From Fig. 3-17,

$$\frac{T_s}{T_c} \approx 0.43$$

Datum = 20°C; therefore,

$$T_i = 250 - 20 = 230$$

Thus,

$$\text{centre temperature} \approx 0.2 \times 230 + 20 \approx 66°C$$

and therefore the surface temperature after 10 hours is:

$$0.43 \times 43 + 20 \approx \textbf{38·5°C}$$

These charts, Figs 3-16–3-19, have been developed for problems in which a body at a constant initial temperature is suddenly placed in a fluid at a different temperature. The main assumption, however, is that the heat transfer coefficient between the body and the fluid is *constant* for all time, and this is not usually the case. Take, for instance, the problem of a heated body placed in still air at a different temperature. The situation is one of natural convection in which the heat transfer coefficient is a strong function of temperature, as shown in Chapters 6 and 7. Another example of considerable practical importance which we have already mentioned is the 'quenching' problem. Here, if a metallic body is heated to over $100°C$ and then suddenly immersed in a bath of water, local boiling on the surface of the body will occur and, as will be shown in Chapter 9, the heat transfer coefficients with boiling can be very much higher than in convection without boiling.

In many practical cases it is only necessary to bound the solution and the assumption of a constant heat transfer coefficient is satisfactory. The charts (Figs 3-16–3-19) then provide a rapid means of solution. However, in some cases accurate solutions are required to problems in which the assumption of a constant heat transfer coefficient is not valid. The cases described above, in which h is a function of temperature, make the conduction problems non-linear and these often have no known analytical solution. These cases, together with the fact that the one-dimensional assumption is often not valid, means that alternative solution procedures such as analogue or numerical methods must be used. It is pertinent to remark here that the present-day availability of high-speed digital computers has moved the emphasis from analogue methods to the numerical approach described in Chapter 4.

3-3   Some operational mathematical methods

The conduction problems treated in Section 3-2 have been solved by the method of separation of variables. This method has been shown to be relatively straightforward and is applicable to a wide range of simple practical problems. However, there are some problems which may be considered simple in formulation but for which the method of separation of variables cannot be used. For instance, transient conduction may occur near the surface of a very large block of material in which, for all practical purposes, one dimension may be considered to extend to infinity. Transient conduction may also occur at the centre of a very long thin rod which, practically, may be considered to extend to infinity in both directions. Each of these problems requires an alternative analytical method for its solution.

No attempt is made in this text to give a complete and rigorous presentation of alternative methods used in the integration of partial differential equations. The purpose of this section is rather to make the reader aware of, and to outline briefly, some alternative methods of analysis which form the basis of many solutions presented in operational mathematical texts on heat conduction, such as that by Carslaw and Jaeger[2].

The methods outlined here may be considered to derive from the important principle that a given arbitrary function may be represented by an infinite trigonometric series. Accordingly, an outline is given of the Fourier sine and cosine series for expansions over a finite range and how the Fourier series takes the form of an integral when the interval is extended to cover an infinite range. The extension of this to the Fourier transform—one of the most powerful tools in modern analysis—is indicated, and its application, together with that of the Laplace transform, is briefly demonstrated.

3-3-1 Fourier series

It has been shown in Section 3-2 that the solution of Laplace's equation in two space coordinates leads to a series which, for a constant position along one coordinate, say y, can express the temperature along the other, x, in the form:

$$f(x) = \tfrac{1}{2}A_0 + \sum_{n=1}^{\infty} A_n \cos(nx) + B_n \sin(nx) \qquad (3\text{-}50)$$

This expansion is known as a Fourier series and the constant coefficients A_n and B_n are known as the amplitudes. The n in Eqn (3-50) may be regarded as consecutive integers in this section, although we have seen how convective boundary conditions produce series solutions in which the n are consecutive roots of trigonometric equations.

We assume, because of the periodicity of the sine–cosine series (2π), that $f(x)$ in Eqn (3-50) is known over the interval $(-\pi, \pi)$ and that we are required to find the constants A_n and B_n to define the series. If we assume that the series is uniformly convergent so that it can be integrated term by term from $-\pi$ to π, we know that

$$\int_{-\pi}^{\pi} \cos(nx)\,dx = \int_{-\pi}^{\pi} \sin(nx)\,dx = 0 \quad \text{for } n = 1, 2, 3, \ldots$$

and that Eqn (3-50) yields

$$\int_{-\pi}^{\pi} f(x)\,dx = A_0\pi$$

If we multiply Eqn (3-50) by $\cos(nx)$ there results

$$f(x)\cos(nx) = \tfrac{1}{2}A_0\cos(nx) + \cdots + A_n\cos^2(nx) + \cdots \qquad (3\text{-}50\text{a})$$

where the omitted terms involve products of the form $\sin(mx)\cos(nx)$ or $\cos(mx)\cos(nx)$ with $m \neq n$. It can quickly be shown that for integral values of m and n:

$$\int_{-\pi}^{\pi} \sin(mx)\cos(nx)\,dx = 0 \quad \text{in general}$$

and

$$\int_{-\pi}^{\pi} \cos(mx)\cos(nx)\,dx = 0 \quad \text{for } m \neq \pm n$$

and hence integration of Eqn (3-50a) yields

$$\int_{-\pi}^{\pi} f(x)\cos(nx)\,dx = A_n \int_{-\pi}^{\pi} \cos^2(nx)\,dx = A_n\pi$$

which gives

$$A_n = \frac{1}{\pi} \int_{-\pi}^{\pi} f(x)\cos(nx)\,dx \quad (n = 1, 2, 3, \ldots)$$

Similarly, multiplying the series (3-50) by $\sin(nx)$ and integrating gives

$$B_n = \frac{1}{\pi} \int_{-\pi}^{\pi} f(x)\sin(nx)\,dx$$

Now, depending on whether $f(x)$ is an *odd* function [that is, $f(-x) \equiv -f(x)$] or an *even* function [that is, $f(-x) \equiv f(x)$] the series is represented by sines alone or cosines alone (for example, x^2 and $\cos x$ are *even* whereas x and $\sin x$ are *odd*). This can be established by examining the integrals of products of odd and even $f(x)$ with $\sin(x)$ or $\cos(x)$. For arbitrary functions which are neither odd nor even both parts of the series are required.

Extension of Fourier series to any range

The foregoing section shows how a function $f(x)$ may be expanded over the interval $-\pi$, π, but in the applications so far considered we require a given function to be expanded over the general interval $-L < x < L$. This is done by changing the scale by putting $z = \pi x/L$. Since $f(x)$ now has a period of $2L$, $f(z)$ has a period of 2π and therefore has a Fourier series in z.

$$f(z) = \tfrac{1}{2}A_0 + \sum_{n=1}^{\infty} A_n \cos(nz) + B_n \sin(nz)$$

and since $z = \pi x/L$, we have

$$f(x) = \tfrac{1}{2}A_0 + \sum_{n=1}^{\infty} A_n \cos\left(\frac{n\pi x}{L}\right) + B_n \sin\left(\frac{n\pi x}{L}\right) \tag{3-51}$$

where, since the limit $-L$ corresponds to $-\pi$ and L to π,

$$A_n = \frac{1}{L} \int_{-L}^{L} f(x) \cos\left(\frac{n\pi x}{L}\right) dx$$

and

$$B_n = \frac{1}{L} \int_{-L}^{L} f(x) \sin\left(\frac{n\pi x}{L}\right) dx$$

These coefficients therefore allow Fourier expansions of a given function to be made over the arbitrary interval $(-L, L)$, where L is finite.

If the function $f(x)$ is given in the half-range $(0, L)$ then, by extending the function into the other half $(-L, 0)$, it can be shown that for a sine series (an *odd* function):

$$f(x) = \sum_{n=1}^{\infty} B_n \sin\left(\frac{n\pi x}{L}\right) \quad \text{with} \quad B_n = \frac{2}{L} \int_{0}^{L} f(x) \sin\left(\frac{n\pi x}{L}\right) dx$$

and for a cosine series (an *even* function):

$$f(x) = \sum_{n=1}^{\infty} A_n \cos\left(\frac{n\pi x}{L}\right) \quad \text{with} \quad A_n = \frac{2}{L} \int_{0}^{L} f(x) \cos\left(\frac{n\pi x}{L}\right) dx$$

The above discussion may be applied directly to the solution of Fourier's one-dimensional transient equation

$$\frac{\partial T}{\partial t} = \alpha \frac{\partial^2 T}{\partial x^2} \quad 0 < x < L$$

for the problem described in Section 3-2-2.

We assume that the initial temperature distribution $T = f(x)$, $t = 0$ $(0 < x < L)$ can be represented by a convergent Fourier series.

The boundary conditions are $T = 0$ at $x = 0$ and L for $t \geqslant 0$, and since $\partial^2 T / \partial x^2$ must exist if T satisfies the differential equation we know that $T(x, t)$ has a Fourier series in x for each fixed $t > 0$:

$$T = \sum_{n=1}^{\infty} B_n(t) \sin\left(\frac{n\pi x}{L}\right) \tag{3-52}$$

where the choice of a sine series automatically satisfies the boundary conditions. Assuming that Eqn (3-52) can be differentiated term by term we have, on substitution into the differential equation,

$$\sum_{n=1}^{\infty} B_n'(t) \sin\left(\frac{n\pi x}{L}\right) = \alpha \sum_{n=1}^{\infty} B_n(t) \left(\frac{n\pi}{L}\right)^2 \left[-\sin\left(\frac{n\pi x}{L}\right)\right]$$

This equation is satisfied if the coefficients of $\sin(n\pi x/L)$ on each side are equated:

$$B'_n(t) = -\alpha \left(\frac{n\pi}{L}\right)^2 B_n$$

Upon integration this gives

$$B_n(t) = C_n \exp\left[-\alpha\left(\frac{n\pi}{L}\right)^2 t\right]$$

where the C_n are constants and hence Eqn (3-52) becomes

$$T = \sum_{n=1}^{\infty} C_n \exp\left[-\alpha\left(\frac{n\pi}{L}\right)^2 t\right] \sin\left(\frac{n\pi x}{L}\right)$$

The initial condition for $t = 0$ gives

$$f(x) = \sum_{n=1}^{\infty} C_n \sin\left(\frac{n\pi x}{L}\right)$$

in which, from the foregoing discussion on the amplitudes of Fourier series, the coefficients are given by

$$C_n = \frac{2}{L} \int_0^L f(x) \sin\left(\frac{n\pi x}{L}\right) dx$$

The final solution, therefore, agrees exactly with the solution obtained by the separation of variables method, Eqn (3-31).

3-3-2 The Fourier transform

The previous section has illustrated how a Fourier series may be used to represent an arbitrary function over a finite interval $(-L, L)$. However, the representation of an arbitrary function over the interval $(-\infty, \infty)$ or $(0, \infty)$ is often required for problems which may be considered, for all practical purposes, to exist in a region extending to infinity in one or both directions. Such a representation leads, in the limit, to a definite integral.

Complex form of a Fourier series

The Fourier series for $f(x)$ in $(-L, L)$:

$$f(x) = \tfrac{1}{2}A_0 + \sum_{n=1}^{\infty} A_n \cos\left(\frac{n\pi x}{L}\right) + B_n \sin\left(\frac{n\pi x}{L}\right)$$

can be written in an equivalent form using the identity $e^{i\theta} = \cos\theta + i\sin\theta$ since

$$\cos\left(\frac{n\pi x}{L}\right) = \tfrac{1}{2}(e^{in\pi x/L} + e^{-in\pi x/L})$$

and

$$\sin\left(\frac{n\pi x}{L}\right) = \frac{1}{2i}\left(e^{in\pi x/L} - e^{-in\pi x/L}\right)$$

then

$$f(x) = \tfrac{1}{2}A_0 + \sum_{n=1}^{\infty}\left(P_n\, e^{in\pi x/L} + Q_n\, e^{-in\pi x/L}\right)$$

where

$$P_n = \tfrac{1}{2}(A_n - iB_n) \quad \text{and} \quad Q_n = \tfrac{1}{2}(A_n + iB_n)$$

We finally obtain

$$f(x) = \sum_{n=-\infty}^{\infty} C_n\, e^{in\pi x/L} \tag{3-53}$$

where

$$C_n = \frac{1}{2L}\int_{-L}^{L} f(x)\, e^{-in\pi x/L}\, dx$$

for the expansion over an arbitrary interval $(-L, L)$.

Generally in a series representation of a function the amplitudes get smaller as the frequencies $\omega_n \equiv n\pi/L$ increase, and the series summation tends to an integral.

From Eqn (3-53) we can write

$$f(x) = \sum_{n=-\infty}^{\infty} \frac{1}{2L}\left[\int_{-L}^{L} f(p)\, e^{-in\pi p/L}\, dp\right] e^{in\pi x/L}$$

and if we let $\omega_n \equiv n\pi/L$ and $\delta\omega_n \equiv \pi/L$, then

$$f(x) = \sum_{n=-\infty}^{\infty} \frac{1}{2\pi}\, e^{i\omega_n x}\left[\int_{-L}^{L} f(p)\, e^{-i\omega_n p}\, dp\right]\delta\omega_n$$

If we let

$$F(\omega_n) = \frac{1}{2\pi}\, e^{i\omega_n x}\left[\int_{-L}^{L} f(p)\, e^{-i\omega_n p}\, dp\right]$$

then, from the fundamental definition of an integral,

$$\sum_{n=-\infty}^{\infty} F(\omega_n)\,\delta\omega_n \to \int_{-\infty}^{\infty} F(\omega)\, d\omega \quad \text{as } \delta\omega_n \to 0$$

and so

$$f(x) = \int_{-\infty}^{\infty} \frac{1}{2\pi}\, e^{i\omega x}\left[\int_{-\infty}^{\infty} f(p)\, e^{-i\omega p}\, dp\right] d\omega$$

or we can write

$$\phi(\omega) = \int_{-\infty}^{\infty} f(p) \, e^{-i\omega p} \, dp$$

and

$$f(x) = \frac{1}{2\pi} \int_{-\infty}^{\infty} \phi(\omega) \, e^{i\omega x} \, d\omega$$

$$\left. \right\} \quad (3\text{-}54)$$

These are known as the Fourier integrals or Fourier transforms. The Fourier transform of a function, say $f(x)$, can be defined as the operation of multiplying the function by e^{ixu} (where u is now the transform variable) and integrating from $-\infty$ to ∞, that is

$$\Phi(u) = \int_{-\infty}^{\infty} e^{ixu} f(x) \, dx$$

and the inverse of this transform is

$$f(x) = \frac{1}{2\pi} \int_{-\infty}^{\infty} \Phi \, e^{-ixu} \, du \qquad (3\text{-}55)$$

As an application of the Fourier transform consider the problem of an infinitely long bar when the initial temperature, $T(x)$ at $t = 0$, is known. The temperature T at any point x and time t is described by the equation

$$\frac{\partial T}{\partial t} = \alpha \frac{\partial^2 T}{\partial x^2}, \quad -\infty < x < \infty \qquad (3\text{-}56)$$

The initial and boundary conditions can be written

$$T = f(x) \quad \text{at } t = 0, \quad -\infty < x < \infty \qquad (3\text{-}57)$$

$$T \to 0 \quad \text{for } t > 0 \text{ as } x \to \pm\infty$$

By multiplying both sides of Eqn (3-56) by e^{ixu} and integrating from $-\infty$ to ∞ we obtain

$$\int_{-\infty}^{\infty} \frac{\partial T}{\partial t} e^{ixu} \, dx = \alpha \int_{-\infty}^{\infty} \frac{\partial^2 T}{\partial x^2} e^{ixu} \, dx$$

Integrating this by parts gives

$$\frac{\partial}{\partial t} \int_{-\infty}^{\infty} T \, e^{ixu} \, dx = \alpha \int_{-\infty}^{\infty} (iu)^2 T \, e^{ixu} \, dx$$

so that we obtain the ordinary differential equation:

$$\frac{d\Phi}{dt} + \alpha u^2 \Phi = 0$$

where $\Phi(u) = \int_{-\infty}^{\infty} T \, e^{ixu} \, dx$ is the Fourier transform of T.

This ordinary differential equation has the solution

$$\Phi = C e^{-\alpha u^2 t}$$

If $F(u)$ is now the Fourier transform of $f(x)$ (the initial condition), that is,

$$F(u) = \int_{-\infty}^{\infty} f(x) e^{ixu} dx$$

then

$$\Phi = F(u) e^{-\alpha u^2 t} \tag{3-58}$$

From the inverse of the transform of T above we have

$$T = \frac{1}{2\pi} \int_{-\infty}^{\infty} \Phi e^{-ixu} du$$

so that by substituting Φ from Eqn (3-58) we have the solution

$$T = \frac{1}{2\pi} \int_{-\infty}^{\infty} F(u) e^{-\alpha u^2 t} e^{-ixu} du \tag{3-59}$$

Although this solution involves two integrations it is an explicit equation for the temperature in terms of the initial condition.

An important class of solutions known as source solutions[2] are obtained by considering the impulsive release of a given amount of heat at a point in an infinite bar. This physical situation can be described by Dirac's delta function $\delta(x)$, which can be defined by

$$\int_{-\infty}^{\infty} f(x) \delta(x - x_0) dx = f(x_0)$$

This function corresponds to a unit amount of heat being released over an infinitesimally small region at x_0 so that an infinite temperature rise results. Similar impulsive situations arise when a switch is closed in an electrical circuit or when a load on a beam is considered to act at a point.

Accordingly, if $f(x) = \delta(x - x_0)$, then $F(u) = e^{ix_0 u}$ and, therefore, in Eqn (3-59) we obtain

$$T = \frac{1}{2\pi} \int_{-\infty}^{\infty} e^{ix_0 u} e^{-\alpha u^2 t} e^{-ixu} du$$

which, by collecting the powers of e and completing the square, gives

$$T = \frac{1}{2\pi} \int_{-\infty}^{\infty} \exp\left\{-\left[\frac{i(x - x_0)}{2\sqrt{(\alpha t)}} + u\sqrt{(\alpha t)}\right]^2\right\} \exp\left\{\left[\frac{i(x - x_0)}{2\sqrt{(\alpha t)}}\right]^2\right\} du$$

$$= \frac{1}{2\pi} \exp\left[-\frac{(x - x_0)^2}{4\alpha t}\right] \int_{-\infty}^{\infty} \exp\left\{-\left[\frac{i(x - x_0)}{2\sqrt{(\alpha t)}} + u\sqrt{(\alpha t)}\right]^2\right\} du$$

The integral in this expression has the standard result $\sqrt{(\pi/\alpha t)}$, so that the solution can be written

$$T = \frac{1}{\sqrt{(4\pi\alpha t)}} \exp\left[-\frac{(x - x_0)^2}{4\alpha t}\right]$$

This solution, known as the *source* solution, is often written more simply as

$$T = t^{-1/2}\, e^{-x^2/4\alpha t} \tag{3-60}$$

and has the properties

$$T \to 0 \text{ as } t \to 0 \text{ for fixed } x \neq 0$$

$$T \to \infty \text{ as } t \to 0 \text{ if } x = 0$$

and

$$\int_{-\infty}^{\infty} T\, dx = 2(\pi\alpha)^{1/2} \quad \text{for all } t > 0$$

This latter expression is, of course, the quantity of heat released per unit area over the plane $x = 0$ at $t = 0$, that is, $2C\rho\sqrt{(\pi\alpha)}$.

Now integration or differentiation of the source solution, Eqn (3-60), also yields a solution of Eqn (3-56) so that this differential equation is also satisfied by

$$\int_0^x t^{-1/2}\, e^{-x^2/4\alpha t}\, dx = 2\alpha^{1/2} \int_0^{x/[2\sqrt{(\alpha t)}]} e^{-\xi^2}\, d\xi$$

where $\xi = x/[2\sqrt{(\alpha t)}]$, and this is known as the *error function* and is usually used with the notation

$$\operatorname{erf} x = \frac{2}{\sqrt{\pi}} \int_0^x e^{-\xi^2}\, d\xi \tag{3-61}$$

We have therefore shown that

$$C \operatorname{erf} \frac{x}{2\sqrt{(\alpha t)}} \tag{3-62}$$

where C is a constant, is a solution of

$$\frac{\partial T}{\partial t} = \alpha \frac{\partial^2 T}{\partial x^2}$$

For given values of x and t Eqn (3-61) has a numerical value and some tables of values, together with further properties of the solution, are given by Carslaw and Jaeger[2].

3-3-3   The Laplace transform

The Laplace transform of a function, $f(x)$—frequently written as the operator $\mathscr{L}(u)$—is defined as the operation of multiplying $f(x)$ by e^{-ux} and integrating from zero to $+\infty$, that is,

$$\mathscr{L}(u) = \int_0^\infty e^{-ux} f(x)\, dx$$

where u may be real or imaginary.

By comparing this with Eqns (3-54) it can be seen to be simply a special case of the Fourier transform. The Laplace transform is widely used for the solution of physical problems and its use can be compared to that of logarithmic tables in which the solution is obtained from tables of anti-logarithms. Tables of Laplace transforms are used in a similar manner except that we deal with functions as opposed to numbers. For example, if $f(x) = x^{1/2}$ then the Laplace transform of $f(x)$ is

$$\mathscr{L}(u) = \int_0^\infty e^{-ux} x^{1/2}\, dx = \left(\frac{\pi}{4u^3}\right)^{1/2}$$

and $x^{1/2}$ is the inverse transform of $(\pi/4u^3)^{1/2}$.

It can be shown by integrating by parts that the Laplace transform of first and second derivatives are

$$\left.\begin{aligned}
\mathscr{L}\left(\frac{df}{dx}\right) &= u\mathscr{L}(u) - f_{x=0} \\[2mm]
\mathscr{L}\left(\frac{d^2 f}{dx^2}\right) &= u^2 \mathscr{L}(u) - uf_{x=0} - \left(\frac{df}{dx}\right)_{x=0}
\end{aligned}\right\} \tag{3-63}$$

where $f_{x=0}$ and $(df/dx)_{x=0}$ are boundary conditions satisfied by $f(x)$.

Application to an ordinary differential equation

We have shown that the temperature $T(x)$ in a simple fin is described by the equation

$$\frac{d^2 T}{dx^2} - m^2 T = 0 \tag{3-3}$$

We assume the boundary conditions $T = T_0$ at $x = 0$ and $dT/dx = 0$ at $x = L$. Applying the Laplace transform to both terms in Eqn (3-3) we obtain

$$\mathscr{L}\left(\frac{d^2 T}{dx}\right) - \mathscr{L}(m^2 T) = 0$$

which, from definitions (3-63), can be expanded to

$$\left[u^2 \mathscr{L}(u) - uT_0 - \left(\frac{dT}{dx} \right)_{x=0} \right] - m^2 \mathscr{L}(u) = 0$$

It is necessary to assume a value for $(dT/dx)_{x=0}$, say D, so that we can write

$$\mathscr{L}(u)(u^2 - m^2) - uT_0 - D = 0$$

which can be rearranged to

$$\mathscr{L}(u) = \frac{uT_0 + D}{u^2 - m^2} = T_0 \left(\frac{u}{u^2 - m^2} \right) + \frac{D}{m} \left(\frac{m}{u^2 - m^2} \right)$$

Now, by consulting tables of Laplace transform pairs[2],[3] we find that the transform $m/(u^2 - m^2)$ corresponds to the function sinh (mx) and $u/(u^2 - m^2)$ corresponds to the function cosh (mx) so that the inverse transformation gives the solution

$$T = T_0 \cosh(mx) + \frac{D}{m} \sinh(mx)$$

To find D we apply the condition $dT/dx = 0$ at $x = L$,

$$\frac{dT}{dx} = 0 = T_0 m \sinh(mx) + D \cosh(mx)$$

$$D = -T_0 m \frac{\sinh(mL)}{\cosh(mL)}$$

and the solution becomes

$$T = T_0 \frac{\cosh[m(L - x)]}{\cosh(mL)}$$

which agrees with Eqn (3-6).

Application to a partial differential equation

The previous section shows that the Laplace transform applied to an ordinary differential equation produced an *algebraic equation* so that, as in the case of the Fourier transform, we expect it to reduce a partial differential equation to an ordinary differential equation.

For example, consider the transient one-dimensional problem of a semi-infinite solid $0 < x < \infty$ (shown in Fig. 3-20) initially at constant temperature T_i which has the surface temperature suddenly reduced to zero at $t = 0$. This is a problem which cannot be solved by 'separation of variables' or by Fourier series since x extends to ∞.

We are required to solve

$$\frac{\partial T}{\partial t} = \alpha \frac{\partial^2 T}{\partial x^2}, \quad 0 < x < \infty$$

with the initial conditions

$$T = T_i \quad \text{at } t = 0, 0 < x < \infty$$

and the boundary conditions

$$T = 0 \quad \text{at } x = 0, t > 0$$
$$T = T_i \quad \text{at } x = \infty, t > 0$$

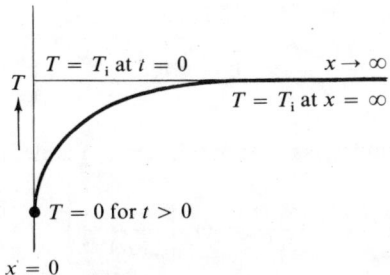

Fig. 3-20 Semi-infinite solid with a sudden change in the surface temperature

Applying the Laplace transform with respect to time, t, to both sides of the equation we have

$$\int_0^\infty e^{-ut} \frac{\partial T}{\partial t} dt = \alpha \int_0^\infty e^{-ut} \frac{\partial^2 T}{\partial x^2} dt$$

We can express the left-hand side in terms of its algebraic transform and interchange the order of integration and differentiation for the right-hand side to obtain

$$u\mathscr{L}(u) - T(x, 0) = \alpha \frac{\partial^2}{\partial x^2} \left[\int_0^\infty e^{-ut} T(x, t) dt \right] \qquad (3\text{-}64)$$

The right-hand side is simply a function of u and x and we can rewrite Eqn (3-64) in the form

$$\frac{d^2}{dx^2} \mathscr{L}(u) - \frac{u}{\alpha} \mathscr{L}(u) = \frac{T_i}{\alpha}$$

subject to the transforms of the boundary conditions

$$\mathscr{L}(0, u) = 0; \qquad \mathscr{L}(\infty, u) = \frac{T_i}{u} \qquad (3\text{-}65)$$

The general solution of this ordinary differential equation is

$$\mathscr{L}(x, u) = \frac{T_i}{u} + A\, e^{-x\sqrt{(u/\alpha)}} + B\, e^{x\sqrt{(u/\alpha)}}$$

The integration constants A and B are determined by considering the transforms of the boundary conditions, Eqn (3-65), which gives

$$\mathscr{L}(x, u) = \frac{T_i}{u} - \frac{T_i}{u}\, e^{-x\sqrt{(u/\alpha)}} \tag{3-66}$$

We can now obtain the inversion of Eqn (3-66) by finding from tables the functions corresponding to the transforms $1/u$ and $(1/u)\, e^{-x\sqrt{(u/\alpha)}}$. These give the solution

$$\frac{T(x, t)}{T_i} = 1 - \text{erfc}\left[\frac{x}{2(\alpha t)^{1/2}}\right]$$

where the complementary error function $\text{erfc}(x)$ is defined as $1 - \text{erf}(x)$.

Hence

$$\frac{T(x, t)}{T_i} = \text{erf}\left[\frac{x}{2(\alpha t)^{1/2}}\right]$$

which is the solution obtained above from the application of Fourier transforms, see Eqn (3-62).

REFERENCES

1. Fourier, J. B. J. *Theorie analytique de la chaleur*. Paris, 1822; translated by A. Freeman, Stechart, New York, 1878.
2. Carslaw, H. S. and J. C. Jaeger. *Conduction of Heat in Solids*. Clarendon Press, Oxford, 1959.
3. Arpaci, V. S. *Conduction Heat Transfer*. Addison-Wesley, Reading, Mass., 1966.
4. Gardner, K. Efficiency of extended surface. *Trans. A.S.M.E.*, 1945, **67** (No. 8).
5. McAdams, W. H. *Heat Transmission* (3rd edn). McGraw-Hill, New York, 1954.
6. Gröber, H. Die Erwärmung und Abkühlung einfacher geometrischer Körper. *Z. Ver. dt. Ing.*, 1925, **67**, 705.
7. Heisler, M. P. Temperature charts for induction and constant-temperature heating. *Trans. A.S.M.E.*, 1947, **69** (No. 3).

PROBLEMS

3-1 A cast iron fin on an engine casing is 0·3 m wide and has a rectangular profile 3 mm thick by 5 cm long. The base temperature of the fin is measured to be 150°C when operating in an ambient temperature of 20°C. If the thermal conductivity of the fin is 52 W/m K and the surface heat transfer coefficient is 50 W/m² K find the temperature at the tip and at a point halfway along the fin. Determine also the power dissipated by 40 such fins (in kW and h.p.). The heat lost from the tips may be neglected.

Ans: 87·9°C, 102·1°C; 5·25 kW, 7·03 h.p.

3-2 A cylindrical rod 2 cm diameter and 0·2 m long protrudes from a heat source at 300°C into air at 40°C. The heat transfer coefficient is known to be 5 W/m² K on all exposed surfaces. Neglecting radial variations of temperature and the heat lost from the tip, find the temperature at the tip and at a point halfway along the rod for rods made from borosilicate glass and mild steel. Take the thermal conductivity of glass and mild steel at this level of temperature to be 1·09 and 48 W/m K respectively. Determine also the fin efficiency of both rods and compare the result with Fig. 3-7.

Ans: Glass 41·2°C, 52·6°C, 0·165; Steel 219·5°C, 238·5°C, 0·79

3-3 By considering a small element of a straight fin of uniform thickness, y, show that the heat flow rate, $\dot{Q}$, past any position x is given by

$$\dot{Q} = kAT_0 m \left[\frac{\sinh m(L - x) + N \cosh m(L - x)}{\cosh mL + N \sinh mL} \right]$$

where $m^2 = 2h/ky$, $N = h_t/mk$, k is the fin thermal conductivity, h and h_t are the heat transfer coefficients over the fin and fin tip respectively, A is the cross-sectional area per unit width, L is the fin length, and T_0 is the difference in temperature between the fin base and the surrounding fluid.

By assuming $h_t = h$ show that the parameter hy/k has a maximum value of 2 for the fin to be useful. If such a fin has a thickness of 1 cm and a thermal conductivity of 25 W/m K determine whether it is useful or not for heat transfer coefficients of 100, 5,000, and 10,000 W/m² K.

3-4 A turbine wheel is cast with the blades and disc integral. If each blade is treated as a one-dimensional fin of uniform cross section conducting heat from the hot gas at T_g to the blade root at $T_{b,r}$ show that the fin can be represented by the metal/fluid interface boundary condition

$$-k \left(\frac{dT}{dn} \right)_{b,r} = h(T_{b,r} - T_g)$$

at its junction with the disc.

A gas turbine wheel has 41 blades and each blade is 0·736 in long and has a cross-sectional area of 0·13 in² in a total blade platform area of 17 in². The periphery of the blade section is 0·15 ft and the thermal conductivity is 15·85 Btu/ft h °R. If the heat transfer coefficient over all the surfaces exposed to the hot gas is 222 Btu/ft² h °R, determine the equivalent mean heat transfer coefficient over the total platform area for estimating the heat transfer into the disc.

Ans: 393 Btu/ft² h °R

3-5 A very wide steel structural member ($k = 33$ Btu/ft h °R) positioned inside the uptake of a furnace may be considered as a truncated triangular fin 1 ft long, 3 inches thick at the base and 1 inch thick at the tip. The temperature at the base is known to be 250°F when the fin is immersed in hot gas at 500°F. If the heat transfer coefficient from gas to fin is estimated to be 10 Btu/ft² h °R, calculate the tip temperature of the fin. Determine the error incurred in estimating this temperature when the fin is considered as a simple straight fin of thickness 2 inches: (a) when the heat transfer to the tip is neglected and (b) when an increment of half thickness is added to the fin length to account for this tip heat transfer.

Ans: 429·4°F; (a) −1·9°F; (b) +8·5°F

3-6 (a) A porous material is cooled by a fluid forced unidirectionally through its interstices. Show that the differential equations satisfied by the temperature of the porous material, T_m, and the temperature of the coolant, T_c, are:

$$k \frac{d^2 T_m}{dx^2} = h_i(T_m - T_c); \qquad GC_p \frac{dT_c}{dx} = h_i(T_m - T_c)$$

where h_i is an internal *volumetric* heat transfer coefficient (that is, it has the units W/m³ K, G is the mass flow per unit surface area, and C_p is the specific heat of the coolant.

(b) By obtaining a single third-order ordinary differential equation from the two differential equations given in (a), show that the solution for the metal and coolant temperatures is given by

$$T_m = \alpha + \left(\frac{1}{2} + \frac{B}{A_1}\right) \beta \exp\left[\left(B - \frac{A_1}{2}\right)x\right] + \left(\frac{1}{2} - \frac{B}{A_1}\right) \gamma \exp\left[-\left(B + \frac{A_1}{2}\right)x\right]$$

$$T_c = \alpha + \beta \exp\left[\left(B - \frac{A_1}{2}\right)x\right] + \gamma \exp\left[-\left(B - \frac{A_1}{2}\right)x\right]$$

where $A_1 = h_i/GC_p$, $B = [(\frac{1}{2}A_1)^2 + h_i/k]^{1/2}$ and α, β, γ are constants.

3-7 If the porous material described in Problem 3-6 generates heat electrically at a rate q_g per unit volume show that the solution of the *appropriate* differential equations gives the metal and coolant temperatures as:

$$T_m = \alpha + \left(\frac{1}{2} + \frac{B}{A_1}\right) \beta \exp\left[\left(B - \frac{A_1}{2}\right)x\right]$$
$$+ \left(\frac{1}{2} - \frac{B}{A_1}\right) \gamma \exp\left[-\left(B + \frac{A_1}{2}\right)x\right] + \frac{A_3}{A_2}\left(\frac{1}{A_1} + x\right)$$

$$T_c = \alpha + \beta \exp\left[\left(B - \frac{A_1}{2}\right)x\right] + \gamma \exp\left[-\left(B + \frac{A_1}{2}\right)x\right] + \frac{A_3}{A_2} x$$

where, in addition to the definitions given in Problem 3-6,

$$A_2 = h_i/k \quad \text{and} \quad A_3 = h_i q_g/kGC_p.$$

3-8 An infinitely long two-dimensional rectangular slab has a thickness L and a constant base temperature T_0. The action of the fluid at temperature T_f surrounding the fin is such that the heat transfer coefficient is infinitely large. Show that the temperature, T, in this two-dimensional region is given by

$$\frac{T - T_f}{T_0 - T_f} = \frac{2}{L} \sum_{n=1}^{\infty} \frac{1 - (-1)^n}{n\pi/L} \exp\left(\frac{-n\pi}{L} x\right) \sin\left(\frac{n\pi}{L} y\right)$$

3-9 A two-dimensional fin of thermal conductivity k, thickness $2L$, and with a base temperature T_0, may be considered to extend to infinity in one direction. The fin experiences a finite heat transfer coefficient h and an ambient fluid temperature T_f.

Show that if the coordinate system is taken at the *centre* of the fin thickness, the temperature of the fin, T, at any point is given by

$$\frac{T - T_f}{T_0 - T_f} = 2 \sum_{n=1}^{\infty} \left(\frac{\sin(p_n L)}{p_n L + \sin(p_n L) \cos(p_n L)} \right) e^{-p_n x} \cos p_n y$$

where the p_n are the roots of the equation $p_n L \tan(p_n L) = hL/k$.

3-10   If a thin rod completely insulated over its length L with the initial temperature distribution $T = f(x)$ suddenly has the temperature of the ends of the rod reduced to zero at $t = 0$, the temperature at any point x in the rod at any time t is given by

$$T = \frac{2}{L} \sum_{n=1}^{\infty} \exp\left[-\left(\frac{n\pi}{L}\right)^2 \alpha t \right] \sin\left(\frac{n\pi x}{L}\right) \int_0^L f(x) \sin\left(\frac{n\pi x}{L}\right) dx$$

where the n are integers and $\alpha\,(=k/\rho C)$ is the thermal diffusivity of the material. If the temperature of the rod is initially constant at $100°C$, determine the first three non-zero terms of the series for the temperature at the centre of the rod as a function of α and t. Comment on the convergence of this series.

3-11   A thin rectangular plate of thickness D loses heat from both flat surfaces into a medium at zero. If the thermal conductivity of the material is k and the surface heat transfer coefficient h, show that the differential equation satisfied by the temperature in the plate is

$$\frac{\partial^2 T}{\partial x^2} + \frac{\partial^2 T}{\partial y^2} - \frac{2hT}{kD} = 0$$

3-12   The thin rectangular plate of Problem 3-11 has the dimensions L_x, L_y and a thickness D. If the edge $y = 0$ is maintained at a temperature $f(x)$ and the other edges are maintained at zero, show that the solution to the differential equation given in Problem 3-11 is given by

$$T = \frac{2}{L_x} \sum_{n=1}^{\infty} \frac{\sin(n\pi x/L_x) \sinh(L_y - y)(N^2 + n^2\pi^2/L_x^2)^{1/2}}{\sinh L_y (N^2 + n^2\pi^2/L_x^2)^{1/2}} \int_0^{L_x} f(x) \sin\left(\frac{n\pi x}{L_x}\right) dx$$

where $N^2 = 2h/kD$.

3-13   A solid circular cylinder of radius r_0 and length L has its two plane ends maintained at a temperature T_E. The temperature of the cylindrical surface $T_R(z)$ has a distribution which is symmetrical about the mid-point of the axis. There is no circumferential variation of temperature. Denoting r and z as radial and axial co-ordinates respectively and with origin at the geometrical centre of the cylinder, show that the steady state solution for the temperature distribution in the cylinder is

$$T - T_E = \frac{2}{L} \sum_{\substack{n=1 \\ (n \text{ odd})}}^{\infty} \frac{\cos(\lambda_n z) I_0(\lambda_n r)}{I_0(\lambda_n r_0)} \int_{-L/2}^{L/2} f(z) \cos(\lambda_n z)\, dz$$

where $\lambda_n = n\pi/L$, $n = 1, 3, 5, 7, \ldots$, and $f(z) = T_R(z) - T_E$.

3-14   A brick wall 13·5 inches thick at an initial temperature of $75°F$ is suddenly exposed to a flow of cold air at $30°F$, on both surfaces, with a heat transfer coefficient of $3·5\ Btu/ft^2\ h\ °R$. Determine the temperature of (a) the centre of the wall, (b) the surface, and (c) a position 3 inches from the surface after 24 hours have passed.

Estimate also the heat transfer per square foot from the wall during this period. (Take the physical properties of the brick to be $\rho = 100$ lb/ft^3, $C = 0\cdot2$ Btu/lb °F, $k = 0\cdot4$ Btu/ft h °R.)

Ans: 33·6°F; 31·0°F; 32·7°F; 950 Btu/ft^2

3-15   An 18 Cr/8 Ni stainless steel rod, 2 cm diameter and very long, initially at a temperature of 50°C is suddenly bathed in hot gas at a temperature of 1,250°C. If the heat transfer coefficient from gas to rod is estimated to be 950 W/m^2 K, estimate the time taken for the centre of the rod to reach a temperature of 650°C and the surface temperature at that time. Determine also the time taken for the centre of the rod to reach 99% of the ultimate temperature change.

Ans: 17 s; 810°C; 100 s

3-16   The core of a nuclear reactor may be considered as a vertical fuel plate of thickness $2L$ surrounded by a coolant maintained at constant temperature. The system is initially at a temperature of zero. Nuclear internal energy, q_g, is then generated in the plate at an exponential decay rate according to $q_g = q_{g,0}\, e^{-bt}$. The heat transfer coefficient between plate and coolant is h. If the temperature variation in the plate is neglected, show that the temperature of the plate is described by the equation

$$\frac{dT}{dt} + \frac{hT}{\rho CL} = \frac{q_{g,0}}{\rho C} e^{-bt}$$

where ρ and C are the density and specific heat of the plate respectively. From first principles show by the method of Laplace transforms that the solution, $T(t)$, is given by

$$T(t) = \frac{q_{g,0}/\rho C}{(h/\rho CL - b)} [e^{-bt} - e^{-ht/\rho CL}]$$

3-17   Solve the differential equations governing the simultaneous flow of heat and mass in a porous material, given in Problem 3-6, by the method of Laplace transforms. Note that

$$\mathscr{L}(e^{bt}\cosh at) = \frac{u - b}{(u - b)^2 - a^2} \quad \text{and} \quad \mathscr{L}\left(\frac{1}{a} e^{bt}\sinh at\right) = \frac{1}{(u - b)^2 - a^2}$$

4

Numerical solution of heat conduction problems

4-1 Introduction

The heat conduction problems treated in Chapter 3 briefly indicate the limited scope of analytical solutions. Only relatively simple geometric shapes can be handled, and these only with the more straightforward type of boundary condition. The vast majority of practical two- and three-dimensional heat conduction problems have no known analytical solution as they usually involve irregular geometries with mathematically inconvenient mixed type boundary conditions.

Analytical solutions sometimes give qualitative information not readily available from numerical solutions, such as the effect of varying certain parameters, but if it is really a numerical solution that is required then evaluating an analytic expression is in itself a numerical process, prone to error, and may take longer than a direct finite difference or a finite element solution.

In many cases it is advisable to use a numerical method which often provides an *adequate* solution more simply and efficiently. Analytical solutions have, of course, the clear advantage that they are *true* solutions, whereas numerical solutions have an inherent error. This is not usually a serious disadvantage as it can often be accurately and quickly estimated. Also, in many practical situations the mathematical model considered is in itself an approximation to the true physical problem so that its true solution is finally subject to uncertain physical interpretation. Numerical solutions, on the other hand, can be made as accurate as the engineering data warrant, as will be demonstrated in this chapter.

With the development of the high-speed digital computer numerical techniques have been developed and extended to handle almost any problem of any degree of complexity. Although the solutions obtained by these techniques give answers only for a given number of discrete points and at discrete time intervals, this need be no disadvantage since use can be made of interpolating functions and graphical output devices. Accordingly, although the numerical techniques described in this chapter are applicable to hand or desk calculators, the emphasis is placed principally on the formulation and procedures for digital computer programming.

Our treatment of numerical heat conduction in this chapter is from the more generalized and more rigorous approach of the numerical analyst as advocated by Fox[1]†, Smith[2], Ames[3], Forsythe and Wasow[4], and Richtmyer and Morton[5]. This procedure is preferred to the 'resistance network' method for reducing the partial differential equations to algebraic form since this physical and often intuitive approach is of limited scope and can sometimes produce approximations of low accuracy and poor consistency within any one problem. It will be shown, however, that this approach can sometimes be of value in difficult cases in which finite difference equations can be constructed more simply from an examination of the heat flow in a control volume.

Of the numerical approximation methods available those employing finite difference approximations are straightforward, are more frequently used, and are generally more applicable than any other. Accordingly, this chapter concentrates on the finite difference approach to both steady-state and transient conduction problems. However, also briefly described is a relatively new approach known as the 'finite element' method, which is essentially a variational approach since the solution of the governing differential equation is obtained by minimizing some integrated quantity (referred to as a functional) which is evaluated over the whole region of interest.

This method is widely used in structural analysis and although it is not as straightforward, conceptually, as the finite difference approach it offers several advantages in the treatment of heat conduction problems, particularly with curved boundaries and normal derivative boundary conditions.

4-1-1 Classification of partial differential equations

It is common in heat transfer to treat steady-state conduction (involving two or more space variables) and transient heat conduction (involving one or more space variables and time) separately, and this is also done in the numerical analysis of these problems since they form separate classes of partial differential equation. These classes are important in that they classify the nature of the procedure used in their solution.

Many conservation principles lead to special forms of the two-dimensional second-order equation

$$A \frac{\partial^2 \phi}{\partial x^2} + B \frac{\partial^2 \phi}{\partial x \, \partial y} + C \frac{\partial^2 \phi}{\partial y^2} + D \frac{\partial \phi}{\partial x} + E \frac{\partial \phi}{\partial y} + F\phi + G = 0$$

where A, B, C, D, E, F, and G can be functions of x, y, and ϕ. Now, by analogy with the conic equation

$$ax^2 + 2bxy + cy^2 = H$$

† Full details of references cited are given at the end of the chapter.

(where a, b, c, and H are constants) which defines an ellipse, a parabola, or a hyperbola according to the sign of $b^2 - ac$, this partial differential equation is said to be:

elliptic   when $B^2 - 4AC < 0$
parabolic   when $B^2 - 4AC = 0$
and   hyperbolic   when $B^2 - 4AC > 0$

Only the first two types of equation are of interest in heat conduction. Elliptic equations usually occur in regions of integration with closed boundaries, and parabolic equations in regions with one boundary at infinity. Accordingly, steady-state problems are elliptic and transient problems are parabolic.

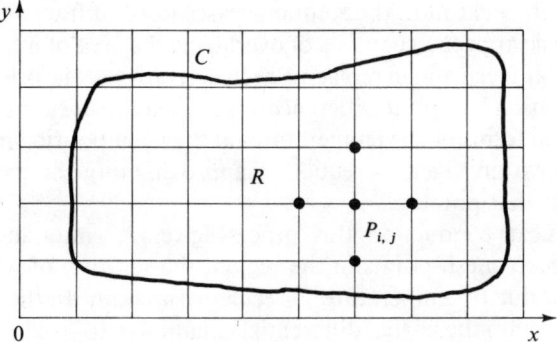

Elliptic (steady-state) heat conduction problem

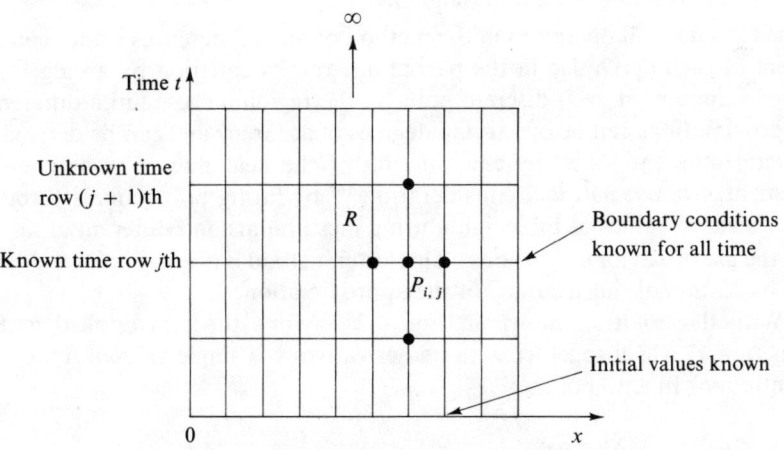

Parabolic (transient) heat conduction problem

Fig. 4-1 Elliptic and parabolic integration regions

4-1-2   Finite difference approach

The application of finite difference methods to both steady-state 'elliptic' and transient 'parabolic' equations are similar in that the integration of the differential equation over a region R is reduced to the solution of a system of algebraic equations. The type of region for both problems in two dimensions is different and is illustrated in Fig. 4-1. The structure of the algebraic equations in these two cases is also different, since with a steady-state 'elliptic' problem, such as Laplace's equation $\partial^2 T/\partial x^2 + \partial^2 T/\partial y^2 = 0$, previously derived as Eqn (2-7), all the algebraic equations are solved together as one system, whereas in transient 'parabolic' problems such as $\partial T/\partial t = \alpha(\partial^2 T/\partial x^2)$, previously derived as Fourier's equation [Eqn (2-5)], they start from initial, known conditions at $t = 0$ and propagate the solution forward from one time row to the next until the solution ceases to be of interest.

In both cases the basic approach consists of overlaying the area of integration R with a rectangular mesh and in replacing each derivative of the original partial differential equation by a finite difference approximation at each mesh point, $P_{i,j}$ in Fig. 4-1, in terms of the temperatures at the neighbouring mesh points and writing down an algebraic equation approximating the partial differential equation at that point.

With steady-state heat conduction this process gives N simultaneous algebraic equations for N mesh points in the region, the solution of which provides a numerical value of temperature for each mesh point. In the case of transient heat conduction the partial differential equation is approximated along one or two rows of mesh points at a time, Fig. 4-1, so that the procedure results in the successive solution of small sets of algebraic equations.

4-1-3   Finite difference approximations

The fundamental operation in the method of finite differences is the replacement of each derivative in the partial differential equation by an algebraic approximation at each discrete point in the region. These finite difference approximations can be of varying degrees of accuracy and can be derived in several different ways: by examination of the heat flow in a differential element, from variational considerations[4] by fitting polynomials through the function values of the neighbouring mesh points and differentiating, or by the use of Taylor expansions. This latter method is brief, simple, and gives an indication of the accuracy of the approximation.

With the notation shown in Fig. 4-2 Taylor's theorem applied to the function T, which together with its derivatives is a single-valued, finite and continuous function of x, gives:

$$T(x + \delta x) = T(x) + \delta x \frac{dT}{dx} + \tfrac{1}{2}\delta x^2 \frac{d^2 T}{dx^2} + \tfrac{1}{6}\delta x^3 \frac{d^3 T}{dx^3}$$
$$+ \tfrac{1}{24}\delta x^4 \frac{d^4 T}{dx^4} + \cdots \quad (4\text{-}1)$$

and

$$T(x - \delta x) = T(x) - \delta x \frac{dT}{dx} + \tfrac{1}{2}\delta x^2 \frac{d^2 T}{dx^2} - \tfrac{1}{6}\delta x^3 \frac{d^3 T}{dx^3}$$

$$+ \tfrac{1}{24}\delta x^4 \frac{d^4 T}{dx^4} - \cdots \quad (4\text{-}2)$$

By adding Eqns (4-1) and (4-2) we obtain

$$T(x + \delta x) + T(x - \delta x) = 2T(x) + \delta x^2 \frac{d^2 T}{dx^2} + o(\delta x^4) \quad (4\text{-}3)$$

where $o(\delta x^4)$ denotes terms containing fourth and higher powers of δx. We may rewrite Eqn (4-3) in the form

$$\left(\frac{d^2 T}{dx^2}\right)_x = \frac{T(x + \delta x) - 2T(x) + T(x - \delta x)}{\delta x^2} + o(\delta x^2) \quad (4\text{-}4)$$

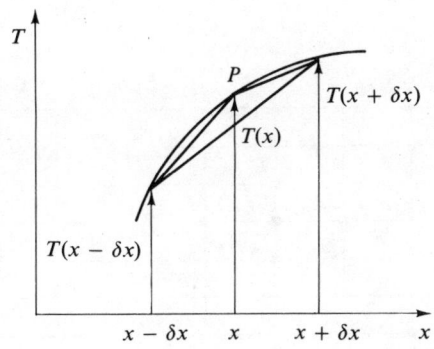

Fig. 4-2 Graphical illustration of finite difference approximations to derivatives

Similarly, by subtracting Eqn (4-2) from Eqn (4-1) and rearranging we obtain the 'central difference' formula

$$\left(\frac{dT}{dx}\right)_x = \frac{T(x + \delta x) - T(x - \delta x)}{2\delta x} + o(\delta x^2) \quad (4\text{-}5)$$

Both these approximations—that is, (4-4) and (4-5)—are of order δx^2, which implies that halving the step length δx approximately quarters the 'discretization' error, which is the error caused by treating the problem in a series of finite or discrete steps. From Fig. 4-2 or from Eqns (4-1) and (4-2) individually, the central difference [Eqn (4-5)] is clearly more accurate than either a forward difference

$$\frac{dT}{dx} = \frac{T(x + \delta x) - T(x)}{\delta x} + o(\delta x) \quad (4\text{-}6)$$

or a backward difference

$$\frac{dT}{dx} = \frac{T(x) - T(x - \delta x)}{\delta x} + o(\delta x) \qquad (4\text{-}7)$$

These two are often used although they have an accuracy only of order δx, so that halving the step length only approximately halves the discretization error.

While it is common in the solution of *ordinary* differential equations to employ an algorithm (a numerical scheme) with a truncation error of the order of δx^4 and δx^5, the finite difference formulae of order δx^2 above are satisfactory for the majority of heat conduction problems. Higher order difference approximations may be used of course, as will be explained later, but the additional programming complexity associated with trying to approximate the boundary conditions to the *same degree of accuracy* is seldom worthwhile.

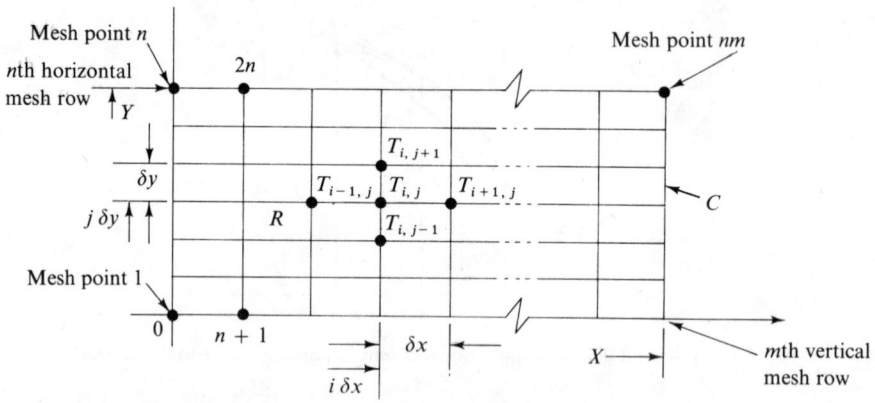

Fig. 4-3 Discretization of a rectangular plate

4-2 Steady-state heat conduction (elliptic equations)

Although numerical techniques are sometimes applied to one-dimensional heat conduction problems such as fins and spines (they can, in many instances, be nonlinear and have no known analytical solution), the value of numerical methods in steady-state conduction is best demonstrated by their application to two-dimensional problems.

The area of integration of a two-dimensional elliptic conduction problem is usually an area R bounded by a closed curve C (which may also have holes inside).

The boundary conditions usually specify either the temperature or the value of its normal derivative, at every point on C, or a mixture of both as would occur with convective heat transfer at the boundary.

4-2-1 Finite difference representation of the main equations

As an illustrative example consider the case of steady conduction in a rectangular plate, Fig. 4-3.

The rectangular area is exactly overlayed with a rectangular mesh with n points in the y direction and m in the x direction, giving a total of nm mesh points at which the solution is to be found. The step lengths are, respectively, $\delta y = Y/(n-1)$ and $\delta x = X/(m-1)$.

For simplicity we assume the thermal conductivity to be independent of temperature so that the partial differential equation obeyed within the region R in Fig. 4-3 is Laplace's equation:

$$\frac{\partial^2 T}{\partial x^2} + \frac{\partial^2 T}{\partial y^2} = 0$$

We approximate this equation at each point *within* the region R by replacing each derivative with its appropriate finite difference equivalent. For the point i, j in Fig. 4-3 we can apply Eqn (4-4) to each second-order derivative in Laplace's equation above, giving the approximation

$$\frac{T_{i+1,j} - 2T_{i,j} + T_{i-1,j}}{\delta x^2} + \frac{T_{i,j+1} - 2T_{i,j} + T_{i,j-1}}{\delta y^2} = 0$$

or

$$T_{i-1,j} + \left(\frac{\delta x}{\delta y}\right)^2 T_{i,j-1} - 2\left[1 + \left(\frac{\delta x}{\delta y}\right)^2\right] T_{i,j}$$
$$+ \left(\frac{\delta x}{\delta y}\right)^2 T_{i,j+1} + T_{i+1,j} = 0 \quad (4\text{-}8)$$

If equal step lengths are used in both x and y directions, that is $\delta x = \delta y$, this equation reduces to the well-known 'five point' formula which, with reference to the coefficients of the labelled points in Fig. 4-3, is often expressed diagrammatically as the 'molecule'

$$\begin{bmatrix} & 1 & \\ 1 & -4 & 1 \\ & 1 & \end{bmatrix} T = 0$$

Finite difference approximations to other partial differential equations in different coordinate systems can be assembled in a similar manner. The important case of Laplace's equation in cylindrical polar coordinates is discussed later in Section 4-4-1.

4-2-2   Representation of boundary conditions

Although we have seen how an algebraic equation can be obtained for the mesh points inside the region R in Fig. 4-3, an equation for each of the points on the boundary C must also be found which satisfies the boundary conditions imposed on the problem. If the temperature is known over the whole boundary the problem of Fig. 4-3 is considerably simplified since the application of Eqn (4-8) to each interior point is sufficient to allow the solution to be found. The known value could either be written into Eqn (4-8) directly so that each equation for a point adjacent to a boundary would have a coefficient on the right-hand side, or each boundary point itself could have a separate equation associated with it.

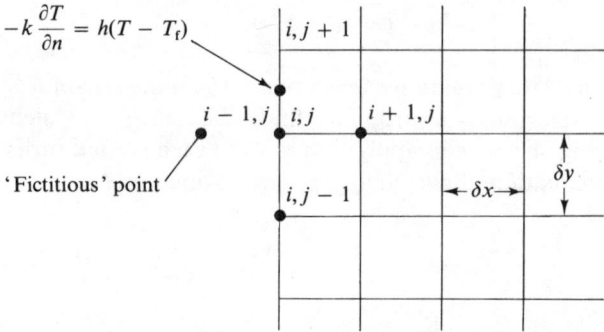

Fig. 4-4   Use of a fictitious point in a derivative boundary condition approximation

Derivative boundary conditions

A common example of this type of boundary condition in practice is when the surface of a heat conducting material is thermally insulated so that there is no heat flow normal to the surface; the boundary condition is then $\partial T/\partial n = 0$. A much more important practical example occurs when a fluid of known temperature, T_f, is in contact with a heat conducting surface. This boundary condition has been given in Section 2-4 as:

$$-k\frac{\partial T}{\partial n} = h(T - T_f) \tag{4-9}$$

where $\partial/\partial n$ denotes differentiation in the direction of the outward normal. If we assume that in Fig. 4-3 this condition is imposed on the surface $x = 0$ we require an algebraic equation to approximate Eqn (4-9). Consider the point i, j in Fig. 4-4. One very obvious way is to approximate $\partial T/\partial n$ with a forward difference [Eqn (4-6)] from the point i, j to $i + 1, j$. Our boundary condition would then be written

$$k\frac{\partial T}{\partial x} = h(T - T_f)$$

and represented by

$$k\left(\frac{T_{i+1,j} - T_{i,j}}{\delta x}\right) = h(T_{i,j} - T_f) \qquad (4\text{-}10)$$

It should be noted that the outward normal in this case is in the negative x direction and therefore a negative sign must be associated with $\partial T/\partial x$. However, this approximation (very commonly used in resistance network formulations) has an error of order δx and is therefore not as accurate as our approximation to the interior points, which is of order δx^2.

One method of overcoming this difficulty is to use a central difference formulation for $\partial T/\partial x$ and to eliminate the 'fictitious' point $i - 1, j$ by also applying the finite difference representation of the main equation [Eqn (4-9)] to the point i, j. Our convective boundary condition is then represented by:

$$k\left(\frac{T_{i+1,j} - T_{i-1,j}}{2\delta x}\right) = h(T_{i,j} - T_f) \qquad (4\text{-}11)$$

which, when the value of the 'fictitious' point $T_{i-1,j}$ from this equation is substituted into Eqn (4-8), gives the algebraic equation for the boundary point i, j in Fig. 4-4 as:

$$\left(\frac{\delta x}{\delta y}\right)^2 T_{i,j-1} - 2\left[1 + \delta x\frac{h}{k} + \left(\frac{\delta x}{\delta y}\right)^2\right] T_{i,j} + \left(\frac{\delta x}{\delta y}\right)^2 T_{i,j+1}$$

$$+ 2T_{i+1,j} = -2\delta x\frac{h}{k} T_f \quad (4\text{-}12)$$

which is an approximation of order δx^2.

If an adiabatic boundary condition, that is, $k(\partial T/\partial x) = 0$, is specified at the point i, j in Fig. 4-4, it is equivalent to the 'coefficient' of heat transfer h in Eqn (4-12) being set equal to zero. The corresponding change to Eqn (4-12) is obvious. It is important to note here that the application of a central difference to the point i, j for this adiabatic boundary condition gives

$$\frac{T_{i+1,j} - T_{i-1,j}}{2\delta x} = 0$$

that is

$$T_{i+1,j} = T_{i-1,j}$$

where $T_{i-1,j}$ is 'fictitious'. This, however, does not mean that $T_{i-1,j}$ and $T_{i+1,j}$ are both equal to $T_{i,j}$ as this would imply that the second derivative is zero, which is not necessarily the case.

If derivative boundary conditions are specified at a corner, as in Fig. 4-5, the use of central difference approximations to the derivatives $\partial T/\partial x$ and $\partial T/\partial y$ means that two fictitious points have to be eliminated with the main approximation to Laplace's equation. The corresponding order δx^2 representation for the corner point i, j in Fig. 4-5 can be shown to be:

$$2T_{i-1,j} + 2\left(\frac{\delta x}{\delta y}\right)^2 T_{i,j-1} - 2\left[1 + \left(\frac{\delta x}{\delta y}\right)^2\left(\delta y\frac{h}{k} + 1\right)\right]T_{i,j}$$
$$= -2\delta y\frac{h}{k}\left(\frac{\delta x}{\delta y}\right)^2 T_f \quad (4\text{-}13)$$

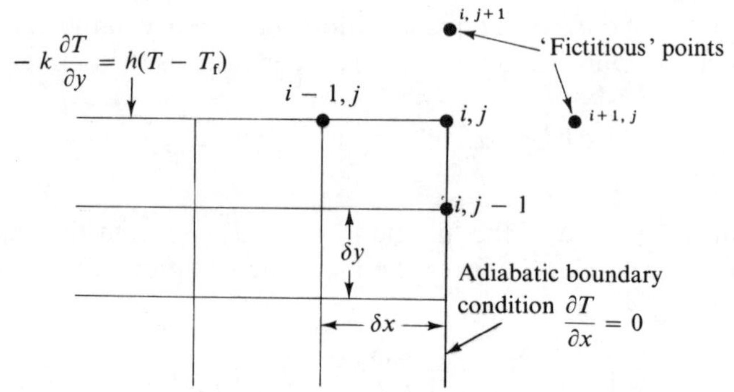

Fig. 4-5 Boundary conditions at a corner point

The use of this 'fictitious' point technique ensures that the accuracy of the finite difference approximation to the whole problem is consistent and that extrapolation procedures associated with improving the accuracy of the solution can be applied more easily. It can be seen from the above that if more accurate difference representations are employed within the region (they inevitably involve more of the neighbouring mesh points), the boundary condition approximation problem is made more difficult, particularly at corners.

4-2-3 Elliptic finite difference equations and their solution

Finite difference methods for solving steady-state heat conduction problems can lead to systems of many thousands of algebraic equations, particularly with irregular geometries, and the solution of the equations is not a trivial matter. In the majority of cases the equations will be linear, so that the coefficients will be constants or functions of the independent variables only, and the mathematical procedures of matrix algebra will apply. However, some problems, such as those involving a temperature-dependent thermal conductivity, produce nonlinear algebraic equations which require iterative methods for their solution.

The methods described in this section can be classed as either *direct* or *iterative*. It has been the practice for the past few years for small systems of equations to be solved by direct matrix elimination methods and large systems by iteration, but the recent development and availability of fast computers and disc storage facilities has meant that direct methods can be used on as many as 6,000–8,000 equations.

Assembly of the matrix coefficients

The system of finite difference equations can be expressed in matrix notation as $A\mathbf{T} = \mathbf{b}$, where A is the matrix of coefficients, $\mathbf{T}$ the required solution vector, and $\mathbf{b}$ a column vector of constants. The matrix equation $A\mathbf{T} = \mathbf{b}$, can be written out as:

$$\left.\begin{array}{c} a_{11}T_1 + a_{12}T_2 + \cdots + a_{1N}T_N = b_1 \\ a_{21}T_1 + a_{22}T_2 + \cdots + a_{2N}T_N = b_2 \\ \vdots \\ a_{N1}T_1 + a_{N2}T_2 + \cdots + a_{NN}T_N = b_N \end{array}\right\} \qquad (4\text{-}14)$$

where N is the order of the matrix (the total number of unknowns) and in which, for the problem of Fig. 4-3, most of the off-diagonal coefficients $a_{i,j}$ would be zero. The coefficients of the finite difference equation for point 1 would be assembled in row 1 of the matrix, and for point n in the matrix row n, and so on, so that for the general point i, j in Fig. 4-3—point k, say, in the ordering sequence—the use of the order δx^2 approximation of Eqn (4-8) would result in the following arrangement of matrix coefficients:

$$\begin{bmatrix} \cdots & a_{kk-n} & \cdots & a_{kk-1} & a_{kk} & a_{kk+1} & \cdots & a_{kk+n} & \cdots \end{bmatrix}\begin{bmatrix} T_k \end{bmatrix} = \begin{bmatrix} 0 \end{bmatrix}$$

$$\begin{array}{ccccc} 1 & & \left(\dfrac{\delta x}{\delta y}\right)^2 & -2\left[1 + \left(\dfrac{\delta x}{\delta y}\right)^2\right] & \left(\dfrac{\delta x}{\delta y}\right)^2 & 1 \end{array}$$

with the coefficient of $T_{i,j}$ placed in the centre diagonal a_{kk}.

Similarly, the arrangement of coefficients for the corner mesh point nm of Fig. 4-3, using the approximation of Eqn (4-13), would be:

$$\begin{bmatrix} \cdots & a_{NN-n} & \cdots & a_{NN-1} & a_{NN} \end{bmatrix}\begin{bmatrix} T_N \end{bmatrix}$$

$$\begin{array}{ccc} 2 & 2\left(\dfrac{\delta x}{\delta y}\right)^2 & -2\left[1 + \left(\dfrac{\delta x}{\delta y}\right)^2\left(\delta y\dfrac{h}{k} + 1\right)\right] \end{array}$$

$$= \begin{bmatrix} b_N \end{bmatrix}$$

$$-2\delta y\dfrac{h}{k}\left(\dfrac{\delta x}{\delta y}\right)^2 T_f$$

where we now have a non-zero constant in the column $\mathbf{b}$.

Most matrix-solving computer programs are assembled to assimilate the matrix coefficients row by row and usually the object of the logic in the main program is to switch to the correct finite difference equation, given the matrix row number.

Ordering of the mesh points

The above section clearly shows the advantage of numbering the mesh points in the grid in a systematic and orderly manner. The mesh points in our illustrative example shown in Fig. 4-3 are ordered from $(0, 0)$, point 1, vertically upwards with m rows of n points per row so that the matrix would have a total order of $N = nm$. This is known as 'pagewise' ordering. Thus the five-point finite difference representation, Eqn (4-8), for the point $T_{i,j}$ in Fig. 4-3 includes only neighbours which are not more than n points backward or forward in the ordering sequence. Consequently, we have a band matrix with a band width of $(2n + 1)$ and advantage of this may be taken with gaussian elimination (see below).

Although *iterative* methods can handle the algebraic equations in any order, it is advantageous to order the mesh points 'pagewise' from the standpoints of the analysis of convergence and stability of the iterative scheme used.

(a)   *Direct methods for simultaneous equations*

Gaussian elimination. This method, sometimes called 'pivotal condensation', is probably the most commonly used direct method for solving systems of linear equations. The basic idea is simple. Referring to the system of Eqns (4-14) we divide the first equation by a_{11} to obtain an expression for T_1 and use the result to eliminate T_1 in all the other equations. We then divide the second equation by the now modified coefficient of T_2 to obtain an expression for T_2 which can then be eliminated in the remaining equations, and so on through the rest of the system. We eventually obtain an equivalent system with an upper triangular matrix of coefficients

$$
\begin{bmatrix}
a_{11} & a_{12} & a_{13} & \cdots & a_{1N} \\
 & a'_{22} & a'_{23} & \cdots & a'_{2N} \\
 & & a'_{33} & \cdots & a'_{3N} \\
 & & & \ddots & \vdots \\
 & & & & a'_{NN}
\end{bmatrix}
\begin{bmatrix}
T_1 \\
T_2 \\
T_3 \\
\vdots \\
T_N
\end{bmatrix}
=
\begin{bmatrix}
b_1 \\
b'_2 \\
b'_3 \\
\vdots \\
b'_N
\end{bmatrix}
$$

where the primes (′) denote modified coefficients. The last unknown in the system can then be found:

$$
T_N = \frac{b'_N}{a'_{NN}}
$$

and the rest follow by back-substitution according to

$$T_k = \frac{1}{a'_{kk}} \left(b'_k - \sum_{i=k+1}^{N} a'_{k,i} T_i \right)$$

where $k = (N - 1), (N - 2), \ldots, 1$.

Row interchanges. Gaussian elimination gives exact results with exact arithmetic, that is, no rounding of significant figures, but the reduction process can sometimes produce 'growth' in the size of the matrix elements. An element can grow only if a large number is added to it during the elimination, and with a fixed number of decimal places this results in the loss of the information originally present in the element which may be vital to the solution. To overcome this, elimination with 'row interchanges' (sometimes called elimination with selection of pivots) is used. The growth is limited by taking the row with the largest coefficient in column 1 and interchanging this row with row 1. By choosing as pivot the largest element in a column, we ensure that the multiples of one row added to another do not exceed unity. Rows and columns are sometimes interchanged as a complete protection against growth, although row interchanges are usually sufficient.

Each of these strategies is based on the assumption that all the elements of A, from row to row, are almost the same size. If there are big differences between rows it will be necessary to *scale* certain rows, a problem which is often important in the solution of systems of two or more partial differential equations.

Recurrence relationships. It is clear that the gaussian elimination process can handle several vectors **b** simultaneously for problems where several calculations are required for different boundary conditions which *only* affect **b**.

Many problems, particularly transient conduction problems, also result in the repeated solution of systems in which only the vectors **b** change. This means that after the first elimination the modified coefficients of A can be stored, and to solve the system for the next vector **b** we need only obtain the modified elements b' of **b** from b'_2 to b'_N until we reach the stage $a'_{NN} T_N = b'_N$, from which point we can back-substitute to obtain **T** using the same relation as before.

For many systems simple recurrence relationships can be obtained which require little storage and which are very economical of computing time. This is particularly true for the tridiagonal matrix of coefficients which results from the transient conduction problems we consider later, and this is described in full by Smith[2].

Matrix inversion. Matrix inversion is *not* generally to be recommended for the solution of finite difference equations since the inverse matrix is rarely required and it usually has all non-zero elements.

This means that no advantage can be taken of any band structure of the original coefficient matrix which would result from sensible ordering of the mesh points and therefore, for a given number of equations, matrix inversion compares very unfavourably with gaussian elimination both in terms of computing time and in storage requirements.

(b) *Iterative methods for simultaneous equations*

Iterative methods are not restricted to linear equations and, therefore, in heat conduction problems the boundary conditions and the thermal conductivity can be made functions of the temperature in the system, as in the case of natural convection and in calculations involving metals at very high temperatures.

For our general system of equations $A\mathbf{T} = \mathbf{b}$ [Eqns (4-14)], any iterative scheme starts with estimates for T_1, T_2, T_3, ..., T_N, denoted by the vector $\mathbf{T}^{(0)}$, and calculates a sequence of approximations $\mathbf{T}^{(1)}$, $\mathbf{T}^{(2)}$, $\mathbf{T}^{(3)}$, ..., $\mathbf{T}^{(n)}$ which converge to the exact solution $\mathbf{T}$ as n increases. When each component of $\mathbf{T}^{(n+1)}$ is calculated individually and explicitly from $\mathbf{T}^{(n)}$ and, possibly, from those components of $\mathbf{T}^{(n+1)}$ that have already been found, the scheme is referred to as a *point iterative method*. When separate subgroups of $\mathbf{T}^{(n)}$ are solved for simultaneously, say twenty at a time by direct methods, the scheme is called a *block* or *line iterative method*.

If an iterative scheme calculates $\mathbf{T}^{(n)}$ using the same cycle of operations for each iteration, it is termed a *stationary method*. The most commonly used processes, and those which we consider here, are stationary, point iterative methods.

Gauss–Seidel method. This procedure is an example of a class of iterative schemes known as *relaxation methods*. Relaxation methods have been used for hand computation for many years, and the basic operation is to take the residual vector

$$\mathbf{r} = \mathbf{b} - A\mathbf{T}$$

(where $\mathbf{T}$ is the approximate solution) and to modify (or relax) one or more components of $\mathbf{T}$ to reduce to zero one or more of the components of $\mathbf{r}$. The equations may be taken in any order, and in the case of hand computation more decimal places may be added as the solution becomes more accurate. In the Gauss–Seidel method, however, we only modify T_k to reduce the kth residual to zero and the equations are cycled in a systematic manner. For the five-point finite difference approximation, Eqn (4-8), with equal step lengths ($\delta x = \delta y$), that is

$$T_{i-1,j} + T_{i,j-1} - 4T_{i,j} + T_{i,j+1} + T_{i+1,j} = 0$$

the Gauss–Seidel iteration is written in the form

$$T_{i,j}^{(n+1)} = \tfrac{1}{4}(T_{i-1,j}^{(n+1)} + T_{i,j-1}^{(n+1)} + T_{i,j+1}^{(n)} + T_{i+1,j}^{(n)}) \qquad (4\text{-}15)$$

In the general case, referring to Eqns (4-14), the Gauss–Seidel method for the kth equation for $T_k^{(n+1)}$ is

$$T_k^{(n+1)} = \frac{b_k}{a_{kk}} - \frac{1}{a_{kk}}(a_{k,1}T_1^{(n+1)} + a_{k,2}T_2^{(n+1)} + \cdots + a_{k,k-1}T_{k-1}^{(n+1)})$$

$$- \frac{1}{a_{k,k}}(a_{k,k+1}T_{k+1}^{(n)} + \cdots + a_{k,N}T_N^{(n)})$$

We illustrate the Gauss–Seidel iteration by a simple example. Consider the three simultaneous equations:

$$\left. \begin{aligned} 5T_1 - 2T_2 - T_3 &= 6 \\ -2T_1 + 8T_2 - 3T_3 &= 7 \\ -T_1 - 3T_2 + 8T_3 &= 5 \end{aligned} \right\}$$

in which the diagonal coefficients dominate. These equations can be written in the form

$$\left. \begin{aligned} T_1 &= \tfrac{1}{5}(6 + 2T_2 + T_3) \\ T_2 &= \tfrac{1}{8}(7 + 2T_1 + 3T_3) \\ T_3 &= \tfrac{1}{8}(5 + T_1 + 3T_2) \end{aligned} \right\} \qquad (4\text{-}16)$$

To obtain T_1, we set $T_2 = T_3 = 0$ and find

$$T_1^{(1)} = \tfrac{6}{5} = 1\cdot 2$$

Inserting this value for T_1, and setting $T_3 = 0$ in the second of Eqns (4-16), we get

$$T_2^{(1)} = \tfrac{1}{8}(7 + 2 \times 1\cdot 2) = 1\cdot 175$$

Finally, $T_3^{(1)}$ is obtained by using $T_1^{(1)}$ and $T_2^{(1)}$ in the third of Eqns (4-16)

$$T_3^{(1)} = \tfrac{1}{8}(5 + 1\cdot 2 + 3 \times 1\cdot 175) = 1\cdot 2156$$

A repetition of the process gives second, third, and higher approximations, that is:

$$T_1^{(2)} = \tfrac{1}{5}(6 + 2 \times 1\cdot 175 + 1\cdot 2156) = 1\cdot 9131$$
$$T_2^{(2)} = \tfrac{1}{8}(7 + 2 \times 1\cdot 9131 + 3 \times 1\cdot 2156) = 1\cdot 8091$$
$$T_3^{(2)} = \tfrac{1}{8}(5 + 1\cdot 9131 + 3 \times 1\cdot 8091) = 1\cdot 5426$$

and so on for $\mathbf{T}^{(3)}, \mathbf{T}^{(4)}, \ldots, \mathbf{T}^{(n)}$. The iteration is normally continued until the changes to $\mathbf{T}$ are sufficiently small. These changes are usually called 'residuals', that is, after the first iteration:

$$r_1 = T_1^{(2)} - T_1^{(1)} = 1\cdot 9131 - 1\cdot 2 = 0\cdot 7131$$

The true solution to the above equations is:

$$T_1 = 2 \cdot 7040, \qquad T_2 = 2 \cdot 4978, \qquad T_3 = 2 \cdot 5247$$

In Eqn (4-15) and in the above example we note that values of **T** already calculated in that cycle, the $(n + 1)$ iterates, are used as soon as they are available. This method is thus a *successive displacement method*. *Simultaneous displacement methods*, such as the Jacobi iteration, calculate the whole new vector $\mathbf{T}^{(n+1)}$ only using the old vector $\mathbf{T}^{(n)}$, and because of their slow rate of convergence they are only of academic interest since the Jacobi iteration provides a comparison of convergence for other iterative methods.

Successive over-relaxation (SOR). The speed at which the Gauss–Seidel iteration for Eqn (4-15) converges can be accelerated by making greater changes to $\mathbf{T}^{(n)}$ than are necessary to reduce the individual residuals to zero, that is by 'over relaxing' the residuals at each point.

The Gauss–Seidel iteration for Eqn (4-15) can be written identically as

$$T_{i,j}^{(n+1)} = T_{i,j}^{(n)} + \tfrac{1}{4}(T_{i-1,j}^{(n+1)} + T_{i,j-1}^{(n+1)} - 4T_{i,j}^{(n)} + T_{i,j+1}^{(n)} + T_{i+1,j}^{(n)})$$

$$= T_{i,j}^{(n)} + r_{i,j}$$

where $r_{i,j}$ is the change in the value of $T_{i,j}$ for one Gauss–Seidel iteration. In the SOR method a change larger than $r_{i,j}$ is made to $T_{i,j}$, defined by

$$T_{i,j}^{(n+1)} = T_{i,j}^{(n)} + \omega r_{i,j}$$

where ω is called the acceleration parameter which, for convergence, has a numerical limit of 2. Some complex systems of algebraic equations (such as those resulting from two-dimensional fluid flow problems) may require under-relaxation, that is $\omega < 1$, but the best value for ω usually lies between 1 and 2.

There is a great deal of literature[3],[6],[7] concerning the best or optimum value of ω (ω_{opt}) that gives the most rapid convergence after a large number of iterations, but unfortunately, except for Laplace's equation and Poisson's equation with simple boundary conditions and simple geometries, no formulae exist for calculating ω_{opt} before starting the calculation.

The value of ω_{opt} depends upon the largest eigenvalue of the iteration matrix, and for any given heat conduction problem the greatest influence on ω_{opt} is exerted by the mesh step length.

For most systems of equations arising from finite difference equations it can be shown that

$$\omega_{\text{opt}} = \frac{2}{1 + (1 - \lambda)^{1/2}} \tag{4-17}$$

where $\lambda = \max |\lambda_i|$ is the maximum eigenvalue (the spectral radius) of the Gauss–Seidel iteration matrix.

If a problem is to be solved many times with no change in the form of the matrix—that is, changing only the value of, say, the boundary conditions—one practical method of estimating ω_{opt} proceeds by using the Gauss–Seidel method to calculate a large set of iterations $\mathbf{T}^{(1)}$, $\mathbf{T}^{(2)}$, $\mathbf{T}^{(3)}$, ..., $\mathbf{T}^{(m)}$ of the N equations $A\mathbf{T} = \mathbf{b}$. Then, since the errors in $\mathbf{T}$ are changing in a progression which is approximately geometric, that is $|e^{(m)}| \approx \lambda |e^{(m-1)}|$, where e is the error in $\mathbf{T}$, λ can be estimated from:

$$\lambda = \lim_{m \to \infty} \frac{\|E_{\mathrm{IT}}^{(m)}\|}{\|E_{\mathrm{IT}}^{(m-1)}\|} \tag{4-18}$$

where $\|E_{\mathrm{IT}}\|$, the norm, is defined as

$$\|E_{\mathrm{IT}}^{(m)}\| = |T_1^{(m)} - T_1^{(m-1)}| + |T_2^{(m)} - T_2^{(m-1)}| + \cdots + |T_N^{(m)} - T_N^{(m-1)}|$$

The value of λ can then be substituted directly into Eqn (4-17) to yield ω_{opt}. Other norms, such as the maximum difference between iterates at any mesh point, may also be used to estimate λ but the result is similar.

The application of this procedure to a practical example is detailed in Section 4-2-5.

With a trivial increase in the amount of arithmetic the computer can be made to calculate the value of ω_{opt} as it proceeds.[7]

Choice of solution procedure for finite difference equations

When a choice of solution procedures for the finite difference equations exists the considerations are primarily the number of storage locations and the computing time.

Direct methods for solving a system of N linear equations with a matrix bandwidth of $2n + 1$ require a minimum of $N(2n + 1)$ storage locations and the computing time is roughly proportional to (Nn^2). Therefore, the time to solve the N equations clearly depends upon the mesh numbering system although in practice, with direct methods, the number of equations that can be solved is limited not by run time but by storage requirements. However, direct methods are finding increasing favour as additional back-up storage facilities, such as disc systems, become more efficient. While for large matrices the computing times with direct methods using magnetic tape auxiliary storage compare unfavourably with efficient iterative methods, the run times with disc storage are quite comparable.

Advantage can be taken by direct methods of the symmetric, positive-definite case, and also of any variation in bandwidth such as occurs in 'difficult regions'.

Iterative procedures inherently require less storage space than direct methods since they only store the non-zero coefficients in the matrix and one or more solution vectors $\mathbf{T}$. They are of particular value where the equations are nonlinear or where the machine storage space is severely limited. However, with three-dimensional steady-state problems, where the matrix bandwidth for direct elimination is necessarily very large, iterative methods are undoubtedly superior.

The convergence rate of any iterative scheme obviously depends upon the specific problem concerned, although with heat conduction equations good convergence is usually obtained. Ames[3] compares the convergence of the three stationary point methods considered above for the five-point finite difference approximations [Eqn (4-8)] with $\delta x = \delta y$ applied to the problem of a square with specified boundary temperatures:

Method	Convergence rate
Jacobi	$\frac{1}{2}\delta x^2$
Gauss–Seidel	δx^2
Optimum SOR	$2\delta x$

It can be seen that, for this simple problem, if $\delta x = 0.1$ the computing time for SOR will be one-twentieth that of the Gauss–Seidel iteration and that the comparison improves as δx decreases. However, it is a feature of all iterative methods that refinement of the finite difference mesh reduces the rate of convergence.

As regards the computing time required, few comparisons between direct elimination and iterative methods have been made. One such comparison is due to Reid[8]. For the simple problem of Laplace's equation in a square giving a matrix of order n^2 and a bandwidth of $2n + 1$ (n mesh points per side), the computing time required for direct elimination is proportional to $\frac{1}{2}n^4$ (or $\frac{1}{2}N^2$). An iterative method such as SOR or Gauss–Seidel requires a computing time roughly proportional to $4n^2$ per iteration so that, for the computer run times to be comparable, the iterative method must converge in $\frac{1}{8}n^2$, that is $\frac{1}{8}N$, iterations. An iterative method can, of course, be stopped when a given accuracy is reached, so that this comparison is somewhat flexible since direct elimination gives a solution whose exactness depends only on the word length in use.

Iterative procedures also exist which sometimes converge faster than optimum SOR. These methods include *alternating direction implicit methods* which employ a combination of horizontal and vertical mesh line iterations, and for such iterations the reader is referred to Varga[6] and others[3, 4, 9]. While testing a computer program in the early stages of development it is common to use as few mesh points as possible and, provided the equations are linear, to solve the finite difference equations by a direct method. Convergence problems and the calculation of acceleration parameters do not then confuse the initial stages of the problem.

4-2-4 Accuracy of steady-state solutions

The total error, E, in a numerical solution, defined as the difference between that solution, T_{NS}, and the true solution to the partial differential equation, T_{TS}, is composed of a discretization error, e_D, and a round-off error, e_R, that is

$$E = T_{TS} - T_{NS} = e_D + e_R$$

The term 'truncation error' is usually reserved for the error in the individual finite difference representations and is consequently the main component in the discretization error e_D.

The round-off error e_R is caused by an inexact solution of the algebraic finite difference equations and is usually small. The dominant error in a stable process is the discretization error e_D, and its estimation is, in general, not simple although decreasing the step size usually decreases the error, that is as $\delta \to 0$, $e_D \to 0$. Clearly, a practical limit to this process is very rapidly reached and methods of estimating e_D are therefore required.

Comparison with analytical solutions. Analytical solutions are sometimes valuable as an indication of the discretization error in a general numerical scheme can be obtained by first taking a simplified geometry and comparing the numerical solution to the true analytical solution. There is, of course, no guarantee that similar errors exist with more complicated geometries.

Extrapolation procedures. These can be powerful techniques for saving computer time and they operate by assuming that a relationship exists between the discretization error and the step length. If T is the extrapolation solution of the differential equation at a particular mesh point and T_1, T_2, and T_3 are the numerical solutions for step sizes δ_1, δ_2, and δ_3 respectively, we can write:

$$\left. \begin{array}{l} T - T_1 = C\delta_1^n \\ T - T_2 = C\delta_2^n \\ T - T_3 = C\delta_3^n \end{array} \right\} \qquad (4\text{-}19)$$

where n is the (unknown) order of the approximation and C is a constant, from which the extrapolated solution, T, can be eliminated.

This is known as 'Richardson's extrapolation'. If the accuracy of the finite difference approximation is known and is consistent throughout the problem, then solutions for only two step sizes are required. For example, if the approximation of Eqn (4-8), which is of order δ^2, is applied to a problem with comparable accuracy on the boundaries, first with the step length equal to δ_1 and then with the step length $\delta_2 = \frac{1}{2}\delta_1$, we have $n = 2$ in Eqns (4-19) and we only require the first two of these equations to show that:

$$T = T_2 + \tfrac{1}{3}(T_2 - T_1) \tag{4-20}$$

Other more elaborate extrapolation formulae employing logarithmic or high-order extrapolation methods are sometimes used.[1, 3, 4]

Higher order finite difference approximations. As we have already shown there is little merit in using more accurate finite difference formulae within the interior of a two-dimensional region if the same accuracy cannot be maintained on the boundaries. With high-order finite difference approximations the truncation error is reduced but the number of mesh points involved in each difference equation increases.

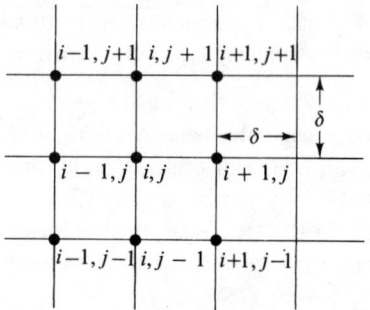

Fig. 4-6   The nine-point approximation to Laplace's equation

One common high-order approximation to Laplace's equation is the nine-point formula[1, 2, 3] (Fig. 4-6) which involves all the points surrounding the node i, j. For equal step sizes the finite difference representation can be shown to be[2, 3]:

$$T_{i-1,j-1} + 4T_{i-1,j} + T_{i-1,j+1} + 4T_{i,j-1} - 20T_{i,j} + 4T_{i,j+1}$$
$$+ T_{i+1,j-1} + 4T_{i+1,j} + T_{i+1,j+1} = 0 \tag{4-21}$$

sometimes expressed as the molecule:

$$\begin{bmatrix} 1 & 4 & 1 \\ 4 & -20 & 4 \\ 1 & 4 & 1 \end{bmatrix} T = 0$$

which has a truncation error of order δ^4. This is clearly very accurate, but there are now more points to be eliminated on the boundary and this order of accuracy is very difficult to achieve with derivative boundary conditions, particularly in corners and with curved geometries.

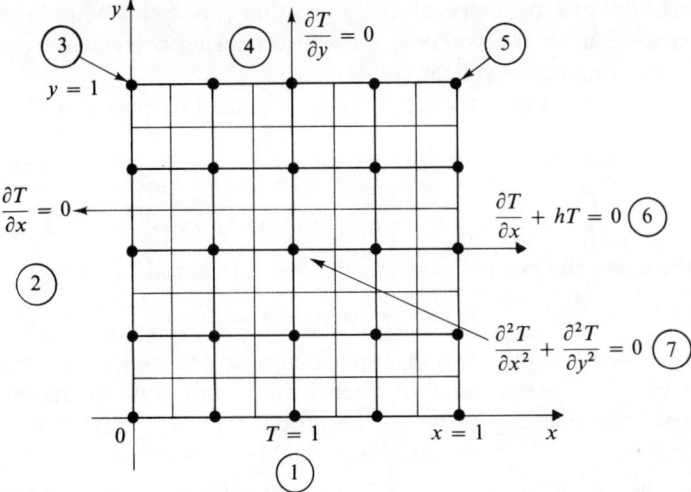

Fig. 4-7 Steady-state problem, Example 1, showing meshes, boundary conditions, and different regions

4-2-5 A steady-state problem

Example 1. As an example of the application of the finite difference approach to steady-state conduction, consider the two-dimensional problem illustrated in Fig. 4-7.

A uniform square section of side unity satisfies Laplace's equation throughout and has the edge $y = 0$ maintained at a temperature of unity, the edges $x = 0$ and $y = 1$ maintained adiabatic, and the edge $x = 1$ convecting heat

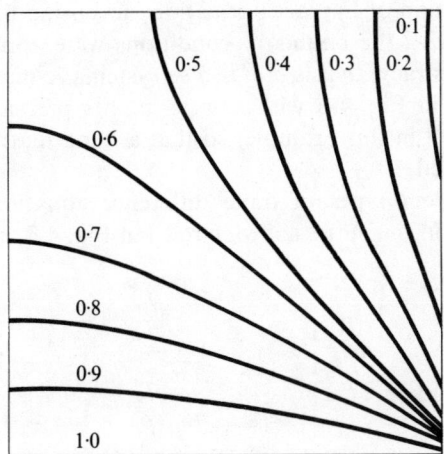

Fig. 4-8 Temperature distribution resulting from the solution of Example 1 (Fig. 4-7)

into a medium at zero temperature. The value of h in this boundary condition is equivalent to h_f/k where h_f is the heat transfer coefficient over $x = 1$ and k is the conductivity of the solid.

The analytical solution to this problem is given by Carslaw and Jaeger[9] as:

$$T = 2h \sum_{i=1}^{\infty} \frac{\cos (p_i x) \cosh [p_i(1 - y)]}{[(p_i^2 + h^2) + h] \cos (p_i) \cosh (p_i)} \qquad (4\text{-}22)$$

where the p_i are the positive roots of the transcendental equation:

$$f(p) = p \tan p - h = 0 \qquad (4\text{-}23)$$

Taking $h = 10$ for our example, this analytical solution is shown graphically in Fig. 4-8. As a point of interest here, this solution required the evaluation of very many terms of the series (over 50), particularly in the region $x = 1$, $y = 0$. The roots of Eqn (4-23), p_i, $i = 1, 2, \ldots, \infty$, had to be obtained numerically, in this case using Newton's iteration, that is $p^{(n+1)} = p^{(n)} - f(p)/f'(p)$, which for Eqn (4-23) is written:

$$p_i^{(n+1)} = p_i^{(n)} - \left\{ \frac{p_i^{(n)} \tan (p_i^{(n)}) - h}{[\tan (p_i^{(n)}) + p_i^{(n)}] \cos^2 (p_i^n)} \right\}$$

taking approximate starting values for each root i.

The first numerical calculation was performed with five mesh points on each side ($\delta x = \delta y = \frac{1}{4}$), so that a total of 25 algebraic equations were employed (Fig. 4-7). The mesh was then halved ($\delta x = \delta y = \frac{1}{8}$), giving 81 equations and Richardson's extrapolation procedure (Section 4-2-4) then used to provide a more accurate solution.

The five-point finite difference representation [Eqn (4-8)] was employed for the approximation of Laplace's equation, and using the procedure outlined in Section 4-2-2 the boundary conditions were approximated to the same degree of accuracy, that is $o(\delta^2)$. The systematic mesh point ordering system illustrated in Fig. 4-3 with n mesh points per side was employed (where $n = 5$ and 9 in this example) so that a band matrix of band width $2n + 1$ was obtained.

The seven different types of finite difference equations for the seven different positions in the mesh, shown circled in Fig. 4-7, are:

Position

1		T_i	$= 1$
2	T_{i-1}	$-4T_i + T_{i+1} + 2T_{i+n}$	$= 0$
3	$2T_{i-1}$	$-4T_i \qquad\quad + 2T_{i+n}$	$= 0$
4	$T_{i-n} + 2T_{i-1}$	$-4T_i \qquad\quad + T_{i+n}$	$= 0$
5	$2T_{i-n} + 2T_{i-1} - (4 + 2h\delta)T_i$		$= 0$
6	$2T_{i-n} + T_{i-1} - (4 + 2h\delta)T_i + T_{i+1}$		$= 0$
7	$T_{i-n} + T_{i-1}$	$-4T_i + T_{i+1} + T_{i+n}$	$= 0$

Table 4-1 Example 1. Comparison between the numerical solutions (N.S.), the extrapolated numerical solution, and the analytical solution (A.S.) for individual mesh points

		$x = 0$	$x = \frac{1}{4}$	$x = \frac{1}{2}$	$x = \frac{3}{4}$	$x = 1$
$y = 1$	N.S. $\delta = \frac{1}{4}$	0·561413	0·531420	0·440448	0·288299	0·084844
	N.S. $\delta = \frac{1}{8}$	0·561501	0·530862	0·438338	0·285457	0·085202
	Extrapolated N.S.	0·561530	0·530676	0·437635	0·284510	0·085321
	A.S.	0·561540	0·530659	0·437577	0·284502	0·085273
	% Error	0·0017	−0·0032	−0·0132	−0·0027	−0·0567
$y = \frac{3}{4}$	N.S. $\delta = \frac{1}{4}$	0·591405	0·561910	0·471037	0·313953	0·093497
	N.S. $\delta = \frac{1}{8}$	0·592014	0·561883	0·469225	0·310505	0·093556
	Extrapolated N.S.	0·592217	0·561874	0·468621	0·309356	0·093576
	A.S.	0·592251	0·561884	0·468559	0·309302	0·093587
	% Error	0·0057	0·0018	−0·0132	−0·0174	0·0121
$y = \frac{1}{2}$	N.S. $\delta = \frac{1}{4}$	0·680387	0·653778	0·567836	0·402977	0·128723
	N.S. $\delta = \frac{1}{8}$	0·682139	0·655129	0·567379	0·398050	0·125069
	Extrapolated N.S.	0·682723	0·655579	0·567227	0·396408	0·123851
	A.S.	0·682789	0·655662	0·567260	0·396148	0·124745
	% Error	0·0097	0·0126	0·0059	−0·0655	0·7167
$y = \frac{1}{4}$	N.S. $\delta = \frac{1}{4}$	0·822586	0·804979	0·743553	0·601397	0·259057
	N.S. $\delta = \frac{1}{8}$	0·824663	0·807275	0·745863	0·597838	0·224114
	Extrapolated N.S.	0·825355	0·808040	0·746633	0·596652	0·212466
	A.S.	0·825375	0·808117	0·746977	0·596835	0·216608
	% Error	0·0024	0·0095	0·0461	0·0307	1·9121
$y = 0$		1·000000	1·000000	1·000000	1·000000	1·000000

where the suffix notation is that of the position of the variable relative to T_i in the solution vector **T**.

The numerical solutions for $\delta = \frac{1}{4}$ and $\frac{1}{8}$ (the difference equations being solved by direct elimination) for five positions in each direction are compared with the extrapolated numerical solution [using Eqn (4-20)] and the analytical solution [Eqn (4-22)] in Table 4-1.

Even with the relatively coarse meshes employed the percentage error between the analytical and the extrapolated numerical solution is shown to be of very small order over the greater part of the region. However, an error approaching 2 per cent is observed near $x = 1$, $y = 0$, but examination of the solution illustrated graphically in Fig. 4-8 shows that rapid changes in temperature occur in this region and that consequently the discretization of the problem in this area is particularly crude. It was mentioned earlier that the convergence of the series in the analytical solution is also slow in this region.

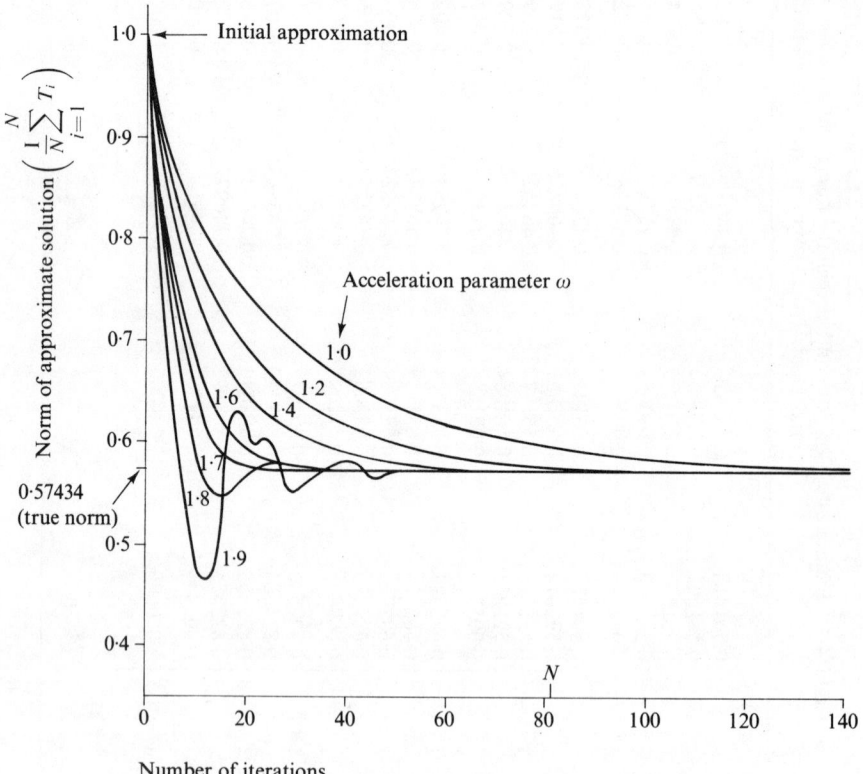

Fig. 4-9   The convergence of SOR for Example 1 ($\delta = \frac{1}{8}$)

The algebraic equations were also solved iteratively and the convergence of the Gauss–Seidel iteration is compared graphically with SOR for various values of the relaxation parameter ω in Fig. 4-9. The convergence to the average of all the mesh values, $(1/N) \sum_{i=1}^{N} T_i$, is plotted against the number of iterations. The optimum value of ω for this system was 1·686, and it can be observed that values of $\omega > 1·686$ cause the solution to oscillate as it converges. The parameter ω, as stated above, has a limit of 2.

The numerical estimation of ω_{opt}

The procedures outlined in Section 4-2-3(b) were applied to the finest mesh used in the above example ($9 \times 9 = 81$ points). The initial approximation of 1·0 was taken for each mesh value and values of

$$\|E_{IT}^{(m)}\| = |T_1^{(m)} - T_1^{(m-1)}| + |T_2^{(m)} - T_2^{(m-1)}| + \cdots + |T_N^{(m)} - T_N^{(m-1)}|$$

after the fifty-ninth and sixtieth iterations of the Gauss–Seidel iteration ($\omega = 1$) were

$$\|E_{IT}^{(59)}\| = 0·122945; \qquad \|E_{IT}^{(60)}\| = 0·118668$$

From Eqn (4-18)

$$\lambda = \frac{0·118668}{0·122945} = 0·96521$$

which, when substituted into Eqn (4-16), yields $\omega_{opt} = 1·686$; that this is in fact the best value for ω would seem reasonable from Fig. 4-9.

4-3 Transient heat conduction (parabolic equations)

Heat conduction problems involving time usually lead to parabolic partial differential equations. The simplest parabolic equation is probably Fourier's one-dimensional transient heat conduction equation as derived in full in Section 2-2

$$\frac{\partial T'}{\partial t'} = \alpha \frac{\partial^2 T'}{\partial x'^2} \tag{4-24}$$

where t', x', and T' are, respectively, dimensional time, distance, and temperature. Equation (4-24), for example, describes the temperature distribution in a uniform rod insulated over the curved surface losing heat from the ends, or, the temperature distribution in a solid bounded by two parallel surfaces from which it is losing heat.

In the first problem, for example, either the temperatures at the ends of the rod or the rate at which heat is lost from the ends may be specified. These are the *boundary* conditions, the temperature distribution in the rod at time $t = 0$ is known as the *initial* condition. With transient conduction, therefore, the area of integration is the infinite area bounded by the x-axis, $t = 0$, and the lines $x = 0$ and $x = L$, where L is length of the rod in this case, and the solution is propagated from $t = 0$ to ∞ in a marching, step-by-step manner.

4-3-1 Dimensionless form of equations

Using the variables in dimensionless form is particularly valuable with the parabolic equations of transient heat conduction as one solution can often be used for similar problems with very different linear dimensions, thermal conductivities, and temperatures. The solution may also be useful for totally different physical problems which obey the same differential equation (the damped oscillation of a pendulum in a viscous fluid, for example).

The transient equation [Eqn (4-24)] applied to a thin uniform rod of length L with a maximum temperature of T_i at zero time can be made dimensionless[2, 5] by using the variables in the form

$$x = \frac{x'}{L}, \quad T = \frac{T'}{T_i}, \quad \text{and} \quad t = \frac{\alpha t'}{L^2}$$

where x', T', and t' are the true distance, true temperature, and true time for the problem concerned. This non-dimensional expression for time is, of course, the Fourier number as discussed in Chapter 3. Equation (4-24), therefore, has the dimensionless form:

$$\frac{\partial T}{\partial t} = \frac{\partial^2 T}{\partial x^2} \tag{4-25}$$

which is the form in which the one-dimensional equation (4-24) will be treated in this section.

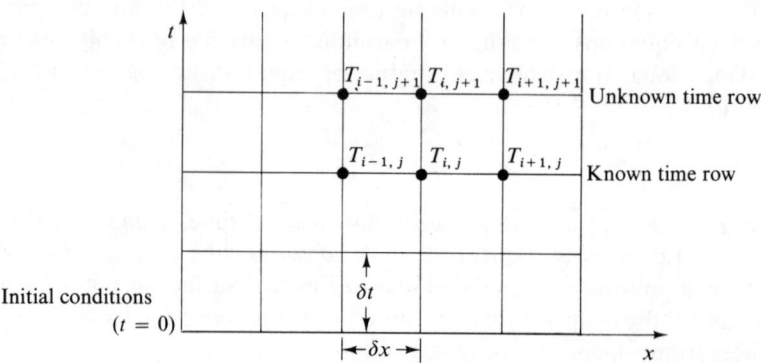

Fig. 4-10   The mesh point system for a one-dimensional parabolic equation

4-3-2 Explicit solutions

An *explicit* method of solution is one in which the temperature of an unknown mesh point is expressed directly in terms of the known temperatures on the previously calculated time row. If the expression for the unknown temperature contains reference to any unknown mesh point values the solution is termed *implicit*.

A straightforward approximation to Eqn (4-25) is to use a forward difference for $\partial T/\partial t$, that is Eqn (4-6), and a central difference for $\partial^2 T/\partial x^2$, that is Eqn (4-4). With the notation of Fig. 4-10 we can then represent Eqn (4-25) by

$$\frac{T_{i,j+1} - T_{i,j}}{\delta t} = \frac{T_{i+1,j} - 2T_{i,j} + T_{i-1,j}}{\delta x^2}$$

which gives, as the formula for the unknown temperature $T_{i,j+1}$,

$$T_{i,j+1} = T_{i,j} + r(T_{i-1,j} - 2T_{i,j} + T_{i+1,j}) \tag{4-26}$$

where $r = \delta t/(\delta x)^2$.

This equation can be used to calculate the values of temperature along the first time row by substituting the initial conditions along $t = 0$ into the right-hand side and then the values along the second and subsequent time rows in a similar manner.

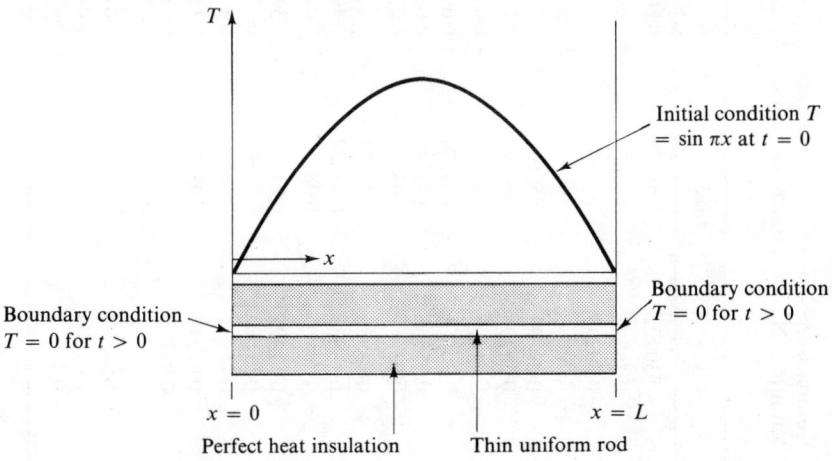

Fig. 4-11 One-dimensional transient conduction, Example 2

Example 2. Consider, as an elementary example (Fig. 4-11), a thin uniform rod completely insulated over its length with its ends maintained at zero and having the initial temperature distribution $T = \sin \pi x$. These conditions can be expressed formally as

Boundary conditions $\quad T = 0$ when $x = 0, 1$ for all t

Initial conditions $\quad\quad T = \sin \pi x$ when $t = 0$ for $0 \leqslant x \leqslant 1$

Table 4-2 Comparison between the numerical (explicit) and the analytical solutions to Example 2, where I.C. = initial condition, N.S. = numerical solution, A.S. = analytical solution, and x and t are dimensionless $(r = 0.20, \delta x = 0.10, \delta t = 0.0020)$

t	$x = 0.0$	0.1	0.2	0.3	0.4	0.5	
0.000	0.0	0.30902	0.58779	0.80902	0.95106	1.00000	I.C.
0.002	0.0	0.30297	0.57628	0.79318	0.93244	0.98042	N.S.
	0.0	0.30298	0.57630	0.79320	0.93247	0.98045	A.S.
	0.0	−0.00	−0.00	−0.00	−0.00	−0.00	% Error
0.004	0.0	0.29704	0.56500	0.77765	0.91418	0.96123	N.S.
	0.0	0.29706	0.56503	0.77770	0.91424	0.96129	A.S.
	0.0	−0.01	−0.01	−0.01	−0.01	−0.01	% Error
0.020	0.0	0.25358	0.48234	0.66388	0.78044	0.82060	N.S.
	0.0	0.25366	0.48249	0.66410	0.78069	0.82087	A.S.
	0.0	−0.03	−0.03	−0.03	−0.03	−0.03	% Error
0.050	0.0	0.18850	0.35855	0.49350	0.58015	0.61000	N.S.
	0.0	0.18865	0.35884	0.49390	0.58062	0.61050	A.S.
	0.0	−0.08	−0.08	−0.08	−0.08	−0.08	% Error
0.100	0.0	0.11499	0.21872	0.30104	0.35389	0.37211	N.S.
	0.0	0.11517	0.21907	0.30153	0.35447	0.37271	A.S.
	0.0	−0.16	−0.16	−0.16	−0.16	−0.16	% Error
0.200	0.0	0.04279	0.08139	0.11202	0.13169	0.13846	N.S.
	0.0	0.04293	0.08165	0.11238	0.13211	0.13891	A.S.
	0.0	−0.32	−0.32	−0.32	−0.32	−0.32	% Error
0.300	0.0	0.01592	0.03028	0.04168	0.04900	0.05152	N.S.
	0.0	0.01600	0.03043	0.04189	0.04924	0.05177	A.S.
	0.0	−0.48	−0.48	−0.48	−0.48	−0.48	% Error

These boundary conditions could be arranged experimentally by keeping the ends of the rod in contact with melting ice, and the initial conditions obtained by heating the rod along its length with a suitably wound electrical coil. If we choose $\delta x = 0.1$ the problem is symmetric about $x = 0.5$ and we need only consider $0 \leqslant x \leqslant 0.5$. The boundary condition equation at $x = 0.5$ will clearly be the adiabatic one, $\partial T/\partial x = 0$, which, with $\delta x = 0.1$, is equivalent to $T_{x=0.4} = T_{x=0.6}$.

If now we choose $\delta t = 0.002$, which gives $r = (\delta t/\delta x^2) = 0.2$, Eqn (4-26) can be simplified to

$$T_{i,j+1} = \tfrac{1}{5}(T_{i-1,j} + 3T_{i,j} + T_{i+1,j})$$

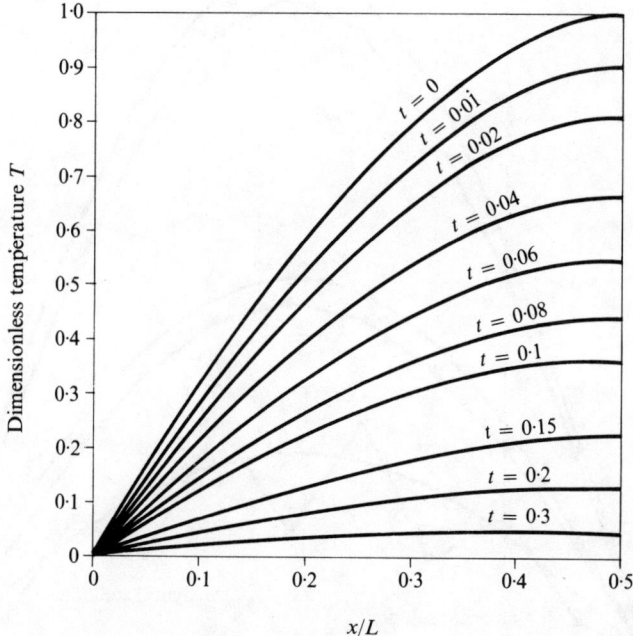

Fig. 4-12 Temperature distribution at various times in an insulated rod, Example 2, with initial temperature distribution $T = \sin \pi x$ and end temperatures zero. The numbers on the curves are values of $(\alpha t/L^2)$

Using this explicit formulation the numerical solution is compared in Table 4-2 with the analytical solution

$$T = e^{-\pi t} \sin \pi x$$

and is shown graphically in Fig. 4-12.

The numerical results in Table 4-2 can be simply checked; for example, for $T_{x=0.4}$ at $t = 0.004$ the simplified form of Eqn (4-26) gives

$$T_{x=0.4} = T_{i,j+1} = \tfrac{1}{5}(0.79318 + 3 \times 0.93244 + 0.98042)$$
$$= 0.91418$$

and for $T_{x=0.5}$ at $t = 0.004$ (employing the adiabatic boundary condition)

$$T_{x=0.5} = T_{i,j+1} = \tfrac{1}{5}(2 \times 0.93244 + 3 \times 0.98042) = 0.96123$$

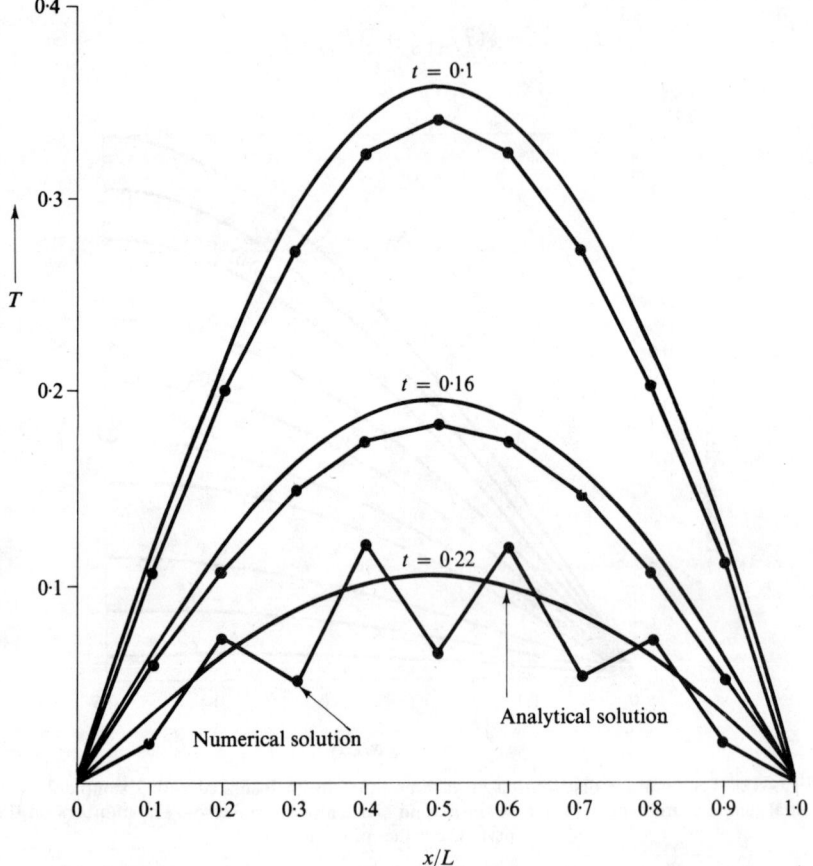

Fig. 4-13   Unstable explicit solution to Example 2, with $r = 1.0$, $\delta t = 0.01$

It can be seen in Table 4-2 that the numerical solution is very accurate indeed. However, it can be shown that as the step length δt is increased the numerical solution becomes less accurate. For values of $r = \delta t/\delta x^2 > 0.5$ the errors in the explicit calculation build up exponentially as t increases so that explicit methods become *unstable* if the forward time step makes $r > 0.5$, with the numerical predictions oscillating with increasing and indeterminable amplitudes about the true solution.

Although this stability condition of $0 < r \leqslant 0.5$ is, in general, true for most explicit problems it is shown by Richtmyer[5] to be slightly affected by derivative boundary conditions.

Consider the same problem above (Example 2) with $\delta t = 0.01$, that is, $r = 1.0$. Although somewhat inaccurate, the resulting solution is not immediately useless as the instability takes time to develop and, as can be seen in Fig. 4-13, it is suddenly manifest, and the solution thereafter useless, after some 20 forward time steps.

The use of the more accurate central difference, Eqn (4-5), for the derivative $\partial T/\partial t$ after the first time step, although seemingly very attractive, has been shown to be unstable for all values of the time step.[2, 3, 10] For an explanation of stability the reader is referred to Mitchell[10].

It is important to note that the problem considered in the above case has the initial function and its derivatives continuous, with no discontinuity between the initial conditions and the boundary conditions at the first time step. Under these circumstances, as can be seen, the finite difference solution can be very accurate. Important practical problems with discontinuous boundary conditions occur, however, and they do not yield such a favourable comparison between the numerical and the true analytical solutions.[2] A common example is the uniform plane slab, bounded by two parallel planes initially at constant temperature, which has the surfaces suddenly reduced to zero temperature at $t = 0$.

For these problems, however, it can be proved that when the boundary values are constant the effect of singularities in the boundary conditions reduces as t increases. If high accuracy is required for small values of t it is possible to remove the singularity by a transformation of both independent variables.[2]

4-3-3   Implicit solutions

Although explicit solutions are mathematically easy to formulate and compute they suffer the limitation on forward time step $\delta t \leqslant 0.5\,\delta x^2$. This limitation clearly becomes more restrictive as δx is reduced (usually to improve accuracy) so that the computing time for even simple problems can become prohibitive.

This restriction can be removed by expressing the derivative $\partial^2 T/\partial x^2$ in terms of the values of T on the $(j + 1)$th time row, instead of on the jth row as in the explicit solutions, with a backward difference for the derivative $\partial T/\partial t$. This is accordingly an implicit method of solution since each time row involves the solution of a system of algebraic equations. In the general implicit case the derivative $\partial^2 T/\partial x^2$ is expressed as a linear combination of finite differences on the jth and $(j + 1)$th time rows. With the notation of Fig. 4-10 the general approximation to $\partial T/\partial t = \partial^2 T/\partial x^2$ can be written simply as

$$\frac{T_{i,j+1} - T_{i,j}}{\delta t} = \frac{1}{\delta x^2} [\theta (T_{i+1,j+1} - 2T_{i,j+1} + T_{i-1,j+1})$$
$$+ (1 - \theta)(T_{i+1,j} - 2T_{i,j} + T_{i-1,j})] \quad (4\text{-}27)$$

where $\theta = 0$ gives the explicit method and $\theta = 1$ gives the fully implicit method.

The approximation is unconditionally stable and convergent for $\frac{1}{2} \leqslant \theta \leqslant 1$, but if $\theta < \frac{1}{2}$ the condition for stability[5] is

$$r = \left(\frac{\delta t}{\delta x^2} \right) \leqslant \frac{1}{2(1 - 2\theta)}$$

The Crank–Nicolson method

This is probably the best-known implicit method for solving $\partial T/\partial t = \partial^2 T/\partial x^2$ and is obtained by setting $\theta = \frac{1}{2}$ in Eqn (4-27) so that the Crank–Nicolson approximation becomes

$$-rT_{i-1,j+1} + (2 + 2r)T_{i,j+1} - rT_{i+1,j+1}$$
$$= rT_{i-1,j} + (2 - 2r)T_{i,j} + rT_{i+1,j} \quad (4\text{-}28)$$

where $r = \delta t/\delta x^2$.

Example 3. As an example of the application of the Crank–Nicolson implicit method consider the previously worked problem:

$$\frac{\partial T}{\partial t} = \frac{\partial^2 T}{\partial x^2} \quad 0 \leqslant x \leqslant 1$$

$$T = 0 \text{ when } x = 0, 1 \text{ for all } t$$

$$T = \sin \pi x \text{ when } t = 0 \text{ for } 0 \leqslant x \leqslant 1$$

By taking $r = \delta t/\delta x^2 = 1$, Eqn (4-28) can be usefully simplified so that with $\delta x = 0 \cdot 1$ it becomes:

$$-T_{i-1,j+1} + 4T_{i,j+1} - T_{i+1,j+1} = T_{i-1,j} + T_{i+1,j} \quad (4\text{-}29)$$

where the left-hand side contains unknown values of T and the right-hand side contains known values. Although this procedure is stable for any value of r, too large a value would obviously give a poor finite difference representation of the derivative $\partial T/\partial t$.

Taking account of the symmetry as before, the initial condition, $T = \sin \pi x$, is numerically:

x	0	0·1	0·2	0·3	0·4	0·5
$T(t = 0)$	0	0·30902	0·58779	0·80902	0·95106	1·0

Table 4-3 Comparison between the numerical (Crank–Nicolson) and the analytical solutions to Example 3 ($r = 1 \cdot 00$, $\delta x = 0 \cdot 10$, $\delta t = 0 \cdot 01$)

t	$x = 0 \cdot 0$	0·1	0·2	0·3	0·4	0·5	
0·00	0·0	0·30902	0·58779	0·80902	0·95106	1·00000	I.C.
0·01	0·0	0·28018	0·53293	0·73352	0·86230	0·90668	N.S.
	0·0	0·27997	0·53254	0·73298	0·86167	0·90602	A.S.
	0·0	0·07	0·07	0·07	0·07	0·07	% Error
0·02	0·0	0·25403	0·48320	0·66507	0·78183	0·82207	N.S.
	0·0	0·25366	0·48249	0·66410	0·78069	0·82087	A.S.
	0·0	0·15	0·15	0·15	0·15	0·15	% Error
0·05	0·0	0·18934	0·36016	0·49571	0·58274	0·61273	N.S.
	0·0	0·18865	0·35884	0·49390	0·58062	0·61050	A.S.
	0·0	0·37	0·37	0·37	0·37	0·37	% Error
0·10	0·0	0·11602	0·22068	0·30374	0·35707	0·37544	N.S.
	0·0	0·11517	0·21907	0·30153	0·35447	0·37271	A.S.
	0·0	0·73	0·73	0·73	0·73	0·73	% Error
0·20	0·0	0·04356	0·08285	0·11404	0·13406	0·14096	N.S.
	0·0	0·04293	0·08165	0·11238	0·13211	0·13891	A.S.
	0·0	1·47	1·47	1·47	1·47	1·47	% Error
0·50	0·0	0·00231	0·00438	0·00603	0·00709	0·00746	N.S.
	0·0	0·00222	0·00423	0·00582	0·00684	0·00719	A.S.
	0·0	3·72	3·72	3·72	3·72	3·72	% Error

where the values of T are denoted by T_i $(i = 1, 2, \ldots, 9)$ as in Fig. (4-14). Using Eqn (4-29) the corresponding values after the first time step, that is $\delta t = 0.01$, must then satisfy the system of simultaneous equations:

$$
\left.
\begin{aligned}
0 + 4T_1 - T_2 &= 0 + 0.58779 \\
-T_1 + 4T_2 - T_3 &= 0.30902 + 0.80902 \\
-T_2 + 4T_3 - T_4 &= 0.58779 + 0.95106 \\
-T_3 + 4T_4 - T_5 &= 0.80902 + 1.0 \\
-2T_4 + 4T_5 &= 0.95106 + 0.95106
\end{aligned}
\right\}
\quad (4\text{-}30)
$$

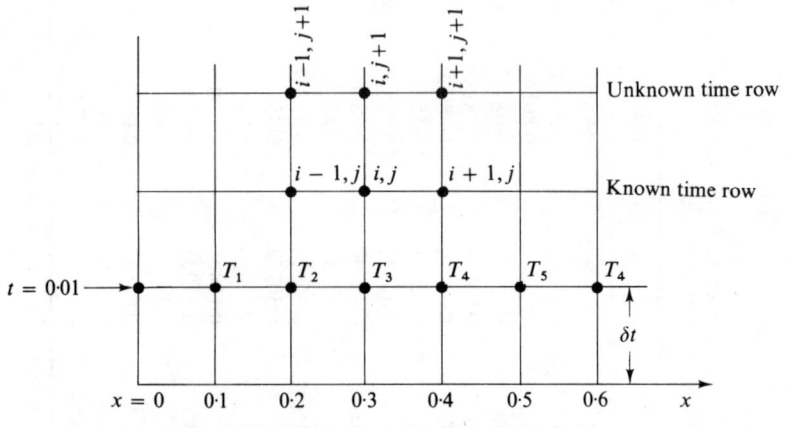

Fig. 4-14 Mesh system for an implicit solution

This linear system is what is known as tridiagonal, and for such systems direct elimination is superior to iteration.

The numerical solution is compared with the analytical one, $T = e^{-\pi t}$ $\sin \pi x$ in Table 4-3, and it is shown by comparison with Table 4-2 that the Crank–Nicolson method of solution is only slightly less accurate than the explicit solution which has five times as many time steps and there is no evidence of instability. Even with this coarse step the error is only 3·72 per cent at $t = 0.5$.

The reader may verify the procedure by substituting the calculated values on the first time row from Table 4-3 into Eqn (4-30).

Example 4. An important practical example is the case of the thin uniform insulated rod (or a uniform slab bounded by two parallel planes), initially at constant temperature, losing heat by convection from both surfaces into a medium at zero temperature.

The problem is therefore to solve the equation

$$
\frac{\partial T}{\partial t} = \frac{\partial^2 T}{\partial x^2} \quad (0 \leqslant x \leqslant 1)
$$

with the initial condition:
$$T = 1 \quad \text{for } 0 \leqslant x \leqslant 1, t = 0$$
and the boundary conditions

$$k\frac{\partial T}{\partial x} = hT \qquad \text{at } x = 0, t > 0$$

$$k\frac{\partial T}{\partial x} = -hT \quad \text{at } x = 1, t > 0$$

Taking $h/k = 1$ simplifies these conditions to $\partial T/\partial x = T$ and $\partial T/\partial x = -T$ at $x = 0$ and 1 respectively.

Using the method of Crank–Nicolson we represent $\partial T/\partial t = \partial^2 T/\partial x^2$ by

$$\frac{T_{i,j+1} - T_{i,j}}{\delta t} = \frac{1}{2}\left[\frac{T_{i+1,j+1} - 2T_{i,j+1} + T_{i-1,j+1}}{\delta x^2}\right.$$
$$\left. + \frac{T_{i+1,j} - 2T_{i,j} + T_{i-1,j}}{\delta x^2}\right]$$

which reduces to Eqn (4-28).

Following the procedures outlined in Section 4-2-2, to be consistent with this approximation we represent the boundary condition at $x = 0$ by the central differences:

$$\frac{T_{i+1,j} - T_{i-1,j}}{2\delta x} = T_{i,j}$$

$$\frac{T_{i+1,j+1} - T_{i-1,j+1}}{2\delta x} = T_{i,j+1}$$

where the values $T_{i-1,j}$ and $T_{i-1,j+1}$ here are 'fictitious'. By applying the main equation to the point $x = 0$ we can eliminate these 'fictitious' values to yield an algebraic equation for $x = 0$ of equivalent accuracy to the main approximation.

As in the previous example we can make use of the symmetry of the problem and apply the adiabatic boundary condition to the point $x = 0.5$, so that with $r = 1$ ($\delta t = 0.01$) and $\delta x = 0.1$ Eqn (4-29) is applicable to the points $x = 0.1, 0.2, 0.3$, and 0.4.

Hence, with the notation of Fig. 4-14 the algebraic equations for the unknown $(j + 1)$th time row are, in matrix notation,

$$
\underset{A}{\begin{bmatrix}
+2\cdot1 & -1 & & & & \\
-1 & +4 & -1 & & & \\
 & -1 & +4 & -1 & & \\
 & & -1 & +4 & -1 & \\
 & & & -1 & +4 & -1 \\
 & & & & -2 & +4
\end{bmatrix}}
\underset{\mathbf{T}_{j+1}}{\begin{bmatrix}
T_0 \\ T_1 \\ T_2 \\ T_3 \\ T_4 \\ T_5
\end{bmatrix}}
=
\underset{\mathbf{b}}{\begin{bmatrix}
-0\cdot1 T_{0,j} + T_{1,j} \\
T_{0,j} + T_{2,j} \\
T_{1,j} + T_{3,j} \\
T_{2,j} + T_{4,j} \\
T_{3,j} + T_{5,j} \\
2T_{4,j}
\end{bmatrix}} \quad (4\text{-}31)
$$

Table 4-4 Comparison between the numerical (Crank–Nicolson) and the analytical solutions to Example 4. Unit initial temperature, convection at ends into a fluid at zero ($r = 1.00, \delta x = 0.10, \delta t = 0.01$)

t		$x = 0.0$	0.1	0.2	0.3	0.4	0.5
0.00	I.C.	1.00000	1.00000	1.00000	1.00000	1.00000	1.00000
0.01	N.S.	0.89083	0.97075	0.99216	0.99789	0.99940	0.99970
	A.S.	0.89646	0.96271	0.99049	0.99835	0.99981	0.99997
	% Error	−0.63	0.84	0.17	−0.05	−0.04	−0.03
0.02	N.S.	0.86238	0.92934	0.97199	0.98998	0.99636	0.99788
	A.S.	0.85848	0.92855	0.96946	0.98911	0.99668	0.99849
	% Error	0.45	0.09	0.26	0.09	−0.03	−0.06
0.10	N.S.	0.71794	0.78335	0.83493	0.87204	0.89437	0.90182
	A.S.	0.71756	0.78276	0.83422	0.87128	0.89360	0.90105
	% Error	−0.05	−0.08	−0.08	0.09	0.09	0.09
0.20	N.S.	0.60414	0.65939	0.70336	0.73532	0.75471	0.76121
	A.S.	0.60402	0.65910	0.70293	0.73479	0.75412	0.76060
	% Error	−0.02	0.04	−0.06	0.07	0.08	0.08
0.50	N.S.	0.36182	0.39491	0.42125	0.44039	0.45201	0.45591
	A.S.	0.36193	0.39494	0.42121	0.44031	0.45189	0.45578
	% Error	−0.03	−0.01	−0.01	0.02	0.03	0.03
1.00	N.S.	0.15397	0.16805	0.17926	0.18740	0.19235	0.19401
	A.S.	0.15415	0.16821	0.17940	0.18753	0.19247	0.19412
	% Error	−0.12	−0.10	−0.08	−0.07	−0.06	−0.06
2.00	N.S.	0.02788	0.03043	0.03246	0.03394	0.03483	0.03513
	A.S.	0.02796	0.03051	0.03254	0.03402	0.03491	0.03521
	% Error	−0.29	−0.27	−0.25	−0.24	−0.24	−0.23

The solution of these equations for several time steps is compared with the analytical solution in Table 4-4 and is illustrated graphically in Fig. 4-15.

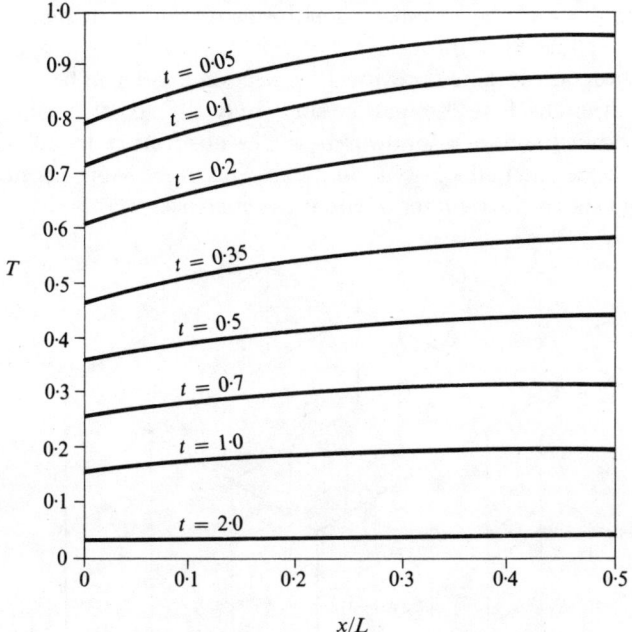

Fig. 4-15 Temperature distribution at various times in an insulated rod, Example 4, with unit initial temperature and convective heat loss from ends

The analytical solution to this problem,[2] as in the first example, requires the evaluation of a series which, in this case, is rapidly convergent:

$$T = 4 \sum_{i=1}^{\infty} \frac{\cos \left[2p_i(x - \tfrac{1}{2})\right]}{e^{4p_i^2 t}(3 + 4p_i^2) \cos (p_i)}, \quad 0 \leqslant x \leqslant 1$$

where the p_i are the positive roots of the equation

$$p \tan (p) = \tfrac{1}{2}$$

Compared with Example 2 the material temperature obviously takes longer to decay and, as Table 4-4 shows that good accuracy is achieved throughout, the example demonstrates that as the total time involved in a transient problem increases (as with small values of h in the above problem), implicit methods become more valuable.

Solution of implicit finite difference equations

As has been demonstrated, compared with steady-state problems, transient heat conduction problems result in the repeated solution of generally smaller systems of algebraic equations. One-dimensional transient equations usually result in a tridiagonal matrix of coefficients which, if linear, is very easily solved by direct elimination methods. In many problems the finite difference approximations and the boundary conditions remain unchanged with time so that the coefficients of the matrix A do not change. In this case the process of elimination, as explained previously in Section 4-2-3, can be considerably simplified after the first elimination since only the vector **b** in Eqn (4-31) need be modified and then followed by back-substitution at each time step. This recurrence method of solution clearly becomes more economical of computing time as the number of equations increases.

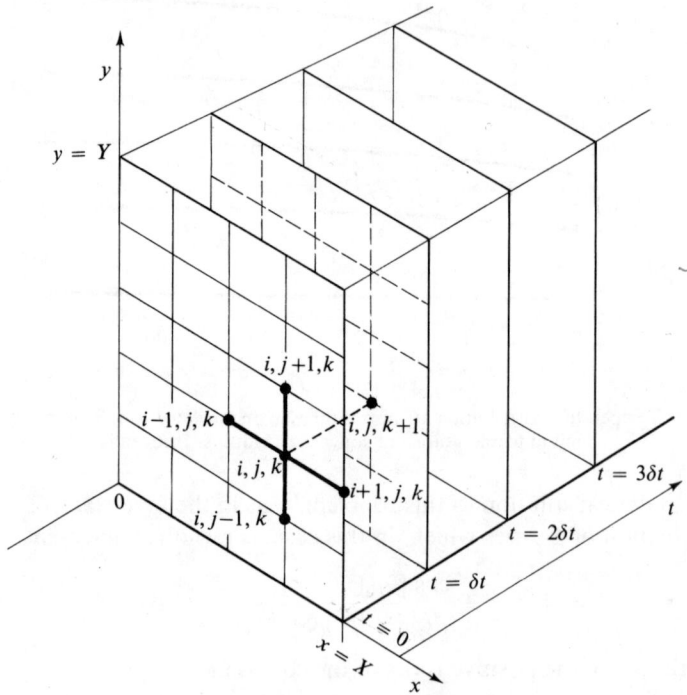

Fig. 4-16   Graphical representation of a two-dimensional transient conduction problem

In general, for one- and two-dimensional transient conduction problems iteration is not recommended unless the equations are nonlinear or unless, for two-dimensional transient cases, the computer storage is severely limited. For large two-dimensional linear transient systems the methods outlined in Section 4-2-3 are applicable, and for each iteration the values at the previous time step would form a close initial approximation.

There has been little information published regarding the iterative solution of nonlinear finite difference equations. The Gauss–Seidel and SOR methods can still be used although the estimation of the best acceleration parameter ω is a complex problem and is best found by trial and error.

4-3-4 Two-dimensional transient heat conduction

These problems, although parabolic in nature, are really three-dimensional since they involve two space coordinates and time. Consider Fourier's equation

$$\frac{\partial T}{\partial t} = \alpha \left(\frac{\partial^2 T}{\partial x^2} + \frac{\partial^2 T}{\partial y^2} \right) \tag{4-32}$$

where $\alpha = k/\rho C$, applied to the rectangular region $0 \leqslant x \leqslant X, 0 \leqslant y \leqslant Y$ shown in Fig. 4-16.

The variables t, T, x, and y must now be used in their dimensional form although for a square-sectioned geometry they can be made dimensionless.

The initial temperature, at $t = 0$, is usually specified at all points on and within the rectangle, and the subsequent conditions on the boundary are usually either derivative, function value known, or a mixture of both.

As in the one-dimensional transient case the solution to Eqn (4-32) can be propagated forward in time explicitly or implicitly.

With the notation of Fig. 4-16, in which we have m mesh points in the x direction and n in the y direction, we define the coordinates of each mesh point (x, y, t) by $x = i\,\delta x$, $y = j\,\delta y$, $t = k\,\delta t$, where $\delta x = X/(m - 1)$ and $\delta y = Y/(n - 1)$. An explicit finite difference representation of Eqn (4-32) is:

$$\frac{T_{i,j,k+1} - T_{i,j,k}}{\delta t} = \alpha \left[\left(\frac{T_{i-1,j,k} - 2T_{i,j,k} + T_{i+1,j,k}}{\delta x^2} \right) \right. $$
$$\left. + \left(\frac{T_{i,j-1,k} - 2T_{i,j,k} + T_{i,j+1,k}}{\delta y^2} \right) \right]$$

As each mesh point value is represented in terms of the known values on previous time steps, this procedure is simple and relatively easy to program. However—as in the case of one-dimensional transient conduction—being explicit, this method is only stable for small forward time steps. For a problem which has the boundary temperatures specified for $t \geqslant 0$ the limit of stability is given by

$$\alpha\,\delta t \left[\frac{1}{\delta x^2} + \frac{1}{\delta y^2} \right] \leqslant \frac{1}{2}$$

The fully implicit method, or the Crank–Nicolson method, in which the derivatives $\partial^2 T/\partial x^2$ and $\partial^2 T/\partial y^2$ are represented wholly or partly on the new time row, will remove this limitation on forward time step but they require the solution of a large number of simultaneous equations (nm in the case above) at each time step.

'Alternating direction implicit' method

This approach, briefly mentioned in Section 4-2-3(b) in connection with the acceleration of iterative solutions of elliptic finite difference equations, has been found to be more efficient than either the explicit or the implicit formulations.

To advance the solution from the kth time row to the $(k + 1)$th row, one of the second-order derivatives, say $\partial^2 T/\partial x^2$, is represented by an implicit difference approximation while the other, $\partial^2 T/\partial y^2$, is represented by an explicit approximation. The solution for the $(k + 1)$th row is then obtained by solving n sets each of m unknowns, since we have n mesh points on $x = 0$ and m on $y = 0$. In going to the $(k + 2)$th time row the procedure is alternated by formulating $\partial^2 T/\partial y^2$ implicitly and $\partial^2 T/\partial x^2$ explicitly and solving m sets each of n unknowns.

With the notation of Fig. 4-16 this scheme has the following representations. From the kth to the $(k + 1)$th time row:

$$\frac{T_{i,j,k+1} - T_{i,j,k}}{\delta t} = \alpha \left[\frac{T_{i-1,j,k+1} - 2T_{i,j,k+1} + T_{i+1,j,k+1}}{\delta x^2} \right.$$
$$\left. + \frac{T_{i,j-1,k} - 2T_{i,j,k} + T_{i,j+1,k}}{\delta y^2} \right]$$

and from the $(k + 1)$th to the $(k + 2)$th time row:

$$\frac{T_{i,j,k+2} - T_{i,j,k+1}}{\delta t} = \alpha \left[\frac{T_{i-1,j,k+1} - 2T_{i,j,k+1} + T_{i+1,j,k+1}}{\delta x^2} \right.$$
$$\left. + \frac{T_{i,j-1,k+2} - 2T_{i,j,k+2} + T_{i,j+1,k+2}}{\delta y^2} \right]$$

and so on.

Provided the step length is constant throughout the problem the method is stable for any forward time step. It is also more efficient than implicit methods since less computing time is required for n sets of m equations than for one set of nm equations.

4-4   Some topics in finite differences

4-4-1   Cylindrical polar coordinates

Curved boundaries with cartesian coordinates necessitate the use of complicated finite difference representations, and circular boundaries (as with tubes and hollow cylinders) can often be more conveniently treated using polar coordinates.

Consider Laplace's equation which, in this coordinate system, can be written:

$$\frac{\partial^2 T}{\partial r^2} + \frac{1}{r} \frac{\partial T}{\partial r} + \frac{1}{r^2} \frac{\partial^2 T}{\partial \theta^2} + \frac{\partial^2 T}{\partial z^2} = 0$$

If there is no variation in T along the z-axis this equation may be represented in the two dimensions r and θ, at the point i, j in Fig. 4-17, by the order δ^2 approximation:

$$\frac{T_{i+1,j} - 2T_{i,j} + T_{i-1,j}}{\delta r^2} + \frac{1}{(i\,\delta r)} \frac{T_{i+1,j} - T_{i-1,j}}{2\delta r}$$

$$+ \frac{1}{(i\,\delta r)^2} \frac{T_{i,j+1} - 2T_{i,j} + T_{i,j-1}}{\delta \theta^2} = 0 \quad (4\text{-}33)$$

which, when applied together with the boundary conditions to every mesh point in the region, results in a system of linear algebraic equations similar to those dealt with in Section 4-2-3.

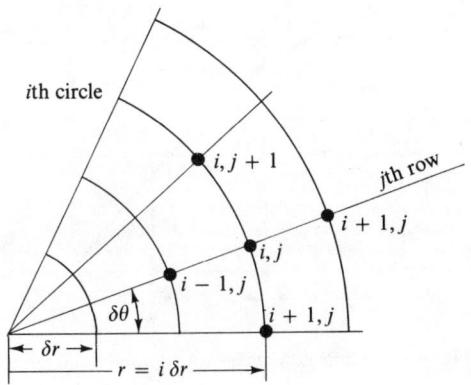

Fig. 4-17 Finite differences in cylindrical polar coordinates

With temperature variations in the z instead of in the θ direction the area of integration is a rectangle, and the last term in Eqn (4-33) would be replaced by a difference equation similar to the first term but involving δz.

A special treatment of Eqn (4-33) is obviously required at $r = 0$ and it is shown by Smith[2] that at $r = 0$, Eqn (4-33) is equivalent to

$$T_0 = T_{\rm m}$$

where $T_{\rm m}$ is the mean of the mesh point values which surround the point $r = 0$, and T_0 is the temperature at $r = 0$.

With circular symmetry, that is $\partial^2 T/\partial \theta^2 = 0$, it can be further shown that at $r = 0$ Laplace's equation is equivalent to:

$$2 \frac{\partial^2 T}{\partial r^2} + \frac{\partial^2 T}{\partial z^2} = 0$$

This has a straightforward finite difference representation so that the possible singularity $0/0$ with $r = 0$ and $\partial T/\partial r = 0$ can be avoided. A similar treatment is, of course, necessary for the equations of transient conduction problems involving radial heat flow.

4-4-2 Curved boundaries and normal derivatives

Curved boundaries with cartesian coordinates present difficulties with finite differences, especially when a normal derivative boundary condition is specified.

With the temperature specified on the boundary it is a straightforward matter to modify the general approximations to derivatives. Consider the geometry in Fig. 4-18 in which the curved boundary intersects the mesh so

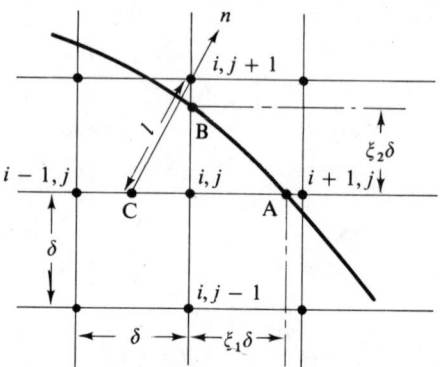

Fig. 4-18 Finite differences with curved boundaries

that OA and OB are fractions of the standard step length δ. By the use of Taylor's theorem, or by fitting quadratic interpolation formulae, it can be shown[1] that for the point i, j the first and second derivatives can be approximated by the equations

$$\left(\frac{\partial T}{\partial x}\right)_{i,j} = \frac{1}{\delta}\left(\frac{1}{\xi_1(1+\xi_1)} T_A - \frac{(1-\xi_1)}{\xi_1} T_{i,j} - \frac{\xi_1}{(1+\xi_1)} T_{i-1,j}\right)$$

$$\left(\frac{\partial^2 T}{\partial x^2}\right)_{i,j} = \frac{1}{\delta^2}\left(\frac{2}{\xi_1(1+\xi_1)} T_A + \frac{2}{(1+\xi_1)} T_{i-1,j} - \frac{2}{\xi_1} T_{i,j}\right)$$

It should be noted that this latter expression, owing to its lack of symmetry, has a greater error (of order δ) than the five-point formula, Eqn (4-8).

For boundary conditions involving the normal derivative of temperature, $\partial T/\partial n$, the choice is either for a simple but inaccurate order δ formula or for a more accurate but much more complex boundary representation. A simple method is to involve points such as $i, j + 1$ and $i + 1, j$ in Fig. 4-18, which are external to the curved boundary, by constructing a normal from $i, j + 1$ to intersect the mesh line from i, j to $i - 1, j$ at C. The normal derivative

$\partial T/\partial n$ at the point $i, j + 1$ can then be represented by $(T_{i,j+1} - T_C)/l$, where l is the distance from $i, j + 1$ to C from which T_C can be eliminated by linear interpolation in terms of $T_{i-1,j}$ and $T_{i,j}$. The resulting finite difference approximation for the point i, j is obviously only of order δ, and for the more desirable but much more complicated order δ^2 approximations the reader is referred to Fox[1].

4-4-3 Interior discontinuities

Heat conduction problems which involve two or more different materials with different thermal conductivities invariably require special approximations for the mesh points which lie on the internal boundaries. These boundaries have a continuous function value, T, but a discontinuous first derivative, $\partial T/\partial x$, across them (Fig. 4-19).

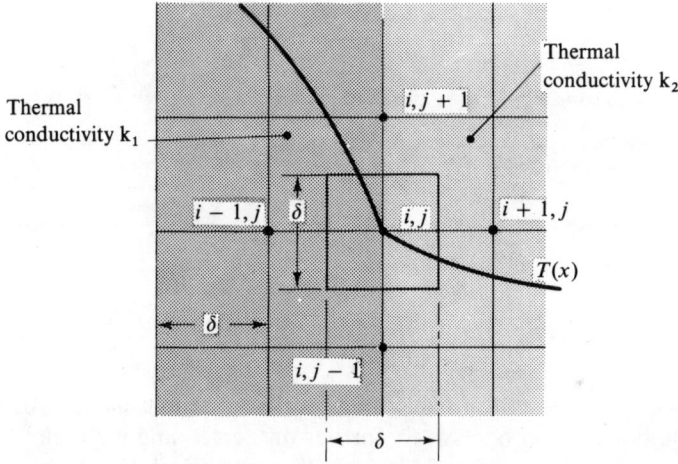

Fig. 4-19 Interior discontinuities in derivatives ($k_2 > k_1$)

A simple and direct method of obtaining a finite difference approximation for the point i, j in Fig. 4-19 is to consider the heat flows into the control element (here a square of side δ) surrounding that point. The sum of the heat flows to a point must be zero, and therefore, using $\dot{Q} = -kA(dT/dn)$, we can write

$$-k_1 \delta \left(\frac{T_{i,j} - T_{i-1,j}}{\delta} \right) - \left(\frac{k_1 + k_2}{2} \right) \delta \left(\frac{T_{i,j} - T_{i,j+1}}{\delta} \right)$$

$$-k_2 \delta \left(\frac{T_{i,j} - T_{i+1,j}}{\delta} \right) - \left(\frac{k_1 + k_2}{2} \right) \delta \left(\frac{T_{i,j} - T_{i,j-1}}{\delta} \right) = 0$$

which reduces to

$$k_1 T_{i-1,j} + \left(\frac{k_1 + k_2}{2}\right) T_{i,j-1} - 2(k_1 + k_2) T_{i,j}$$

$$+ \left(\frac{k_1 + k_2}{2}\right) T_{i,j+1} + k_2 T_{i-1,j} = 0$$

If $k_1 = k_2$ this further reduces to the standard five-point formula for Laplace's equation.

The same procedure can be extended to mesh points on interior corners and on junctions between interior boundaries and outer surfaces.

4-4-4 Infinite boundaries

Problems frequently arise in which one boundary should strictly be at an infinite distance from the region of interest, as in Fig. 4-20.

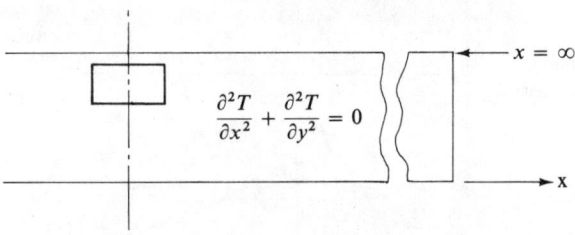

Fig. 4-20 Problem with an infinite boundary

This, of course, is not possible numerically and the usual approach is to move the boundary progressively further out, extending the region of integration, until further movement has no effect on the solution. This involves either an increase in the number of mesh points or the use of an expanding step length.

The true boundary condition at infinity for Fig. 4-20 is, of course, $\partial T/\partial x = 0$, the finite difference representation of which has been dealt with in Section 4-2-2. However, for this problem $\partial^2 T/\partial x^2 = 0$ at $x = \infty$, and Laplace's equation thus reduces to $\partial^2 T/\partial y^2 = 0$ on this boundary; this often has an analytical solution. Thus, the finite difference equations on this boundary can be written in terms of the mesh points along that boundary line only.

4-4-5 Simultaneous partial differential equations

Although steady-state heat conduction problems do not usually result in the application of more than one differential equation, systems of partial differential equations occur in the boundary layer prediction of convective heat transfer (see Chapters 6 and 7) and in simultaneous heat and mass transfer in porous materials. To illustrate the numerical method of solution of systems of partial differential equations, consider an elliptic problem with the cooling method known as 'transpiration cooling' which involves forcing a coolant unidirectionally through a porous metal as shown in Fig. 4-21.

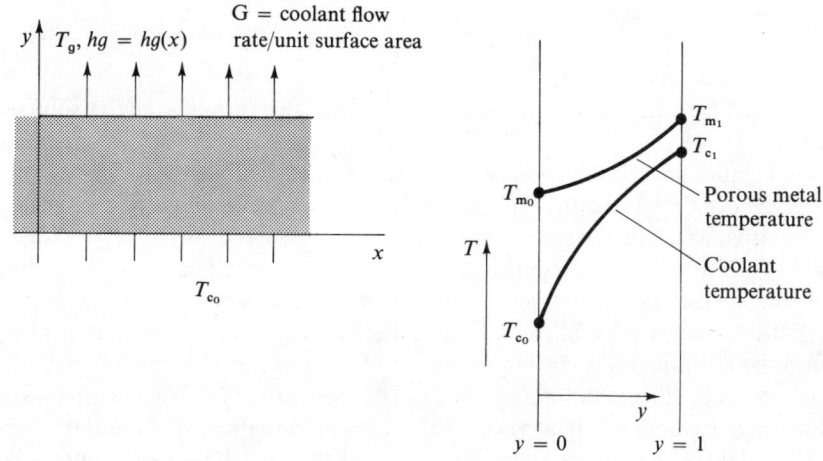

Fig. 4-21 Two-dimensional transpiration cooling

If the gas temperature, T_g, or the heat transfer coefficient, h_g, in Fig. 4-21 is a function of the x coordinate the system will be two-dimensional with two independent variables—the metal temperature T_m and the coolant temperature T_c. The governing differential equations are:

$$k\frac{\partial^2 T_m}{\partial x^2} + k\frac{\partial^2 T_m}{\partial y^2} = h_i(T_m - T_c) \qquad (4\text{-}34a)$$

$$GC_p\frac{\partial T_c}{\partial y} = h_i(T_m - T_c) \qquad (4\text{-}35a)$$

where conduction in the coolant fluid has been neglected and where h_i is the volumetric heat transfer coefficient from metal to coolant $[h_i(G)]$, k is the thermal conductivity of the porous metal, G is the coolant mass flow rate per unit surface area and C_p is the specific heat of the coolant.

With the mesh notation illustrated in Fig. 4-3, with n mesh points across the metal thickness shown in Fig. 4-21, two variables are associated with each mesh point so that the algebraic equations substituted in the matrix run, alternately, one for the metal temperature and one for the coolant temperature so that a total of nm mesh points yields a matrix of order $2nm$. Accordingly, a finite difference representation of Eqns (4-34a) and (4-35a), with step lengths δx and δy in the x and y directions respectively, is:

$$\left(\frac{T_{i-2n} - 2T_i + T_{i+2n}}{\delta x^2}\right) + \left(\frac{T_{i-2} - 2T_i + T_{i+2}}{\delta y^2}\right) - \frac{h_i}{k}(T_i - T_{i+1}) = 0$$

(4-34b)

$$\left(\frac{T_{i+2} - T_{i-2}}{2\delta y}\right) - \frac{h_i}{GC_p}(T_{i-1} - T_i) = 0 \qquad (4\text{-}35b)$$

The suffix notation is that of the position of the variables in the solution vector $T = T_{m_1}, T_{c_1}, T_{m_2}, T_{c_2}, \ldots, T_{m_N}, T_{c_N}$ where 'm' denotes the metal temperature and 'c' the coolant temperature.

The boundary conditions for the metal temperature equations are usually derivative and are related to the heating rates on the surfaces $y = 0$ and $y = Y$. The coolant temperature, T_{co}, is usually specified on $y = 0$ while on the surface $y = Y$, which usually has no boundary condition specified for the coolant temperature, a finite difference equation can be obtained by fitting quadratic interpolation formulae to the three mesh points nearest the surface and differentiating to obtain the derivative $\partial T/\partial y$ in Eqn (4-35). Fitting polynomial functions through mesh point values, or boundary condition data, can be conveniently accomplished using Lagrange's interpolation formula which gives an nth degree polynomial which passes through the $n + 1$ points whose coordinates are (x_i, y_i):

$$y = \frac{(x - x_1)(x - x_2) \ldots (x - x_n)}{(x_0 - x_1)(x_0 - x_2) \ldots (x_0 - x_n)} y_0$$

$$+ \frac{(x - x_0)(x - x_2) \ldots (x - x_n)}{(x_1 - x_0)(x_1 - x_2) \ldots (x_1 - x_n)} y_1$$

$$+ \cdots + \frac{(x - x_0)(x - x_1) \ldots (x - x_{n-1})}{(x_n - x_0)(x_n - x_1) \ldots (x_n - x_{n-1})} y_n$$

4-5 The finite element method

4-5-1 Introduction

The finite element approach was developed and has been extensively used for structural analysis.[11,12] It is becoming of increasing importance in other engineering fields such as fluid mechanics and heat conduction.[11,13] While the finite difference method represents a direct approach by approximating individual derivatives in the differential equation, the finite element procedure is an approximation applied to the variational form of the partial differential equation so that, in general, the problem becomes one of finding a function which minimizes a 'variational functional' over the field of interest.

The region is divided into a number of elements, usually triangular as shown in an actual example in Fig. 4-22,[13] and the unknown function is uniquely specified throughout the field by a discrete number of values associated with the node points of the elements. The value of the function at a particular node point influences only the function at the nodes of the adjacent elements so that the minimization of the functional throughout the whole region results, as with finite differences, in a system of simultaneous equations. Although the finite difference and finite element methods occasionally result in identical equations, the finite element method has advantages in the freedom of choice of the shape of elements (that is, the choice of placement of node points in complex regions) and in the ease with which derivative and mixed boundary conditions on irregularly shaped boundaries can be handled.

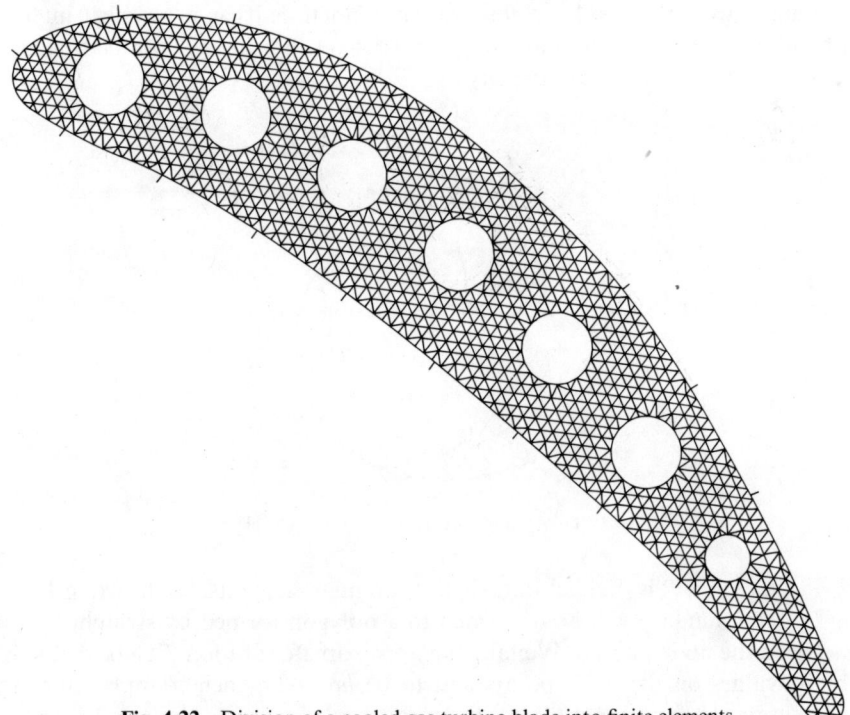

Fig. 4-22 Division of a cooled gas turbine blade into finite elements

4-5-2 The mathematical problem

To describe the procedure we consider the solution of Laplace's equation:

$$\frac{\partial^2 T}{\partial x^2} + \frac{\partial^2 T}{\partial y^2} = 0 \qquad (4\text{-}36)$$

over the region R with the boundary condition:

$$\frac{\partial T}{\partial n} + \sigma T + p = 0, \quad \sigma \geqslant 0 \tag{4-37}$$

on the boundary C, where $\partial/\partial n$ denotes differentiation of the outward normal. This equation can easily be interpreted in terms of the convective boundary condition equation (2-12).

By applying the Euler conditions of the calculus of variations[11] it can be shown that the problem defined by Eqns (4-36) and (4-37) corresponds to that of finding a function T which minimizes the functional

$$\int_R \int \frac{1}{2}\left[\left(\frac{\partial T}{\partial x}\right)^2 + \left(\frac{\partial T}{\partial y}\right)^2\right] dx \, dy + \int_C (pT + \tfrac{1}{2}\sigma T^2) \, ds \tag{4-38}$$

evaluated over the whole region R. This functional has a physical interpretation in terms of thermal energy, but for our purposes it can be treated simply as a mathematical quantity.

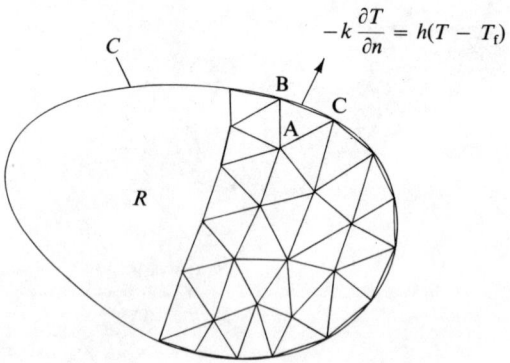

Fig. 4-23   Division of a region into triangular elements

If the region R is divided into small triangular elements, as shown in Fig. 4-23, the boundary will be deformed to a polygon formed by straight lines between the node points. We take the approximate solution T to be defined by its values on the node points and to be *linear* on each triangle. When evaluating the functional (4-38), there is a contribution to the area integral from each triangle and if the triangle forms a part of the boundary (for instance, BC on the triangle ABC in Fig. 4-23), there will also be a contribution to the line integral.

The resulting quadratic function by which we approximate (4-38) can be written in matrix form[13] as:

$$\tfrac{1}{2}\mathbf{T}^T A \mathbf{T} - \mathbf{T}^T \mathbf{b}$$

where **T** is the vector of values at mesh points inside R and on C and **b** is a vector of the same order. By differentiating this expression with respect to each of the nodal values in **T**, and equating the result to zero, it follows that the minimization of the functional (4-38) corresponds to the solution of the system of equations:

$$A\mathbf{T} = \mathbf{b} \qquad (4\text{-}39)$$

The matrix of coefficients A will be symmetric, positive definite, and, if the grid points are sensibly ordered, will have a band structure so that the direct and iterative methods outlined in Section 4-2-3 can be used for its solution.

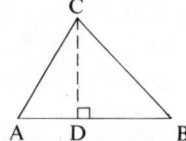

Fig. 4-24   A triangular element

Assembly of the matrix equations

If ABC in Fig. 4-24 is a typical grid triangle, where T is *linear* on ABC and has the values T_A, T_B, and T_C at the vertices ABC, then, by taking the derivatives of T in the functional (4-38) to be differences in terms of the nodal values T_A, T_B, and T_C, the contribution to the area integral can be determined as follows:

Let D be the foot of the perpendicular from C to AB. The contribution of interest is:

$$\int_{ABC}\int\left[\left(\frac{\partial T}{\partial x}\right)^2 + \left(\frac{\partial T}{\partial y}\right)^2\right] dx\,dy$$

$$= \tfrac{1}{2}(AB)(CD)\left[\left(\frac{T_B - T_A}{AB}\right)^2 + \left(\frac{T_C - T_D}{CD}\right)^2\right]$$

$$= \frac{1}{2}\left[\frac{CD}{AD}(T_B - T_A)^2 + \frac{AB}{CD}\left(T_C - \frac{BD.T_A + AD.T_B}{AB}\right)^2\right]$$

It is clear that this is a quadratic function of T_A, T_B, and T_C and it remains to find the coefficients. The coefficient of T_C^2 is

$$\tfrac{1}{2}(AB/CD) = \tfrac{1}{2}(\cot A + \cot B)$$

and the coefficient of $T_A T_C$ is

$$- BD/CD = - \cot B$$

The values of the other coefficients follow by symmetry, so that the contribution is:

$$\tfrac{1}{2}(T_A T_B T_C) = \begin{bmatrix} \cot B + \cot C & -\cot C & -\cot B \\ -\cot C & \cot C + \cot A & -\cot A \\ -\cot B & -\cot A & \cot A + \cot B \end{bmatrix} \begin{bmatrix} T_A \\ T_B \\ T_C \end{bmatrix} \tag{4-40}$$

If the triangle ABC has two vertices (say B and C as in Fig. 4-22) on the boundary C, then the line integral will contribute to the functional (4-38):

$$\int_{BC} (pT + \tfrac{1}{2}\sigma T^2)\, ds \approx \frac{\text{length of BC}}{2} [T_B T_C]$$

$$+ \frac{\text{length of BC}}{4} [T_B T_C] \begin{bmatrix} \sigma(B) & \\ & \sigma(C) \end{bmatrix} \begin{bmatrix} T_B \\ T_C \end{bmatrix} \tag{4-41}$$

in which the trapezium rule has been used and the arc length approximated by the length of the straight line BC.

The first part of expression (4-41) will contribute to the vector **b** and the second part to the coefficients in the matrix A, as will be shown later.

4-5-3 Finite elements with equiangular triangles

If, for example, a region was systematically divided into equilateral triangles with n mesh points along each vertical mesh row, then we can illustrate how the algebraic equation for the nodal value T_i is assembled by considering the triangular elements surrounding the point i in Fig. 4-25.

For triangle I, which is taken to be wholly within the region, Eqn (4-40) can be simplified to give:

$$A\mathbf{T} \equiv \frac{\sqrt{3}}{2} \begin{bmatrix} 2 & -1 & -1 \\ -1 & 2 & -1 \\ -1 & -1 & 2 \end{bmatrix} \begin{bmatrix} T_i \\ T_{i+n} \\ T_{i+n+1} \end{bmatrix} \tag{4-42}$$

so that when these coefficients are added to the successive contributions of the five other triangular elements (triangles II to VI) surrounding T_i, the coefficients of A at that stage will have the values:

Row position

$$\frac{\sqrt{3}}{2}
\begin{array}{ccccccc|c}
 i-n & i-n+1 & i-1 & i & i+1 & i+n & i+n+1 & \mathbf{T} \\
 4 & -1 & -1 & -2 & & & & i-n \\
 -1 & 4 & & -2 & -1 & & & i-n+1 \\
 -1 & & 4 & -2 & & -1 & & i-1 \\
 -2 & -2 & -2 & 12 & -2 & -2 & -2 & i \\
 & -1 & & -2 & 4 & & -1 & i+1 \\
 & & -1 & -2 & & 4 & -1 & i+n \\
 & & & -2 & -1 & -1 & 4 & i+n+1
\end{array}$$

$$(4\text{-}43)$$

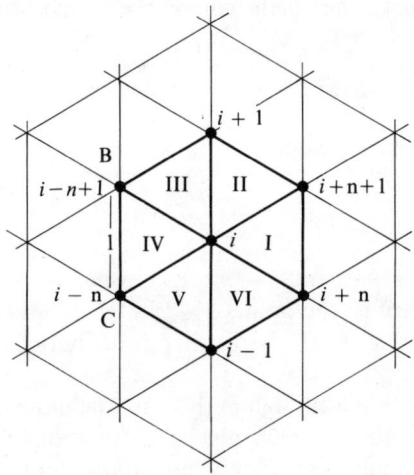

Fig. 4-25   An equiangular finite element mesh

Now, since all the elements surrounding the point i have been considered there will be no further contribution to the row i, and since $b_i = 0$ it is clear that the equation for the node i in Fig. 4-25 can be illustrated as the following molecule, which is in fact equivalent to a finite difference approximation for a regular hexagonal mesh.

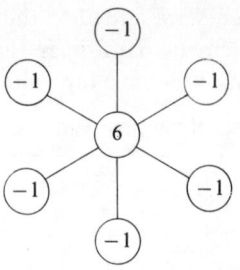

If triangle IV in Fig. 4-24 were to have one side, say, BC of length l, on the boundary over which the applied boundary condition was

$$-k \frac{\partial T}{\partial n} = h(T - T_f) \tag{4-44}$$

then, provided h, k, and T_f are fixed constants, the parameters σ and p will have the values:

$$\sigma(B) = \sigma(C) = \frac{h}{k} \quad \text{and} \quad p(B), p(C) = -\frac{h}{k} T_f$$

and the contribution to the coefficients in the matrix A (4-43) would be:

$$\frac{l}{2} \frac{h}{k} \begin{bmatrix} 1 & \\ & 1 \end{bmatrix} \begin{bmatrix} T_{i-n+1} \\ T_{i-n} \end{bmatrix}$$

with the vector **b** having the contributions

$$-l \frac{h}{k} T_f$$

in the positions $i - n + 1$ and $i - n$ respectively.

If elements adjacent to the points T_{i-n} and T_{i-n+1} also form part of the boundary there would clearly be further contributions to A and **b** to be added in.

We may conclude that although in this case the finite difference and finite element methods result in similar algebraic equations for the interior grid points, the algebraic equations for the boundary points are not similar and are in fact more easily obtained using finite elements since finite differences would involve complicated interpolation inside the region.

In this section we have only discussed triangular elements with nodes placed at the vertices, which implies that although the function is continuous the first and higher derivatives are not. To overcome this, more complex elements can be used such as triangles with nodes placed at midsides so that the describing function on an element can be nonlinear. This means that fewer elements are required for the same overall accuracy.

A consequence of using larger elements is that the representation of curved boundaries is not as accurate and, accordingly, a further development is the use of elements with curved sides.

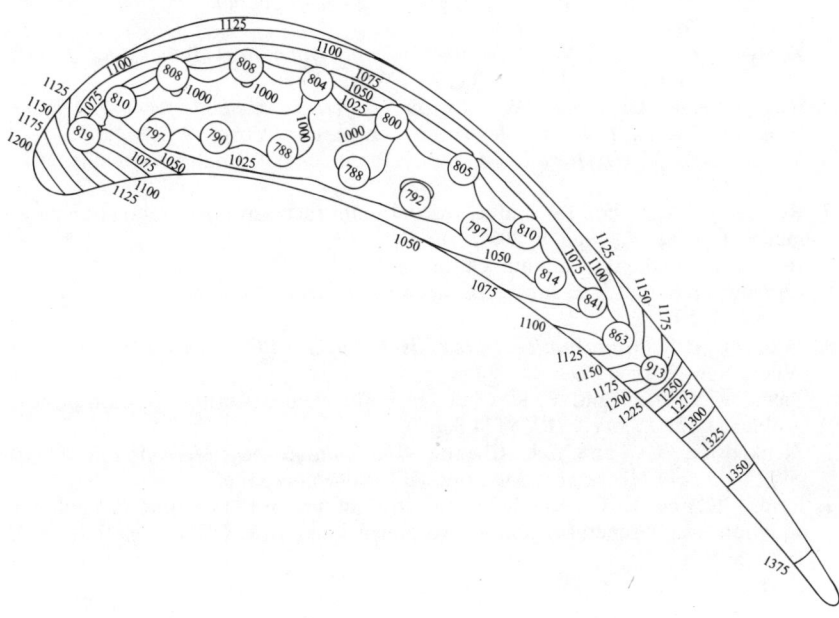

Fig. 4-26 A graphical numerical solution to a gas turbine blade problem

Result of an actual problem

A numerical solution[13] for the temperature contours in a two-dimensional section of an air cooled gas turbine blade is illustrated in Fig. 4-26 as a practical example of the application of this technique. This solution was obtained using the finite element method described above, with a total of over 5,000 mesh points within the region, and has an accuracy estimated at better than 0·5 per cent. This graphical solution illustrates the comments made in the introduction to this chapter regarding the effectiveness of graphical output devices since this solution, except for the numerals, was drawn automatically by the computer linked to a graph plotter.

The specification for the aerofoil shape, the size and position of the air-cooling holes, the boundary conditions for the holes, and the boundary conditions around the aerofoil surface were sufficient to allow the program to compute and plot the solution.

REFERENCES

1. Fox, L. *Numerical Solution of Ordinary and Partial Differential Equations.* Pergamon Press, Oxford, 1962.
2. Smith, G. D. *Numerical Solution of Partial Differential Equations.* Oxford University Press, London, 1965.
3. Ames, W. F. *Numerical Methods for Partial Differential Equations.* Nelson, London, 1969.
4. Forsythe, G. E. and W. R. Wasow. *Finite Difference Methods for Partial Differential Equations.* John Wiley, New York, 1960.
5. Richtmyer, R. D. and K. W. Morton. *Difference Methods for Initial Value Problems* (2nd edn). John Wiley (Interscience), New York, 1967.
6. Varga, R. S. *Matrix Iterative Analysis.* Prentice-Hall, Englewood Cliffs, N.J., 1962.
7. Reid, J. K. A method for finding the optimum successive over relaxation parameter. *Comput. J.*, 1966, **9**, 200.
8. Reid, J. K. Personal Communication.
9. Carslaw, H. S. and J. C. Jaeger. *Conduction of Heat in Solids.* Clarendon Press, Oxford, 1959.
10. Mitchell, A. R. *Computational Methods in Partial Differential Equations.* John Wiley, New York, 1969.
11. Zienkiewicz, O. C. and Y. K. Cheung. Finite elements in the solution of field problems. *The Engineer*, 1965 (24 Sept.).
12. Zienkiewicz, O. C. and Y. K. Cheung. *The Finite Element Method in Structural and Continuum Mechanics.* McGraw-Hill, New York, 1967.
13. Reid, J. K. and A. B. Turner. Fortran subroutines for the solution of Laplace's equation over a general region in two dimensions. *U.K.A.E.A. Rept T. P. 422*, 1970 (Sept.).

PROBLEMS

4-1 A steel spine of thermal conductivity 54 W/m K, 1 cm diameter and 20 cm long protrudes from a base at 100°C into a fluid at zero. The heat transfer coefficient is 10 W/m^2 K and the tip of the fin is thermally insulated. A one-dimensional numerical solution is required for the temperature of the fin. If nodes are placed at both ends of the fin and at equal spacings along it show by (a) considering the heat flow to and from each node point and (b) representing the differential equation for a simple fin, Eqn (3-3), by finite differences, that one finite difference scheme is

$$T_i = 0.4576\,(T_{i-1} + T_{i+1}) \quad \text{and} \quad T_i = 0.9152\,T_{i-1}$$

for the inner nodes and the end node respectively.

Take a step length of 5 cm, that is five nodes, an initial assumption of 100, 75, 50, 25, 0°C for these nodes, and solve the resulting four simultaneous equations by the Gauss–Seidel iteration. Show that after ten iterations the numerical solution compares with the analytical one as follows:

Distance along fin (m)	0	0·05	0·1	0·15	0·2
Numerical solution (°C)	100	67·55	47·92	37·49	34·31
Analytical solution (°C)	100	67·75	48·32	37·88	34·63

4-2 Recalculate the fin problem described in Problem 4-1 using the Jacobi iteration (which uses only the nth iterates to calculate the $(n + 1)$th); show that a reduced rate of convergence results and that after ten iterations the numerical solution is:

Distance along fin (m)	0	0·05	0·1	0·15	0·2
Numerical solution (°C)	100	67·34	46·47	36·65	31·96

Calculate also the heat dissipated by the fin, from the heat conducted through the base, with the temperature derivative approximated (a) by a forward difference from $x = 0$ and (b) by differentiating the quadratic fitted through the first three node values (use Lagrange's interpolation formula, Section 4-4-5). Compare both results with the analytical value given by Eqn (3-7)

Ans: 2·78 W, 3·27 W, 3·42 W

4-3 The temperature in a square sectioned slab, of side 1 m and thermal conductivity 10 W/m K satisfies Laplace's equation

$$\frac{\partial^2 T}{\partial x^2} + \frac{\partial^2 T}{\partial y^2} = 0$$

and has the sides $x = 0$ and $y = 0$ maintained at a temperature of zero. The side $y = 1$ has the surface maintained at the temperature $T = 80x°C$ whilst the side $x = 1$ convects heat into a medium at zero with a heat transfer coefficient of 10 W/m^2 K. Using a step length $\delta x = \delta y$ of $\frac{1}{2}$ and central difference approximations for Laplace's equation and the boundary condition along $x = 1$, show that the temperatures of the centre of the slab and the centre of the side $x = 1$ have the values $15\frac{5}{9}$ and $22\frac{2}{9}°C$ respectively.

4-4 Heat is generated internally in a thin square metal plate such that the temperature within the plate satisfies Poisson's equation

$$\frac{\partial^2 T}{\partial x^2} + \frac{\partial^2 T}{\partial y^2} + 2 = 0$$

If the square is bounded by $x = \pm 1$, $y = \pm 1$ and the temperature is maintained at zero on the boundary, calculate two finite difference solutions using a square mesh with step lengths of $h = 1$ and $h = \frac{1}{2}$ respectively. Taking the discretization error in the finite difference equations to be proportional to h^2 use Richardson's extrapolation, Eqn (4-20), to calculate an extrapolated value for the point (0, 0) and compare the result with the true value of 0·589.

4-5 The temperature T in a square of side 4 units satisfies Laplace's equation for two-dimensional steady heat flow. The temperatures along the boundaries are given by $T = 0$ along $x = 0$ and $y = 0$, $T = x^3$ along $y = 4$, and $T = 16y$ along $x = 4$. Use a square mesh of side unity and calculate the temperatures (to ± 0.05) at the nine internal mesh points of the square. Use either the Gauss–Seidel iteration (Eqn 4-15) or a relaxation method (in which the largest residual is usually relaxed first). Compare the results with the following values: T along $x = 1$, 2·60, 4·45, 4·49; T along $x = 2$, 5·95, 10·73, 12·52; T along $x = 3$, 10·49, 20·02, 26·88. Resolve the problem using a sequence of systematic SOR iterations taking $\omega = 1.1716$ so that the iteration becomes

$$T_{i,j}^{(n+1)} = \frac{\omega}{4}\left[T_{i-1,j}^{(n+1)} + T_{i,j-1}^{(n+1)} + T_{i+1,j}^{(n)} + T_{i,j+1}^{(n)}\right] + (1 - \omega)T_{i,j}^{(n)}$$

and show that only seven iterations are necessary to give the solution correct to 2 decimals.

4-6 Derive the finite difference representation of Laplace's equation (Section 4-4-2) for a mesh point which has both the perpendicular mesh lines which pass through it intersected by a curve within one step length. A thin, semicircular metal plate of uniform thermal conductivity and radius 2 units has the diameter maintained at a temperature of zero and the circumference maintained at unity. Using a square mesh with a step length of unity show that the three internal mesh points have values of 0·705, 0·602, and 0·705.

4-7 It is required to solve Laplace's equation for the temperatures in a disc which has a truncated triangular profile, and rather than use an orthogonal mesh system a tapered mesh system is proposed. The step in the radial direction r is constant but the step in the z direction reduces towards the tip of the disc such that the number of mesh points in the z direction remains constant. The system is axisymmetric, that is no θ variation, and the angle of taper of the disc is small. Show that Laplace's equation takes the form, in the (r, z') coordinate system,

$$\frac{\partial^2 T}{\partial r^2} + \left(\frac{1}{r} - \frac{1}{R_0 - r}\right)\frac{\partial T}{\partial r} + \frac{\partial^2 T}{\partial z'^2} = 0$$

where R_0 is the radial distance from the centre-line of the extrapolated intersection of the two faces of the disc and z' is the axial distance from the disc *surface*.

4-8 An electrical resistance-wire analogue is required for the two-dimensional steady state problem illustrated as Example 1, Section 4-2-5. Simulation of the problem with the step length $\delta x = \delta y = \frac{1}{4}$ is required with the resistance wires joining the node points shown in Fig. 4-7 and with external resistances at the boundary points along $x = 1$ to account for the surface/fluid resistance to heat transfer. Show, by considering the heat flow to particular points, or otherwise, that if the resistance between the internal nodes of the mesh is R ohms then the resistance wires forming the outer boundary will have a resistance of $2R$ ohms between nodes, and that the surface heat transfer resistances will have values of $0·4R$ ohms along $x = 1$ and a value of $0·8R$ ohms at $x = 1$, $y = 1$ when the ratio h_f/k is 10.

4-9 Show that Fourier's one-dimensional transient heat flow equation

$$\frac{\partial T'}{\partial t'} = \alpha \frac{\partial^2 T'}{\partial x'^2}$$

where $\alpha = k/\rho C$ can be written non-dimensionally as

$$\frac{\partial T}{\partial t} = \frac{\partial^2 T}{\partial x^2}$$

where $x = x'/L$, $T = T'/T_i$ and $t = \alpha t'/L^2$ and where T_i and L are an initial temperature and a characteristic length of the problem, respectively.

4-10 The temperature T of a thin rod, insulated along its length, obeys Fourier's equation

$$\frac{\partial T}{\partial t} = \frac{\partial^2 T}{\partial x^2}$$

where the variables are non-dimensional, and has the initial temperature distribution, symmetrical about the mid point $x = 0.5$, given by $T = 2x, 0 \leqslant x \leqslant \frac{1}{2}$. Formulate an explicit method of solution and show that with step lengths $\delta x = 0.1$ and $\delta t = 0.001$, one finite difference approximation to the differential equation is given by

$$T_{i,j+1} = \tfrac{1}{10}(T_{i-1,j} + 8T_{i,j} + T_{i+1,j})$$

Calculate five steps of the solution and show that by comparison with the analytical solutions at $t = 0.005$ for $x = 0.3$ and 0.5, which are 0.5966 and 0.8404 respectively, the discontinuity in the initial condition gives a greater error at $x = 0.5$.

4-11 Recalculate the problem described in Problem 4-10 with the step length $\delta x = 0.1$ as before but with the step length $\delta t = 0.01$ so that the explicit finite difference formulation $T_{i,j+1} = T_{i-1,j} - T_{i,j} + T_{i+1,j}$ is *unstable*. Show that, although it is the correct solution for the conditions given, it is of no value.

4-12 An infinitely wide slab of finite thickness, initially at constant temperature, loses heat by convection from both faces to two fluids at different temperatures. The temperature within the slab satisfies the equation

$$\frac{\partial T}{\partial t} = \frac{\partial^2 T}{\partial x^2}, \qquad 0 \leqslant x \leqslant 1$$

and the boundary conditions $k\,\partial T/\partial x = h_1(T - T_{f_1})$ at $x = 0$ and $-k\,\partial T/\partial x = h_2(T - T_{f_2})$ at $x = 1$ where t, x and the temperatures are non-dimensional. Approximate the boundary conditions by central differences and obtain explicit finite difference equations for the three different types of node point. Compare the result with the boundary conditions approximated by forward (at $x = 0$) and backward (at $x = 1$) differences.

4-13 The temperature, T, in a thin rod, insulated along its length, obeys Fourier's equation

$$\frac{\partial T}{\partial t} = \frac{\partial^2 T}{\partial x^2}$$

where the variables are non-dimensional. The rod, of unit length, has the initial temperature distribution symmetrical about the mid point $x = 0.5$ and given by $T = x, 0 \leqslant x \leqslant \frac{1}{2}$, and the boundary condition $T = 0$ at $x = 0$ and $x = L$. Show that the Crank–Nicolson implicit finite difference scheme with the step lengths $\delta x = 0.1, \delta t = 0.01$ gives the formulation

$$-T_{i-1,j+1} + 4T_{i,j+1} - T_{i+1,j+1} = T_{i-1,j} + T_{i+1,j}$$

where the $j + 1$th values are unknown. Calculate a numerical solution of the problem and show that the above formulation gives a stable solution. Evaluate also the analytical solution to the problem

$$T = \frac{4}{\pi^2} \sum_{n=1}^{\infty} \frac{1}{n^2} \sin\left(\tfrac{1}{2}n\pi\right) \sin\left(n\pi x\right) e^{-n^2\pi^2 t}$$

and show that at the time step $t = 0.1$ both solutions compare as follows:

x	0	0·1	0·2	0·3	0·4	0·5
Numerical solution	0	0·0474	0·0902	0·1241	0·1459	0·1535
Analytical solution	0	0·0467	0·0888	0·1222	0·1437	0·1511

5

The basic equations of convection

5-1 Basic concepts of convective heat transfer

Convection is the mode of heat transfer that occurs in a fluid owing to the combination of molecular conduction within the fluid itself and the energy transport created by the motion of the fluid particles. If this motion is produced predominantly by density variations, generated from the temperature gradients within the fluid, the energy transport is said to be due to *free convection* or *natural convection*: if the motion is produced principally by a superimposed velocity field, the energy transport is said to be due to *forced convection*. While, in general, both free and forced convection phenomena exist together, in many practical problems they can be treated as separate effects.

In most engineering situations the fluid is bounded on one or more sides by a solid 'wall'. In the previous chapters attention has been focused on the calculation of temperature within such a wall, and an important class of conduction problems specifies a boundary condition in terms of the heat flux normal to the wall. In the present nomenclature, the heat flux through the wall, q_s, is determined from Newton's law of cooling, as shown in Chapter 1, by

$$q_s = h(T_s - T_\infty) \qquad (5\text{-}1)$$

T_s being the wall temperature, T_∞ being the fluid temperature at some as yet unspecified distance from the wall, and h being the heat transfer coefficient, the value of which is prescribed. In general, the thermal boundary condition on the solid wall will influence the heat transported by convection processes, and will thus affect the value of the heat transfer coefficient itself, but in many problems h can be estimated solely from a knowledge of the fluid dynamics of the flow system.

For flow past a stationary wall, fluid viscosity causes the molecules adjacent to the wall to adhere to the surface (the so-called *no-slip* condition) causing a variation in velocity from zero at the wall to U_∞ at the *free stream*, as shown in Fig. 5-1(a). The region in which the velocity changes from zero to U_∞ is termed the *boundary layer*. Although the velocity u approaches U_∞ asymptotically, it is customary to define a finite boundary layer thickness, δ, for example as the distance from the wall to the point where u is equal to an arbitrary value of U_∞ (say, $u = 0.99 U_\infty$). For low-viscosity fluids such as air or water, the value of δ is usually very small compared with the basic dimensions of the bounding geometry for an *external flow* system, such as flow over plates, turbine blades, aircraft wings, and so on. For *internal flow* systems such as flow in pipes, ducts, diffusers, and so on, the boundary layers are only thin compared with the duct width for a limited distance (the so-called *entry length*) and eventually the wall boundary layers merge, leaving no free stream region. The flow in the latter case is termed *fully developed* flow.

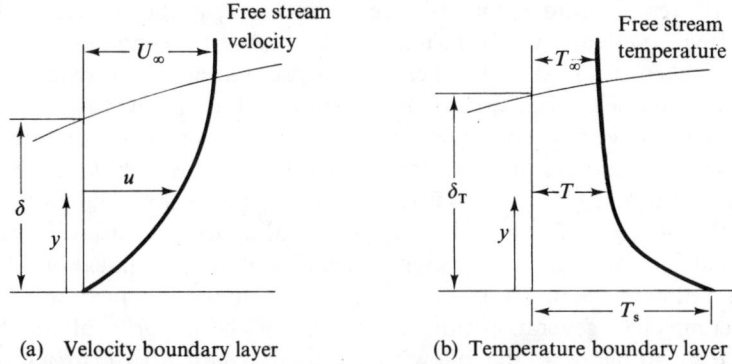

(a)   Velocity boundary layer   (b) Temperature boundary layer

Fig. 5-1   Flow past a stationary wall

In an analogous manner to the velocity behaviour, the temperature of a fluid flowing past a wall varies from the surface temperature of the wall to the free stream temperature, as shown in Fig. 5-1(b). The regions of large thermal change are confined within a thermal boundary layer of thickness δ_T which, as we shall see in Section 5-3, is related to the thickness of the velocity boundary layer according to the fluid properties. It is the thermal boundary layer which contributes the main resistance to heat transfer from the free stream to the wall. Through the fluid particles adhering to the wall, heat can only be transferred by molecular conduction, where Fourier's equation, as defined in Chapters 1 and 2, applies such that

$$q_s = -k\left(\frac{\partial T}{\partial n}\right)_s \tag{5-2}$$

k being the thermal conductivity of the fluid, and $(\partial T/\partial n)_s$ being the fluid temperature gradient normal to the surface. If the temperature field can be evaluated, then the heat transfer coefficient can be calculated by

$$h = -k\,\frac{(\partial T/\partial n)_s}{T_s - T_\infty} \tag{5-3}$$

For external flow systems, T_∞ is the uniform free stream temperature outside the thermal boundary layer, but for internal flow systems T_∞ is taken as a representative temperature, usually the average or *bulk temperature*, of the fluid. In only a few cases can h be determined directly from Eqn (5-3), and in most problems of engineering interest h is evaluated from approximate or semi-empirical solutions of the fluid behaviour.

In the general case of flow in three dimensions, where the flow and temperature fields interact, an adequate description of the fluid behaviour can only be obtained from evaluation of the three velocity components u, v, and w, the temperature T, the pressure p, the density ρ, the viscosity μ, and the thermal conductivity k of the fluid. This involves the simultaneous solution of eight equations: the law of conservation of momentum provides three of these equations, while conservation of mass, the energy equation, the equation of state, and empirical relations for viscosity and conductivity provide the remaining equations. In order to simplify this complex problem, attention in this chapter will be restricted to two-dimensional flow systems.

The flow is assumed to be *steady*, such that the dependent variables at any given spatial location do not vary with time. The above simplification allows the derivation of the fundamental equations of convection which, while not being completely general, permit description of the majority of tractable engineering problems. In the following section the *continuity equation* will be derived from the principle of the conservation of mass; the derivation of the momentum equations will be based upon the application of Newton's second law; and the energy equation will be derived by applying the first law of thermodynamics.

The fluid motion itself can either be *laminar*, where the streamlines behave in a well-ordered manner, or *turbulent*, where irregular velocity fluctuations are superimposed on the mean fluid motion. The turbulence itself consists of three-dimensional eddies of varying velocity and size, and energy is transported from the main flow, principally by means of the larger eddies. This energy is dissipated as heat by the action of viscosity, due predominantly to the smaller eddies existing near the bounding walls of the flow system. The structure of turbulence is a complex phenomenon which is still not fully understood, and the mathematics that has so far been developed is so limited that semi-empirical methods are invariably used for the solution of engineering problems. Initially the basic equations will be limited to laminar flow, but in Section 5-4 the steady flow equation will be extended to turbulent flow by time-averaging the turbulent fluctuations.

5-2   The basic equations

5-2-1   Conservation of mass

The *continuity equation* is simply the mathematical statement of the conservation of mass. For a two-dimensional steady flow system the mass flow into an infinitesimal volume must be equal to the mass flow out of the volume. In cartesian coordinates it is convenient to consider flow through an element in the x–y plane with unit depth in the z direction (as illustrated in Fig. 5-2).

The rate of mass flow into the element from the x direction is equal to $\rho u\,\delta y$, and the rate of mass flow out of the element in the x direction is $\rho u\,\delta y + \partial/\partial x(\rho u)\,\delta x\,\delta y$, where ρ is the fluid density. Similar expressions hold in the y direction, and to satisfy the continuity requirement it is necessary that

$$\rho u\,\delta y + \rho v\,\delta x = \left[\rho u\,\delta y + \frac{\partial}{\partial x}(\rho u)\,\delta x\,\delta y\right] + \left[\rho v\,\delta x + \frac{\partial}{\partial y}(\rho v)\,\delta y\,\delta x\right] ,$$

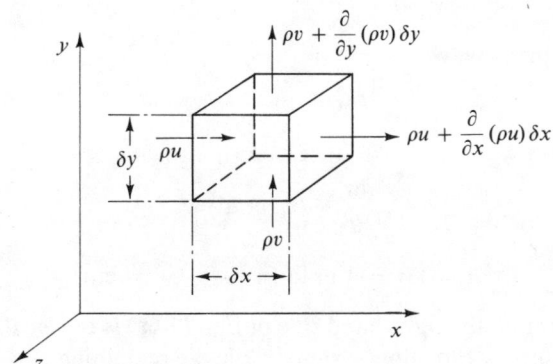

Fig. 5-2 Mass flux through a control volume

The *continuity equation* can simply be written as

$$\frac{\partial}{\partial x}(\rho u) + \frac{\partial}{\partial y}(\rho v) = 0 \tag{5-4a}$$

or

$$u\frac{\partial \rho}{\partial x} + v\frac{\partial \rho}{\partial y} = -\rho\left(\frac{\partial u}{\partial x} + \frac{\partial v}{\partial y}\right) \tag{5-4b}$$

In order to transform into any two- or three-dimensional coordinate system, Eqn (5-4a) can be written in the vector notation

$$\text{div } \rho\mathbf{w} = 0 \tag{5-5}$$

where the velocity vector is

$$\mathbf{w} = \mathbf{e}_1 w_1 + \mathbf{e}_2 w_2 + \mathbf{e}_3 w_3$$

$\mathbf{e}_1$, $\mathbf{e}_2$, and $\mathbf{e}_3$ being the unit vectors in the three orthogonal directions, and w_1, w_2, and w_3 the velocity components in these directions.

5-2-2 Equations of motion

Newton's second law, which states that for a constant mass the product of mass and acceleration of a body is equal to the sum of the external forces acting on that body, is used to derive the equations of motion. It is convenient at this stage to introduce the *substantial derivative* notation D/Dt. For a scalar quantity ϕ, where $\phi = \phi(x, y, t)$,

$$\frac{D\phi}{Dt} = \frac{\partial \phi}{\partial t} + \frac{\partial \phi}{\partial x}\frac{\partial x}{\partial t} + \frac{\partial \phi}{\partial y}\frac{\partial y}{\partial t}$$

or

$$\frac{D\phi}{Dt} = \frac{\partial \phi}{\partial t} + u\frac{\partial \phi}{\partial x} + v\frac{\partial \phi}{\partial y} \tag{5-6a}$$

For a vector quantity, $\mathbf{w}$,

$$\frac{D\mathbf{w}}{Dt} = \frac{\partial \mathbf{w}}{\partial t} + \frac{d\mathbf{w}}{dt} \tag{5-6b}$$

where

$$\frac{d\mathbf{w}}{dt} = (\mathbf{w}.\mathrm{grad})\,\mathbf{w}$$

and

$$(\mathbf{w}.\mathrm{grad})\,\mathbf{w} = \tfrac{1}{2}\,\mathrm{grad}\,(\mathbf{w}.\mathbf{w}) - \mathbf{w}\times\mathrm{curl}\,\mathbf{w}$$

The first term on the right-hand side of Eqn (5-6a) is the *local acceleration*, which is zero for a steady flow system, while the remaining two terms are the *convective acceleration* due to the motion between two regions at differing velocities.

The external forces acting on a fluid particle are of two types: *body forces* which act on the fluid element due to an external force field (for example, gravitational forces), and *surface forces* which are caused by the stresses acting on the fluid element.

Consider a fluid element subjected to a body force $\mathbf{F}$ per unit volume, where

$$\mathbf{F} = \mathbf{i}F_x + \mathbf{j}F_y \tag{5-7}$$

and a surface force, $\mathbf{T}$, per unit volume, where

$$\mathbf{T} = \mathbf{i}T_x + \mathbf{j}T_y \tag{5-8}$$

$\mathbf{i}$ and $\mathbf{j}$ being the unit vectors in the x and y directions. The equations of motion of the fluid element of mass $\rho(\delta x\,\delta y)$ can then be written as

$$\left.\begin{array}{l}\rho(\delta x\,\delta y)\dfrac{Du}{Dt} = F_x(\delta x\,\delta y) + T_x(\delta x\,\delta y) \\[2mm] \rho(\delta x\,\delta y)\dfrac{Dv}{Dt} = F_y(\delta x\,\delta y) + T_y(\delta x\,\delta y)\end{array}\right\} \tag{5-9}$$

Or in vector notation

$$\rho \frac{Dw}{Dt} = F + T \qquad (5\text{-}10)$$

5-2-3   Relationship between stress and rate of strain of a fluid element

When the stress system, generated by the surface forces, is linearly related to the *rate of strain* of the fluid element, the fluid is said to be *newtonian* and obeys *Stokes' law* of friction (which is analogous to Hooke's law for elastic solids). The fluid is said to be *isotropic* if the relation between the components of stress and the components of the rate of strain is the same in all directions. As most fluids of engineering interest are newtonian and isotropic, attention will be limited to those fluids. The reader interested in non-newtonian fluids is referred to specialist books in that area, for example, the work of Skelland[1]†.

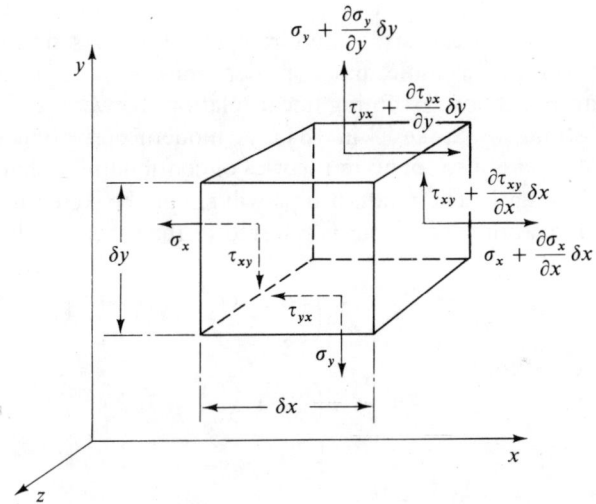

Fig. 5-3   Stress system on a fluid element

A description of the surface forces in terms of the stress system can be gained by considering a fluid element, of unit depth in the z direction, shown in Fig. 5-3.

In the x and y direction σ_x and σ_y are the normal stresses, respectively; τ_{xy} is the shear stress acting (along the y direction) on a plane normal to the x direction; and τ_{yx} is the shear stress acting (along the x direction) on a plane normal to the y direction. A simple force balance in the x and y directions

† Full details of references cited are given at the end of the chapter.

shows that the surface forces can be replaced by the normal and shear stresses, such that

$$T_x \, \delta x \, \delta y = \frac{\partial \sigma_x}{\partial x} \delta x \, \delta y + \frac{\partial \tau_{yx}}{\partial y} \delta y \, \delta x$$

$$T_y \, \delta x \, \delta y = \frac{\partial \sigma_y}{\partial y} \delta y \, \delta x + \frac{\partial \tau_{xy}}{\partial x} \delta x \, \delta y$$

For steady flow, Eqns (5-9) can now be expressed as

$$\left.\begin{aligned}
\rho\left(u \frac{\partial u}{\partial x} + v \frac{\partial u}{\partial y}\right) &= F_x + \frac{\partial \sigma_x}{\partial x} + \frac{\partial \tau_{yx}}{\partial y} \\
\rho\left(u \frac{\partial v}{\partial x} + v \frac{\partial v}{\partial y}\right) &= F_y + \frac{\partial \sigma_y}{\partial y} + \frac{\partial \tau_{xy}}{\partial x}
\end{aligned}\right\} \quad (5\text{-}11)$$

The relationship between stress and rate of strain was first derived by Navier[2] in 1827, from molecular considerations, but was later based on Newton's law of friction, assuming linear relations between the stresses and the rate of strain, by Stokes[3] in 1845. A modern derivation is given by Schlichting[4], and is also found in theories of continuum mechanics such as the work of Yung[5]. The relationships will simply be stated here, and for the formal derivations the reader is referred to the works quoted above:

$$\left.\begin{aligned}
\sigma_x &= -p - \tfrac{2}{3}\mu\left(\frac{\partial u}{\partial x} + \frac{\partial v}{\partial y}\right) + 2\mu \frac{\partial u}{\partial x} \\
\sigma_y &= -p - \tfrac{2}{3}\mu\left(\frac{\partial u}{\partial x} + \frac{\partial v}{\partial y}\right) + 2\mu \frac{\partial v}{\partial y} \\
\tau_{xy} &= \tau_{yx} = \mu\left(\frac{\partial v}{\partial x} + \frac{\partial u}{\partial y}\right)
\end{aligned}\right\} \quad (5\text{-}12)$$

where p is the *thermodynamic pressure* and μ is the constant of proportionality between shear stress and rate of strain called the *absolute or dynamic viscosity* of the fluid. For liquids, μ is nearly independent of pressure, but it decreases with increasing temperature. For gases, to a first approximation, μ can be regarded as independent of pressure, but it increases with temperature.

5-2-4 Navier–Stokes equations

Substitution of relationships (5-12) into Eqns (5-11) produces the *Navier–*

Stokes equations for a two-dimensional, compressible, steady flow system:

$$\rho\left(u\frac{\partial u}{\partial x}+v\frac{\partial u}{\partial y}\right)$$
$$= F_x-\frac{\partial p}{\partial x}+\frac{\partial}{\partial x}\left[2\mu\frac{\partial u}{\partial x}-\tfrac{2}{3}\mu\left(\frac{\partial u}{\partial x}+\frac{\partial v}{\partial y}\right)\right]+\frac{\partial}{\partial y}\left[\mu\left(\frac{\partial v}{\partial x}+\frac{\partial u}{\partial y}\right)\right]$$
$$\rho\left(u\frac{\partial v}{\partial x}+v\frac{\partial v}{\partial y}\right)$$
$$= F_y-\frac{\partial p}{\partial y}+\frac{\partial}{\partial y}\left[2\mu\frac{\partial v}{\partial y}-\tfrac{2}{3}\mu\left(\frac{\partial u}{\partial x}+\frac{\partial v}{\partial y}\right)\right]+\frac{\partial}{\partial x}\left[\mu\left(\frac{\partial v}{\partial x}+\frac{\partial u}{\partial y}\right)\right]$$

(5-13)

A gas can be treated as incompressible if its velocity is small compared with the velocity of sound. From a study of gas dynamics, for which the reader is referred to Shapiro[6], it can be shown that in isentropic flow (that is, neglecting heating and the effects of friction)

$$(\rho_0 - \rho_\infty)/\rho_0 = 1 - [1 + \tfrac{1}{2}(\gamma - 1)\mathbf{M}^2]^{1/(1-\gamma)}$$
$$\approx \tfrac{1}{2}\mathbf{M}^2 + \cdots$$

where the subscript ∞ refers to the free stream, and the subscript 0 refers to fluid brought to rest isentropically. The Mach number, $\mathbf{M}$, is defined as the ratio of the magnitude of the fluid velocity to the velocity of sound at the same conditions, and $\gamma(\equiv C_p/C_v)$ is the ratio of the specific heats at constant pressure and constant volume. The flow may be regarded as incompressible if $\tfrac{1}{2}\mathbf{M}^2 \ll 1$; for example for $\mathbf{M} = 0.3$ (which corresponds approximately to an air velocity of 330 ft/s or 96 m/s at atmospheric conditions) the error in ρ, neglecting density changes, is of order 5 per cent. For an *incompressible fluid*, where ρ is constant, Eqn (5-4a) reduces to

$$\frac{\partial u}{\partial x} + \frac{\partial v}{\partial y} = 0 \tag{5-14}$$

Changes in the viscosity of a gas can be neglected for temperature variations of less than 50°C (122°F). With liquids the variation of viscosity with temperature is greater, and for temperature changes above 10°C (50°F) an empirical viscosity law should be used.

Considerable simplifications to Eqns (5-13) can be made for an incompressible fluid with a constant viscosity, and under these conditions the system reduces to

$$\rho\left(u\frac{\partial u}{\partial x} + v\frac{\partial u}{\partial y}\right) = F_x - \frac{\partial p}{\partial x} + \mu\left(\frac{\partial^2 u}{\partial x^2} + \frac{\partial^2 u}{\partial y^2}\right)$$
$$\rho\left(u\frac{\partial v}{\partial x} + v\frac{\partial v}{\partial y}\right) = F_y - \frac{\partial p}{\partial y} + \mu\left(\frac{\partial^2 v}{\partial x^2} + \frac{\partial^2 v}{\partial y^2}\right)$$

(5-15)

or, in vector notation,

$$\rho \frac{D\mathbf{w}}{Dt} = \mathbf{F} - \mathrm{grad}\, p + \mu \nabla^2 \mathbf{w} \qquad (5\text{-}16)$$

where ∇^2 is the Laplacian operator, and for a cartesian coordinate system, $\nabla^2 = \partial^2/\partial x^2 + \partial^2/\partial y^2 + \partial^2/\partial z^2$.

Although the derivation has been for a two-dimensional cartesian co-ordinate system, Eqn (5-16) is valid for any three-dimensional coordinate system. For later use in this text it is convenient also to state Eqn (5-16) in terms of axisymmetric (all tangential derivatives being zero) cylindrical polar coordinates, which are useful for dealing with flow in pipes, annuli, and so on. The equations for the radial and axial components are stated below, but can readily be derived from Eqn (5-16). In polar coordinates, where $\mathbf{w} = v_r \mathbf{e}_r + v_z \mathbf{e}_z$ ($\mathbf{e}_r$ and $\mathbf{e}_z$ being unit vectors in the r and z directions), Eqn (5-16) can be expressed in component form as

$$\left.\begin{aligned}
\rho\left(v_r \frac{\partial v_r}{\partial r} + v_z \frac{\partial v_r}{\partial z} \right) &= F_r - \frac{\partial p}{\partial r} + \mu\left(\frac{\partial^2 v_r}{\partial r^2} + \frac{1}{r}\frac{\partial v_r}{\partial r} - \frac{v_r}{r^2} + \frac{\partial^2 v_r}{\partial z^2} \right) \\
\rho\left(v_r \frac{\partial v_z}{\partial r} + v_z \frac{\partial v_z}{\partial z} \right) &= F_z - \frac{\partial p}{\partial z} + \mu\left(\frac{\partial^2 v_z}{\partial r^2} + \frac{1}{r}\frac{\partial v_z}{\partial r} + \frac{\partial^2 v_z}{\partial z^2} \right)
\end{aligned}\right\} \quad (5\text{-}17)$$

From Eqn (5-5), the continuity equation—remembering that the fluid is incompressible—becomes

$$\frac{\partial v_r}{\partial r} + \frac{v_r}{r} + \frac{\partial v_z}{\partial z} = 0 \qquad (5\text{-}18)$$

For compressible flow problems it is necessary to introduce a relationship for evaluating the density. One such relationship which emerges from thermo-dynamic consideration is the equation of state for a perfect gas

$$p - \rho RT = 0 \qquad (5\text{-}19)$$

where R is the gas constant and T the absolute temperature of the gas. Except for fluid problems involving isothermal flow processes the fluid temperature must be calculated, and having derived the momentum equations for a fluid it is now necessary to formulate the energy equation. This draws up a balance of heat, work, and energy using the first law of thermodynamics.

5-2-5   Energy equation

In order to establish an energy balance for a fluid element it is necessary to enumerate the various energy forms. For simplicity it will be assumed that there are no chemical reactions or other heat sources within the fluid, and radiation will be neglected since it is normally very small in fluids, as shown in Chapter 8. A fluid particle will possess *internal energy* comprising an innate energy due to the random motion of the molecules, and a kinetic energy due to the motion of the particle. Across the particle boundaries heat will be transferred by conduction due to the temperature gradient (or molecular energy differences), and mechanical energy will be dissipated by viscous forces to cause generation of heat by friction. There will be work done by body forces, if these are significant, and for compressible fluids there will be an additional work term due to the changes of volume of the fluid element.

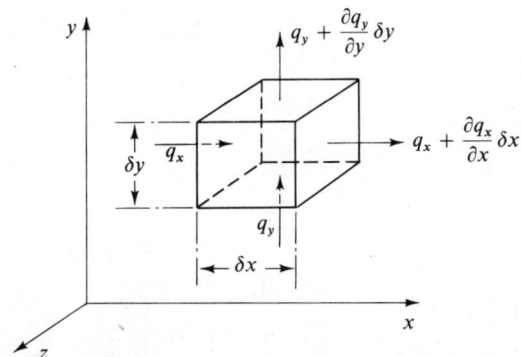

Fig. 5-4 Heat transfer by conduction into a fluid element

From the first law of thermodynamics an energy balance can be established on a fluid element of area $\delta x \, \delta y$ and unit depth, with a mass $\rho \, \delta x \, \delta y$. A quantity of heat δQ added to the particle in time δt will increase its internal energy by an amount δE while performing an amount of work δW. In differential terms

$$\frac{dQ}{dt} = \frac{dE}{dt} + \frac{dW}{dt} \tag{5-20}$$

Applying Fourier's law to the fluid element illustrated in Fig. 5-4, in an analogous way to that used for conduction in a solid, shows that the net rate of heat transfer per unit area by conduction in the x direction (that is δq_x *from* the element) is

$$\delta q_x = -\frac{\partial}{\partial x}\left(k \frac{\partial T}{\partial x}\right)\delta x$$

similarly for the y direction

$$\delta q_y = -\frac{\partial}{\partial y}\left(k \frac{\partial T}{\partial y}\right)\delta y$$

As radiation is absent, the only transfer of heat through the boundaries of the fluid particle occurs by conduction, and so the rate of heat flow *into* the element is

$$\frac{dQ}{dt} = \frac{\partial}{\partial x}\left(k\,\frac{\partial T}{\partial x}\right)\delta x\,\delta y + \frac{\partial}{\partial y}\left(k\,\frac{\partial T}{\partial y}\right)\delta y\,\delta x \qquad (5\text{-}21)$$

The internal energy consists of the specific internal energy, e per unit mass, and the kinetic energy, $\frac{1}{2}(u^2 + v^2)$ per unit mass so that

$$\frac{dE}{dt} = \rho\,\delta x\,\delta y\left[\frac{de}{dt} + \frac{1}{2}\frac{d}{dt}(u^2 + v^2)\right] \qquad (5\text{-}22)$$

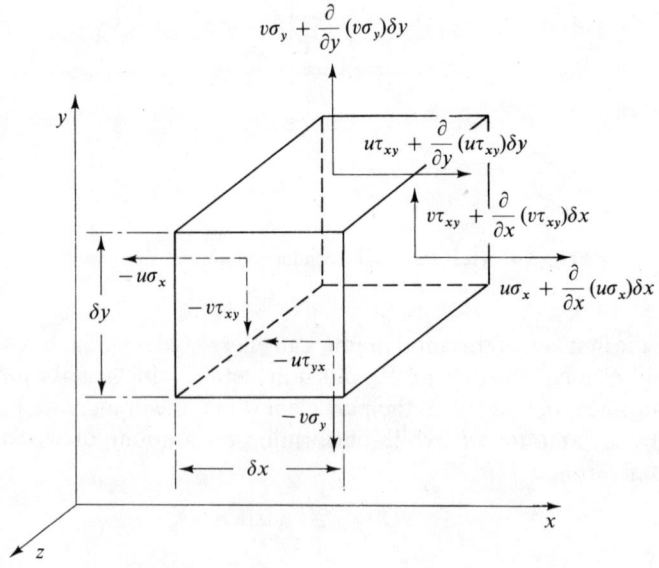

Fig. 5-5   Rate of frictional work on a fluid element

In order to represent the rate of frictional work, it is convenient to consider Fig. 5-5. The rate of work performed *on* the element due to the normal stress in the x direction is simply

$$\left[-u\sigma_x + u\sigma_x + \frac{\partial}{\partial x}(u\sigma_x)\,\delta x\right]\delta y$$

which is

$$\frac{\partial}{\partial x}(u\sigma_x)\,\delta x\,\delta y$$

Thus, the total rate of work due to friction, dW_f/dt, applied *to* the element is

$$\frac{dW_f}{dt} = \left[\frac{\partial}{\partial x}(u\sigma_x + v\tau_{xy}) + \frac{\partial}{\partial y}(v\sigma_y + u\tau_{yx})\right]\delta x\,\delta y \qquad (5\text{-}23)$$

The rate of work done by the body forces acting on the fluid particle, represented by dW_F/dt, is simply

$$\frac{dW_F}{dt} = (F_x u + F_y v)\,\delta x\,\delta y \qquad (5\text{-}24)$$

Hence, the total rate of work done *by* the particle is given by

$$\frac{dW}{dt} = -\frac{dW_f}{dt} - \frac{dW_F}{dt} \qquad (5\text{-}25)$$

From Eqns (5-11), using Eqns (5-23) and (5-24), Eqn (5-25) becomes

$$-\frac{dW}{dt} = \rho\left[u\left(u\frac{\partial u}{\partial x} + v\frac{\partial u}{\partial y}\right) + v\left(u\frac{\partial v}{\partial x} + v\frac{\partial v}{\partial y}\right)\right]\delta x\,\delta y$$

$$+ \left[\sigma_x\frac{\partial u}{\partial x} + \tau_{xy}\frac{\partial v}{\partial x} + \sigma_y\frac{\partial v}{\partial y} + \tau_{yx}\frac{\partial u}{\partial y}\right]\delta x\,\delta y \qquad (5\text{-}26)$$

The expression within the first set of square brackets in Eqn (5-26) can be simplified by remembering that, for a steady flow system,

$$\frac{d}{dt}(u^2) = u\frac{\partial}{\partial x}(u^2) + v\frac{\partial}{\partial y}(u^2)$$

The expression within the second set of square brackets in Eqn (5-26) can be simplified by using Eqns (5-12), and Eqn (5-26) can be rewritten as

$$-\frac{dW}{dt} = \left\{\tfrac{1}{2}\rho\frac{d}{dt}(u^2 + v^2)\right\}\delta x\,\delta y + \left[\mu\Phi - p\left(\frac{\partial u}{\partial x} + \frac{\partial v}{\partial y}\right)\right]\delta x\,\delta y \qquad (5\text{-}27)$$

where Φ represents the *dissipation function* given by

$$\Phi = 2\left[\left(\frac{\partial u}{\partial x}\right)^2 + \left(\frac{\partial v}{\partial y}\right)^2\right] + \left(\frac{\partial v}{\partial x} + \frac{\partial u}{\partial y}\right)^2 - \frac{2}{3}\left(\frac{\partial u}{\partial x} + \frac{\partial v}{\partial y}\right)^2 \qquad (5\text{-}28)$$

Substitution of Eqns (5-21), (5-22), and (5-27) into Eqn (5-20) produces the result

$$\rho\frac{de}{dt} = \frac{\partial}{\partial x}\left(k\frac{\partial T}{\partial x}\right) + \frac{\partial}{\partial y}\left(k\frac{\partial T}{\partial y}\right) - p\left(\frac{\partial u}{\partial x} + \frac{\partial v}{\partial y}\right) + \mu\Phi \qquad (5\text{-}29)$$

It now remains to express the specific internal energy in terms of convenient thermodynamic parameters. [The reader not familiar with thermodynamics is advised to refer to Keenan[7] or van Wylen[8].] The specific enthalpy, i, is defined as

$$i \equiv e + p/\rho \tag{5-30}$$

hence
$$\frac{de}{dt} = \frac{di}{dt} - \frac{1}{\rho}\left(\frac{dp}{dt} - \frac{p}{\rho}\frac{dp}{dt}\right) \tag{5-31}$$

For a pure homogeneous substance with specific entropy s and absolute temperature T,

$$di = T\,ds + dp/\rho \tag{5-32}$$

where
$$ds = \left(\frac{\partial s}{\partial T}\right)_p dT + \left(\frac{\partial s}{\partial p}\right)_T dp \tag{5-33}$$

and from Maxwell's relations

$$\left(\frac{\partial s}{\partial T}\right)_p = \frac{1}{T}\left(\frac{\partial i}{\partial T}\right)_p \tag{5-34}$$

$$\left(\frac{\partial s}{\partial p}\right)_T = \frac{1}{\rho^2}\left(\frac{\partial \rho}{\partial T}\right)_p \tag{5-35}$$

Defining the *specific heat at constant pressure*, C_p, by

$$C_p \equiv \left(\frac{\partial i}{\partial T}\right)_p \tag{5-36}$$

and the *volume expansion coefficient*, β, by

$$\beta \equiv -\frac{1}{\rho}\left(\frac{\partial \rho}{\partial T}\right)_p \tag{5-37}$$

Eqn (5-32) can be rewritten in the form

$$di = C_p\,dT + dp\,\frac{1 - T\beta}{\rho} \tag{5-38}$$

From Eqn (5-4b)

$$\frac{d\rho}{dt} = -\rho\left(\frac{\partial u}{\partial x} + \frac{\partial v}{\partial y}\right) \tag{5-39}$$

and substituting Eqns (5-38) and (5-39) into (5-31) produces

$$\frac{de}{dt} = C_p\frac{dT}{dt} - \frac{T\beta}{\rho}\frac{dp}{dt} - \frac{p}{\rho}\left(\frac{\partial u}{\partial x} + \frac{\partial v}{\partial y}\right) \tag{5-40}$$

Equation (5-29) can now be written

$$\rho C_p \frac{dT}{dt} = T\beta \frac{dp}{dt} + \frac{\partial}{\partial x}\left(k \frac{\partial T}{\partial x}\right) + \frac{\partial}{\partial y}\left(k \frac{\partial T}{\partial y}\right) + \mu\Phi \qquad (5\text{-}41)$$

which is the *energy equation for the two-dimensional steady flow of a compressible, variable property fluid*. In general, C_p, β, and k will be temperature dependent, but simplifications can be made for a *perfect gas*, where Eqn (5-19) is valid, or for an *incompressible fluid*, where ρ is assumed to be a constant independent of temperature and pressure. If k is also assumed to be constant then for a perfect gas, where $\beta = 1/T$,

$$\rho C_p\left(u \frac{\partial T}{\partial x} + v \frac{\partial T}{\partial y}\right) = \left(u \frac{\partial p}{\partial x} + v \frac{\partial p}{\partial y}\right) + k\left(\frac{\partial^2 T}{\partial x^2} + \frac{\partial^2 T}{\partial y^2}\right) + \mu\Phi \qquad (5\text{-}42a)$$

and for an incompressible fluid where $\beta = 0$,

$$\rho C_p\left(u \frac{\partial T}{\partial x} + v \frac{\partial T}{\partial y}\right) = k\left(\frac{\partial^2 T}{\partial x^2} + \frac{\partial^2 T}{\partial y^2}\right) + \mu\Phi \qquad (5\text{-}42b)$$

For many flows frictional dissipation, represented by $\mu\Phi$, is small and can be neglected. Under these conditions Eqn (5-42b) can be simplified and expressed in vector notation as

$$\rho C_p(\mathbf{w}\cdot\text{grad } T) = k\nabla^2 T$$

or in axisymmetric cylindrical polar coordinates

$$\rho C_p\left(v_r \frac{\partial T}{\partial r} + v_z \frac{\partial T}{\partial z}\right) = k\left(\frac{\partial^2 T}{\partial r^2} + \frac{1}{r}\frac{\partial T}{\partial r} + \frac{\partial^2 T}{\partial z^2}\right) \qquad (5\text{-}43)$$

Despite the mathematical complexity of the energy equation, the physical structure can be elicited from consideration of the form of Eqn (5-41). In words, the convective heat transfer due to the fluid motion is balanced by the work due to volumetric changes, the heat conducted through the fluid, and the viscous dissipation of the kinetic energy of the fluid.

While the energy equation, the Navier–Stokes equations and the continuity equation, together with the appropriate relationships for the fluid properties, provide a comprehensive description of energy transfer in a moving fluid, the resulting system of partial differential equations presents formidable mathematical difficulties in solution. The number of solutions of the full equations is limited to situations of mathematical interest rather than of engineering importance. In subsequent sections it will be shown how the *boundary layer assumptions* can simplify the equations, and solutions will be presented in Chapter 6, but it is now convenient to use the theory of similarity to derive the dimensionless parameters that give a deeper physical insight into the problems of fluid flow and heat transfer.

5-3   Theory of similarity

It is convenient to rewrite the basic relationship to permit a complete presentation of the system of equations under discussion.

To make the form suitable for free and forced convection, the body force will be considered due to a gravitational field, $\mathbf{g} = \mathbf{i}g \cos \phi + \mathbf{j}g \sin \phi$, at an incidence of ϕ to the x direction. Using θ to denote the temperature difference between the fluid particle and its surroundings, that is, $\theta = T - T_\infty$ where the subscript ∞ refers to a temperature datum (usually that of the free stream or any other convenient reference), the relative volumetric change is simply $\beta\theta$. The gravitational force per unit volume is $\rho\mathbf{g}$, and so the body force resulting from a relative change in volume is simply

$$\mathbf{F} = \rho\beta\theta\mathbf{g} \tag{5-44}$$

Rewriting Eqns (5-4a), (5-13), and (5-42) for a perfect gas with constant properties, and introducing Eqn (5-44), the equations become

$$\frac{\partial}{\partial x}(\rho u) + \frac{\partial}{\partial y}(\rho v) = 0 \tag{5-45}$$

$$\rho\left(u\frac{\partial u}{\partial x} + v\frac{\partial u}{\partial y}\right) = \rho g\beta\theta \cos \phi - \frac{\partial p}{\partial x}$$
$$+ \mu\left[\left(\frac{\partial^2 u}{\partial x^2} + \frac{\partial^2 u}{\partial y^2}\right) + \frac{1}{3}\frac{\partial}{\partial x}\left(\frac{\partial u}{\partial x} + \frac{\partial v}{\partial y}\right)\right] \tag{5-46}$$

$$\rho\left(u\frac{\partial v}{\partial x} + v\frac{\partial v}{\partial y}\right) = \rho g\beta\theta \sin \phi - \frac{\partial p}{\partial y}$$
$$+ \mu\left[\left(\frac{\partial^2 u}{\partial x^2} + \frac{\partial^2 u}{\partial y^2}\right) + \frac{1}{3}\frac{\partial}{\partial y}\left(\frac{\partial u}{\partial x} + \frac{\partial v}{\partial y}\right)\right] \tag{5-47}$$

$$\rho C_\mathrm{p}\left(u\frac{\partial T}{\partial x} + v\frac{\partial T}{\partial y}\right) = \left(u\frac{\partial p}{\partial x} + v\frac{\partial p}{\partial y}\right) + k\left(\frac{\partial^2 T}{\partial x^2} + \frac{\partial^2 T}{\partial y^2}\right) + \mu\Phi \tag{5-48}$$

$$p - \rho RT = 0 \tag{5-49}$$

The five dependent variables u, v, p, ρ, and T are described by the above equations together with the relevant boundary conditions. The equations can be made dimensionless by referring the variables to characteristic quantities, that is,

$$\tilde{x} = x/l, \qquad \tilde{y} = y/l$$

where l is a characteristic linear dimension—for example, the diameter of a pipe or the length of a plate. Similarly,

$$\tilde{u} = u/U_\infty, \qquad \tilde{v} = v/U_\infty$$

where U_∞ is a convenient reference velocity,

$$\tilde{p} = p/\rho_\infty U_\infty^2 \quad \text{and} \quad \tilde{\rho} = \rho/\rho_\infty$$

where ρ_∞ is the density at the same conditions as U_∞ and $\frac{1}{2}\rho_\infty U_\infty^2$ is the *dynamic head* at these conditions.

For the temperature,

$$\tilde{\theta} = (T - T_\infty)/\Delta T$$

where ΔT is a suitable temperature difference, for example, $\Delta T = T_s - T_\infty$, in which T_s is the temperature of a heated surface cooled by a fluid at temperature T_∞.

Substitution of these dimensionless quantities into Eqns (5-46) and (5-48) gives, for the case of $\phi = 0$,

$$\frac{\rho_\infty U_\infty^2}{l}\left\{\tilde{\rho}\left(\tilde{u}\frac{\partial \tilde{u}}{\partial \tilde{x}} + \tilde{v}\frac{\partial \tilde{u}}{\partial \tilde{y}}\right)\right\} = \frac{\rho_\infty U_\infty^2}{l}\frac{\partial \tilde{p}}{\partial \tilde{x}} + \rho_\infty g \beta_\infty \Delta T \tilde{\rho}\tilde{\theta}\tilde{\beta}$$

$$+ \frac{\mu U_\infty}{l^2}\left\{\frac{\partial^2 \tilde{u}}{\partial \tilde{x}^2} + \frac{\partial^2 \tilde{v}}{\partial \tilde{y}^2} + \frac{1}{3}\frac{\partial}{\partial \tilde{x}}\left(\frac{\partial \tilde{u}}{\partial \tilde{x}} + \frac{\partial \tilde{v}}{\partial \tilde{y}}\right)\right\}$$

$$(5\text{-}50)$$

$$\frac{\rho_\infty C_p U_\infty \Delta T}{l}\left\{\tilde{\rho}\left(\tilde{u}\frac{\partial \tilde{\theta}}{\partial \tilde{x}} + \tilde{v}\frac{\partial \tilde{\theta}}{\partial \tilde{y}}\right)\right\} = \frac{\rho_\infty U_\infty^3}{l}\left\{\tilde{u}\frac{\partial \tilde{p}}{\partial \tilde{x}} + \tilde{v}\frac{\partial \tilde{p}}{\partial \tilde{y}}\right\} + \frac{k\,\Delta T}{l^2}\left\{\frac{\partial^2 \tilde{\theta}}{\partial \tilde{x}^2} + \frac{\partial^2 \tilde{\theta}}{\partial \tilde{y}^2}\right\}$$

$$+ \frac{\mu U_\infty^2}{l^2}\tilde{\Phi}$$

$$(5\text{-}51)$$

where

$$\tilde{\Phi} = 2\left\{\left(\frac{\partial \tilde{u}}{\partial \tilde{x}}\right)^2 + \cdots\right\} = \frac{l^2}{U_\infty^2}\Phi$$

Dividing Eqn (5-50) by $(\rho_\infty U_\infty^2/l)$ produces the two dimensionless groups $[\mu/(\rho_\infty U_\infty l)]$ and $[g\beta_\infty \Delta T l/U_\infty^2]$. The first group is the ratio of viscous to inertial forces, and its reciprocal is familiar as the Reynolds number **R** where

$$\mathbf{R} \equiv U_\infty l/v_\infty \qquad (5\text{-}52)$$

and $v_\infty \equiv \mu/\rho_\infty$ is called the *kinematic viscosity* of the fluid.

The second group, which is the ratio of buoyancy forces to inertial forces, can be written as

$$g\beta_\infty \Delta T l/U_\infty^2 = \mathbf{G}/\mathbf{R}^2$$

where **G**, the Grashof number, is defined as

$$\mathbf{G} \equiv g\beta_\infty \Delta T l^3/v_\infty^2 \qquad (5\text{-}53)$$

Dividing Eqn (5-51) by $[\rho_\infty C_p U_\infty \Delta T/l]$ produces the three dimensionless groups $[U_\infty^2/(C_p \Delta T)]$, $[k/(\rho_\infty C_p U_\infty l)]$, and $[\mu U_\infty/(\rho_\infty C_p l\, \Delta T)]$. The first is the ratio of the *dynamic temperature* $\frac{1}{2}U_\infty^2/C_p$, due to fluid motion, to the *static temperature* difference, ΔT. The ratio is termed the Eckert number **E**, where

$$\mathbf{E} \equiv U_\infty^2/(C_p \, \Delta T) \tag{5-54}$$

Inverting the second group produces the result

$$\rho_\infty C_p U_\infty l/k = \mathbf{PR}$$

where the Prandtl number, **P**, is given by

$$\mathbf{P} \equiv \mu C_p/k \tag{5-55}$$

The Prandtl number is solely a function of the fluid properties and can be regarded as the ratio of kinematic viscosity, v_∞, to the *thermal diffusivity*, α_∞, where

$$\alpha_\infty = k/(\rho_\infty C_p) \tag{5-56}$$

The final group is not an independent parameter but is merely a function of the Reynolds and Eckert numbers, such that

$$\mu U_\infty/(\rho_\infty C_p l \, \Delta T) = \mathbf{E/R}$$

It can now be seen that solutions of Eqns (5-50) and (5-51) depend on the four independent dimensionless groups defined by Eqns (5-52)–(5-55). The units of the quantities contained within these groups can be found from Table 5-1, but care must be exercised in the conversion of work and heat units.

Table 5-1   Conversion factors for British and SI units

Quantity	BS units	SI units	Conversion factor[a]
Specific heat, C_p	Btu/lb °R	kJ/kg K	4·1868
Thermal conductivity, k	Btu/ft h °R	W/m K	1·7307
Volume expansion coefficient, β	°R^{-1}	K^{-1}	1·8000
Absolute viscosity, μ	lb/ft h	kg/m s	4·1333 × 10^{-4}
	lb/ft s	kg/m s	1·4882
Kinematic viscosity, v	ft^2/h	m^2/s	2·5806 × 10^{-5}
	ft^2/s	m^2/s	0·09290
Density, ρ	lb/ft^3	kg/m^3	16·023

Note: 1 W = 1 J/s = 1 N m/s.

[a] To convert from British to SI units, multiply by factor. To convert from SI to British units, divide by factor.

Example. Calculate the Eckert number for the flow of air, with a free stream temperature and velocity of 100°C (212°F) and 200 m/s (656 ft/s), respectively, over an aerofoil at a temperature of 40°C (104°F). The specific heat of air is given as 1·01 kJ/kg K (0·241 Btu/lb °R).

In SI units 1 kJ/kg K is equivalent to 1,000 m²/s² K, and so

$$\frac{U_\infty^2}{C_\mathrm{p}} = \frac{200^2}{1\cdot01 \times 1,000} = 39\cdot6°C$$

$$\Delta T = 100 - 40 = 60°C$$

hence

$$\mathbf{E} = \frac{U_\infty^2}{C_\mathrm{p}\,\Delta T} = \frac{39\cdot6}{60} = 0\cdot66$$

In the British system the unit for specific heat is Btu/lb °R, and we must convert this to the unit ft²/s² °R by multiplying by g_c (32·2 lb ft/s² lbf) and Joules equivalent, J (778 ft lbf/Btu). Thus,

$$\frac{U_\infty^2}{C_\mathrm{p}} = \frac{656^2}{0\cdot241 \times 32\cdot2 \times 778} = 71\cdot4°F$$

$$\Delta T = 212 - 104 = 108°F$$

hence

$$\mathbf{E} = \frac{71\cdot4}{108} = 0\cdot66$$

In many practical problems, it is required to calculate the heat flux through the wall, or to estimate the heat transfer coefficient, h, given in Eqn (5-3). The dimensionless parameter describing the wall heat flux can be found by writing the temperature gradient at the wall as

$$\frac{l}{T_s - T_\infty}\left(\frac{\partial T}{\partial n}\right)_s = -\frac{l}{T_s - T_\infty}\frac{q_s}{k}$$

from Eqn (5-2).

The Nusselt number, **N**, is defined by

$$\mathbf{N} \equiv \frac{lq_s}{k(T_s - T_\infty)} \tag{5-57}$$

and from Eqn (5-1) it can be seen that

$$h = k\mathbf{N}/l$$

A similar argument can be employed for the skin friction, where the wall shear stress, τ_s, is given by

$$\tau_s = \mu\left(\frac{\partial u}{\partial y}\right)_s$$

such that

$$\frac{l}{U_\infty}\left(\frac{\partial u}{\partial y}\right)_s = \frac{\tau_s l}{\mu U_\infty} = \tfrac{1}{2}\mathbf{R}\mathbf{C}_\mathrm{f}$$

where the skin friction coefficient, C_f, is defined by

$$C_f \equiv \frac{\tau_s}{\frac{1}{2}\rho_\infty U_\infty^2} \tag{5-58}$$

It will be shown in Chapter 6 that a strong correlation exists between N and C_f.

In general, for a given spatial location,

$$N = N(R, P, G, E)$$

but special solutions can be found where one or more of the independent dimensionless groups can be neglected. For many practical systems the dynamic temperature is much less than the static temperature difference, and the effect of Eckert number can be ignored. Further, if the ratio of buoyancy forces to inertial forces is small, the effect of Grashof number can be ignored and heat is principally transferred by *forced convection*. Conversely, if the buoyancy forces are relatively large, the flow becomes independent of the Reynolds number and heat is transferred by *free* or *natural convection*. Hence,

$$\text{forced convection:} \quad N = N(R, P) \tag{5-59a}$$

$$\text{free convection:} \quad N = N(G, P) \tag{5-59b}$$

The above dimensionless parameters can be derived without reference to the system of Eqns (5-45)–(5-49) purely by *dimensional analysis*. Also, other groups can be formed from the four basic independent parameters, and any reader wishing to pursue this particular topic is recommended to read the books on dimensional analysis by Buckingham[9], Bridgeman[10], and Langhaar[11].

The object of using the theory of similarity was to identify the parameters that dominate heat transfer theory and control the solution of the mathematical equations. Now that this has been done it is necessary to simplify the equations before attempting to find solutions. Before the introduction of the simplifying assumptions, it is expedient to introduce the physical phenomenon that occurs in the majority of engineering situations: turbulence.

5-4 Turbulence and the time-average equations

Equations (5-45)–(5-48) were derived for steady *laminar* flow, where the velocity at any point does not vary with time; however, real fluids exhibit a characteristic feature called *turbulence*. Osborne Reynolds[12] showed, by injecting dye into the water flowing through a transparent pipe, that at low flow rates the streamlines were parallel, that is, the dye travelled in straight lines. As the flow rate was increased the streamlines became sinuous and distorted, and the well-ordered laminar flow gave way to turbulent motion.

Velocity traverses across the pipe also revealed that the velocity profiles of laminar and turbulent flow were significantly different.

The onset of turbulence is caused basically by fluid instability. A small fluctuation will be damped out by viscous effects if the flow is stable and the resulting flow will continue to be laminar. However, under certain conditions a small fluctuation will be amplified, and the laminar flow will then break down into three-dimensional eddying motion superimposed on the original streamline flow. The *transition* from laminar to turbulent flow depends on a number of factors (see Schlichting[4] for a discussion of these factors)—for example fluid velocity, pressure gradients, body forces, system geometry, surface roughness, fluid properties, free stream turbulence, and so on— but Reynolds was able to find empirical relationships for the onset of turbulence in pipes and channels.

The critical value of Reynolds number for which the flow becomes turbulent, $\mathbf{R}_{d, crit}$, was found to be

$$\mathbf{R}_{d, crit} = (\bar{U}d/v)_{crit} \approx 2{,}300$$

d being the pipe diameter and $\bar{U}$ being the bulk mean velocity through the pipe, that is, the volumetric flow rate divided by the pipe area. It was later found that the critical Reynolds number can vary between 2,000 and 40,000, the latter case applying to quiescent inlet conditions.

Experiments by later investigators[13, 14] showed that for flow over a smooth flat plate

$$\mathbf{R}_{x, crit} = (U_{\infty}x/v)_{crit} \approx 3 \times 10^5 \text{ to } 10^6$$

where x denotes the distance from the leading edge of the plate, and the subscript ∞ refers to conditions in the 'free stream' away from the plate surface. For Reynolds numbers below the critical value, disturbances due, for example, to surface irregularities or free stream turbulence are damped out by viscous action. Above the critical values the disturbances propagate causing the main flow to become unstable, and the laminar flow breaks down into eddy motion.

Equations (5-45)–(5-48) were derived for steady flow, whereas in turbulent flow the fluctuations superimposed upon the mean velocity vary continuously with time. The instantaneous velocity components u and v can be represented by

$$u = \bar{u} + u', \qquad v = \bar{v} + v' \tag{5-60}$$

where the bars represent time-average velocities at a point, and the primes denote the time-dependent fluctuating components. Similarly, for the pressure, density, and temperature:

$$p = \bar{p} + p', \qquad \rho = \bar{\rho} + \rho', \qquad T = \bar{T} + T' \tag{5-61}$$

Thus a thermocouple, with a sufficiently small time constant, placed at a fixed location in a turbulent fluid would record an output similar to that shown in Fig. 5-6.

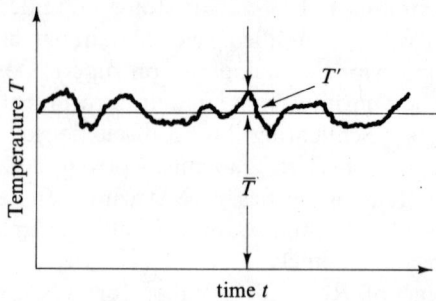

Fig. 5-6 Turbulent temperature variation with time

The time-average components, $\bar{\phi}$, are obtained by integrating the component ϕ at a fixed spatial location over a period of time, Δt. Δt is sufficiently long compared with the period of the fluctuations such that $\bar{\phi}$ is independent of t:

$$\bar{\phi} = \frac{1}{\Delta t} \int_t^{t+\Delta t} \phi \, dt \tag{5-62}$$

and, as a consequence of this definition,

$$\overline{\phi'} = \overline{u'} = \overline{\rho'} = \overline{p'} = \overline{T'} = 0 \tag{5-63}$$

and it is left as an exercise for the reader to prove that, for any two dependent variables ϕ and ψ, where $\phi = \bar{\phi} + \phi'$ and $\psi = \bar{\psi} + \psi'$,

$$\left. \begin{array}{c} \overline{\bar{\phi}} = \bar{\phi}, \qquad \overline{\phi + \psi} = \bar{\phi} + \bar{\psi}, \qquad \overline{\bar{\phi}\psi} = \bar{\phi}\bar{\psi} \\ \\ \overline{\partial\phi/\partial n} = (\partial/\partial n)\bar{\phi} \end{array} \right\} \tag{5-64}$$

and

It now remains to express the equation of motion in terms of the time-average of the velocity components. For incompressible flow, the continuity equation [Eqn (5-14)] becomes

$$\frac{\partial u}{\partial x} + \frac{\partial v}{\partial y} = \frac{\partial \bar{u}}{\partial x} + \frac{\partial \bar{v}}{\partial y} + \frac{\partial u'}{\partial x} + \frac{\partial v'}{\partial y} \tag{5-65}$$

and averaging over a period of time

$$\frac{\overline{\partial u'}}{\partial x} = \frac{\overline{\partial v'}}{\partial y} = 0$$

hence

$$\frac{\partial \bar{u}}{\partial x} + \frac{\partial \bar{v}}{\partial y} = 0$$

and

$$\frac{\partial u'}{\partial y} + \frac{\partial v'}{\partial y} = 0$$

(5-66)

Ignoring body forces, Eqns (5-15) can be written

$$\rho\left[\frac{\partial}{\partial x}(u^2) + \frac{\partial}{\partial y}(uv)\right] = -\frac{\partial p}{\partial x} + \mu\left[\frac{\partial^2 u}{\partial x^2} + \frac{\partial^2 u}{\partial y^2}\right]$$

$$\rho\left[\frac{\partial}{\partial x}(uv) + \frac{\partial}{\partial y}(v^2)\right] = -\frac{\partial p}{\partial y} + \mu\left[\frac{\partial^2 v}{\partial x^2} + \frac{\partial^2 v}{\partial y^2}\right]$$

(5-67)

and a time average of Eqns (5-67), using the relations given by Eqns (5-60)–(5-64), yields the following result

$$\rho\left[\bar{u}\frac{\partial \bar{u}}{\partial x} + \bar{v}\frac{\partial \bar{u}}{\partial y}\right] = -\frac{\partial \bar{p}}{\partial x} + \frac{\partial}{\partial x}\left(\mu\frac{\partial \bar{u}}{\partial x} - \rho\overline{u'^2}\right) + \frac{\partial}{\partial y}\left(\mu\frac{\partial \bar{u}}{\partial y} - \rho\overline{u'v'}\right)$$

$$\rho\left[\bar{u}\frac{\partial \bar{v}}{\partial x} + \bar{v}\frac{\partial \bar{v}}{\partial y}\right] = -\frac{\partial \bar{p}}{\partial y} + \frac{\partial}{\partial x}\left(\mu\frac{\partial \bar{v}}{\partial x} - \rho\overline{u'v'}\right) + \frac{\partial}{\partial y}\left(\mu\frac{\partial \bar{v}}{\partial y} - \rho\overline{v'^2}\right)$$

(5-68)

The turbulence terms, generated from the convective accelerations, have been taken over to the right-hand side of Eqns (5-68) in order to emphasize their role as *apparent stresses*, which are usually termed Reynolds stresses, and Eqns (5-68) are often termed the *Reynolds equations*.[15] Comparing Eqns (5-68) with Eqns (5-11), the resultant stresses are given by

$$\sigma_x = -p + 2\mu\frac{\partial \bar{u}}{\partial x} - \rho\overline{u'^2}$$

$$\sigma_y = -p + 2\mu\frac{\partial \bar{v}}{\partial y} - \rho\overline{v'^2}$$

$$\tau_{xy} = \tau_{yx} = \mu\left(\frac{\partial \bar{u}}{\partial y} + \frac{\partial \bar{v}}{\partial x}\right) - \rho\overline{u'v'}$$

(5-69)

The Reynolds stresses σ'_x, σ'_y, and τ'_{xy} are $-\rho\overline{u'^2}$, $-\rho\overline{v'^2}$, and $-\rho\overline{u'v'}$ respectively.

The physical significance of these apparent stresses can be seen by considering a plane of area δA normal to the x direction through which, in time δt, a mass of fluid $\rho u\,\delta A\,\delta t$ passes. The momentum transfer in the x direction, δJ_x, is therefore $\rho u^2\,\delta A\,\delta t$ and in the y direction $\delta J_y = \rho uv\,\delta A\,\delta t$. The time-averages per unit time of the momenta are given by $\overline{\delta J_x} = \overline{\rho u^2}\,\delta A$ and $\overline{\delta J_y} = \overline{\rho uv}\,\delta A$ and, from Eqns (5-60)–(5-64), $\overline{\delta J_x} = \rho(\bar{u}^2 + \overline{u'^2})\,\delta A$ and $\overline{\delta J_y} = \rho(\overline{uv} + \overline{u'v'})\,\delta A$. There must be an equal and opposite force to the rate of change of momentum exerted on the area by its surroundings, and the forces per unit area (or stresses) are equal to $-\rho(\bar{u}^2 + \overline{u'^2})$ in the x direction and $-\rho(\bar{u}\bar{v} + \overline{u'v'})$ in the y direction. Thus, the fluctuations give rise to the apparent stresses, on the area considered, of $\sigma'_x = -\rho\overline{u'^2}$ and $\tau'_{xy} = -\rho\overline{u'v'}$. By similar arguments for the y direction, it can be seen that $\sigma'_y = -\rho\overline{v'^2}$ and $\tau'_{yx} = -\rho\overline{v'u'}$.

A similar argument can be applied to the energy equation which, for laminar incompressible flow (rewriting Eqn (5-43) and neglecting dissipative effects), is

$$\rho C_p\left(u\frac{\partial T}{\partial x} + v\frac{\partial T}{\partial y}\right) = k\left(\frac{\partial^2 T}{\partial x^2} + \frac{\partial^2 T}{\partial y^2}\right) \tag{5-70}$$

Applying Eqns (5-60)–(5-64) to Eqn (5-70) yields the time-average energy equation for incompressible turbulent flow:

$$\rho C_p\left(\bar{u}\frac{\partial \bar{T}}{\partial x} + \bar{v}\frac{\partial \bar{T}}{\partial y}\right) = \frac{\partial}{\partial x}\left(k\frac{\partial \bar{T}}{\partial x} - \rho C_p\overline{u'T'}\right)$$

$$+ \frac{\partial}{\partial y}\left(k\frac{\partial \bar{T}}{\partial y} - \rho C_p\overline{v'T'}\right) \tag{5-71}$$

In practice the term $\rho\overline{u'v'}$ is of opposite sign to the velocity gradient and can be several orders of magnitude greater than the laminar shear stress. Its effect, therefore, is to increase greatly the shear stress acting on the fluid particle. As the stresses in the momentum equation are increased by the fluctuating velocities, the heat fluxes in the energy equation are increased in a similar manner by the transport of energy due to the transverse mixing of fluid elements. The increase in fluxes due to turbulence can be compared to the increase in fluxes due to an increase in the transport properties, μ and k of the fluid.

While few exact solutions of the Navier–Stokes equations have been discovered, no solutions of the Reynolds equations exist owing to the lack of knowledge concerning the structure of turbulence itself and its relationship to the mean values. This is not to say that problems involving turbulent flow are intractable, and many meaningful predictions have been made by making judicious assumptions and by using semi-empirical relationships for the turbulence terms. Attention will now be focused on the boundary layer approximations and the resulting equations that can be used for laminar or turbulent flow.

5-5 Boundary layer equations

5-5-1 Boundary layer concept

Exact solutions of the Navier–Stokes and energy equations are few in number, and for most practical problems it is necessary either to idealize the physical system in order to formulate tractable mathematical problems or to simplify the mathematics in order to obtain physically realistic answers. It is the object of this chapter to reduce the basic equations—employing physical arguments and using the results of solutions of the exact equations— to a form that can be readily solved by analytical, approximate, or numerical techniques.

The Navier–Stokes equations represent a balance of inertia, viscous, pressure and body forces, and simplifications can be introduced by ignoring one or more of these force terms. For low-speed flow, or *creeping motion*, which occurs for example in oil-lubricated journal bearings, the inertia and body forces can be neglected in comparison with the viscous and pressure forces. In aerodynamics the flow is often considered *inviscid* or *potential* so that only the pressure and inertia forces are important, and the resulting equations are known as the Euler equations. However, for flow over a solid body, irrespective of how small the viscosity of the fluid, the transition from zero velocity at the wall to the free stream value away from the wall is associated with velocity gradients that make the viscous forces significant.

free stream velocity

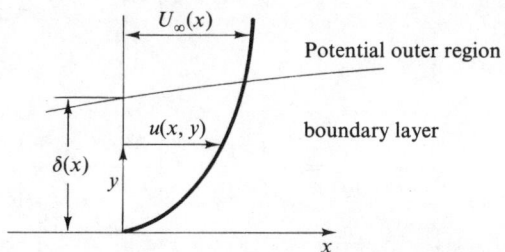

Fig. 5-7 Boundary layer flow over a solid body

In 1904 Prandtl[16] simplified the Navier–Stokes equations by the concept of a boundary layer, on the surface of a body, in which all the significant viscous effects could be contained. Outside the boundary layer the velocity gradients were small enough for viscous effects to be negligible, and the flow in this outer region was considered to be potential. In real flows the influence of the wall can be found throughout the fluid, but for large Reynolds numbers (where inertia forces are much greater than viscous forces) the velocity gradients are only significant very close to the wall.

5-5-2 Differential boundary layer equations

Exact solutions of the Navier–Stokes equations (see Schlichting[4]) show that the boundary layer thickness, δ, increases with increasing viscosity, or more precisely

$$\delta \sim v^{1/2}$$

From consideration of Fig. 5-7, which shows the flow of fluid over a solid body of unit length along the x-axis, it can be seen that the velocity component, u, must change from zero at the wall to U_∞ (the free stream velocity, U_∞, will be assumed to have a value of order unity) in a distance δ from the wall. As u is of order U_∞ and x is of order 1, then $\partial u/\partial x$ must be of order 1.

From the continuity equation, $\partial u/\partial x$ is of order $\partial v/\partial y$, and as y is of order δ, v must also be of order δ. An order of magnitude evaluation of the Navier–Stokes equations can now be made, and to simplify the argument Eqns (5-14) and (5-15)—for incompressible flow ignoring buoyancy effects—are written below. The order of magnitude of every term is written beneath each of the equations.

$$\frac{\partial u}{\partial x} + \frac{\partial v}{\partial y} = 0 \tag{5-72}$$

$$\frac{1}{1} \qquad \frac{\delta}{\delta}$$

$$u\frac{\partial u}{\partial x} + v\frac{\partial u}{\partial y} = -\frac{1}{\rho}\frac{\partial p}{\partial x} + v\left[\frac{\partial^2 u}{\partial x^2} + \frac{\partial^2 u}{\partial y^2}\right] \tag{5-73}$$

$$1\,\frac{1}{1} \qquad \delta\,\frac{1}{\delta} \qquad\qquad\qquad \delta^2\left[\frac{1}{1^2} \quad \frac{1}{\delta^2}\right]$$

$$u\frac{\partial v}{\partial x} + v\frac{\partial v}{\partial y} = -\frac{1}{\rho}\frac{\partial p}{\partial y} + v\left[\frac{\partial^2 v}{\partial x^2} + \frac{\partial^2 v}{\partial y^2}\right] \tag{5-74}$$

$$1\,\frac{\delta}{1} \qquad \delta\,\frac{\delta}{\delta} \qquad\qquad\qquad \delta^2\left[\frac{\delta}{1^2} \quad \frac{\delta}{\delta^2}\right]$$

From Eqn (5-73) it can be seen that for thin boundary layers, where $\delta \ll 1$,

$$\partial^2 u/\partial x^2 \ll \partial^2 u/\partial y^2$$

From Eqn (5-74) it can be deduced that $(1/\rho)(\partial p/\partial y)$ is of order δ, or the pressure difference across the boundary layer, obtained by integration of $\partial p/\partial y$ over a distance δ, is of order δ^2. Thus, an important result of the boundary layer assumptions is that the pressure is practically constant in the y direction (the *cross-stream* direction) but can vary in the x direction (or *streamwise* direction). Hence, for boundary layer flows $p = p(x)$, and as the velocity gradients at the edge of the boundary layer (where $u = U_\infty$) are small enough to regard the flow as inviscid, Eqn (5-73) in the limit as $y \to \delta$ reduces to

$$U_\infty \frac{dU_\infty}{dx} = -\frac{1}{\rho}\frac{dp}{dx} \tag{5-75a}$$

for the outer flow region. It can be seen that Eqn (5-75a) is of order 1, and is hence in accord with Eqn (5-73). Alternatively, Eqn (5-75a) can be integrated to give

$$p + \tfrac{1}{2}\rho U_\infty^2 = \text{constant} \qquad (5\text{-}75b)$$

which is known as *Bernoulli's equation*.

Equations (5-72), (5-73), and (5-74) can now be reduced to two partial differential equations,

$$u\frac{\partial u}{\partial x} + v\frac{\partial u}{\partial y} = -\frac{1}{\rho}\frac{dp}{dx} + v\frac{\partial^2 u}{\partial y^2} \qquad (5\text{-}76)$$

and

$$\frac{\partial u}{\partial x} + \frac{\partial v}{\partial y} = 0 \qquad (5\text{-}77)$$

with the boundary conditions

$$y = 0: \quad u = v = 0; \qquad y = \delta: \quad u = U_\infty$$

and the pressure gradient specified by Eqn (5-75a). In addition, at the start of integration of Eqn (5-76), at $x = x_0$, initial velocity distributions must be known or assumed.

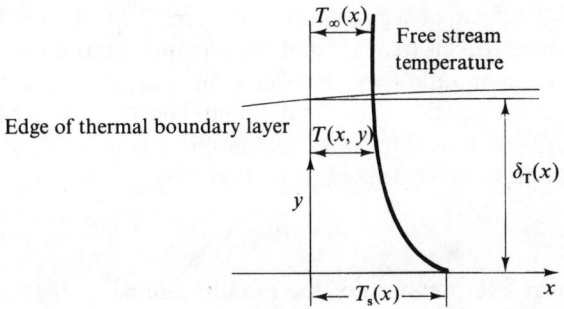

Fig. 5-8 Thermal boundary layer on a solid body

As well as avoiding the solution of Eqn (5-74), the boundary layer assumptions have completely changed the nature of the partial differential equations. Equation (5-73) is *elliptic* while Eqn (5-76) is *parabolic* and from the discussion on these classifications given in the preceding chapters the reader will be aware that in elliptic equations the behaviour of a dependent variable at any location is influenced by its behaviour on all surrounding locations. Thus, the term 'elliptic' can be thought of as implying a completely bounded problem. Truncation of elliptic equations to parabolic form by removal of the second derivative in the streamwise direction (or x direction), implies that the dependent variable is not bounded on the downstream side.

A physical explanation is that the flow at a point in a boundary layer is insensitive to the fluid behaviour downstream of that point, whereas in elliptic flows downstream effects can be propagated upstream. The boundary layer equations therefore require boundary conditions on both sides of the flow system (the wall and the free stream), and also specification of initial values is necessary at the outset of the problem.

Similar arguments can be used for the energy equation, and Fig. 5-8 shows the thermal boundary layer formed by flow over a solid body. An order of magnitude evaluation of the energy equation can be made using Eqn (5-43), for incompressible flow. The temperature is assumed of order unity, and the thermal boundary layer has a thickness δ_T. In a similar manner to the momentum equations, where the inertial and viscous forces are of the same order if v is of order δ^2, the conduction and convection terms will be of order unity if α, the thermal diffusivity, is of order δ_T^2. Equation (5-43) is rewritten as

$$u \frac{\partial T}{\partial x} + v \frac{\partial T}{\partial y} = \alpha \left[\frac{\partial^2 T}{\partial x^2} + \frac{\partial^2 T}{\partial y^2} \right] + v \left(\frac{\Phi}{C_p} \right) \qquad (5\text{-}78)$$

$$1 \; \frac{1}{1} \qquad \delta \, \frac{1}{\delta_T} \qquad \delta_T^2 \left[\frac{1}{1^2} \quad \frac{1}{\delta_T^2} \right] \qquad \delta^2 \left(\frac{1}{\delta^2} \right)$$

From Eqn (5-28) it can be seen that the only term in the dissipation function, Φ, that is of order $1/\delta^2$ is $(\partial u/\partial y)^2$ and the remaining terms can be neglected from order of magnitude considerations. In general, the boundary layer thickness, δ, is not equal to the thermal boundary layer thickness, δ_T, but a consequence of the boundary layer assumptions is that $\delta^2/\delta_T^2 \sim v/\alpha$ or, by using the Prandtl number defined in Section 5-3,

$$\delta \sim \mathbf{P}^{1/2} \, \delta_T \qquad (5\text{-}79)$$

Thus a physical interpretation of the Prandtl number is that it governs the relative thickness of the velocity and thermal boundary layers. For gases, $\mathbf{P}$ is of order unity, so the edges of the boundary layers are nearly coincident. However, the thermal boundary layers of liquids, with Prandtl numbers from 10 to 1,000, is much smaller than the velocity boundary layer. Conversely, for liquid metals, with Prandtl numbers between 0·003 and 0·03, the velocity boundary layer is much thinner than the thermal boundary layers. For fluids where $\mathbf{P} \to \infty$ or $\mathbf{P} \to 0$ special consideration is necessary, but for a large variety of liquids and all gases the boundary layer approximations can be used to rewrite Eqn (5-78) as

$$\rho C_p \left[u \frac{\partial T}{\partial x} + v \frac{\partial T}{\partial y} \right] = k \frac{\partial^2 T}{\partial y^2} + \mu \left(\frac{\partial u}{\partial y} \right)^2 \qquad (5\text{-}80)$$

In laminar boundary layers it is customary to write the shear stress component τ_{xy} as τ, where

$$\tau = \mu \frac{\partial u}{\partial y} \tag{5-81a}$$

Similarly, the heat flux in the cross-stream direction, q_y, is written as q, where

$$q = -k \frac{\partial T}{\partial y} \tag{5-81b}$$

The boundary layer approximations are also valid for turbulent flow, but the turbulence terms must be included in the shear stress and heat flux such that

$$\tau = \mu \frac{\partial u}{\partial y} - \rho \overline{u'v'} \tag{5-82a}$$

and

$$q = -\left(k \frac{\partial T}{\partial y} - \rho C_p \overline{T'v'} \right) \tag{5-82b}$$

where u and T are assumed to be time-average components. The buoyancy forces may also be included in the equations, so that for laminar or turbulent boundary layer flow of an incompressible† fluid

$$\frac{\partial u}{\partial x} + \frac{\partial v}{\partial y} = 0 \tag{5-83a}$$

$$\rho \left[u \frac{\partial u}{\partial x} + v \frac{\partial u}{\partial y} \right] = \rho \beta g \cos \phi (T - T_\infty) - \frac{dp}{dx} + \frac{\partial \tau}{\partial y} \tag{5-83b}$$

$$\rho C_p \left[u \frac{\partial T}{\partial x} + v \frac{\partial T}{\partial y} \right] = -\frac{\partial q}{\partial y} + \tau \frac{\partial u}{\partial y} \tag{5-83c}$$

For laminar flow τ and q are given by Eqns (5-81). For turbulent flow the components in Eqns (5-83) are time-average terms and Eqns (5-82), supplemented by assumptions for the turbulence terms, are used. Equations (5-83) represent a system of equations for u, v, and T, the pressure gradient being prescribed by the potential flow equation, Eqn (5-75a).

It is left as an exercise for the reader to prove to himself that the boundary layer approximations are valid for compressible flow with variable transport

† Strictly speaking for incompressible flow $\beta = 0$, but for many free convection problems the flow velocity may be low enough to allow density changes caused by pressure variation to be neglected while still retaining the buoyancy term due to thermal variations. For forced convection the buoyancy term is neglected.

properties. The system of six equations is written below, again the bars over time-average values have been omitted as their inclusion is now unnecessary:

$$\frac{\partial}{\partial x}(\rho u) + \frac{\partial}{\partial y}(\rho v) = 0 \tag{5-84a}$$

$$\rho\left[u\frac{\partial u}{\partial x} + v\frac{\partial u}{\partial y}\right] = \rho\beta g \cos\phi(T - T_\infty) - \frac{dp}{dx} + \frac{\partial\tau}{\partial y} \tag{5-84b}$$

$$\rho C_p\left[u\frac{\partial T}{\partial x} + v\frac{\partial T}{\partial y}\right] = u\frac{dp}{dx} - \frac{\partial q}{\partial y} + \tau\frac{\partial u}{\partial y} \tag{5-84c}$$

$$\rho = \rho(p, T), \qquad \mu = \mu(T), \qquad k = k(T) \tag{5-84d}$$

Empirical relations for μ and k are necessary for the calculation of the shear stress and heat flux, and for turbulent flow these must be supplemented by assumed forms of the turbulence terms.

Equations (5-83) and (5-84) are valid for a large range of fluids and flow conditions, and although the equations were derived from Prandtl's boundary layer concept they are valid for internal as well as external flow. If the flow has a predominant direction, such that cross-stream gradients are much larger than streamwise gradients, the truncation of the Navier–Stokes equations is valid. The assumptions are often valid for free jets (a stream of fluid moving at a velocity different from that of its environment), wall jets (jets flowing along a solid wall), flow in ducts, but they are not valid in regions of jet impingement, sudden changes in geometry, *separation*†, or in any situation where the flow is recirculating or where there is no dominant flow direction. Equations (5-83) and (5-84) are valid for flow over curved surfaces, such as aerofoils, cylinders, and so on, provided that the boundary layer is thin compared with the radius of curvature of the body surface. A more detailed discussion of the range of application of the boundary layer equations is given in Chapter 6.

The differential equations of the boundary layer permit the detailed distribution of velocity and temperature within the boundary layer to be predicted either analytically or numerically. However, for many convection problems of engineering interest it is sufficient to be able to calculate the Nusselt number, and details of the flow are not necessary. The integral equations of the boundary layer will now be presented as a means to the end of predicting gross or overall fluid and thermal behaviour.

---

† Separation is the condition, usually caused by the presence of an *adverse pressure gradient* (dp/dx positive) or a sudden change in body geometry, when the shear stress at the body surface tends to zero, and flow reversal occurs causing the boundary layer to leave the surface or separate. Thickening of a boundary layer approaching separation invalidates the assumption of relatively large cross-stream gradients.

5-5-3   Integral boundary layer equations

The integral equations can be derived without reference to the Navier–Stokes and energy equations from consideration of the momentum and energy transport into and out of an elemental strip of the boundary layer. The integral momentum equation was originally derived by von Karman[17] in 1912 by balancing the momentum transport to an element of the boundary layer with the pressure force acting across the element and the wall shear force acting on the element.

Having already derived the differential boundary layer equations, a more direct approach is to integrate these equations in the cross-stream direction. Using Eqn (5-75a), Eqn (5-83b) can be integrated across the boundary layer such that

$$\int_0^\delta \left[u \frac{\partial u}{\partial x} + v \frac{\partial u}{\partial y} \right] dy = \int_0^\delta \left[U_\infty \frac{dU_\infty}{dx} + \frac{1}{\rho} \frac{\partial \tau}{\partial y} \right] dy$$
$$+ \int_0^\delta \beta g \cos \phi (T - T_\infty) \, dy \qquad (5\text{-}85)$$

It should be noted that the second term on the left-hand side can be written, with the aid of Eqn (5-83a),

$$\int_0^\delta v \frac{\partial u}{\partial y} \, dy = uv \Big|_0^\delta - \int_0^\delta u \frac{\partial v}{\partial y} \, dy$$
$$= -U_\infty \int_0^\delta \frac{\partial u}{\partial x} \, dy + \int_0^\delta u \frac{\partial u}{\partial x} \, dy \qquad (5\text{-}86)$$

Substitution of Eqn (5-86) into Eqn (5-85), and rearranging the terms, yields:

$$\int_0^\delta \left[2u \frac{\partial u}{\partial x} - U_\infty \frac{\partial u}{\partial x} - U_\infty \frac{dU_\infty}{dx} \right] dy$$
$$= -\frac{\tau_s}{\rho} + \int_0^\delta \beta g \cos \phi \, (T - T_\infty) \, dy \qquad (5\text{-}87)$$

alternatively,

$$\frac{d}{dx} \int_0^\delta u(U_\infty - u) \, dy + \frac{dU_\infty}{dx} \int_0^\delta (U_\infty - u) \, dy$$
$$= \frac{\tau_s}{\rho} - \int_0^\delta \beta g \cos \phi \, (T - T_\infty) \, dy \qquad (5\text{-}88)$$

Equation (5-88) is the *momentum integral equation* for steady, laminar or turbulent, incompressible boundary layers. For solution, U_∞ is prescribed and a variation of u as a function of y is assumed. In laminar flow the value of τ_s, the wall shear stress, is controlled by the assumed form of the velocity distribution, whereas in turbulent flow an empirical value is used. It is therefore usual to regard Eqn (5-88) as a first-order differential equation for δ.

A similar relationship can be obtained by integrating the energy equation over a cross-stream distance, Δ, that exceeds the thickness of both the velocity and thermal boundary layers. Ignoring dissipation, Eqn (5-83c) can be written in the integral form

$$\int_0^\Delta \rho C_p \left[u \frac{\partial T}{\partial x} + v \frac{\partial T}{\partial y} \right] dy = -\int_0^\Delta \frac{\partial q}{\partial y} dy \qquad (5\text{-}89)$$

Using Eqn (5-83a), the second term on the left-hand side can be expanded, to give

$$\int_0^\Delta v \frac{\partial T}{\partial y} dy = vT \Big|_0^\Delta - \int_0^\Delta T \frac{\partial v}{\partial y} dy$$

$$= -\int_0^\Delta T_\infty \frac{\partial u}{\partial x} dy + \int_0^\Delta T \frac{\partial u}{\partial x} dy \qquad (5\text{-}90)$$

Substituting Eqn (5-90) into (5-89) yields the result

$$\int_0^\Delta \rho C_p \left[u \frac{\partial T}{\partial x} + \frac{\partial u}{\partial x} (T - T_\infty) \right] dy = q_s \qquad (5\text{-}91)$$

where q_s is the heat flux through the solid wall. If T_∞ is constant in value, corresponding to an isothermal free stream, Eqn (5-91) can be simplified and, noting that for $y \geqslant \delta_T$ the integrand on the left-hand side vanishes, Eqn (5-91) can be written as

$$\frac{d}{dx} \int_0^{\delta_T} u(T - T_\infty) \, dy = \frac{q_s}{\rho C_p} \qquad (5\text{-}92)$$

This equation is known as the *energy integral equation* for steady, laminar or turbulent, incompressible boundary layers with an isothermal free stream. For assumed cross-stream distributions of u and T, Eqn (5-92) can be regarded as a first-order ordinary differential equation for δ_T. As for the momentum integral equation, the wall flux for laminar flow is implicitly controlled by the assumed temperature distribution, whereas in turbulent flow an empirical value must be used. It should be pointed out that errors between assumed and actual profiles are not so significant when these profiles are integrated, and reasonable answers can be obtained with relatively crude inputs.

Having derived the governing equations of convective heat transfer—the continuity equation, Navier–Stokes equations, and the energy equation—and having extended them for turbulent flow and simplified them by the boundary layer assumptions, it now remains to find solutions. Chapter 6 is concerned with solutions for laminar and turbulent boundary layers in external and internal flow systems with either free or forced convection. Despite the simplifications used most practical problems can only be solved approximately or by numerical methods, and Chapter 7 will present formulae for use in engineering convection systems and will provide an introduction to the numerical techniques useful for the solution of a wide range of practical problems.

Before turning to the solution of the heat transfer equations it is appropriate to consider the analogy between the transfer of heat and the transfer of matter, or mass, in a moving fluid. The similarity between Fourier's law of conduction and Fick's law of diffusion has already been shown in Chapter 1; the similarity that also exists between convective heat and mass transfer can now be demonstrated.

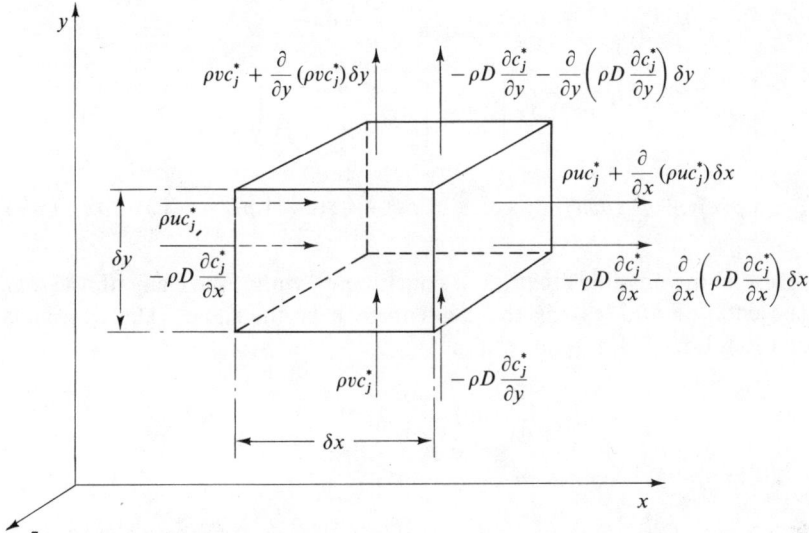

Fig. 5-9   Mass transfer of component j into a control volume

5-5-4   Mass transfer in boundary layers

Fick's law, as stated in Chapter 1, is strictly valid only for a *binary system*, for example, the evaporation of water into air, in the absence of a temperature gradient. For a two-dimensional cartesian coordinate system, Fick's law can be generalized for compressible flow (see Kays[18]) as

$$\dot{m}'_{j,x} = -\rho D \frac{\partial c_j^*}{\partial x} \tag{5-93}$$

where $\dot{m}'_{j,x}$ is the mass flux of constituent j (for example, water) transported by diffusion in the x direction, and c_j^* is the mass fraction of that constituent, that is the ratio of the mass of the constituent j to the mass of the mixture (such as water vapour and air). The mass fraction c^* is related to the mass concentration c by $c^* = c/\rho$. In general, the mixture density ρ and the diffusion coefficient D will be variable. A similar relationship to Eqn (5-93) will obtain for $\dot{m}'_{j,y}$, the mass flux of constituent j transported by diffusion in the y direction.

As well as the diffusion due to a concentration gradient, mass will be transferred by convection if the fluid is moving. There may also be a creation of the constituent j with a mass rate of creation per unit volume R_j due to a chemical reaction. Thus, considering the control volume shown in Fig. 5-10, with sides δx, δy, and unity in the z direction, a simple mass balance for steady laminar flow reveals that

$$\rho u c^*_j \, \delta y - \rho D \frac{\partial c^*_j}{\partial x} \, \delta y + \rho v c^*_j \, \delta x - \rho D \frac{\partial c^*_j}{\partial y} \, \delta x + R_j \, \delta x \, \delta y$$

$$= \left[\rho u c^*_j + \frac{\partial}{\partial x} (\rho u c^*_j) \, \delta x \right] \delta y - \left[\rho D \frac{\partial c^*_j}{\partial x} + \frac{\partial}{\partial x} \left(\rho D \frac{\partial c^*_j}{\partial x} \right) \delta x \right] \delta y$$

$$+ \left[\rho v c^*_j + \frac{\partial}{\partial y} (\rho v c^*_j) \, \delta y \right] \delta x - \left[\rho D \frac{\partial c^*_j}{\partial y} + \frac{\partial}{\partial y} \left(\rho D \frac{\partial c^*_j}{\partial y} \right) \delta y \right] \delta x \quad (5\text{-}94)$$

Despite the possible creation of constituent j within the element, the *total* mass entering and leaving the element must be the same. The continuity equation, Eqn (5-4a), requires that

$$\frac{\partial}{\partial x} (\rho u) + \frac{\partial}{\partial y} (\rho v) = 0$$

thus, Eqn (5-94) becomes

$$\rho u \frac{\partial c^*_j}{\partial x} + \rho v \frac{\partial c^*_j}{\partial y} = \frac{\partial}{\partial x} \left(\rho D \frac{\partial c^*_j}{\partial x} \right) + \frac{\partial}{\partial y} \left(\rho D \frac{\partial c^*_j}{\partial y} \right) + R_j \quad (5\text{-}95)$$

For boundary layer flows, where $\partial^2/\partial x^2 \ll \partial^2/\partial y^2$,

$$\rho u \frac{\partial c^*_j}{\partial x} + \rho v \frac{\partial c^*_j}{\partial y} = -\frac{\partial \dot{m}'_j}{\partial y} + R_j \quad (5\text{-}96)$$

where, for laminar flow,

$$\dot{m}'_j = \dot{m}'_{j,y} = -\rho D \frac{\partial c^*_j}{\partial y} \quad (5\text{-}97a)$$

For turbulent flow, by analogy with Eqns (5-82),

$$\dot{m}'_j = -\left(\rho D \frac{\partial c^*_j}{\partial y} - \overline{\rho c^{*\prime}_j v'} \right) \quad (5\text{-}97b)$$

For incompressible flow with no chemical reaction, Eqn (5-96) simplifies to

$$u \frac{\partial c_j}{\partial x} + v \frac{\partial c_j}{\partial y} = -\frac{\partial \dot{m}'_j}{\partial y} \tag{5-98}$$

where, for laminar flow,

$$\dot{m}'_j = -D \frac{\partial c_j}{\partial y} \tag{5-99a}$$

and for turbulent flow,

$$\dot{m}'_j = -\left(D \frac{\partial c_j}{\partial y} - \overline{c'_j v'} \right) \tag{5-99b}$$

It can be seen that Eqn (5-98) is very similar to Eqn (5-83c) and the diffusion coefficient, D, resembles the thermal diffusivity, α. By analogy with the Prandtl number, where $\mathbf{P} = v/\alpha$, we can define a Schmidt number, **Sc**, based on the mixture properties, such that

$$\mathbf{Sc} = \frac{v}{D} = \frac{\mu}{\rho D} \tag{5-100}$$

The Schmidt number affects the transfer of mass in an analogous way to that in which the Prandtl number affects the transfer of heat.

In order to calculate the mass flux at the surface of a solid body it is convenient to use a *mass transfer coefficient*, K, analogous to the heat transfer coefficient, such that

$$\dot{m}'_{j,s} = -D \left(\frac{\partial c_j}{\partial y} \right)_s = K(c_{j,s} - c_{j,\infty}) \tag{5-101}$$

where the subscript 's' refers to the surface conditions and ∞ refers to the free stream conditions for external flow. For internal flows it is necessary to use a convenient bulk mean concentration for $c_{j,\infty}$.

Just as the Nusselt number can be regarded as a dimensionless heat transfer coefficient ($\mathbf{N} = hl/k$), so we can define the Sherwood number, **Sh**, as

$$\mathbf{Sh} = \frac{Kl}{D} \tag{5-102}$$

As would be expected from the similarity between the equations of heat and mass transfer, the Nusselt number and the Sherwood number behave in an analogous fashion. This analogy will be demonstrated in Chapter 7, where an empirical correlation for **N** and **Sh** is given. Also, as the boundary layer equations for the transfer of mass, momentum, and heat are so similar in form, they are equally amenable to the numerical method described in

Chapter 7. In Chapter 9 the analogy between heat and mass transfer will be demonstrated in problems involving the condensation of a liquid from a moving gas stream.

REFERENCES

1. Skelland, A. *Non-Newtonian Flow and Heat Transfer*. John Wiley, New York, 1967.
2. Navier, M. Mémoire sur les lois du mouvement des fluides. *Mém. Acad. Sci.*, 1827, **6**, 389.
3. Stokes, G. G. On the theories of internal friction of fluids in motion. *Trans. Camb. Phil. Soc.*, 1845, **8**, 287.
4. Schlichting, H. *Boundary Layer Theory* (6th edn). McGraw-Hill, New York, 1968.
5. Yung, Y. C. *A First Course in Continuum Mechanics*. Prentice-Hall, Englewood Cliffs, N.J., 1969.
6. Shapiro, A. H. *The Dynamics and Thermodynamics of Compressible Fluid Flow*, vol 1. Ronald Press Company, New York, 1953.
7. Keenan, J. H. *Thermodynamics*. John Wiley, New York, 1963.
8. van Wylen, G. J. *Thermodynamics*. John Wiley, New York, 1959.
9. Buckingham, E. On physically similar systems; illustrations of the use of dimensional equations. *Phys. Rev.*, 1914, **4**, 345.
10. Bridgeman, P. W. *Dimensional Analysis*. Yale University Press, New Haven, Conn., 1931.
11. Langhaar, H. L. *Dimensional Analysis and Theory of Models*. John Wiley, New York, 1951.
12. Reynolds, O. On the experimental investigation of the circumstances which determine whether the motion of water shall be direct or sinuous, and the law of resistance in parallel channels. *Phil. Trans. Roy. Soc.*, 1883, **174**, 935.
13. Burgess, J. M. The motion of a fluid in the boundary layer along a plane smooth surface. *Proc. First International Congress on Applied Mechanics*, Delft, 1924.
14. van der Hegge Zignen, B. G. Measurements of the velocity distribution in the boundary layer along a plane surface. Thesis, Delft, 1924.
15. Reynolds, O. On the dynamic theory of incompressible viscous fluids and the determination of the criterion. *Phil. Trans. Roy. Soc.*, 1895, T**186** A, 123.
16. Prandtl, L. Über Flüssigkeits bewegung bei sehr kleiner Reibung. *Proc. Third International Mathematics Congress*, Heidelburg, 1904.
17. von Karman, Th. Über laminare und turbulente Reibung. *ZAMM*, 1912, **1**, 233.
18. Kays, W. M. *Convective Heat and Mass Transfer*. McGraw-Hill, New York, 1966.

PROBLEMS

Note: where not stated, fluid property values should be obtained from the Appendix.

5-1   By considering the axisymmetric steady flow of mass and energy into an annular element of radius, r, thickness δr, and length, δz, show, stating your assumptions, that for an incompressible fluid

$$\frac{\partial}{\partial r}(rV_r) + \frac{\partial}{\partial z}(rV_z) = 0$$

and, if frictional heating is negligible,

$$\rho C_p \left(V_r \frac{\partial T}{\partial r} + V_z \frac{\partial T}{\partial z} \right) = k \left(\frac{\partial^2 T}{\partial r^2} + \frac{1}{r} \frac{\partial T}{\partial r} + \frac{\partial^2 T}{\partial z^2} \right)$$

V_r and V_z are the velocity components in the r and z directions, respectively, and other symbols have their usual meaning.

5-2 Calculate the appropriate Reynolds number for the following cases: (a) water with a bulk velocity of 10 m/s and a bulk temperature of 20°C flowing in a pipe of 0·1 m bore; (b) air with a relative velocity of 900 ft/s, a pressure of 56 lbf/in², and a temperature of 300°F, flowing over a compressor blade with a 2-inch chord; (c) hydrogen at 27°C, 0·1 MN/m², flowing with a velocity of 10 m/s over a hot-wire probe with a wire diameter of 5×10^{-6} m; (d) engine oil at 140°F with a velocity of 50 ft/s at a distance of 3 inches from the leading edge of a flat plate.

Ans: (a) 9.96×10^5; (b) 1.83×10^6; (c) 0.45; (d) 1.38×10^4

5-3 Calculate the Prandtl number of fluids with the following properties: (a) a gas with $k = 0.0398$ W/m K, $\mu = 2.569 \times 10^{-5}$ kg/m s, $C_p = 1.055$ kJ/kg K; (b) a gas with $\alpha = 0.510$ ft²/h, $v = 10.22 \times 10^{-5}$ ft²/s; (c) a liquid with $\rho = 888$ kg/m³, $C_p = 1,880$ J/kg K, $v = 900 \times 10^{-6}$ m²/s, $k = 0.145$ W/m K; (d) a liquid metal with $\rho = 847$ lb/ft³, $\mu = 3.90$ lb/ft h, $\alpha = 0.17$ ft²/h.

Ans: (a) 0.68; (b) 0.72; (c) $10,400$; (d) 0.027

5-4 Calculate the appropriate Grashof number for the following cases: (a) a radiator 0·5 m high with a temperature of 80°C in the room of a house with a temperature of 27°C; (b) a cubic tank, the sides of which are 3 feet long and are maintained at 140°F, filled with engine oil at 68°F; (c) a tube of 0·01 m diameter and a surface temperature of $-10°$C in a container of liquid ammonia at 20°C; (d) a hot wire probe with a wire diameter of 2×10^{-4} inches and a temperature of 740°F in a stream of nitrogen with a pressure of 15 lbf/in², a temperature of 260°F and a velocity of 300 ft/s.

Ans: (a) 8.78×10^8; (b) 2.59×10^5; (c) 5.6×10^6; (d) 1.3×10^{-6}

5-5 Calculate the appropriate Nusselt number for the following cases: (a) a square plate of 0·1 m side and a temperature of 90°C releasing 0·5 kW of heat into a tank of water at 20°C; (b) a 4-inch diameter pipe through which a liquid metal, with a thermal conductivity of 41·8 Btu/ft h °R, is pumped, such that the heat transfer coefficient is 1,200 Btu/ft² h °R; (c) a brick wall 3 m high, 0·1 m thick, with a thermal conductivity of 0·4 W/m K through which the temperature drops linearly between 20°C and 18°C when the air temperature on the colder side is 2°C; (d) an isothermal square duct of 3-inch side and 100 ft long with a temperature of 90°F through which water with bulk velocity of 5 ft/s enters at 200°F and leaves at 190°F.

Ans: (a) 118; (b) 9.56; (c) 61.7; (d) 41.2

5-6 A plate 1 m long has a constant temperature of 80°C and is cooled by air flowing lengthwise over both surfaces with a free-stream velocity of 10 m/s. Calculate the Prandtl number of the fluid and the overall Reynolds number of the flow. If the plate is vertical calculate the overall Grashof number, and discuss whether or not free convection is negligible. For the air the free-stream conditions are 0°C and 10^5 N/m² pressure, and the following properties can be used: $C_p = 1.01$ kJ/kg K, $k = 0.0242$ W/m K, $\mu = 17.1 \times 10^{-6}$ kg/m s.

Ans: 0.71, 7.5×10^5, 1.6×10^{10}

5-7   Repeat Problem 5-6 for a plate 32·8 ft long at a temperature of 176°F cooled by air at 14·6 lbf/in² and 32°F with a velocity of 32·8 ft/s. For the air take $C_\mathrm{p} = 0.24$ Btu/lb °R, $k = 0.014$ Btu/ft h °R, $v = 1.46 \times 10^{-4}$ ft²/s. Is free convection still negligible?

$$Ans: 0.71, \ 7.5 \times 10^6, \ 1.6 \times 10^{13}$$

5-8   The flat plate of Problem 5-6 is placed with its length along the x-axis and its leading edge at $x = 0$. If the local Nusselt number, $\mathbf{N}_x$, is related to the local Reynolds number, $\mathbf{R}_x$, and the Prandtl number by

$$\mathbf{N}_x = 0.03\mathbf{P}^{1/3}\mathbf{R}_x^{0.8}$$

calculate the value of the local heat transfer coefficient, h, at a position of $x = 0.5$ m. If the plate is 0·5 m wide what power must be supplied to the entire plate to maintain it at the stated temperature of 80°C?

$$Ans: 37.2 \ \mathrm{W/m^2 \ K}, \ 3.28 \ \mathrm{kW}$$

5-9   By considering the balance of mass, momentum, and energy into an infinitesimal element within the boundary layer, where

$$\frac{\partial^2 u}{\partial y^2} \gg \frac{\partial^2 u}{\partial x^2}, \qquad \frac{\partial^2 T}{\partial y^2} \gg \frac{\partial^2 T}{\partial x^2}, \quad \text{etc.,}$$

derive Eqns (5-83) from first principles, stating all the assumptions made.

5-10   Consider the boundary layer flow between two coaxial circular discs of radius R placed an axial distance Z apart such that $Z \ll R$. Fluid is fed from the centre of the discs and flows radially outward. By considering the balance of momentum and energy for steady laminar incompressible flow of an element of radius r, radial thickness dr, and axial thickness dz, subtending an angle $d\phi$ at the axis, show from first principles, stating your assumptions, that

$$\rho\left(V_r \frac{\partial V_r}{\partial r} + V_z \frac{\partial V_r}{\partial z}\right) = -\frac{dp}{dr} + \mu \frac{\partial^2 V_r}{\partial z^2}$$

$$\rho C_\mathrm{p}\left(V_r \frac{\partial T}{\partial r} + V_z \frac{\partial T}{\partial z}\right) = k \frac{\partial^2 T}{\partial z^2}$$

5-11   Using the above technique, show that for steady laminar axisymmetric incompressible flow in a long circular pipe

$$\rho\left(V_r \frac{\partial V_z}{\partial r} + V_z \frac{\partial V_z}{\partial z}\right) = -\frac{dp}{dz} + \frac{\mu}{r}\frac{\partial}{\partial r}\left(r \frac{\partial V_z}{\partial r}\right)$$

$$\rho C_\mathrm{p}\left(V_r \frac{\partial T}{\partial r} + V_z \frac{\partial T}{\partial z}\right) = \frac{k}{r}\frac{\partial}{\partial r}\left(r \frac{\partial T}{\partial r}\right)$$

5-12   By considering the balance of mass and momentum into a boundary layer element of length dx and width, δ, where δ is the boundary layer thickness, show from first principles that for the case of a zero pressure gradient, and negligible buoyancy forces,

$$\frac{d}{dx}\int_0^\delta u(U_\infty - u)\, dy = \frac{\tau_s}{\rho}$$

where all symbols have their usual meanings.

5-13 Ignoring viscous dissipation use the technique of the above problem to show that for a thermal boundary layer of thickness δ_T with an isothermal free stream

$$\frac{d}{dx} \int_0^{\delta_T} u(T - T_\infty) \, dy = \frac{q_s}{\rho C_p}$$

where all symbols have their usual meanings.

6

Boundary layer analysis of convection problems for incompressible flow

6-1 Classification of flow systems

In Chapter 5 the equations governing the transfer of heat by free and forced convection were derived for the general situation, and by making the boundary layer assumptions these equations were truncated to a simpler form. While the boundary layer equations cannot be regarded as a general description of all convection problems, they can be applied to a large variety of practical situations. It is not possible to apply a rigorous classification to the convection problems that can be treated by boundary layer analysis: all that can be stated is that if the velocity and temperature gradients in the streamwise direction are always much smaller than the gradients in the cross-stream direction, then the boundary layer equations will provide a good description of the flow behaviour. If, however, there is no principal flow direction such that the velocity gradients or thermal gradients in one direction are not dominant, then the boundary layer assumptions will be invalid.

The simple geometries of Fig. 6-1 illustrate flow situations that are common in engineering practice. Figure 6-1(a) shows the flow past a plate, wing, turbine blade, and so on, where the adverse pressure gradient imposed by the external flow field causes the flow to separate. While the boundary layer region is shown extending to the point of separation, in practice—owing to relative weakening of cross-stream gradients and the corresponding inaccuracy of shear stress calculations—predictions made from the solution of the boundary layer equations will become increasingly less accurate as this point is approached. After separation it may occur that the main flow reattaches itself to the wall, forming a new boundary layer. This often occurs at transition from laminar to turbulent flow.

Figure 6-1(b) shows the flow in a pipe or duct where initially separate boundary layers form on the walls while the flow external to these boundary layers can be treated as inviscid. While the flow in many situations—such as flow between the walls of a duct—is confined in the geometric sense, the boundary layers do not significantly interact and the problem can be treated as external flow, that is, the pressure gradient is prescribed by the inviscid flow external to the boundary layers. Where the boundary layers interact,

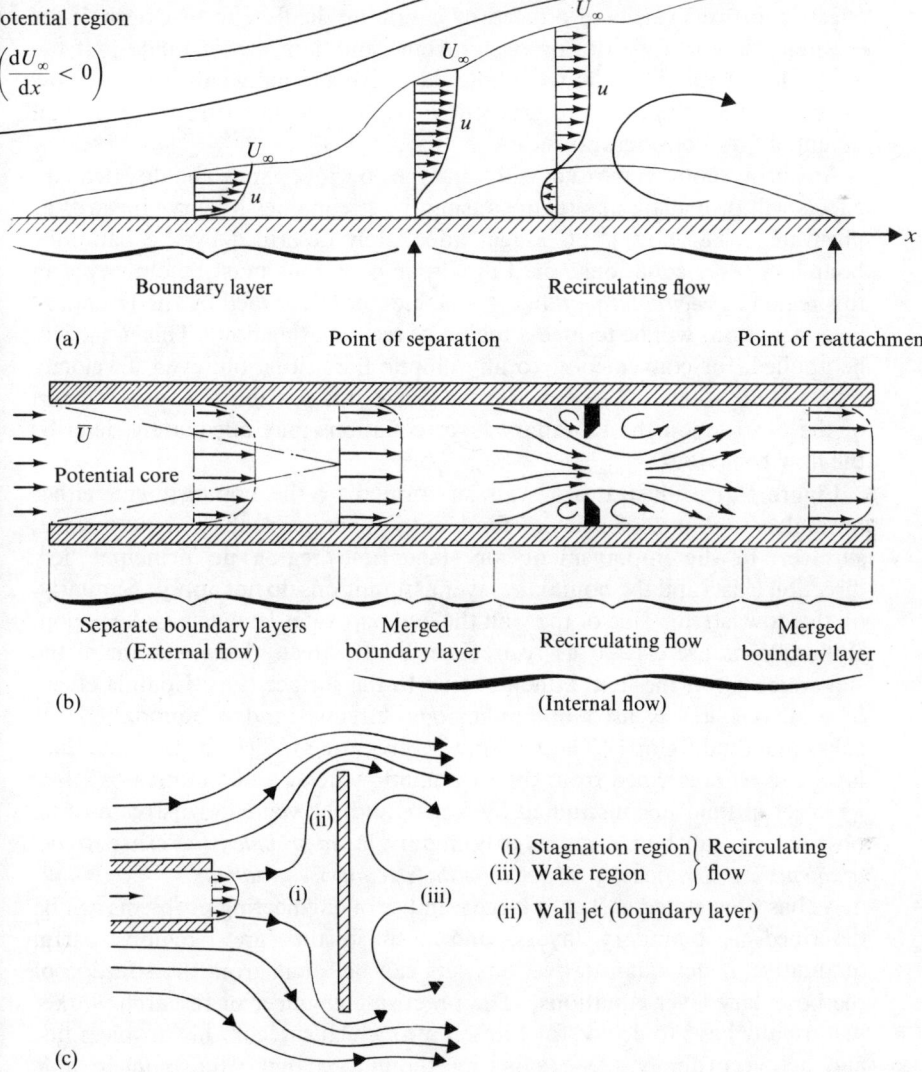

Fig. 6-1 (a) External flow with an adverse pressure gradient; (b) internal flow with an obstruction; (c) wall jet with external flow field

or where a potential core does not exist, the whole space between the walls of the duct must be considered in its entirety. In these circumstances the continuity equation supplies additional information to enable the pressure gradient to be calculated. It is thus apparent that external flows are merely a subset of internal flow problems, and the distinction between these classifications is often blurred. However, in many engineering problems effective answers can be produced by assuming the flow to be external; for example, flow in cascades or over turbine and compressor blades. If the duct is long enough to allow the boundary layers to merge, the ensuing flow is internal and the pressure gradient must be calculated from other than potential flow considerations.

Any obstruction in the duct will cause the flow to separate and downstream effects will propagate upstream causing the streamlines to curve inwards as illustrated. The flow is no longer adequately described by the parabolic boundary layer equations: the full elliptic equations must be employed in this region of *recirculating flow*. Any region not described by the boundary layer equations will be termed a region of recirculating flow. This name will be applied, for convenience, to any elliptic flow situation, even if velocity reversals do not exist. As for external flows, it is possible that downstream of the obstruction the boundary layer equations may adequately describe the flow behaviour.

Figure 6-1(c) illustrates a wall jet formed by the impingement, either normal or oblique, of a jet of fluid onto an impermeable, or permeable, surface. In the impingement, or stagnation, region no principal flow direction exists and the boundary layer assumptions do not apply. Similarly, on the downstream side of the wall the flow separates leaving a wake region. Both regions are classed as recirculating flow areas. Downstream of the stagnation point the fluid attaches itself to the surface (this 'Coanda effect' of attaching jets is used in fluidic logic circuits) and a boundary layer subsequently develops. The streamwise pressure gradient in the boundary layer can be prescribed from the external flow field. In certain cases a free jet (a jet of fluid unconstrained by walls), and the wake downstream of an obstruction, can be treated as a boundary layer *provided the cross-stream gradients are significantly larger than the streamwise gradients*.

As has been stated earlier, a large number of engineering problems can be described as boundary layers, and even in the 'grey' regions useful qualitative, if not quantitative, answers can be found from the solution of the boundary layer equations. The practising engineer or research worker will usually have to decide for himself into which category his problem fits, and act accordingly. One sobering thought is that while considerable mathematical and physical expertise has been gained from a study of boundary layers, the state of the art in recirculating flows—particularly for the turbulent case—is considerably less developed. For turbulent re-circulating flow, problems are usually treated by the use of semi-empirical formulae, although in recent years numerical solutions of the elliptic flow equations have been obtained by using finite difference techniques. However, the rest of this chapter will be devoted to the solution of a selected number of boundary layer problems.

The chapter will progress from laminar to turbulent flow, and a range of problems involving free and forced convection will be examined. The particular examples chosen have been selected to illustrate the application of classical techniques to the solution of a representative class of boundary layer equations, to elicit and discuss the significant physical features of these solutions, and to form a basis for the semi-empirical formulae and the numerical techniques described in Chapter 7.

6-2 Laminar forced flow between parallel plates

Despite its simplicity, this system provides a model from which can be elicited significant facts that are important to the understanding of heat transfer by forced convection (Fig. 6-2). Parallel flows constitute a particularly simple fluid motion where there is a velocity component in only one

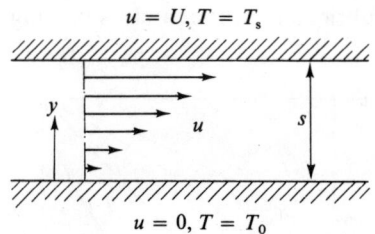

Fig. 6-2 Flow between a stationary and a moving plate

direction. From the continuity equation, Eqn (5-83a), it follows that if $v = 0$ then

$$\frac{\partial u}{\partial x} = \frac{\partial v}{\partial y} = 0 \tag{6-1}$$

thus $u = u(y)$. For laminar forced convection, the boundary layer momentum equation, Eqn (5-83b), can be expressed as

$$\rho\left[u\frac{\partial u}{\partial x} + v\frac{\partial u}{\partial y}\right] = -\frac{dp}{dx} + \mu\frac{\partial^2 u}{\partial y^2} \tag{6-2}$$

hence, from Eqn (6-1),

$$0 = -\frac{dp}{dx} + \mu\frac{d^2 u}{dy^2} \tag{6-3}$$

For the case of the bottom wall stationary and the top wall moving with a velocity U, that is, $u(0) = 0$, $u(s) = U$, the solution of Eqn (6-3) is

$$u = \frac{y}{s}U - \frac{s^2}{2\mu}\frac{dp}{dx}\left[\frac{y}{s}\left(1 - \frac{y}{s}\right)\right] \tag{6-4}$$

This case is known as *Couette flow*, and for a zero pressure gradient

$$u = \frac{y}{s} U \tag{6-5}$$

If both plates are stationary,

$$u = -\frac{s^2}{2\mu}\frac{dp}{dx}\left[\frac{y}{s}\left(1 - \frac{y}{s}\right)\right] \tag{6-6}$$

which is known as *Poiseuille flow* between parallel plates. Thus the general solution comprises a linear velocity distribution between the two plates, due to simple shear flow, superimposed on which is the quadratic distribution caused by the pressure parameter, P, where $P \equiv -(s^2/2\mu U)\,(dp/dx)$. The result is shown in Fig. 6-3, and it can be observed that reverse flow occurs if $P < -1$. The flow is basically the same as that existing in the narrow clearance between a lubricated journal and its bearing where inertial effects are negligible.

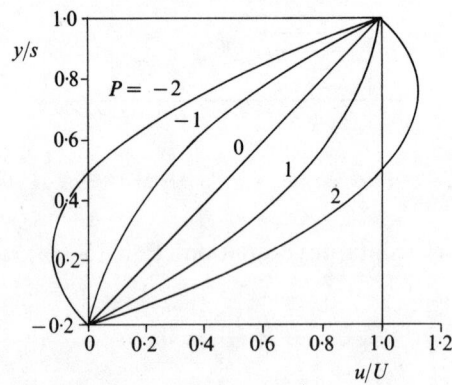

Fig. 6-3 Velocity distribution in Couette flow illustrating the effect of pressure gradient
$$\left(P = \frac{-s^2}{2\mu U}\frac{dp}{dx}\right)$$

The thermal problem can be solved by considering Eqn (5-83c) which, for laminar flow, can be written

$$\rho C_p\left[u\frac{\partial T}{\partial x} + v\frac{\partial T}{\partial y}\right] = k\frac{\partial^2 T}{\partial y^2} + \mu\left(\frac{\partial u}{\partial y}\right)^2 \tag{6-7}$$

For isothermal plates we postulate $T = T(y)$, and as $v = 0$, we can write

$$0 = k\frac{d^2 T}{dy^2} + \mu\left(\frac{du}{dy}\right)^2 \tag{6-8}$$

For the boundary conditions $T(0) = T_0$, $T(s) = T_s$, $T_s > T_0$, and for the case of a zero pressure gradient Eqn (6-8) can be integrated, using Eqn (6-5), to give the solution

$$\frac{T - T_0}{T_s - T_0} = \frac{y}{s} + \frac{\mu U^2}{2k(T_s - T_0)} \frac{y}{s}\left(1 - \frac{y}{s}\right) \tag{6-9}$$

or

$$\theta = \eta[1 + \tfrac{1}{2}\mathbf{PE}(1 - \eta)] \tag{6-10}$$

where $\eta \equiv y/s$, $\theta \equiv (T - T_0)/(T_s - T_0)$, and $\mathbf{P}$ and $\mathbf{E}$ are equivalent to the Prandtl and Eckert numbers defined by Eqns (5-55) and (5-54), respectively, which for convenience are written below as

$$\mathbf{P} \equiv \mu C_p/k \quad \text{and} \quad \mathbf{E} \equiv U^2/[C_p(T_s - T_0)]$$

Inspection of Eqn (6-10) reveals that the linear temperature profile for $\mathbf{PE} = 0$ (that is, U or $\mu = 0$) is supplemented by a parabolic distribution due to heat generated by friction (that is, U and $\mu \neq 0$). Although for $T_s > T_0$ the upper plate will only be cooled by the fluid between the plates if $\partial T/\partial y|_{y=s} = 0$. From Eqn (6-9) it can be seen that the heat flow *from* the top plate *to* the fluid, q_s, is

$$q_s = k \frac{\partial T}{\partial y}\bigg|_{y=s} = k \frac{T_s - T_0}{s}[1 - \tfrac{1}{2}\mathbf{PE}] \tag{6-11}$$

Thus, for $T_s > T_0$ the upper plate will be

cooled if $\mathbf{PE} < 2$

heated if $\mathbf{PE} > 2$

This very important result illustrates how frictional heating can prevent, and indeed reverse, the flow of heat from a hot plate even though the *ostensible* temperature difference is favourable for cooling to occur.

The concept of the adiabatic wall temperature, discussed in Chapter 5, can now be graphically illustrated as $\partial T/\partial y|_{y=s} = 0$ (that is, the upper wall is adiabatic) if $\mathbf{PE} = 2$. Hence, heat transfer from the top plate is proportional to $T_s - T_{ad,s}$ where

$$T_{ad,s} = T_0 + \tfrac{1}{2}\mathbf{P}U^2/C_p \tag{6-12}$$

$T_{ad,s}$ is the *adiabatic wall temperature*; that is, the temperature assumed by the top plate when the heat flux through it is zero. Thus,

for cooling the top plate $T_s > T_{ad,s}$

for heating the top plate $T_s < T_{ad,s}$

The resulting temperature distributions for a range of **PE**, based on Eqn (6-10), is shown in Fig. 6-4.

It should be noted that Eqn (6-3) for parallel flow can be produced from the Navier–Stokes equations, and Eqn (6-8) can be derived from the full energy equation. Hence Eqns (6-5) and (6-9) constitute simple, but exact, solutions of the Navier–Stokes and energy equations. It is left as an exercise for the reader to verify this fact for himself.

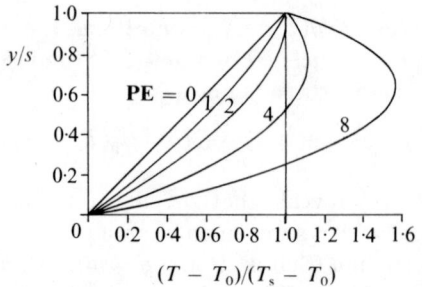

Fig. 6-4   Temperature distribution in Couette flow illustrating the effect of frictional heating

6-3   Laminar forced flow in a circular pipe

As stated in Section 6-1, when fluid flows through a pipe a boundary layer builds up on the pipe wall leaving a potential core in the centre. The boundary layer may start as a laminar layer which eventually thickens to fill the pipe, eliminating the potential core. When this happens the flow is known as *fully developed laminar flow*, while the region preceding the point of merger is known as the *hydrodynamic entry length*. The entry length, l_e, can be calculated for laminar flow from the formula

$$l_e/d = 0.0575 \mathbf{R}_d \qquad (6\text{-}13)$$

where d is the pipe diameter and the Reynolds number $\mathbf{R}_d$ is based on this diameter, such that $\mathbf{R}_d \equiv \rho \overline{U} d / \mu$.

For incompressible flow the bulk mean velocity $\overline{U}$ is calculated by

$$\overline{U} = \frac{1}{\pi r_0^2} \int_0^{r_0} 2\pi r v_z \, dr = \text{constant} \qquad (6\text{-}14)$$

where v_z is the axial velocity component, r is the radial distance from the pipe centre-line, and $r_0 \,(=\tfrac{1}{2}d)$ is the pipe radius.

If the Reynolds number, based on diameter as the characteristic dimension, is sufficiently high (for a pipe $\mathbf{R}_{d,\,\text{crit}} \approx 2{,}300$, although this value depends on many other parameters as discussed in Section 5-4), then the fully developed flow will be turbulent. Also, it is possible that even in the entry region, transition from laminar to turbulent flow will occur if $\mathbf{R}_z$ is large enough. [$\mathbf{R}_z \equiv \rho U_\infty z / \mu$, where U_∞ is the velocity of the potential core and z the distance from the pipe entrance. The critical value for a flat plate ($\mathbf{R}_{x,\,crit} \approx 3 \cdot 5 \times 10^5$) can serve as a crude guide to transition.] If the boundary layer becomes turbulent, the entry region will be shorter than that predicted by Eqn (6-13) owing to the increased thickening of the turbulent boundary layer.

In addition to a hydrodynamic entry region, there will also be a *thermal entry region*, unless the flow is isothermal, owing to the development of thermal boundary layers on the pipe wall. Since the relative thicknesses of the thermal and hydrodynamic boundary layers depend on the Prandtl number of the fluid, the thermal entry length will be longer than the hydrodynamic entry length for $\mathbf{P} > 1$, and vice versa. In the majority of engineering applications flow in pipes is mainly turbulent, apart from very viscous fluids such as oil. For viscous fluids, $\mathbf{P} \gg 1$, and so the thermal entry length can be very great. Another problem encountered in the calculation of convection in laminar pipe flow is that the transport properties, μ and k, are, in viscous fluids, strongly temperature dependent.

One of the simplest cases to consider—from which the mathematical techniques used to produce solutions, and the physical significance of these solutions, can be learned—is the case of *laminar flow, with fully developed velocity and temperature profiles, with constant heat flux through the pipe wall.*

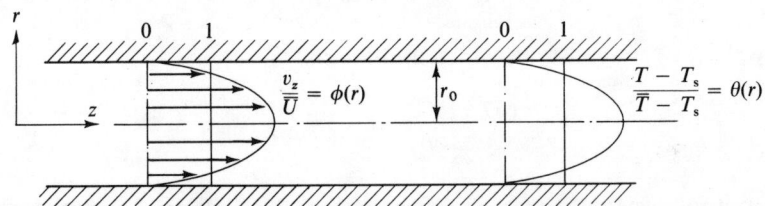

Fig. 6-5 Fully developed velocity and temperature profiles in laminar pipe flow

Before attempting to solve the problem, it is necessary to define the system in mathematical terms. For fully developed profiles there are no changes in cross-stream distribution in the streamwise direction. Thus, referring to Fig. 6-5 where the pipe centre-line is along the z-axis, fully developed velocity and temperature profiles are defined by

$$\frac{\partial \phi}{\partial z} = \frac{\partial \theta}{\partial z} = 0 \qquad (6\text{-}15)$$

where $\qquad \phi \equiv \dfrac{v_z}{U} \quad \text{and} \quad \theta \equiv \dfrac{T - T_s}{\overline{T} - T_s} \qquad (6\text{-}16)$

The bulk velocity $\bar{U}$ is calculated from Eqn (6-14), and the bulk temperature $\bar{T}$ is calculated from

$$\bar{T} = \frac{1}{\pi r_0^2 \bar{U}} \int_0^{r_0} 2\pi r v_z T \, dz \qquad (6-17)$$

and T_s is the temperature of the pipe wall.

Typical velocity and temperature profiles are shown in Fig. 6-5, and as a consequence of the fact that $\theta = \theta(r)$ it follows that $(\partial\theta/\partial r)_{r=\text{const}}$ is constant; that is, the profiles at any z station are said to be *similar*. Not only is a fully developed profile invariant with z, but its radial gradient—at any radius—is also invariant with z. We can deduce further implications by considering the case when a heat flux, q_s, passes through the pipe wall into the fluid. Now we can define a heat transfer coefficient, h, such that the heat flux is proportional to h and a convenient temperature difference, which we shall arbitrarily take as $T_s - \bar{T}$. Thus

$$q_s = h(T_s - \bar{T}) \qquad (6-18)$$

also

$$q_s = -k \left(\frac{\partial T}{\partial r} \right)_{r=r_0} \qquad (6-19)$$

Hence

$$\frac{k}{h} = -\frac{(\partial T/\partial r)_{r=r_0}}{T_s - \bar{T}} \qquad (6-20)$$

But $(\partial\theta/\partial r)_{r=r_0} = \text{constant}$, which implies that

$$\frac{(\partial T/\partial r)_{r=r_0}}{\bar{T} - T_s} = \text{constant} \qquad (6-21)$$

and so it follows that (k/h) is invariant with z. Also, $\partial\theta/\partial z = 0$ implies that

$$\frac{T - T_s}{\bar{T} - T_s} \frac{d}{dz} (\bar{T} - T_s) = \frac{\partial}{\partial z} (T - T_s) \qquad (6-22)$$

For constant heat flux $(\partial T/\partial r)_{r=r_0}$ is constant, and as a consequence of Eqn (6-21) $(\bar{T} - T_s)$ is constant. A further constraint is revealed from a heat balance over an elemental pipe length, δz, where

$$2\pi r_0 q_s \, \delta z = \frac{d}{dz} \left[C_p \int_0^{r_0} 2\pi \rho r v_z T \, dr \right] \delta z \qquad (6-23)$$

Using Eqn (6-17) for $\bar{T}$ it follows from Eqn (6-23) that

$$q_s = \tfrac{1}{2} r_0 \rho C_p \bar{U} \frac{d\bar{T}}{dz} \qquad (6-24)$$

which, for q_s constant, implies that $d\bar{T}/dz$ is constant. From Eqns (6-21), (6-22), and (6-24) the implication of fully developed pipe flow with constant heat flux is that

$$\frac{\partial T}{\partial z} = \frac{d\bar{T}}{dz} = \frac{dT_s}{dz} = \text{constant} \tag{6-25}$$

Using the above constraints we can now solve the forced convection problem. The continuity equation, Eqn (5-18), can be written

$$\frac{\partial v_r}{\partial r} + \frac{v_r}{r} = -\frac{\partial v_z}{\partial z} \tag{6-26}$$

As $\partial v_z/\partial z$ is zero for developed flow

$$\frac{\partial}{\partial r}(rv_r) = 0 \tag{6-27}$$

which, in order to preserve the no-slip condition at the wall, implies that $v_r = 0$.

The Navier–Stokes equations, Eqns (5-17), reduce to the single component form which neglecting body forces and putting $v_r = 0$ is simply

$$\rho v_z \frac{\partial v_z}{\partial z} = -\frac{dp}{dz} + \mu\left(\frac{\partial^2 v_z}{\partial r^2} + \frac{1}{r}\frac{\partial v_z}{\partial r} + \frac{\partial^2 v_z}{\partial z^2}\right)$$

As

$$\frac{\partial v_z}{\partial z} = \frac{\partial^2 v_z}{\partial z^2} = 0$$

then

$$0 = -\frac{dp}{dz} + \frac{\mu}{r}\frac{d}{dr}\left(r\frac{dv_z}{dr}\right) \tag{6-28}$$

Using the boundary conditions that $dv_z/dr = 0$ at $r = 0$, and $v_z = 0$ at $r = r_0$, Eqn (6-28) can be integrated to give the result that

$$v_z = -\frac{r_0^2}{4\mu}\frac{dp}{dz}\left[1 - \left(\frac{r}{r_0}\right)^2\right]$$

Using Eqn (6-14),

$$\bar{U} = -\frac{r_0^2}{8\mu}\frac{dp}{dz} \tag{6-29}$$

hence

$$\frac{v_z}{\bar{U}} = 2\left[1 - \left(\frac{r}{r_0}\right)^2\right] \tag{6-30}$$

In pipe flow it is common to use a friction factor, f, defined by

$$f \equiv -\frac{dp/dz}{\frac{1}{2}\rho \bar{U}^2/d} \tag{6-31}$$

which can be evaluated from Eqn (6-29) to give the result

$$f = 64/\mathbf{R}_d \tag{6-32}$$

where $d = 2r_0$ is the pipe diameter. This result is often referred to as the Hagen–Poiseuille law of friction, as Hagen[1]† obtained experimental values that were later given theoretical verification by Poiseuille[2].

The energy equation, neglecting viscous dissipation, can be written from Eqn (5-43) as

$$\rho C_p \left(v_r \frac{\partial T}{\partial r} + v_z \frac{\partial T}{\partial z} \right) = k \left(\frac{\partial^2 T}{\partial r^2} + \frac{1}{r}\frac{\partial T}{\partial r} + \frac{\partial^2 T}{\partial z^2} \right) \tag{6-33}$$

which, using Eqn (6-25), and the fact that $v_r = 0$, reduces to

$$\rho C_p \left(v_z \frac{d\bar{T}}{dz} \right) = \frac{k}{r}\frac{\partial}{\partial r}\left(r \frac{\partial T}{\partial r} \right) \tag{6-34}$$

This equation can be integrated at any z location using the boundary conditions that $(\partial T/\partial r) = 0$ at $r = 0$ and $T = T_s$ at $r = r_0$, and using Eqn (6-30) for v_z, such that

$$r\frac{\partial T}{\partial r} = \frac{2\bar{U}}{\alpha}\frac{d\bar{T}}{dz}\left[\frac{r^2}{2} - \frac{r^4}{4r_0^2} \right] \tag{6-35}$$

and

$$T = T_s - \frac{2\bar{U}}{\alpha}\frac{d\bar{T}}{dz}\left[\frac{3r_0^2}{16} - \frac{r^2}{4} + \frac{r^4}{16r_0^2} \right] \tag{6-36}$$

where α, the thermal diffusivity, is defined by $\alpha = k/\rho C_p$. If q_s is specified, Eqn (6-36) can be simplified using Eqn (6-24) for q_s such that

$$T_s - T = \frac{4q_s}{kr_0}\left[\frac{3r_0^2}{16} - \frac{r^2}{4} + \frac{r^4}{16r_0^2} \right] \tag{6-37}$$

Integration of Eqn (6-37) provides the result that

$$T_s - \bar{T} = \frac{11}{24}\frac{q_s r_0}{k} \tag{6-38}$$

and defining a Nusselt number,

$$\mathbf{N}_d \equiv \frac{q_s d}{k(T_s - \bar{T})}$$

it follows that

$$\mathbf{N}_d = 4{\cdot}364$$

† Full details of references cited are given at end of chapter.

The Nusselt number for fully developed laminar pipe flow and constant heat flux is therefore a constant, independent of the Reynolds and Prandtl numbers.

It should be noted that while the velocity distribution given by Eqn (6-30) attributed to Poiseuille constitutes an exact solution of the Navier–Stokes equations, Eqn (6-37) is not an exact solution of the energy equation as viscous dissipation is not identically zero; it has been assumed to be negligible. For pipe flow, unlike the Couette flow experienced in lubrication problems, frictional heating is usually small enough to be neglected. The boundary condition of specified heat flux is not artificial as the flux is often prescribed for flow in pipes subject to electrical heating or thermal radiation. The only artificiality is in the fact stated at the beginning of this section that the transport properties are likely to be temperature dependent for fully developed laminar flow.

The energy equation for a constant surface temperature requires an iterative solution which yields the result that $N_d = 3·658$, showing that the Nusselt number, and the heat transfer coefficient, is dependent on the thermal boundary conditions—an obvious consequence that is often over-looked in solving practical problems.

The above example has been chosen for its simplicity to illustrate some important features of laminar pipe flow. It is not proposed to devote more time to this particular topic as the problems that can be solved are of an artificial rather than a practical nature. The reader interested in pursuing this subject is referred to Kays[3], where the flow in pipes and annuli are considered for a variety of boundary conditions.

6-4 Laminar forced flow over a flat plate : exact solutions

The flat plate is the traditional model for boundary layer flows as it provides a simple geometry that is capable of revealing most of the important aspects of the fluid dynamics and heat transfer characteristics of all external flows. It has been used as the basis of many calculations for the friction and heat loss of curvilinear surfaces such as turbine and compressor blades, aircraft wings, and so on, and as long as the boundary layer thickness is small compared with the radius of curvature of the surface ($\delta/r < 1/300$), then the solutions of the cartesian boundary layer equations provide useful answers to the flow problems. From the theoretical work of Murphy[4] it would appear that curvature can be important when $\delta/r > 0·05$.

In the general case the potential flow external to the boundary layer will prescribe the streamwise pressure gradient, and the temperature distribution over the surface will be arbitrary. However, while both of these effects will modify the answers to the problem, considerable insight can be gained by considering the simplest model: an isothermal plate with zero pressure gradient. The differential boundary layer equations will first be solved exactly in order to familiarize the reader with some of the mathematical techniques employed in external laminar flow problems and to introduce the concept of *recovery factors*. The integral equations will then be solved by an approximate method to illustrate the power of this technique which can yield satisfactory answers for relatively little mathematical complexity.

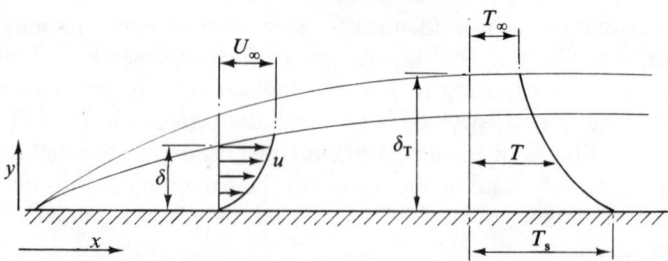

Fig. 6-6   Velocity and temperature boundary layers for forced flow over a heated plate

6-4-1   Fluid dynamics

Figure 6-6 illustrates the flow of fluid with a uniform velocity, U_∞, parallel to a plate aligned with the x-axis and with its leading edge at $x = 0$. As the potential velocity is constant, the pressure gradient is zero and so the boundary layer equations, Eqns (5-83a) and (5-83b), can be rewritten for forced laminar flow as

$$u\frac{\partial u}{\partial x} + v\frac{\partial u}{\partial y} = v\frac{\partial^2 u}{\partial y^2} \tag{6-39}$$

$$\frac{\partial u}{\partial x} + \frac{\partial v}{\partial y} = 0 \tag{6-40}$$

with the boundary conditions

$$y = 0 : u = v = 0; \qquad y = \infty : u = U_\infty$$

It is heuristically assumed that the velocity profiles are *similar* such that $u/U_\infty = \phi(y/\delta)$, where $\delta(x)$ is the thickness of the hydrodynamic boundary layer at any x location, and the function ϕ is independent of x. Exact solutions of the Navier–Stokes equations reveal that $\delta \sim \sqrt{(vx/U_\infty)}$, so we introduce a new coordinate, η, where

$$\eta \equiv y\sqrt{\frac{U_\infty}{vx}} \sim \frac{y}{\delta} \tag{6-41}$$

The continuity equation can be satisfied by a stream function, ψ, where

$$u = \frac{\partial \psi}{\partial y} \quad \text{and} \quad v = -\frac{\partial \psi}{\partial x}$$

so that

$$\frac{\partial u}{\partial x} + \frac{\partial v}{\partial y} = 0$$

Further, we introduce the transform

$$\psi = \sqrt{(vxU_\infty)}f(\eta) \qquad (6\text{-}42)$$

such that

$$u = \frac{\partial \psi}{\partial \eta}\frac{\partial \eta}{\partial y} = \sqrt{(vxU_\infty)}f'(\eta)\sqrt{\frac{U_\infty}{vx}}$$

that is,

$$\frac{u}{U_\infty} = f'(\eta) \qquad (6\text{-}43)$$

which satisfies the similarity condition. Equation (6-39) can now be expressed in terms of ψ, with the result that

$$\frac{\partial \psi}{\partial y}\frac{\partial^2 \psi}{\partial x\,\partial y} - \frac{\partial \psi}{\partial x}\frac{\partial^2 \psi}{\partial y^2} = v\frac{\partial^3 \psi}{\partial y^3} \qquad (6\text{-}44)$$

Table 6-1 Laminar flow over a flat plate: the Blasius solution as computed by Howarth[6] from Eqn (6-45)

η	f	f'	f''
0	0	0	0·33206
0·2	0·00664	0·06641	0·33199
0·4	0·02656	0·13277	0·33147
0·6	0·05974	0·19894	0·33008
0·8	0·10611	0·26471	0·32739
1·0	0·16557	0·32979	0·32301
1·4	0·32298	0·45627	0·30787
1·8	0·52952	0·57477	0·28293
2·2	0·78120	0·68132	0·24835
2·6	1·07252	0·77246	0·20646
3·0	1·39682	0·84605	0·16136
3·4	1·74696	0·90177	0·11788
3·8	2·11605	0·94112	0·08013
4·2	2·49806	0·96696	0·05052
4·6	2·88826	0·98269	0·02948
5·0	3·28329	0·99155	0·01591
5·4	3·68094	0·99616	0·00793
5·8	4·07990	0·99838	0·00365
6·2	4·47948	0·99937	0·00155
6·6	4·87931	0·99977	0·00061
7·0	5·27926	0·99992	0·00022
7·4	5·67924	0·99998	0·00007
7·8	6·07923	1·00000	0·00002
8·2	6·47923	1·00000	0·00001
8·6	6·87923	1·00000	0·00000

Using Eqns (6-41) and (6-42) we can write Eqn (6-44) as

$$ff'' + 2f''' = 0 \tag{6-45}$$

with the boundary conditions that $f(0) = f'(0) = 0$, and $f'(\infty) = 1$. It is left as an exercise for the reader to prove these results.

The third-order nonlinear ordinary differential equation, Eqn (6-45), was originally solved by Blasius[5] who employed series expansions, and later by Howarth[6] whose numerical solutions for f, f', and f'' are given in Table 6-1.

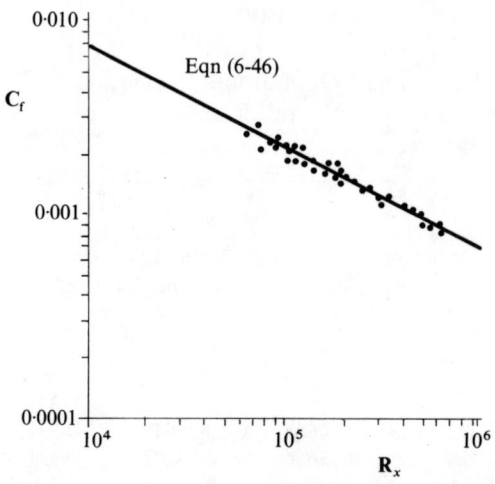

Fig. 6-7   Comparison between the theory of Blasius and the experiments of Liepmann and Dhawan for laminar flow over a flat plate

For later reference, the skin friction coefficient, C_f, defined as $C_f \equiv \tau_s / (\frac{1}{2}\rho U_\infty^2)$, where τ_s is the shear stress on the plate surface, will be calculated:

$$\tau_s = \mu\left(\frac{\partial u}{\partial y}\right)_{y=0} = \mu U_\infty \sqrt{\frac{U_\infty}{vx}} f''(0)$$

Hence, from Table 6-1,

$$C_f = 0.664 R_x^{-1/2} \tag{6-46}$$

where $R_x \equiv U_\infty x/v$.

Figure 6-7 shows the excellent agreement between Eqn (6-46) and the measurements of Liepmann and Dhawan[7]. Having obtained a solution for the velocity distribution, the thermal boundary layer problem can now be considered.

6-4-2 The isothermal plate

The thermal boundary layer is described by Eqn (5-83c), which for laminar flow can be written

$$u\frac{\partial T}{\partial x} + v\frac{\partial T}{\partial y} = \alpha\frac{\partial^2 T}{\partial y^2} + \frac{\mu}{\rho C_p}\left(\frac{\partial u}{\partial y}\right)^2 \tag{6-47}$$

with the boundary conditions: $y = 0$, $T = T_s$; $y = \infty$, $T = T_\infty$. Again, using the similarity transforms (6-41) and (6-42), the reader is left to verify for himself that Eqn (6-47) can be expressed as

$$\frac{d^2 T}{d\eta^2} + \tfrac{1}{2}\mathbf{P}f\frac{dT}{d\eta} = -\mathbf{P}\frac{U_\infty^2}{C_p}(f'')^2 \tag{6-48}$$

where $\mathbf{P} \equiv v/\alpha$. As the equation is linear, the general solution can be presented in the form

$$T(\eta) - T_\infty = c\theta_1(\eta) + \frac{U_\infty^2}{2C_p}\theta_2(\eta) \tag{6-49}$$

where T_∞ is the temperature outside of the thermal boundary layer, c is an arbitrary constant, and θ_1 and θ_2 are the solution of the homogeneous equation and a particular solution, respectively. It is convenient to choose boundary conditions such that θ_1 is the solution of the cooling problem when frictional heating is ignored, and θ_2 is the solution for the case of an adiabatic plate, that is for $(\partial T/\partial y)_{y=0} = 0$, when frictional heating is considered. Hence, the homogeneous equation can be expressed as

$$\theta_1'' + \tfrac{1}{2}\mathbf{P}f\theta_1' = 0 \tag{6-50}$$

where $\theta_1(0) = 1$ and $\theta_1(\infty) = 0$, and the non-homogeneous equation becomes

$$\theta_2'' + \tfrac{1}{2}\mathbf{P}f\theta_2' = -2\mathbf{P}(f'')^2 \tag{6-51}$$

where $\theta_2'(0) = \theta_2(\infty) = 0$. It can now be seen from Eqn (6-49), by putting $\eta = 0$, that

$$c = T_s - T_\infty - \frac{U_\infty^2}{2C_p}\theta_2(0)$$

6-4-3 The cooling problem: ignoring frictional heating

Putting $\phi_1 = \theta_1'$ in Eqn (6-50) produces the first-order equation

$$\frac{d\phi_1}{d\eta} = -\tfrac{1}{2}\mathbf{P}f\phi_1 \tag{6-52}$$

with the solution

$$\ln \phi_1 = -\tfrac{1}{2}\mathbf{P}\int f\, d\eta + c_1 \tag{6-53}$$

where c_1 is an arbitrary constant. This quadrature can be solved by substituting for f from Eqn (6-45) where $-\tfrac{1}{2}f = f'''/f''$. Hence, Eqn (6-53) simplifies to

$$\ln \phi_1 = \mathbf{P}\ln f'' + c_1$$

or

$$\phi_1 = \frac{d\theta_1}{d\eta} = c_2(f'')^{\mathbf{P}} \tag{6-54}$$

Integration of Eqn (6-54) yields the result

$$\theta_1 = c_2 \int (f'')^{\mathbf{P}}\, d\eta + c_3$$

where the arbitrary constants can be removed by using the boundary conditions $\theta_1(0) = 1$ and $\theta_1(\infty) = 0$ such that $c_2 = [-\int_0^\infty (f'')^{\mathbf{P}}\, d\eta]^{-1}$ and $c_3 = 1$, and the final result is expressed as the quadrature

$$\theta_1(\eta) = 1 - \frac{\int_0^\eta (f'')^{\mathbf{P}}\, d\eta}{\int_0^\infty (f'')^{\mathbf{P}}\, d\eta} \tag{6-55}$$

This answer was first produced by Pohlhausen[8] and the numerical solutions of Eqn (6-55) are shown plotted in Fig. 6-8 for a range of Prandtl numbers.

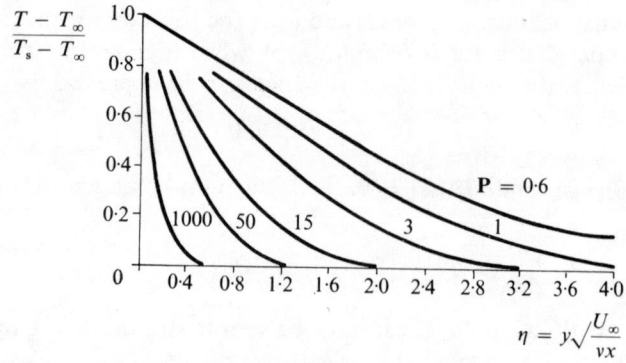

Fig. 6-8 Temperature distribution for laminar flow over a flat plate, neglecting frictional heating effects (after Pohlhausen[8])

For the case of $\mathbf{P} = 1$, Eqn (6-55) can be readily integrated to give

$$\theta_1(\eta) = 1 - \frac{f'(\eta) - f'(0)}{f'(\infty) - f'(0)} = 1 - \frac{u}{U_\infty}$$

or

$$\frac{T_s - T}{T_s - T_\infty} = \frac{u}{U_\infty} \tag{6-56}$$

This result shows that, ignoring frictional heating, the velocity and temperature distributions are similar if the Prandtl number is unity. We turn now to the consideration of heat transfer from the flat plate. In order to calculate the temperature gradient at the plate surface, we note from Eqn (6-54) that

$$\frac{d\theta_1}{d\eta} = c_2(f'')^P$$

and from Table 6-1, $f''(0) = 0.332$, hence

$$-\left(\frac{d\theta_1}{d\eta}\right)_{\eta=0} = -0.332^P c_2 = a_1(P) \tag{6-57}$$

where $c_2 = -1/\int_0^\infty (f'')^P d\eta$. The function $a_1(P)$ has been evaluated numerically for a range of Prandtl numbers by Pohlhausen and the results are shown in Table 6-2. The results can be approximated by

$$\left. \begin{array}{ll} a_1(P) \approx 0.332P^{1/3} & 0.6 < P < 10 \\ 0.564P^{1/2} & P \to 0 \\ 0.339P^{1/3} & P \to \infty \end{array} \right\} \tag{6-58}$$

$a_1(P)$ will be discussed again during the calculation of the Nusselt numbers after we have considered the effect of frictional heating.

Table 6-2   Heat transfer from a flat plate: the dimensionless constant, $a_1(P)$, computed by Pohlhausen[8] from Eqn (6-57), and the recovery factor $\mathscr{R}$ computed from Eqn (6-65) for $\theta_2(0)$

P	0·6	0·7	0·8	0·9	1·0	1·1	7	10	15
a_1	0·276	0·293	0·307	0·320	0·332	0·344	0·645	0·730	0·835
$\mathscr{R}$	0·770	0·835	0·895	0·950	1·000	1·050	2·515	2·965	3·535

6-4-4   The adiabatic plate: including frictional heating

The non-homogeneous second-order equation, Eqn (6-51), can be turned into a first-order equation by putting $\phi_2 = \theta_2'$ such that

$$\phi_2' + \tfrac{1}{2}Pf\phi_2 = -2P(f'')^2 \tag{6-59}$$

Using the standard mathematical technique of variation of the parameter we look for a solution of Eqn (6-59) in the form

$$\phi_2 = \gamma\phi_1 + \varepsilon\phi_1 \tag{6-60}$$

where ϕ_1 is a solution of the homogeneous equation, and from Eqn (6-54) $\phi_1 = (f'')^P$ while γ is an arbitrary constant and ε is an arbitrary function. Substituting $\varepsilon\phi_1$ for ϕ_2 in Eqn (6-59) gives

$$\varepsilon'\phi_1 + \varepsilon(\phi_1' + \tfrac{1}{2}Pf\phi_1) = -2P(f'')^2$$

hence

$$\varepsilon' = -\frac{2P(f'')^2}{\phi_1} = -2P(f'')^{2-P} \tag{6-61}$$

and

$$\varepsilon = -2P \int (f'')^{2-P}\, d\eta \tag{6-62}$$

Substitution of Eqn (6-62) into Eqn (6-60) yields

$$\phi_2 = \gamma(f'')^P - 2P(f'')^P \int (f'')^{2-P}\, d\eta \tag{6-63}$$

Using the boundary condition that $\phi_2(0) = 0$ implies $\gamma = 0$,

$$\phi_2 = -2P(f'')^P \int_0^\eta (f'')^{2-P}\, d\eta \tag{6-64}$$

As $\theta_2(\infty) = 0$, then

$$\theta_2(\eta) = \int_\infty^\eta \frac{d\theta_2}{d\eta}\, d\eta = -\int_\eta^\infty \phi_2\, d\eta$$

Hence, using Eqn (6-64), $\theta_2(\eta)$ can be expressed as

$$\theta_2(\eta) = 2P \int_\eta^\infty \left[(f'')^P \int_0^\eta (f'')^{2-P}\, d\eta \right] d\eta \tag{6-65}$$

Again the quadrature can only, in general, be evaluated by numerical methods, but for a unity Prandtl number direct integration is possible, and it is left to the reader to show that, for $P = 1$,

$$\theta_2(\eta) = 1 - (f')^2$$

Referring to Eqn (6-49) it can be seen that the solution to the adiabatic plate problem can be expressed as

$$T_{ad}(\eta) - T_\infty = \frac{U_\infty^2}{2C_p}\, \theta_2(\eta)$$

and the temperature of the adiabatic plate surface, $T_{ad, s}$ will be given by

$$T_{ad, s} = T_\infty + \frac{U_\infty^2}{2C_p} \theta_2(0) \qquad (6\text{-}66)$$

where $\theta_2(0)$ is evaluated numerically from Eqn (6-65) and is solely a function of the Prandtl number, **P**. For convenience, Eqn (6-66) is rewritten in the form

$$T_{ad, s} = T_\infty + \mathscr{R} \frac{U_\infty^2}{2C_p} \qquad (6\text{-}67)$$

where $\mathscr{R}$ is termed the *recovery factor* and is, for laminar flow, dependent only on the Prandtl number of the fluid and is unity when **P** = 1. Values of $\mathscr{R}$, which is equal to $\theta_2(0)$, are given in Table 6-2 but the following approximations can be used with reasonable accuracy:

$$\begin{array}{ll} \text{Air, water (moderate } \mathbf{P}) & \mathscr{R} \approx \mathbf{P}^{1/2} \\ \text{Oils (large } \mathbf{P}) & \mathscr{R} \approx 1{\cdot}9\mathbf{P}^{1/3} \end{array} \right\} \qquad (6\text{-}68)$$

It should be repeated, however, that just as for laminar pipe flows the above flat plate theory takes no account of property variations and will only be applicable to oils if the temperature differences are small.

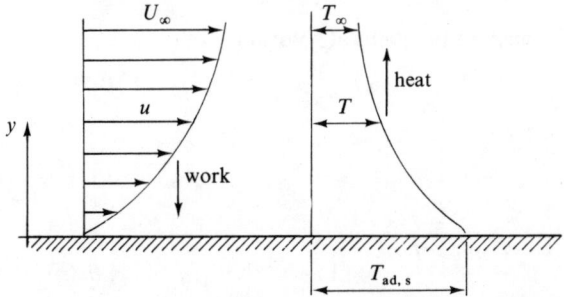

Fig. 6-9 Velocity and temperature distribution in the boundary layer on an adiabatic plate

A certain appreciation as to the physical significance of the recovery factor can be gained by considering the *total temperature*, T_t, of a moving fluid, which is given by

$$T_t = T_\infty + \frac{U_\infty^2}{2C_p}$$

This would be the temperature recorded by a thermometer inserted at the stagnation point of a body placed in the fluid, and for this reason T_t is sometimes called the *stagnation temperature*. Thus $\mathscr{R}$ is the ratio of the frictional, or viscid, temperature rise to the frictionless, or inviscid, temperature rise—that is

$$\mathscr{R} = \frac{T_{ad,s} - T_\infty}{T_t - T_\infty} \qquad (6\text{-}69)$$

From Table 6-2 it can be seen that for $\mathbf{P} < 1$ the viscid temperature rise is less than the inviscid rise, and for $\mathbf{P} > 1$ the reverse is true. The physical reason for this can be appreciated if it is recalled that the Prandtl number is the ratio of the kinematic viscosity to the thermal diffusivity of the fluid. Figure 6-9 illustrates the fact that the wall to free stream temperature difference must be large enough to balance the heat flow away from the plate with the work transport towards the plate. If $v > \alpha$ ($\mathbf{P} > 1$) the temperature difference will be greater than if $v < \alpha$ ($\mathbf{P} < 1$). For $v = \alpha$ the balance is achieved by complete recovery of the kinetic energy of the fluid, that is, $\mathscr{R} = 1$ when $\mathbf{P} = 1$.

It is apparent that if $T_s > T_{ad,s}$ the fluid will cool the plate, and if $T_s < T_{ad,s}$ the fluid will heat the plate. Therefore, the adiabatic temperature difference $T_s - T_{ad,s}$ is a useful criterion of heat transfer rates. It will obviously be advisable to base heat transfer coefficients and Nusselt numbers on the adiabatic temperature difference rather than $T_s - T_\infty$ as the wall heat flux can go to zero even when the latter temperature difference is non-zero. We shall now look more closely at heat transfer from the flat plate.

6-4-5   Heat transfer: the general solution

The general solution of Eqn (6-47) can now be expressed in the form of Eqn (6-49):

$$T(\eta) - T_\infty = [(T_s - T_\infty) - (T_{ad,s} - T_\infty)]\theta_1(\eta) + \frac{U_\infty^2}{2C_p}\theta_2(\eta) \quad (6\text{-}70)$$

or

$$\frac{T(\eta) - T_\infty}{T_s - T_\infty} = [1 - \tfrac{1}{2}\mathscr{R}\mathbf{E}]\theta_1(\eta) + \tfrac{1}{2}\mathbf{E}\theta_2(\eta) \qquad (6\text{-}71)$$

where $\mathbf{E} \equiv U_\infty^2/[C_p(T_s - T_\infty)]$. For heat transfer

$$q_s = -k\left(\frac{\partial T}{\partial y}\right)_{y=0} = -k\sqrt{\frac{U_\infty}{vx}}\left(\frac{\partial T}{\partial \eta}\right)_{\eta=0}$$

which, employing Eqn (6-70) and remembering that $(\partial\theta_2/\partial\eta)_{\eta=0} = 0$, yields the result

$$q_s = -k\sqrt{\frac{U_\infty}{vx}}(T_s - T_{ad,s})\left(\frac{\partial\theta_1}{\partial\eta}\right)_{\eta=0}$$

Hence, defining the local Nusselt number, N_x, as

$$N_x \equiv \frac{q_s x}{k(T_s - T_{ad,s})} \qquad (6\text{-}72a)$$

and using Eqn (6-57) we find that

$$N_x = a_1(P)R_x^{1/2} \qquad (6\text{-}73a)$$

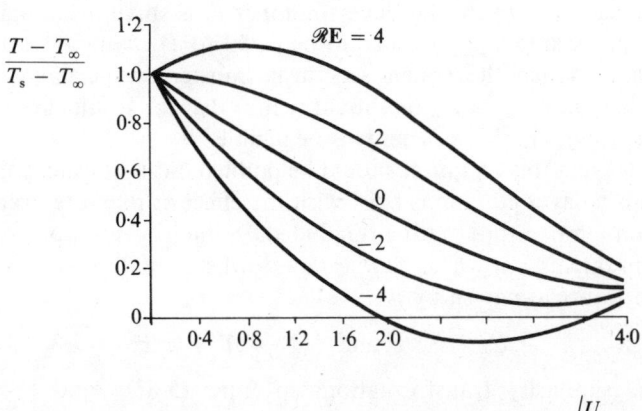

$$\eta = y\sqrt{\frac{U_\infty}{vx}}$$

Fig. 6-10   Temperature distribution for laminar flow over a flat plate, including frictional heating effects for $P = 0.7$

It is convenient to mention here that overall heat transfer rates can be calculated from an overall Nusselt number, $\overline{N}$ defined as

$$\overline{N} \equiv \frac{\bar{q}_s l}{k(T_s - T_{ad,s})} \qquad (6\text{-}72b)$$

where $\bar{q}_s$ is the average heat transfer rate over a plate of length l, that is

$$\bar{q}_s = \frac{1}{l}\int_0^l q_s\, dx$$

It is therefore apparent that

$$\overline{N} = 2N_x \qquad (6\text{-}73b)$$

Thus, unlike the Couette and Pouiseuille flows, heat transfer from a flat plate is governed by the Prandtl *and* the Reynolds numbers. It is interesting to observe that the plate will only be cooled by the fluid if $\mathscr{R}E < 2$, that is if $T_s > T_{ad,s}$, which should be compared with the condition that $PE < 2$ for cooling in Couette flow: for $P = 1$ these conditions are identical. The temperature distribution for a range of values of $\mathscr{R}E$ calculated from Eqn (6-71) is shown in Fig. 6-10.

In conclusion, after a certain amount of mathematical manipulation we have been able to produce exact solutions of the boundary layer equations, and we have been able to bring out the effect of frictional heating and the concept of recovery factors and adiabatic wall temperatures. These are important concepts that can still be retained for high-speed applications, even when the flow is turbulent and compressible, although under these conditions the answers will obviously be quantitatively different. For low-speed applications, where the Eckert number E, is small, frictional heating is often ignored and $T_{ad, s}$ is taken to be equal to T_∞, which is exactly true in the limit. When the reader sees heat transfer coefficients or Nusselt numbers based on $T_s - T_\infty$ he should realize that the results are only valid when viscous dissipation of energy is negligible.

Before leaving this section it should be pointed out that exact solutions of the boundary layer equations exist with the effect of pressure gradient and free stream temperature gradient included. Similarity solutions exist for a class of problems known as *wedge flows*† where the free stream velocity and temperature are given by

$$U_\infty \propto x^m \quad \text{and} \quad (T_s - T_\infty) \propto x^n$$

Using the similarity transformations in Eqns (5-83b) and (5-83c), the boundary layer equations can be written

$$f''' + \tfrac{1}{2}(m + 1)ff'' + m[1 - (f')^2] = 0 \qquad (6\text{-}74a)$$

with $f(0) = f'(0) = 0$ and $f'(\infty) = 1$, and

$$\theta'' + \tfrac{1}{2}(m + 1)\mathbf{P}f\theta' - n\mathbf{P}f'\theta = -\mathbf{P}\mathbf{E}x^{2m-n}(f'')^2 \qquad (6\text{-}74b)$$

with $\theta(0) = 1$, $\theta(\infty) = 0$ where $\theta \equiv (T - T_\infty)/(T_s - T_\infty)$ which, for $m = n = 0$, reduce to Eqns (6-45) and (6-48).

Similarity solutions for Eqn (6-74b) will only exist if $2m - n = 0$ or if dissipation is neglected. The latter case has been considered by a number of authors, and Eckert's[9] solution for the effect of pressure gradient (that is, the effect of m) on the Nusselt number with $n = 0$ for a range of Prandtl numbers is shown graphically in Fig. 6-11. Attention is drawn to Eqn (5-75) where the prescribed pressure gradient is given by

$$\frac{1}{\rho}\frac{dp}{dx} = -U_\infty \frac{dU_\infty}{dx}$$

hence, $m > 0$ corresponds to a negative (or favourable) pressure gradient and $m < 0$ corresponds to a positive (or adverse) pressure gradient. The latter has the effect of 'retarding' the fluid and, hence, thickening the boundary layer with the possibility of separation. Detailed examination of the

---

† The free stream velocity is that due to the potential flow past a wedge, the included angle of which is equal to $2\pi m/(m + 1)$.

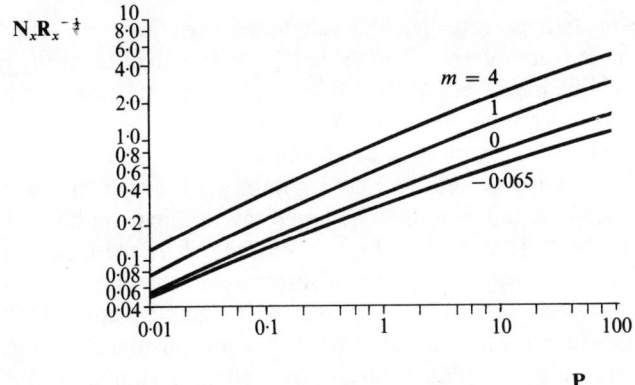

Fig. 6-11 Local Nusselt number as a function of Prandtl number for laminar flow over a flat plate where $U_\infty \propto x^m$ (after Eckert[9])

momentum equation by Hartree[10] revealed the existence of point of inflexions for $m < 0$ with separation at $m = -0.091$. Thus the effect of an adverse pressure gradient is to reduce heat transfer and eventually to cause separation of the boundary layer. In practice, laminar separation is usually followed by reattachment of a turbulent boundary layer, as the higher energy level of turbulent flow renders the boundary layer more resistant to separation by an adverse pressure gradient. A favourable gradient has the reverse effect as the boundary layer is stabilized and is less likely to become turbulent. The resulting laminar boundary layer, which is thinner than that associated with an adverse gradient, assists the transfer of heat, as can be seen by the higher Nusselt numbers for $m > 0$ in Fig. 6-11.

The effect of temperature boundary conditions can also be important, as is illustrated in Fig. 6-12 which shows the effect of the exponent n on the

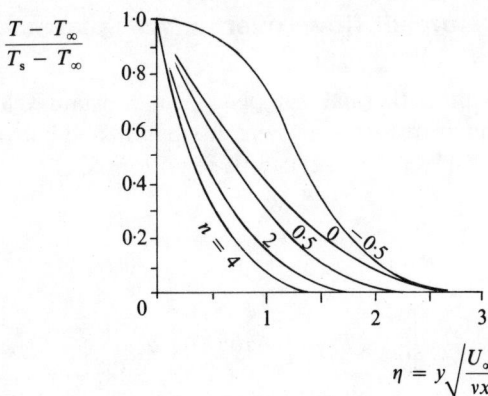

Fig. 6-12 Temperature distribution for laminar flow over a flat plate with $P = 0.7$, illustrating the effect of plate temperatures of the form: $T - T_\infty \propto x^n$ (after Levy[11])

cross-stream temperature profile calculated by Levy[11], who ignored dissipation. It is instructive to note that for $n > 0$ the heat transfer from a hot plate will be increased, but for $n < -\frac{1}{2}$ the hot plate cannot be cooled by convection to the cooler fluid. We have already seen how a hot plate can be heated by a cooler fluid due to frictional heating, but the phenomenon of a hot plate that is heated by a cooler fluid when dissipation is negligible is more difficult to understand. The apparent paradox can be explained by considering the fluid particles close to the hot wall: these particles move downstream from hotter upstream regions, and as the plate is being cooled (by some external means) at a greater rate than the adjacent fluid, the temperature gradients on the plate surface become positive, transferring heat from the fluid into the plate. The inverse effect is similar to *film cooling* where a wall jet of cold fluid acts as a thermal buffer between a wall and a hot gas stream: heat is transferred from both the wall and the gas stream into the coolant.

It should now be clear to the reader that viscous dissipation, pressure gradients, and arbitrary wall temperatures can have a significant effect on heat transfer from a flat plate, and indeed from any other surface. The foregoing has been an introduction to, rather than a complete treatment of, flat plates, and for calculation of the completely general boundary layer problem the reader is referred to the numerical techniques described in Chapter 7.

It is now a convenient time to reintroduce the integral equations, and in order to emphasize the comparative simplicity of this technique we shall take a second look at the flat plate problem. As the object is to illustrate the power of this approximate solution technique, and not to elicit new physical aspects of the problem, we shall only consider the isothermal plate and ignore the effects of pressure gradient and viscous dissipation.

6-5 Laminar forced flow over a flat plate: approximate solutions

For the case of an isothermal flat plate in a constant velocity, constant temperature, laminar free stream where dissipation and buoyancy forces are negligible, Eqns (5-88) and (5-92) can be rewritten as

$$\frac{d}{dx} \int_0^\delta u(u - U_\infty) \, dy = v\left(\frac{\partial u}{\partial y}\right)_{y=0} \tag{6-75}$$

and

$$\frac{d}{dx} \int_0^{\delta_T} u(T - T_\infty) \, dy = -\alpha\left(\frac{\partial T}{\partial y}\right)_{y=0} \tag{6-76}$$

The technique used to solve these equations is based on the method of Pohlhausen[12] where the velocity and temperature distributions are represented by polynomials. First, assuming similar profiles, let us define $\eta \equiv y/\delta$ and $g(\eta) \equiv u/U_\infty$, then Eqn (6-75) can be expressed as

$$\frac{d}{dx}\left[\delta \int_0^1 g(g - 1)\, d\eta\right] = -\frac{v}{U_\infty \delta}\left(\frac{dg}{d\eta}\right)_{\eta=0} \tag{6-77}$$

and we approximate $g(\eta)$ by

$$g(\eta) = a + b\eta + c\eta^2 + d\eta^3 + \cdots$$

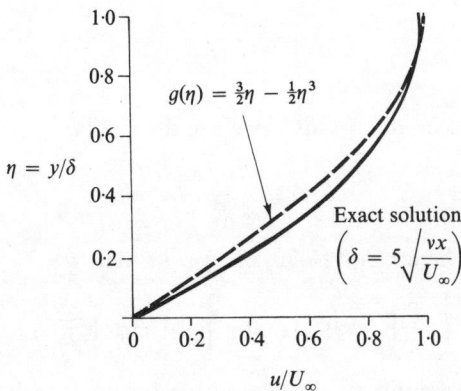

Fig. 6-13 Comparison between approximate and exact velocity distributions for laminar flow over a flat plate

For simplicity we shall consider only these first four terms and eliminate the coefficients a, b, c, and d from the boundary conditions that $g(0) = 0$ and $g(1) = 1$, together with the compatibility requirements† that $g''(0) = 0$ and $g'(1) = 0$. Other terms could be introduced, and the coefficients removed by further compatibility conditions, but it is the object of this exercise to show the reader that reasonable answers are produced by relatively crude assumptions. Using these constraints, we find that

$$g(\eta) = \tfrac{3}{2}\eta - \tfrac{1}{2}\eta^3$$

which is shown plotted in Fig. 6-13 against the exact solution taken from Table 6-1 with $\delta = 5\sqrt{(vx/U_\infty)}$, which corresponds to $u/U_\infty \approx 0.99$. Obviously the addition of more terms to the polynomial would improve the agreement, but these will have little effect on the heat transfer problem.

† Considering Eqn (5-83b) for laminar forced flow, as $y \to 0$, $-dp/dx + \mu(\partial^2 u/\partial y^2) \to 0$. If $dp/dx = 0$ then $(\partial^2 u/\partial y^2)_{y=0}$ and higher derivatives are zero. It is also apparent that as $y \to \delta$, $u \to U_\infty$ and $(\partial u/\partial y)_{y=\delta}$ and all higher derivatives are zero if $U_\infty = U_\infty(x)$.

Using the results that $\int_0^1 g(g - 1) \, d\eta = -\frac{39}{280}$ and $(dg/d\eta)_{\eta=0} = \frac{3}{2}$ we get from Eqn (6-77) the following first-order ordinary differential equation for δ:

$$-\frac{39}{280} \frac{d\delta}{dx} = -\frac{3}{2} \frac{v}{U_\infty \delta}$$

Hence

$$\delta \frac{d\delta}{dx} = \frac{140}{13} \frac{v}{U_\infty}$$

and as $\delta = 0$ at $x = 0$,

$$\tfrac{1}{2}\delta^2 = \frac{140}{13} \frac{vx}{U_\infty}$$

or

$$\delta/x = 4 \cdot 64 \mathbf{R}_x^{-1/2} \tag{6-78}$$

where $\mathbf{R}_x \equiv U_\infty x/v$.

For the thermal boundary layer we use the similarity profile

$$\theta(\eta_T) \equiv \frac{T(\eta_T) - T_\infty}{T_s - T_\infty}$$

where $\eta_T \equiv y/\delta_T$, hence Eqn (6-76) can be written

$$\frac{d}{dx}\left[\delta_T \int_0^1 g(\eta)\theta(\eta_T) \, d\eta_T\right] = \frac{\alpha}{U_\infty \delta_T}\left(\frac{d\theta}{d\eta_T}\right)_{\eta_T = 0} \tag{6-79}$$

Again we assume a polynomial distribution for $\theta(\eta_T)$ such that

$$\theta(\eta_T) = a_T + b_T \eta_T + c_T \eta_T^2 + d_T \eta_T^3 + \cdots$$

and use the boundary conditions that $\theta(0) = 1$, $\theta(1) = 0$ and the compatibility requirements that $\theta''(0) = 0$ and $\theta'(1) = 0$ to show that

$$\theta(\eta_T) = 1 - \tfrac{3}{2}\eta_T + \tfrac{1}{2}\eta_T^3 \tag{6-80}$$

We also introduce a new variable ζ as the ratio of the thermal boundary layer to the hydrodynamic boundary layer thickness, such that

$$\zeta \equiv \delta_T/\delta = \eta/\eta_T \tag{6-81}$$

Introducing Eqns (6-80) and (6-81), Eqn (6-79) becomes

$$\frac{d}{dx}\left[\delta_T \zeta \left(\frac{3}{20} - \frac{3}{280}\zeta^2\right)\right] = \frac{3}{2}\frac{\alpha}{U_\infty \delta_T} \tag{6-82}$$

For moderate Prandtl numbers $\delta \approx \delta_T$, that is ζ is of order unity, and so the term involving $\frac{3}{280}\zeta^2$ can be neglected to give

$$\frac{3}{20}\frac{d}{dx}(\zeta^2 \delta) = \frac{3}{2}\frac{\alpha}{U_\infty \zeta \delta}$$

or
$$\zeta^3 \delta \frac{d\delta}{dx} + 2\zeta^2 \delta^2 \frac{d\zeta}{dx} = \frac{10\alpha}{U_\infty}$$

Using Eqn (6-78),

$$\zeta^3 + \frac{4}{3} x \frac{d}{dx} (\zeta^3) = \frac{13}{14} \frac{\alpha}{v}$$

hence, for the case of an unheated starting length x_0 (where $\delta_T = 0$ if $x \leqslant x_0$),

$$\zeta = 0.975 P^{-1/3} [1 - (x_0/x)^{3/4}]$$

For the case of $x_0 = 0$, or as $x \to \infty$,

$$\zeta = \delta_T/\delta \approx P^{-1/3} \tag{6-83}$$

Defining the Nusselt number as

$$N_x \equiv \frac{q_s x}{k(T_s - T_\infty)}$$

then as
$$q_s = -\frac{k}{\delta_T} \left(\frac{\partial\theta}{\partial\eta_T}\right)_{\eta_T=0} (T_s - T_\infty)$$

and $(\partial\theta/\partial\eta_T)_{\eta_T=0} = -\frac{3}{2}$ from Eqn (6-80), it follows that

$$N_x = 0.331 P^{1/3} R_x^{1/2} \tag{6-84}$$

The agreement between this approximate answer and the exact solution given by Eqn (6-73), which for moderate Prandtl numbers is

$$N_x = 0.332 P^{1/3} R_x^{1/2}$$

is very good.

Having seen the simplicity of this approximate method, we shall now carry out a similar analysis of a natural convection problem: laminar free convection from a vertical plate.

6-6 Laminar free convection from a vertical plate

In forced convection problems the buoyancy forces due to density gradients in the fluid are ignored as they are usually small-order terms. When there is no external flow field, or the buoyancy forces are much greater than the inertial forces—that is $G_x \gg R_x^2$ where the local Grashof number G_x is defined as

$$G_x \equiv g\beta x^3 (T_s - T_\infty)/v^2 \tag{6-85}$$

where $T_s > T_\infty$—then, as we saw in Section 5-3, the Nusselt number becomes independent of the Reynolds numbers (and effectively independent of the Eckert number as dissipation is usually negligible) such that

$$\mathbf{N}_x = \mathbf{N}_x(\mathbf{G}_x, \mathbf{P})$$

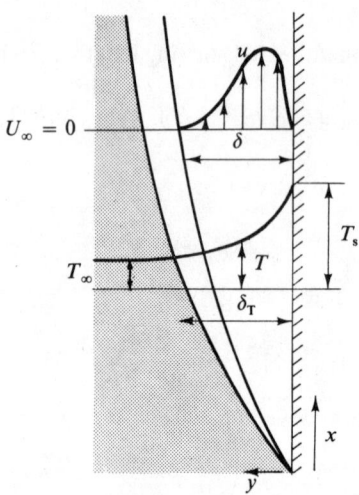

Fig. 6-14   Temperature and velocity distributions due to free convection from a vertical heated plate

For the case of the vertical plate illustrated in Fig. 6-14, we shall consider the simplest case of an isothermally heated plate at temperature T_s in a quiescent fluid ($U_\infty = 0$) with a constant temperature† T_∞. Dissipation will be neglected, and we shall restrict the analysis to fluids with a Prandtl number near unity so that we can assume $\delta \approx \delta_T$. In these circumstances for laminar flow, Eqns (5-88) and (5-92) can be written, assuming β to be constant,

$$\frac{d}{dx}\int_0^\delta u^2\, dy = g\beta \int_0^\delta (T - T_\infty)\, dy - v\left(\frac{\partial u}{\partial y}\right)_{y=0} \qquad (6\text{-}86a)$$

$$\frac{d}{dx}\int_0^\delta u(T - T_\infty)\, dy = -\alpha \left(\frac{\partial T}{\partial y}\right)_{y=0} \qquad (6\text{-}86b)$$

It is readily obvious that free convection problems involve the solution of coupled equations, where in forced convection the momentum and energy equation can be solved separately. However, as in Section 6-5, we shall use

---

† Note that in a quiescent fluid a pressure gradient can be generated owing to density variations in the 'free stream' outside the boundary layer. For an isothermal 'free stream' this effect can be ignored, and the pressure gradient neglected.

polynomial distributions for u and T to satisfy the boundary condition that $u(0) = u(\delta) = 0$, $T(0) = T_s$ and $T(\delta) = T_\infty$.

It is apparent that the velocity profile will have a turning point inside the boundary layer, but we shall assume that the temperature decays from the wall to the free stream. The following simple profiles will be assumed

$$u = u_1\eta(1 - \eta)^2 \tag{6-87a}$$

and

$$\theta = (1 - \eta)^2 \tag{6-87b}$$

where $\eta \equiv y/\delta$ and $\theta \equiv (T - T_\infty)/(T_s - T_\infty)$. These profiles have the properties

$$u(0) = u(1) = u'(1) = 0, \qquad u'(0) = u_1$$

$$\theta(0) = 1, \qquad \theta(1) = \theta'(1) = 0, \qquad \theta'(0) = -2$$

Also, differentiation of Eqn (6-87a) reveals that the maximum velocity, u_{max}, is given by

$$u_{max} = 4u_1/27$$

Also

$$\int_0^1 u^2 \, d\eta = u_1^2/105, \qquad \int_0^1 \theta \, d\eta = \tfrac{1}{3}, \qquad \int_0^1 u\theta \, d\eta = u_1/30 \quad (6\text{-}88)$$

hence Eqns (6-86a) and (6-86b) can be expressed as

$$\frac{1}{105}\frac{d}{dx}(u_1^2\delta) = \frac{1}{3}g\beta(T_s - T_\infty)\delta - v\frac{u_1}{\delta} \tag{6-89a}$$

$$\frac{1}{30}\frac{d}{dx}(u_1\delta) = 2\frac{\alpha}{\delta} \tag{6-89b}$$

As we are looking for similarity solutions, we assume answers of the form

$$u_1 = c_1 x^{m_1} \quad \text{and} \quad \delta = c_2 x^{m_2}$$

where c_1, c_2, m_1, and m_2 are constants, and substitution into Eqns (6-89a) and (6-89b) yields

$$\frac{1}{105}c_1^2 c_2(2m_1 + m_2)x^{2m_1 + m_2 - 1} = \frac{1}{3}g\beta(T_s - T_\infty)c_2 x^{m_2} - v\frac{c_1}{c_2}x^{m_1 - m_2} \tag{6-90a}$$

$$\frac{1}{30}c_1 c_2(m_1 + m_2)x^{m_1 + m_2 - 1} = 2\frac{\alpha}{c_2}x^{-m_2} \tag{6-90b}$$

Similarity solutions only exist if both sides of these equations are independent of x, thus the exponents m_1 and m_2 must be related by

$$2m_1 + m_2 - 1 = m_2 = m_1 - m_2$$

and

$$m_1 + m_2 - 1 = -m_2$$

or

$$m_2 = \tfrac{1}{4} \quad \text{and} \quad m_1 = \tfrac{1}{2}$$

Simultaneous solution of Eqns (6-90a) and (6-90b) for the coefficients c_1 and c_2 provides the results

$$c_1 = 5 \cdot 17 v \left(\frac{20}{21} + \frac{v}{\alpha}\right)^{-1/2} \left(\frac{g\beta(T_s - T_\infty)}{v^2}\right)^{1/2} \tag{6-91}$$

$$c_2 = 3 \cdot 93 \left(\frac{20}{21} + \frac{v}{\alpha}\right)^{1/4} \left(\frac{g\beta(T_s - T_\infty)}{v^2}\right)^{-1/4} \left(\frac{v}{\alpha}\right)^{-1/2} \tag{6-92}$$

Consequently,

$$\delta/x = c_2 x^{m_2 - 1} = 3 \cdot 93 (0 \cdot 952 + \mathbf{P})^{1/4} \mathbf{P}^{-1/2} \mathbf{G}_x^{-1/4} \tag{6-93}$$

If we define the local Nusselt number as

$$\mathbf{N}_x \equiv \frac{q_s x}{k(T_s - T_\infty)}$$

where

$$q_s = -k\left(\frac{\partial T}{\partial y}\right)_{y=0} = 2k \frac{T_s - T_\infty}{\delta}$$

then

$$\mathbf{N}_x = 2x/\delta$$

and, using Eqn (6-93),

$$\mathbf{N}_x = 0 \cdot 508 \mathbf{P}^{1/2} (0 \cdot 952 + \mathbf{P})^{-1/4} \mathbf{G}_x^{1/4} \tag{6-94}$$

This relationship is valid for gases ($\mathbf{P} < 1$), and in order to evaluate the volume expansion coefficient, β, which from Eqn (5-37) is given by

$$\beta \equiv -\frac{1}{\rho}\left(\frac{\partial \rho}{\partial T}\right)_p$$

we can use the perfect gas law (where, for constant pressure, the density is inversely proportional to the absolute temperature of the gas) such that $\beta = 1/T$. The temperature T is normally taken as the absolute free stream temperature, and all properties should be evaluated at this temperature. In Chapter 7 it is shown that for compressible flow problems, where property variations with temperature are important, the film temperature will be evaluated in a slightly different manner.

For air, with $\mathbf{P} = 0.71$, Eqn (6-94) simplifies to

$$\mathbf{N}_x = 0.378\mathbf{G}_x^{1/4}$$

which is only 5 per cent different from the exact solution calculated numerically by Schmidt and Beckmann[13] as

$$\mathbf{N}_x = 0.360\mathbf{G}_x^{1/4}$$

The numerically calculated velocity and temperature profiles, which agree well with the experimental data of Schmidt and Beckmann, are shown in Fig. 6-15 together with the approximate profiles, used in the above approximate analysis, which are attributed to Squire[14]. The reader interested in pursuing this topic is referred to the paper by Ostrach[15], who solved the boundary layer equations numerically for a range of Prandtl numbers. It is interesting to note that the approximate solution, Eqn (6-94), is within ten per cent of the exact solutions for $0.01 < \mathbf{P} < 1,000$, which is remarkable considering the assumption made that $\delta = \delta_T$.

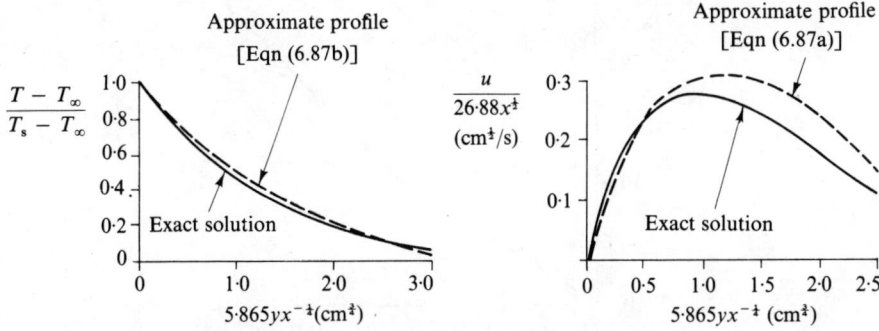

Fig. 6-15   Comparison between approximate and exact temperature and velocity distributions for laminar free convection from a vertical plate

Having demonstrated that the apparently gross assumptions of the integral techniques of analysis yield reasonably accurate predictions for laminar flows, attention will now be turned to the more common turbulent regime. It is, however, necessary to consider what governs the transition from laminar to turbulent flow in free convection. Figure 6-16 shows the experimental values of mean Nusselt number, $\overline{\mathbf{N}}$, against the product $\mathbf{G}_l\mathbf{P}$ as measured by Eckert and Jackson[16], where

$$\overline{\mathbf{N}} \equiv \frac{\overline{h}l}{k} \quad \text{and} \quad \mathbf{G}_l \equiv \frac{g\beta l^3(T_s - T_\infty)}{v^2}$$

and for a plate of length, l, the mean heat transfer coefficient, $\bar{h}$, is given by

$$\bar{h} = \frac{1}{l(T_s - T_\infty)} \int_0^l q_s(x)\, dx$$

For laminar free convection, where the local heat transfer coefficient, h, is proportional to $x^{-1/4}$, $\bar{h} = 4h/3$, or the mean coefficient over a length x is one and one-third times the local value at x. It can be seen from Fig. 6-16 that there is a change of slope around $\mathbf{G}_l\mathbf{P} = 10^9$: this corresponds to transition from laminar to turbulent flow. For laminar flow the results are correlated by $\overline{\mathbf{N}} = 0{\cdot}555(\mathbf{G}_l\mathbf{P})^{1/4}$ which for air with $\mathbf{P} = 0{\cdot}71$ reduces to $\overline{\mathbf{N}} = 0{\cdot}512\mathbf{G}_l^{1/4}$. Equation (6-94) can be rewritten as

$$\overline{\mathbf{N}} = 0{\cdot}677\mathbf{P}^{1/2}(0{\cdot}952 + \mathbf{P})^{-1/4}\mathbf{G}_l^{1/4} \qquad (6\text{-}95)$$

which, for $\mathbf{P} = 0{\cdot}71$, reduces to $\overline{\mathbf{N}} = 0{\cdot}504\mathbf{G}_l^{1/4}$ agreeing quite well with the correlation.

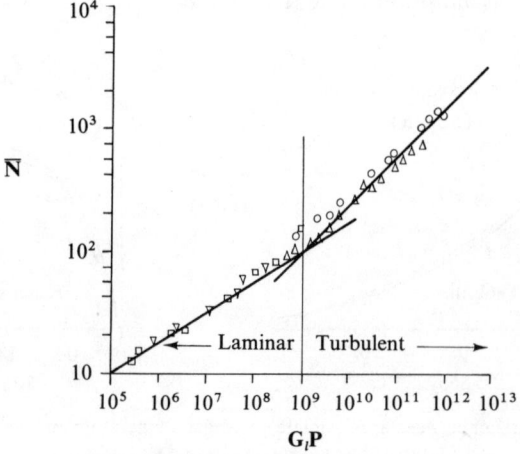

Fig. 6-16   Mean Nusselt number as a function of Grashof number for the case of free convection from a vertical surface (after Eckert and Jackson[16])

The effect of a non-zero free stream velocity is to reduce this transition value, and for a classification of convection into free, forced, and mixed laminar and turbulent regimes the reader is referred to the paper by Eckert and Diaguila[17]. Sparrow and Gregg[18] have analysed free convection over a vertical plate for very small Prandtl numbers, and Le Fevre[19] has studied the asymptotic values for $\mathbf{P} \to 0$ and $\mathbf{P} \to \infty$. Sparrow et al.[20] have also obtained solutions of the differential boundary layer equation for velocity and surface temperature distributions of the form $U_\infty \propto x^m$ and $(T_s - T_\infty) \propto x^{2m-1}$. Free convection from a horizontal heated cylinder was studied by

Hermann[21] whose results agree well with the experiments conducted in air by Jodlbauer[22], and a number of correlations for free convection from various geometries are included in Chapter 7.

It is now a convenient time to turn our attention to turbulent flows. We shall look more closely at the structure of turbulent boundary layers and introduce some of the techniques that are useful for the solution of turbulent free and forced convection problems.

6-7 Analogy between transfer of heat and momentum in turbulent flow

Similarities between velocity and temperature distributions in the preceding laminar examples have already become evident; as we shall shortly see, the similarities are more than coincidental and are true, under certain circumstances, in turbulent as well as in laminar flow. Before proceeding further in that direction it may be profitable to look more closely at some of the hypotheses that have been postulated concerning the structure of a turbulent boundary layer.

Despite the large amount of time devoted by eminent research workers, engineers, physicists, and mathematicians into the theoretical study of turbulent flow, very little has been of use to the practising engineer or designer. At the present time something of a dichotomy exists between the mathematical models of the theoreticians and the semi-empirical methods that are used by the designers. As we shall see in Chapter 7 there is hope that new numerical prediction methods, making use of the digital computer, will be able to act as a proving ground for turbulent hypotheses, but for the present we must resign ourselves to relatively crude assumptions regarding the exchange of heat and momentum in turbulent flow. We shall confine ourselves to relatively simple, but very useful, turbulence concepts, and for the rest of this section we shall revert to the bar and prime notation of Section 5-4.

As early as 1877 Boussinesq[23] used the concept of a mixing coefficient, A_τ, for turbulent flow analogous to the coefficient of viscosity, μ, in laminar flow. Hence for turbulent flow, recalling Eqn (5-82a), the shear stress can be represented by†

$$\tau = \mu \frac{\partial \bar{u}}{\partial y} - \rho \overline{u'v'} = (\mu + A_\tau) \frac{\partial \bar{u}}{\partial y} \tag{6-96}$$

In 1925 Prandtl proposed his well-known *mixing length hypothesis* in which he assumed, from physical considerations, that

$$\overline{u'v'} = -l^2 |u'| \, |v'| = -l^2 \left| \frac{\partial \bar{u}}{\partial y} \right| \frac{\partial \bar{u}}{\partial y}$$

† In this section, bars will be used to denote time-average values.

where l is termed the *mixing length* and is similar to the mean free path used in the kinetic theory of gases. Hence, Eqn (6-96) can be expressed as

$$\tau = \left(\mu + \rho l^2 \left| \frac{\partial \bar{u}}{\partial y} \right| \right) \frac{\partial \bar{u}}{\partial y} \tag{6-97}$$

where the unknown coefficient A_τ has been replaced by a term involving an unknown length scale, l, and the modulus of the velocity gradient. However, experimental observations and physical insight allow values of l to be assumed, and despite the fact that the mixing length model has obvious weaknesses (experimental measurements have shown that τ does not always vanish when $\partial \bar{u}/\partial y = 0$), it has been used as the basis of many successful calculation procedures.

As the turbulent eddies must be damped out at the wall, owing to the no-slip condition, the mixing length must be zero at the wall. The region where the turbulent shear stress is small compared with the laminar shear stress is known as the *laminar sublayer* or, more correctly, the *viscous sublayer*. Close to a solid wall, if the pressure gradient is zero, the shear stress will be constant owing to the fact that if the inertial terms are negligible, $\partial \tau/\partial y = 0$. Hence, in the viscous sublayer $\partial^2 \bar{u}/\partial y^2 = 0$, implying a linear velocity distribution, and hence the term laminar sublayer.

Despite the fact that outside this region the inertial effects are not necessarily negligible, Prandtl assumed that for fully developed flow past a flat plate, $\tau = \tau_s$, and that the mixing length varied linearly with y, that is,

$$l = \kappa y \tag{6-98}$$

where κ is a mixing length constant.

Away from the wall, where viscous effects are negligible, Eqn (6-97) reduces to

$$\tau_s = \rho \kappa^2 y^2 \left(\frac{d\bar{u}}{dy} \right)^2$$

or

$$\frac{d\bar{u}}{dy} = \frac{u^*}{\kappa y}, \tag{6-99}$$

where

$$u^* \equiv \sqrt{(\tau_s/\rho)} \tag{6-100}$$

is known as the *friction velocity*.

Equation (6-99) can be integrated to give

$$\frac{\bar{u}}{u^*} = \frac{1}{\kappa} \ln y + A \tag{6-101}$$

where A is an arbitrary constant. For a pipe of radius r_0, where $\bar{u} = U$ at $y = r_0$, Eqn (6-101) becomes

$$\frac{U - \bar{u}}{u^*} = \frac{1}{\kappa} \ln\left(\frac{r_0}{y}\right) \tag{6-102}$$

which is known as the *universal velocity-defect law*. For pipe flow, despite the fact that $dp/dx \neq 0$, a value of $\kappa = 0.4$ gives a very good fit with experimental data for rough and smooth pipes.[24,25] For large Reynolds numbers it is found that a value of $A = 5.5$ in Eqn (6-101) gives a good fit with experimental data outside the wall region.

While many more fundamental analyses have been performed and many measurements made to show that a logarithmic law of the wall is applicable for most boundary layer flows, the complete structure is still not completely understood. The simple model of a viscous region and a law-of-the-wall region has been extended to include a 'buffer' region between the two (where viscous and turbulence effects are significant), and wake regions to allow for free stream effects have been considered. Universal laws to include many or all of these effects have been postulated, but for clarity we shall focus our attention on relatively simple turbulent boundary layer models.

Returning to heat transfer in turbulent flow, the heat flux—it may be recalled from Eqn (5-82b)—is given by

$$q = -\left(k \frac{\partial \bar{T}}{\partial y} - \rho C_p \overline{v'T'}\right) \tag{6-103}$$

We postulate a turbulent law analogous to Fourier's law of thermal conduction, and rewrite Eqn (6-103) as

$$q = -(k + C_p A_q) \frac{\partial \bar{T}}{\partial y} \tag{6-104}$$

In general, the turbulent exchange coefficient for momentum and heat, A_τ and A_q are similar but not identical, but by analogy with the laminar Prandtl number $(\mathbf{P} \equiv \mu C_p / k)$ we introduce a turbulent Prandtl number, $\mathbf{P}_t$, where

$$\mathbf{P}_t \equiv A_\tau / A_q \tag{6-105}$$

Unlike the laminar Prandtl number, which is solely a function of the fluid properties, $\mathbf{P}_t$ depends upon the velocity and thermal distributions. Many attempts have been made to indirectly measure the distribution of $\mathbf{P}_t$ across boundary layers, and a summary of these results can be found in Kestin and Richardson's paper[26]. Despite considerable experimental uncertainty and conflicting evidence, a value of $\mathbf{P}_t = 1$ is often used in calculation procedures, principally for the computational simplicity that this assumption brings.

It is now informative to introduce the above postulates into the momentum and energy boundary layer equations, and for the case of a zero pressure gradient and negligible buoyancy forces, Eqns (5-83b) and (5-83c) are written below as

$$\rho\left(\bar{u}\frac{\partial \bar{u}}{\partial x} + \bar{v}\frac{\partial \bar{u}}{\partial y}\right) = \frac{\partial \tau}{\partial y} \tag{6-106a}$$

$$\rho C_p\left(\bar{u}\frac{\partial \bar{T}}{\partial x} + \bar{v}\frac{\partial \bar{T}}{\partial y}\right) = -\frac{\partial q}{\partial y} + \tau\frac{\partial \bar{u}}{\partial y} \tag{6-106b}$$

Now

$$\tau\frac{\partial \bar{u}}{\partial y} = \frac{\partial}{\partial y}(\bar{u}\tau) - \bar{u}\frac{\partial \tau}{\partial y}$$

hence, using Eqn (6-106a), Eqn (6-106b) can be written

$$\rho C_p\left(\bar{u}\frac{\partial \bar{T}}{\partial x} + \bar{v}\frac{\partial \bar{T}}{\partial y}\right) = -\frac{\partial q}{\partial y} + \frac{\partial}{\partial y}(\bar{u}\tau) - \rho\bar{u}\left(\bar{u}\frac{\partial \bar{u}}{\partial x} + \bar{v}\frac{\partial \bar{u}}{\partial y}\right) \tag{6-107}$$

A total enthalpy, $\tilde{h}$, is now defined as

$$\tilde{h} \equiv C_p\bar{T} + \tfrac{1}{2}\bar{u}^2 \tag{6-108}$$

and using the turbulent exchange coefficients for τ and q such that

$$\tau = (\mu + A_\tau)\frac{\partial \bar{u}}{\partial y} \tag{6-109a}$$

and

$$q = \left(\frac{\mu}{P} + \frac{A_\tau}{P_t}\right)\frac{\partial \bar{T}}{\partial y} \tag{6-109b}$$

Eqn (6-107) can be expressed as

$$\rho\left(\bar{u}\frac{\partial \tilde{h}}{\partial x} + \bar{v}\frac{\partial \tilde{h}}{\partial y}\right) = \frac{\partial}{\partial y}\left\{C_p\left(\frac{\mu}{P} + \frac{A_\tau}{P_t}\right)\frac{\partial \bar{T}}{\partial y} + \bar{u}(\mu + A_\tau)\frac{\partial \bar{u}}{\partial y}\right\} \tag{6-110}$$

For the case of $P = P_t = 1$, Eqn (6-110) simplifies to

$$\rho\left(\bar{u}\frac{\partial \tilde{h}}{\partial x} + \bar{v}\frac{\partial \tilde{h}}{\partial y}\right) = \frac{\partial}{\partial y}(\mu + A_\tau)\frac{\partial \tilde{h}}{\partial y} \tag{6-111a}$$

and is analogous to Eqn (6-106a), which is written as

$$\rho\left(\bar{u}\frac{\partial \bar{u}}{\partial x} + \bar{v}\frac{\partial \bar{u}}{\partial y}\right) = \frac{\partial}{\partial y}(\mu + A_\tau)\frac{\partial \bar{u}}{\partial y} \tag{6-111b}$$

As the equations are analogous, then, for similar initial and boundary conditions, the answers must be analogous. If we consider the isothermal flat plate with a temperature T_s, and a constant free stream velocity and temperature U_∞ and T_∞, respectively, the solutions to Eqns (6-111a) and (6-111b) must be related by

$$\frac{\tilde{h} - \tilde{h}_s}{\tilde{h}_\infty - \tilde{h}_s} = \frac{\bar{u}}{U_\infty} \qquad (6\text{-}112)$$

Now

$$q_s = -k\left(\frac{\partial \bar{T}}{\partial y}\right)_{y=0} = \frac{k}{C_p}\left(\frac{\partial \tilde{h}}{\partial y}\right)_{y=0}$$

hence, from Eqn (6-112),

$$\frac{q_s}{C_p \tau_s} = \frac{T_s - T_\infty - \frac{1}{2}U_\infty^2/C_p}{U_\infty} \qquad (6\text{-}113)$$

Thus, for an adiabatic wall, where $q_s = 0$,

$$T_{ad,\,s} = T_\infty + \frac{U_\infty^2}{2C_p} \qquad (6\text{-}114)$$

which is identical with the result for laminar flow given by Eqn (6-66), where for $\mathbf{P} = 1$, $\theta_2(0) = 1$. Equation (6-113) is usually generalized for non-unity Prandtl numbers by the introduction of a recovery factor, $\mathscr{R}$, analogous with that of Eqn (6-67), such that

$$T_{ad,\,s} = T_\infty + \mathscr{R}\,\frac{U_\infty^2}{2C_p} \qquad (6\text{-}115)$$

Figure 6-17 shows some values of $\mathscr{R}$ measured by Eckert and Weise[27] showing the transition from laminar to turbulent flow for air. As the turbulent recovery factor will depend on both the laminar and turbulent Prandtl numbers, assumed or experimental values are used in the calculations.

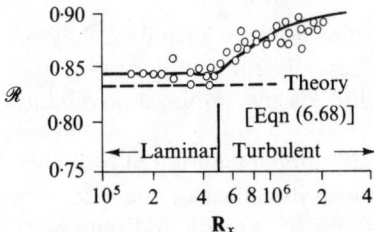

Fig. 6-17   Recovery factor as a function of Reynolds number (after Eckert and Weise[27])

Experimental values for high-speed flow past cones have been conducted by Mack[28], while Rotta[29] obtained a theoretical equation for the recovery factor such that $\mathscr{R} = \mathscr{R}(\mathbf{P}, \mathbf{P}_t, C_f)$. Many authors use the simple expression $\mathscr{R} = \mathbf{P}^{1/3}$ for moderate Prandtl numbers (cf. $\mathscr{R} = \mathbf{P}^{1/2}$ for laminar flow), and for air with $\mathbf{P} = 0.72$, this yields $\mathscr{R} = 0.896$, which is in reasonable accord with measured values. As with laminar flow, the adiabatic wall temperature is only significantly different from the free stream temperatures in high-speed flow where aerodynamic heating is important.

If we keep the same definition of Nusselt number that we used in Eqn (6-72), where $\mathbf{N}_x \equiv q_s x/[k(T_s - T_{ad,s})]$, and if we recall the definition of skin friction coefficient used in Eqn (5-58), where $\mathbf{C}_f \equiv \tau_s/(\frac{1}{2}\rho U_\infty^2)$, then as $\mathbf{P} = 1$, Eqn (6-113) can be expressed as

$$\mathbf{N}_x = \tfrac{1}{2}\mathbf{R}_x\mathbf{C}_f \tag{6-116}$$

This is a dimensionless representation of the *Reynolds analogy* which was discovered in a simpler form by Reynolds[30] in 1874. It should be remembered that it is only strictly valid in boundary layer flows if:

   (i) $dp/dx = 0$;
   (ii) $\mathbf{P} = \mathbf{P}_t = 1$;
   (iii) the initial and boundary conditions for the temperature and velocity fields are similar.

For the case of laminar flow over a flat plate, Eqn (6-116) applied to the skin friction coefficient given by Eqn (6-46) yields the result

$$\mathbf{N}_x = 0.332\mathbf{R}_x^{-1/2}$$

which is identical with the exact solution given by Eqn (6-73) (where, for $\mathbf{P} = 1$, $a_1(\mathbf{P}) = 0.332$). As the hydrodynamic effects are uncoupled from thermal effects in this example, it is convenient for moderate Prandtl numbers [where $a_1(\mathbf{P}) = 0.332\mathbf{P}^{1/3}$ from Eqn (6-58)] to generalize the Reynolds analogy to

$$\mathbf{N}_x = \tfrac{1}{2}\mathbf{P}^{1/3}\mathbf{R}_x\mathbf{C}_f \tag{6-117}$$

This relationship is also used for turbulent flow problems, although—as we shall see in later pages—the non-similarity between the temperature and the velocity fields in the viscous sublayer invalidates the analogy in this region.

In order to gain a better understanding of heat transfer in turbulent flows it is necessary to consider the effect of the viscous sublayer. Figure 6-18 illustrates the distributions of velocity and temperature within a turbulent boundary layer which is assumed to consist of two regimes: a viscous sublayer and an outer turbulent region. All the restrictions stated above for the Reynolds analogy apply, with the exception that the laminar Prandtl number is not necessarily unity. As the laminar effects are only assumed to be significant within the viscous layer, the analogy between Eqns (6-111a) and (6-111b) in the turbulent region is still apparent (putting $\mu = 0$). Neglecting dissipation (that is, ignoring terms involving $\frac{1}{2}U_\infty^2/C_p$), Eqn (6-113) can be applied in the turbulent region as

$$\frac{q_l}{C_p\tau_l} = \frac{\overline{T}_l - T_\infty}{U_\infty - \bar{u}_l} \qquad (6\text{-}118)$$

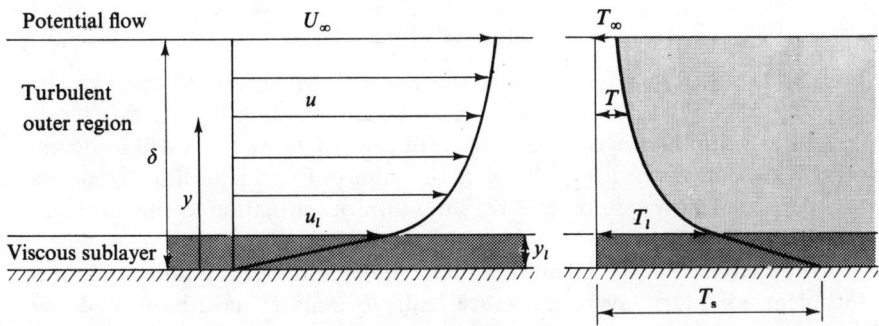

Fig. 6-18 Velocity and temperature profiles in the Prandtl–Taylor boundary layer model

where the subscript l refers to conditions at the edge of the sublayer. In the viscous sublayer, where $A_\tau = 0$ and the inertial and convection terms are negligible, close examination of Eqns (6-106) reveals that if dissipation is ignored

$$\frac{\partial\tau}{\partial y} = \frac{\partial q}{\partial y} = 0$$

hence
$$\tau_l = \tau_s = \mu\bar{u}_l/y_l \qquad (6\text{-}119a)$$

and
$$q_l = q_s = k(T_s - \overline{T}_l)/y_l \qquad (6\text{-}119b)$$

Now
$$T_s - \overline{T}_l = (T_s - T_\infty) - (\overline{T}_l - T_\infty)$$

or, using Eqn (6-118),

$$T_s - \overline{T}_l = (T_s - T_\infty) - \frac{q_l}{C_p\tau_l}(U_\infty - \bar{u}_l) \qquad (6\text{-}120)$$

Dividing Eqn (6-119b) by (6-119a) gives the result that

$$\frac{q_s}{\tau_s} = \frac{q_l}{\tau_l} = \frac{k(T_s - \overline{T}_l)}{\mu\bar{u}_l}$$

and, using Eqn (6-120),

$$\frac{q_s}{\tau_s} = \frac{C_p}{\bar{u}_l\mathbf{P}}\left[(T_s - T_\infty) - \frac{q_s}{C_p\tau_s}(U_\infty - \bar{u}_l)\right]$$

Hence
$$\frac{q_s}{\tau_s} = \frac{C_p(T_s - T_\infty)}{U_\infty[(\mathbf{P} - 1)\bar{u}_l/U_\infty + 1]} \tag{6-121}$$

or, if we replace $T_{ad,s}$ by T_∞ in the definition of our Nusselt number, we find that

$$\mathbf{N}_x = \frac{\frac{1}{2}\mathbf{P}\mathbf{R}_x\mathbf{C}_f}{[1 + (\mathbf{P} - 1)\bar{u}_l/U_\infty]} \tag{6-122}$$

Equation (6-122) is referred to as the *Prandtl–Taylor modification* of the Reynolds analogy after Prandtl[31] and Taylor[32]. For $\mathbf{P} = 1$, this result is identical with the Reynolds analogy. In general, however, it will be necessary to calculate the ratio of $\bar{u}_l/U_\infty$ and the value of $\mathbf{C}_f$ from the fluid dynamics of a particular flow system, and we shall turn our attention to this problem in subsequent sections.

It is apparent that the Prandtl–Taylor model is an oversimplification of a turbulent boundary layer, and more realistic analyses have been made by dividing the boundary layer into a viscous sublayer, a buffer region, and an outer turbulent region. By matching conditions at the edge of these regions an expression for $\mathbf{N}_x$, more elaborate but more accurate than Eqn (6-122), was found by von Karman[33], Boelter et al.[34] and Martinelli[35]. Martinelli's analysis is applicable to liquid metals, where the low Prandtl numbers prevent the thermal conductivity from being ignored in the outer turbulent region, as $\mu/\mathbf{P}$ is of order $A_\tau/\mathbf{P}_t$. All the above analyses were conducted for $\mathbf{P}_t = 1$, but Reichardt[36] derived a relationship to include both the effects of laminar and turbulent Prandtl numbers. Rotta[29] and van Driest[37] studied the effect of an arbitrarily varying turbulent Prandtl number in the boundary layer on a flat plate, and it was concluded, by Rotta, that only the value of $\mathbf{P}_t$ close to the wall significantly affects the heat transfer. Hence, a constant value of $\mathbf{P}_t$ (a suitable value appears to be $\mathbf{P}_t = 0{\cdot}9$) is usually assumed in most calculation procedures.

It should also be pointed out that free stream turbulence and pressure gradient can have a significant effect on heat transfer.[38] An adverse pressure gradient ($d\mathbf{P}/dx > 0$) tends to cause a thickening of the boundary layer, and hence reduces the heat transfer. However with a strong favourable pressure gradient ($d\mathbf{P}/dx < 0$ it is possible for a turbulent boundary layer to be *laminarized*, that is, reverse transition can occur.[39]

The heat transfer–momentum analogy is often employed for pipe flow, and for systems where the geometry and boundary conditions do not satisfy the requirements for a mathematical analogy, as reasonable predictions of heat transfer rates can be made from an examination of the fluid dynamics of the system. In the remaining sections of this chapter, the Reynolds analogy and the Prandtl–Taylor modification will be applied to turbulent convection problems involving pipes and flat plates.

6-8   **Turbulent forced flow in a circular pipe**

Similar to the case of laminar pipe flow discussed in Section 6-3, an entry length exists before fully developed profiles are formed. The turbulent entry region is considerably shorter than that found in laminar flow, and Nikuradse[40, 41]—who conducted an extensive experimental study of pipe flows—found that fully developed turbulent velocity profiles existed after an inlet length of 25 to 40 diameters.

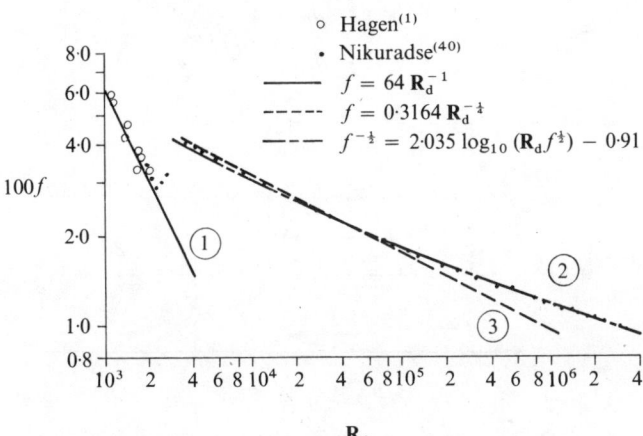

Fig. 6-19   Frictional resistance in a smooth pipe

Some experimental measurements of the friction factor, f, conducted by Hagen[1] and Nikuradse, are shown plotted against the Reynolds number, $\mathbf{R}_d$, in Fig. 6-19. f has been defined in Eqn (6-31) as

$$f \equiv -\frac{dp/dz}{\frac{1}{2}\rho \overline{U}^2/d}$$

where $\overline{U}$ is the bulk mean velocity† given by Eqn (6-14). It can be seen that the Hagen–Poiseuille law for laminar friction is invalid for $\mathbf{R}_d > 2 \times 10^3$ where transition from laminar to turbulent flow commences. In 1911, some years before Nikuradse's experiments, Blasius[42] made a survey of the then existing data and proposed the following empirical relationship for the friction factor

$$f = 0.3164 \mathbf{R}_d^{-1/4} \tag{6-123}$$

This is known as the Blasius formula, valid for moderate turbulent Reynolds number up to $\mathbf{R}_d \approx 10^5$, and it has had important and far-reaching

---

† For the rest of this chapter all velocity and temperature components will be time-averaged values, and bars will only be used to denote bulk mean values.

consequences leading to the establishment of *the one-seventh power law* velocity profiles. In order to discuss this result further it is necessary to remember that in fully developed pipe flow inertial terms are zero, hence a momentum balance on the elemental length of pipe, δz, shown in Fig. 6-20 reveals the result that

$$2\pi r_0 \tau_s \, \delta z = -\pi r_0^2 \frac{dp}{dz} \delta z$$

hence

$$\tau_s = -\frac{1}{2} r_0 \frac{dp}{dz} = \frac{1}{8} f \rho \bar{U}^2 \qquad (6\text{-}124)$$

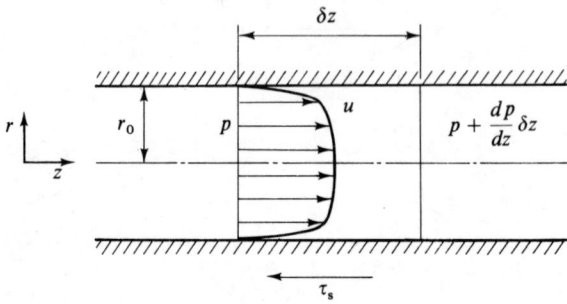

Fig. 6-20 Forces acting on a fluid in an elemental length of pipe

Substitution of the Blasius formula, Eqn (6-123), into Eqn (6-124) yields the result

$$\tau_s = 0.03325 \rho \bar{U}^{7/4} \nu^{1/4} r_0^{-1/4} \qquad (6\text{-}125)$$

Introducing the friction velocity u^*, where $u^* \equiv \sqrt{(\tau_s/\rho)}$, Eqn (6-125) can be rewritten in the form

$$\frac{\bar{U}}{u^*} = 6.99 \left(\frac{u^* r_0}{\nu} \right)^{1/7} \qquad (6\text{-}126)$$

It is proposed to generalize Eqn (6-126) for the distribution of velocity across the pipe, and if we assume that outside the viscous sublayer, $u \propto y^{1/7}$ (where, for convenience, $u = v_z$, the axial velocity component, and $y = r_0 - r$, the distance from the pipe wall), then applying Eqn (6-14) we find that $\bar{U}/U \approx 0.8$. From Eqn (6-126) we find that

$$U/u^* = 8.74(u^* r_0/\nu)^{1/7}$$

and

$$u/u^* = 8.74(u^* y/\nu)^{1/7} \qquad (6\text{-}127)$$

where U is the velocity at $y = r_0$, that is, the centre-line velocity. Thus the Blasius resistance formula is consistent with the one-seventh power law velocity profiles, and from Eqn (6-127)

$$u^* = 0\cdot15u^{7/8}(v/y)^{1/8}$$

or
$$\tau_s = 0\cdot0225\rho U^{7/4}(v/r_0)^{1/4} \tag{6-128}$$

This relationship, while only strictly valid for pipe flow at moderate Reynolds number, has been used as the basis of many turbulent boundary layer calculation procedures, often with very good results.

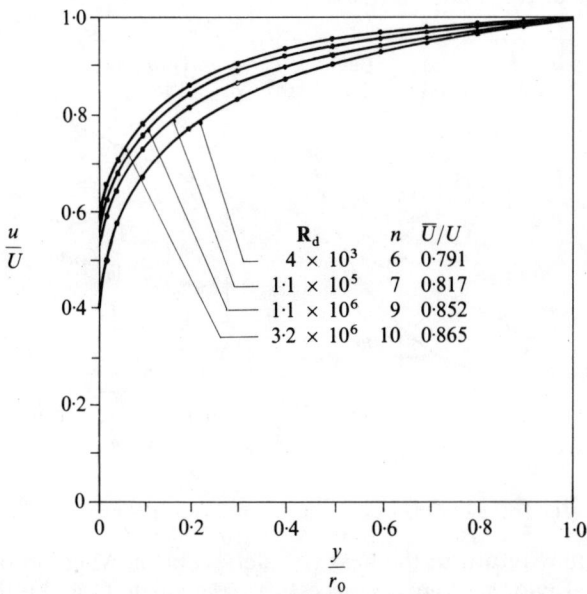

Fig. 6-21 Velocity profiles in turbulent pipe flow for a range of Reynolds numbers (after Nikuradse[40, 41])

It is possible to fit the experimentally measured velocities of Nikuradse by power law profiles of the type

$$u/U = (y/r_0)^{1/n}$$

where the index $n = 7$ for $\mathbf{R}_d \approx 10^5$, but increases with increasing Reynolds numbers, as shown in Fig. 6-21 where the values of n, selected to give a good fit to the data, are given together with the ratio of $\overline{U}/U$ for these n [which, as the reader can confirm, is given by $\overline{U}/U = 2n^2(n + 1)^{-1}(2n + 1)^{-1}$]. Prandtl's universal velocity defect law, Eqn (6-102), is in good agreement, outside the sublayer, with measured profiles for $\kappa = 0{\cdot}4$, and Eqn (6-101) provides a good fit to the data if it is written in the form

$$u/u^* = 2{\cdot}5 \ln (yu^*/v) + 5{\cdot}5 \tag{6-129}$$

This equation, usually referred to as the *law of the wall*, is shown plotted in Fig. 6-22 and is often used outside its pipe flow application. From Eqns (6-124) and (6-129) it is possible to show that

$$f^{-1/2} = 2{\cdot}035 \log_{10} (\mathbf{R}_{\mathrm{d}} f^{1/2}) - 0{\cdot}91 \tag{6-130}$$

which is shown as curve 2 in Fig. 6-19, and can be seen to give a good fit over a wide range of Reynolds numbers.

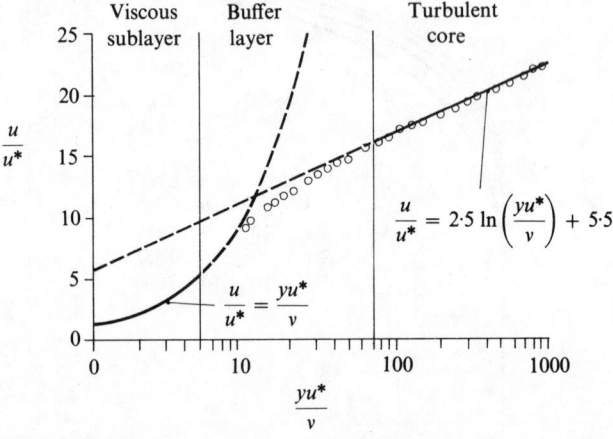

Fig. 6-22   Law of the wall for smooth pipes (after Nikuradse [40,41])

We shall now return to the heat transfer problem, where in order to use the Prandtl–Taylor analogy it is necessary to calculate $\mathbf{C}_{\mathrm{f}}$ and u_l/U_∞ for pipe flow. For pipe flow it is convenient to rewrite Eqn (6-122) in terms of the bulk values defined in Section 6-3 as

$$\mathbf{N}_{\mathrm{d}} = \frac{\tfrac{1}{2}\mathbf{P}\mathbf{R}_{\mathrm{d}}\mathbf{C}_{\mathrm{f}}}{1 + (\mathbf{P} - 1)u_l/\overline{U}} \tag{6-131}$$

where
$$\mathbf{N}_{\mathrm{d}} \equiv \frac{q_{\mathrm{s}}d}{k(T_{\mathrm{s}} - \overline{T})} \quad \text{and} \quad \mathbf{C}_{\mathrm{f}} \equiv \frac{\tau_{\mathrm{s}}}{\tfrac{1}{2}\rho\overline{U}^2}$$

It is apparent from Eqn (6-124) that $\mathbf{C}_{\mathrm{f}} = \tfrac{1}{4} f$ hence using the Blasius formula, Eqn (6-123),

$$\mathbf{C}_{\mathrm{f}} = 0{\cdot}0791 \mathbf{R}_{\mathrm{d}}^{-1/4} \tag{6-132}$$

In the viscous sublayer, where $\tau_s = \mu u_l / y_l$, Eqn (6-125) can be used to give

$$\mu u_l / y_l = 0.03325 \rho \bar{U}^{7/4} v^{1/4} r_0^{-1/4} \qquad (6\text{-}133)$$

and in order to join the sublayer to the outer region it is necessary that

$$u_l / U = (y_l / r_0)^{1/7} \qquad (6\text{-}134)$$

Elimination of y_l between Eqns (6-133) and (6-134), with $\bar{U} \approx 0.8U$, gives the result

$$u_l / \bar{U} = 2.5 \mathbf{R_d}^{-1/8} \qquad (6\text{-}135)$$

Hence, from Eqn (6-131) the Nusselt number can be expressed as

$$\mathbf{N_d} = \frac{0.0396 \mathbf{PR_d^{3/4}}}{1 + 2.5 \mathbf{R_d^{-1/8}(P - 1)}} \qquad (6\text{-}136)$$

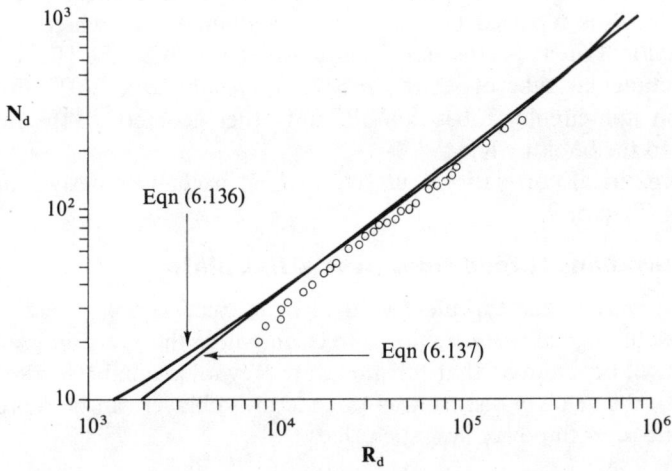

Fig. 6-23 Comparison between the calculated Nusselt numbers and the experiments of Deissler and Eian[43] for turbulent pipe flow

The above expression provides satisfactory estimates of the Nusselt number for turbulent pipe flow with moderate Reynolds numbers. Figure 6-23 shows the measured results of Deissler and Eian[43] for air, with $\mathbf{P} = 0.73$, together with the Nusselt numbers calculated from Eqn (6-136), and from the modified Reynolds analogy of Eqn (6-117) where

$$\mathbf{N_d} = 0.0396 \mathbf{P}^{1/3} \mathbf{R_d^{3/4}} \qquad (6\text{-}137)$$

The calculated curves are both slightly higher than the measured values, and it can be seen that over this range of Reynolds numbers, and for $\mathbf{P} = 0.73$, Eqns (6-136) and (6-137) yield very similar results.

The above analysis was conducted for *hydraulically smooth pipes*, that is, pipes in which the surface roughness of the walls does not affect the boundary layer structure. The ratio of the characteristic roughness dimension, y_r, to the pipe diameter, d, is known as the *relative roughness*, and pipes are re-garded as hydraulically smooth if $y_r < y_l$ such that $y_r u^*/v < 5$.

Nikuradse[41] and Moody[44] obtained a large number of experimental data for rough pipes, and the *Moody diagram*, in which the friction factor f is plotted against Reynolds number for a large range of relative roughness values, is a useful design tool. While for small values of roughness ratio, and for fluids of moderate Prandtl numbers, the analogy between N_d and C_f can be used, it becomes unreliable for high Prandtl numbers, where heat transfer resistance is in the viscous sublayer, and for large values of roughness. The reader is referred to the paper by Dipprey and Sabersky[45] for more details of the effect of roughness on heat transfer.

For non-circular tubes, Eqn (6-136) can be used if the characteristic dimension, d, is replaced by the equivalent diameter, d_e where $d_e = 4 \times$ cross-sectional area $\div$ cross-sectional perimeter of the tube [for example, for a rectangular tube of section $a \times b$, $d_e = 4ab/(2a + 2b)$]. For more details on non-circular tubes, annuli, and other geometries, the reader is referred to the book by Kays[3].

Semi-empirical correlations, useful for heat exchanger design, are pre-sented in Chapter 7.

6-9   Turbulent forced flow over a flat plate

We shall consider the turbulent equivalent of Section 6-4, where we con-sider a plate aligned with the flow direction where the pressure gradient is zero. It will be assumed that for moderate Reynolds numbers, $5 \times 10^5 < R_x < 10^7$, the velocity distribution outside the sublayer can be represented by the one-seventh power law, such that

$$u/U_\infty = (y/\delta)^{1/7} \tag{6-138}$$

and $\delta(x)$ is the thickness of the boundary layer where the velocity, u, is 99 per cent of the free stream value, U_∞. Experimental support for this assumption is shown in Fig. 6-24, based on the measurements of Dhawan[46].

We shall also use the Blasius expression for shear stress, given by Eqn (6-128) for pipe flows, but modified for external flows as

$$\tau_s = 0.0225\rho U_\infty^{7/4}(v/\delta)^{1/4} \tag{6-139}$$

We are now in a position to solve the integral momentum equation for forced turbulent flow, rewritten from Eqn (5-88) for the zero pressure gradient case $(dU_\infty/dx = 0)$ as

$$\frac{d}{dx} \int_0^\delta u(U_\infty - u)\, dy = \frac{\tau_s}{\rho} \tag{6-140}$$

From Eqns (6-138) and (6-139) it follows that

$$\frac{7}{12}\frac{d\delta}{dx} = 0.0225\left(\frac{\nu}{U_\infty\delta}\right)^{1/4} \tag{6-141}$$

which, for $\delta = 0$ at $x = 0$, has the solution

$$\delta(x) = 0.37x\mathbf{R}_x^{-1/5} \tag{6-142}$$

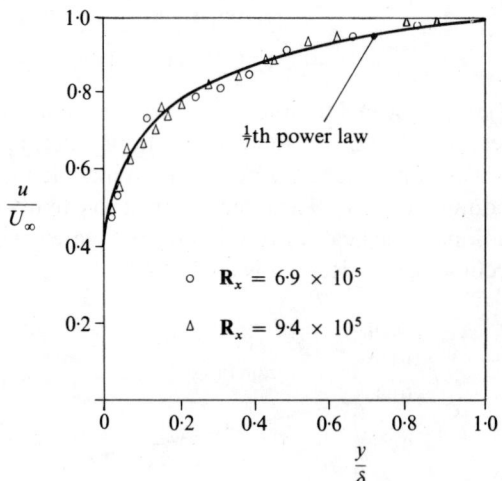

Fig. 6-24 Comparison between $\frac{1}{7}$th power law velocity profile and the experimental data of Dhawan[46] for turbulent flow over a flat plate

It can be seen from Eqn (6-139) that

$$\mathbf{C_f} \equiv \frac{\tau_s}{\frac{1}{2}\rho U_\infty^2} = 0.045\left(\frac{\nu}{U_\infty\delta}\right)^{1/4} \tag{6-143}$$

or, using Eqn (6-142),

$$\mathbf{C_f} = 0.0576\mathbf{R}_x^{-1/5} \tag{6-144}$$

which agrees reasonably well with Dhawan's results shown in Fig. 6-25.

Agreement over a wider range of Reynolds numbers can be obtained by using a logarithmic velocity profile, similar to Eqn (6-129) for pipes. The added complexity is not warranted for this example, and we shall now apply the Prandtl–Taylor analogy to the flat plate, using the above relationships based on the one-seventh power law.

Just as for pipe flow, we assume that the velocity at the edge of the viscous sublayer, u_l, is related to the sublayer thickness, y_l, by the equation

$$u_l/U_\infty = (y_l/\delta)^{1/7} \tag{6-145}$$

Also, using eqn (6-139),

$$\mu u_l / y_l = 0{\cdot}0225 \rho U_\infty^{7/4} (v/\delta)^{1/4} \qquad (6\text{-}146)$$

With the aid of Eqn (6-142), Eqns (6-145) and (6-146) can be solved to give

$$u_l / U_\infty = 2{\cdot}12 \mathbf{R}_x^{-1/10} \qquad (6\text{-}147)$$

Substitution of Eqns (6-144) and (6-147) into Eqn (6-122) provides the result

$$\mathbf{N}_x = \frac{0{\cdot}0292 \mathbf{PR}_x^{0{\cdot}8}}{1 + 2{\cdot}12 \mathbf{R}_x^{-0{\cdot}1}(\mathbf{P} - 1)} \qquad (6\text{-}148)$$

Equation (6-148) does not include the effect of the laminar boundary layer, which will be present before transition occurs. In practice, in order to calculate the heat loss from a flat plate it is necessary to include Eqn (6-73) for the initial laminar boundary layer. The effect of arbitrary temperature or arbitrary flux distribution on the heat transfer from a flat plate has been calculated using integral techniques, by Reynolds *et al.*[47]

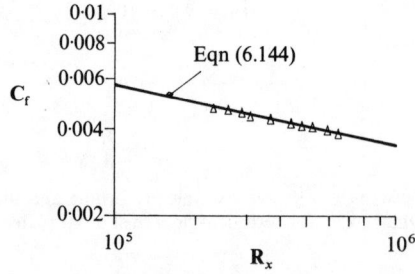

Fig. 6-25   Skin friction coefficient for turbulent flow over a flat plate (after Dhawan[46])

Chapter 7 includes some design formulae for the calculation of heat transfer from a flat plate, and the numerical technique which will be discussed in the next chapter is capable of solving the boundary layer equations for the case of arbitrary boundary conditions, variable properties, and mass transfer through the surface. Although semi-empirical data are used for the turbulent flux terms, the technique will be shown to be a very powerful tool that is capable of reasonable predictions for a very wide range of external, and internal, flows.

6-10   Turbulent free convection from a vertical plate

We saw in Section 6-6 that for $\mathbf{G}_l\mathbf{P} > 10^9$ heat transfer from a vertical heated plate would be controlled by turbulent flow. It is proposed here to use the same method of analysis as we used in Section 6-6, with the exception that we shall assume that the velocity and temperature profiles can be represented by the one-seventh power laws that we used in Section 6-9. For the case of an isothermal plate with temperature T_s and a quiescent free stream with constant temperature T_∞, we can rewrite Eqns (5-88) and (5-92) for turbulent free convection, neglecting viscous dissipation, as

$$\frac{d}{dx} \int_0^\delta u^2 \, dy = g\beta \int_0^\delta (T - T_\infty) \, dy - \frac{\tau_s}{\rho} \tag{6-149a}$$

$$\frac{d}{dx} \int_0^\delta u(T - T_\infty) \, dy = \frac{q_s}{\rho C_p} \tag{6-149b}$$

It has been implicitly assumed that $\delta_T \approx \delta$ and we now postulate the one-seventh power law profiles [see reference (16)]

$$u = u_1 \eta^{1/7} (1 - \eta)^4 \tag{6-150a}$$

and $$\theta = (1 - \eta^{1/7}) \tag{6-150b}$$

where $$\eta \equiv y/\delta \quad \text{and} \quad \theta \equiv (T - T_\infty)/(T_s - T_\infty)$$

For small values of η Eqn (6-150a) tends to $u = u_1 \eta^{1/7}$ (cf. $u = U_\infty \eta^{1/7}$ for forced convection over a flat plate), and it is proposed to modify Eqn (6-139) for the shear stress, such that

$$\tau_s/\rho = 0{\cdot}0225 u_1^2 \left(\frac{\nu}{u_1 \delta}\right)^{1/4} \tag{6-151a}$$

As we are considering Prandtl numbers of order unity (as $\delta_T \approx \delta$) we shall make use of the modified Reynolds analogy, Eqn (6-117), where,

$$N_x = \tfrac{1}{2} P^{1/3} R_x C_f \tag{6-152}$$

which implies that

$$\frac{q_s x}{k(T_s - T_\infty)} = \frac{1}{2} P^{1/3} \left(\frac{u_1 x}{\nu}\right) \left(\frac{\tau_s}{\tfrac{1}{2}\rho u_1^2}\right)$$

or $$\frac{q_s}{\rho C_p} = P^{-2/3} \frac{T_s - T_\infty}{u_1} \left(\frac{\tau_s}{\rho}\right) \tag{6-151b}$$

Substitution of Eqns (6-150) and (6-151) into Eqns (6-149) produces the results

$$0{\cdot}0523 \frac{d}{dx} (u_1^2 \delta) = 0{\cdot}125 g\beta(T_s - T_\infty)\delta - 0{\cdot}0225 u_1^2 \left(\frac{\nu}{u_1 \delta}\right)^{1/4} \tag{6-153a}$$

$$0{\cdot}0366 \frac{d}{dx} (u_1 \delta) = 0{\cdot}0225 P^{-2/3} u_1 \left(\frac{\nu}{u_1 \delta}\right)^{1/4} \tag{6-153b}$$

By analogy with the laminar free convection case it is proposed to solve these coupled differential equations by similarity substitutions of the form

$$u_1 = c_1 x^{m_1} \quad \text{and} \quad \delta = c_2 x^{m_2}$$

Using these substitutions, Eqns (6-153) become

$$0 \cdot 0523 \frac{d}{dx}(c_1^2 c_2 x^{2m_1 + m_2})$$

$$= 0 \cdot 125 g \beta (T_s - T_\infty) c_2 x^{m_2} - 0 \cdot 0225 (c_1 x^{m_1})^{7/4} (c_2 x^{m_2})^{-1/4} v^{1/4} \quad \text{(6-154a)}$$

$$0 \cdot 0366 \frac{d}{dx}(c_1 c_2 x^{m_1 + m_2}) = 0 \cdot 0225 \mathbf{P}^{-2/3} (c_1 x^{m_1})^{3/4} (c_2 x^{m_2})^{-1/4} v^{1/4} \quad \text{(6-154b)}$$

In order that similarity solutions exist it is obviously necessary that

$$2m_1 + m_2 - 1 = m_2 = \tfrac{7}{4}m_1 - \tfrac{1}{4}m_2$$

and
$$m_1 + m_2 - 1 = \tfrac{3}{4}m_1 - \tfrac{1}{4}m_2$$

hence
$$m_1 = 0 \cdot 5 \quad \text{and} \quad m_2 = 0 \cdot 7$$

After some straightforward, but tedious, manipulation of Eqns (6-154) it can be shown that

$$c_1 = 0 \cdot 0689 c_2^{-5} v \mathbf{P}^{-8/3} \quad \text{(6-155a)}$$

and
$$c_2 = \left\{ 0 \cdot 00338 \frac{v^2}{g\beta(T_s - T_\infty)} (1 + 0 \cdot 494 \mathbf{P}^{2/3}) \mathbf{P}^{-16/3} \right\}^{1/10} \quad \text{(6-155b)}$$

Using the fact that $u_1 = c_1 x^{0.5}$ and remembering that the local Grashof number, $\mathbf{G}_x$, is defined by $\mathbf{G}_x \equiv g\beta(T_s - T_\infty)x^3/v$, we can write u_1 in the form

$$u_1 = 1 \cdot 185 \frac{v}{x} \mathbf{G}_x^{1/2} (1 + 0 \cdot 494 \mathbf{P}^{2/3})^{-1/2} \quad \text{(6-156a)}$$

Similarly, using the fact that $\delta = c_2 x^{0.7}$, we can express the boundary layer thickness as

$$\delta/x = 0 \cdot 565 \mathbf{G}_x^{-1/10} \mathbf{P}^{-8/15} (1 + 0 \cdot 494 \mathbf{P}^{2/3})^{1/10} \quad \text{(6-156b)}$$

We are now in a position to solve Eqn (6-152) for the Nusselt number, as

$$\mathbf{R}_x = u_1 x/v = 1 \cdot 185 \mathbf{G}_x^{1/2} (1 + 0 \cdot 494 \mathbf{P}^{2/3})^{-1/2} \quad \text{(6-157)}$$

and
$$\tfrac{1}{2} \mathbf{C}_f = \frac{\tau_s}{\rho u_1^2} = 0 \cdot 0225 \left(\frac{v}{u_1 x} \right)^{1/4} \left(\frac{x}{\delta} \right)^{1/4} \quad \text{(6-158)}$$

Hence, using Eqns (6-156), (6-157), (6-158) we find that

$$\mathbf{N}_x = 0 \cdot 0295 \mathbf{G}_x^{2/5} \mathbf{P}^{7/15} (1 + 0 \cdot 494 \mathbf{P}^{2/3})^{-2/5} \quad \text{(6-159)}$$

As in Section 6-4, we can define a mean Nusselt number, $\overline{N} \equiv \overline{h}l/k$, and an overall Grashof number, $G_l \equiv g\beta l^3(T_s - T_\infty)/\nu^2$, where l is the height of the vertical plate and

$$\overline{h} = \frac{1}{l(T_s - T_\infty)} \int_0^l q_s \, dx$$

It can be seen from Eqn (6-159) that $h \propto x^{1/5}$ (as $h = N_x x/k$). Hence, as $q_s = h(T_s - T_\infty)$, it follows that $\overline{h} = 5h/6$, or

$$\overline{N} = 0\cdot0246 G_l^{2/5} P^{7/15}(1 + 0\cdot494 P^{2/3})^{-2/5} \qquad (6\text{-}160)$$

For air, with $P = 0\cdot71$, Eqn (6-160) simplifies to

$$\overline{N} = 0\cdot0183 G_l^{2/5} \qquad (6\text{-}161)$$

which almost agrees with the experimental results, shown plotted in Fig. 6-16, correlated by $\overline{N} = 0\cdot0210(G_l P)^{2/5}$. Not surprisingly the agreement is less satisfactory for Prandtl numbers very different from unity, and for $P > 5$ the error soon exceeds 10 per cent, although the analysis is quite satisfactory for gases. Correlations for free convection in turbulent flow for a number of geometries will be included in the design formulae of Chapter 7.

6-11 Concluding remarks

In this chapter consideration has been given to representative classes of boundary layer convection problems. Analytical techniques have been applied to laminar problems where the solution procedure is not unduly complicated or where the complexity is justified by the physical insight that is elicited. The reader has been introduced to some approximate solution procedures for the integral equations which, for laminar and turbulent, free and forced convection problems, provide a powerful tool for producing realistic estimates of Nusselt numbers. The analogy between the transfer of heat and the transport of momentum has been shown to apply, within stated limits, to laminar and turbulent flow, and this fact has been used to advantage in the calculation of heat transfer in turbulent boundary layer problems.

Values of Nusselt numbers calculated from the above examples can provide the heat transfer coefficient, h, which has been used in the derivative boundary conditions of the conduction problems mentioned in earlier chapters. In many engineering situations, however, geometry and boundary conditions often make analytical techniques extremely tedious. Any solution that is found is usually in an inconvenient form for subsequent calculation purposes, and the engineer who is designing equipment involving heat transfer calculations has, in the past, relied primarily on semi-empirical formulae or the correlation of experimental data. Examples on what is believed to be useful working formulae—and their range of validity—are given in Chapter 7 for a range of convection problems.

Engineers know that it is dangerous to extrapolate results beyond the stated range of validity, yet often—when no other data exist—this is the only course open. However, with the availability of large high-speed digital computers it is more reliable, and often as convenient, to use numerical solution procedures to calculate the Nusselt numbers for new problems. If a solution procedure has been experimentally confirmed for a range of flow conditions then the answers obtained for new conditions are, if the solution procedure is satisfactory, less error-prone than the extrapolation of an experimental correlation. We shall, in the next chapter, present a solution procedure which is believed to be satisfactory for a very large range of research and design problems.

REFERENCES

1. Hagen, G. Über die Bewegung des Wassers in engen zlindrischen Röhren. *Pogg. Ann.*, 1839, **46**, 423.
2. Poiseuille, J. Recherches expérimentelles sur le mouvement des liquides dans les tubes de très petits diametres. *Compte Rendus*, 1840, **11**, 961.
3. Kays, W. M. *Convective Heat and Mass Transfer*. McGraw-Hill, New York, 1966.
4. Murphy, J. S. Some effects of surface curvature on laminar boundary layer flow. *J. Aero. Sci.*, 1953, **20**, 338.
5. Blasius, H. Grenzschichten in Flüssigkeiten mit kleiner Reibung. *ZAMP*, 1908, **56**, 1.
6. Howarth, L. On the solution of the laminar boundary layer equations. *Proc. Roy. Soc. A*, 1938, **164**, 547.
7. Liepmann, H. W. and S. Dhawan. Direct measurements of local skin friction in low-speed and high-speed flow. *Proc. First U.S. National Congress on Applied Mechanics*, 1951, 869.
8. Pohlhausen, E. Der Wärmeaustausch zwischen festen Körpern und Flüssigkeiten mit kleiner Reibung und kleiner Wärmeleitung. *ZAMM*, 1921, **1**, 115.
9. Eckert, E. R. G. Die Berechnung des Wärmeübergangs in der laminaren Grenzschicht umströmter Körper. *VDI-ForschHft*, 1942, **416**.
10. Hartree, D. R. On an equation occurring in Falkner and Skan's approximate treatment of the equations of the boundary layer. *Proc. Camb. Phil. Soc.*, 1937, **33**, 223.
11. Levy, S. Heat transfer to constant property laminar boundary layer flows with power-function free-stream velocity and wall temperature variation. *J. Aero. Sci.*, 1952, **19**, 341.
12. Pohlhausen, K. Zur naherungsweisen Integration der Differentialgleichung der laminaren Reibungsschicht. *ZAMM*, 1921, **1**, 252.
13. Schmidt, E. and W. Beckmann. Das Temperatur- und Geschwindigkeitsfeld von einer licher Wandtemperatur. *Forsch. Geb. IngWes.*, 1930, **1**, 391.
14. Squire, H. B. *Modern Developments in Fluid Dynamics*, vol. II (S. Goldstein, ed.). Oxford University Press, London, 1938.
15. Ostrach, S. An analysis of laminar free convection flow and heat transfer about a flat plate parallel to the direction of the generating body force. *NACA Rept 1111*, 1953.

16. Eckert, E. R. G. and T. W. Jackson. Analysis of turbulent free convection boundary layer on a flat plate. *NACA Rept 1015*, 1951.

17. Eckert, E. R. G. and A. J. Diaguila. Convective heat transfer for mixed, free and forced flow through tubes. *Trans. A.S.M.E.*, 1954, **76**, 497.

18. Sparrow, E. M. and J. L. Gregg. Details of exact low Prandtl number boundary layer solutions for forced and for free convection. *NASA Memo. 2-27-59E*, 1959.

19. Le Fevre, E. J. Laminar free convection from a vertical plane surface. *Mech. Engng Res. Lab.*, Heat 113 (Great Britain), 1956.

20. Sparrow, E. M., R. Eichhorn and J. L. Gregg. Combined forced and free convection in a boundary layer flow. *Phys. Fluids*, 1959, **2**, 319.

21. Hermann, R. Wärmeübertragung bei freier Strömung am Waagerechten Zylinder in zweiatomigen Gasen. *VDI-ForschHft*, 1936, **379**.

22. Jodlbauer, K. Das Temperatur- und Geschwindigkeitsfeld um ein geheiztes Rohr bei freier Konvektion. *Forsch. Geb. IngWes.*, 1933, **4**, 157.

23. Boussinesq, J. Théorie de l'écoulement tourbillant. *Mém. prés. Acad. Sci.*, 1877, **XXIII**, 46.

24. Prandtl, L. Über die ausgebildet Turbulenz. *ZAMM*, 1925, **5**, 136.

25. Nikuradse, J. Gesetzmäßigkeit der turbulenten Strömung in glatten Röhren. *Forsch. Arb. IngWes.*, 1932 (No. 356).

26. Kestin, J. and P. D. Richardson. Heat transfer across turbulent incompressible boundary layer. *Int. J. Heat Mass Transfer*, 1963, **6**, 147.

27. Eckert, E. and W. Weise. Messung der Temperatur aus der Oberfläche schnell angeströmter unbeheizter Körper. *Forsch. Geb. IngWes.*, 1942, **13**, 246.

28. Mack, L. M. An experimental investigation of the temperature recovery factor. *Rept 20–80*, Jet Propulsion Laboratory, Calif. Inst. Tech., Pasadena, 1954.

29. Rotta, J. C. Temperatur verteilungen in der turbulenten Grenzschicht an der ebenen Platte. *Int. J. Heat Mass Transfer*, 1964, **7**, 215.

30. Reynolds, O. On the extent and action of the heating surface for steam boilers. *Proc. Manchester Lit. Phil. Soc.*, 1874, **14**, 7.

31. Prandtl, L. Eine Beziehung zwischen Wärmeaustausch und Strömung zwiderstand der Flüssigkeiten. *Phys. Z.*, 1910, **11**, 1072.

32. Taylor, G. I. Conditions at the surface of a hot body exposed to the wind. *A.R.C. R. and M. 272*, 1919.

33. von Karman, Th. The analogy between fluid friction and heat transfer. *Trans. A.S.M.E.*, 1939, **61**, 705.

34. Boelter, L. M. K., R. C. Martinelli and F. Jonassen. Remarks on the analogy between heat transfer and momentum transfer. *Trans. A.S.M.E.*, 1941, **63**, 447.

35. Martinelli, R. C. Heat transfer to molten metals. *Trans. A.S.M.E.*, 1947, **69**, 947.

36. Reichardt, H. Der Einfluß der wandnähen Strömung auf den turbulenten Wärmeübergang. *Rept Max-Planck-Institut für Strömungsforschung*, 1950 (No. 3), 1.

37. van Driest, E. R. The turbulent boundary layer with variable Prandtl number, in *Fifty Years of Boundary Layer Research*, 257–271. Braunschweig, 1955.

38. Junkhan, G. H. and G. K. Serovy. Effects of free stream turbulence and pressure gradient on flat plate boundary layer velocity profiles and on heat transfer. *J. Heat Transfer*, 1967, **89**, 169.

39. Launder, B. E. and W. P. Jones. On the prediction of laminarization. *A.R.C. CP 1036*, 1968.

40. Nikuradse, J. Gesetzmäßigkeit der turbulenten Strömung in glatten Röhren. *Forsch. Arb. IngWes.*, 1932 (No. 356).

41. Nikuradse, J. Strömungsgesetze in rauhen Röhren. *Forsch. Arb. IngWes.*, 1933 (No. 361).

42. Blasius, H. Das Ahnlichkeitsgesetz bei Reibungsvorgängen in Flüssigkeiten. *Forsch. Arb. IngWes.*, 1913 (No. 131).
43. Deissler, R. G. and C. S. Eian. Analytical and experimental investigation of fully developed turbulent flow of air in a smooth tube with heat transfer with variable fluid properties. *NACA Tech. Note 2629*, 1952.
44. Moody, L. F. Friction factors for pipe flow. *Trans. A.S.M.E.*, 1944, **66**, 671.
45. Dipprey, D. F. and R. H. Sabersky. Heat and momentum transfer in smooth and rough tubes at various Prandtl numbers. *Int. J. Heat Mass Transfer*, 1963, **6**, 329.
46. Dhawan, S. Direct measurements of skin friction. *NACA Tech. Note 2567*, 1952.
47. Reynolds, O., W. M. Kays and S. J. Kline. Heat transfer in the turbulent incompressible boundary layer.
   I. Constant wall temperature. *NASA Memo. 12-1-58W*, 1958.
   II. Step wall temperature distribution. *NASA Memo. 12-2-58W*, 1958.
   III. Arbitrary wall temperature and heat flux. *NASA Memo. 12-3-58W*, 1958.
   IV. Effect of location of transition and prediction of heat transfer in a known transition region. *NASA Memo. 12-4-58W*, 1958.

PROBLEMS

Note: where not stated, fluid property values should be obtained from the Appendix.

6-1 Consider the flow between two plates placed a distance s apart parallel to the x-axis, along $y = 0$ and $y = s$, and infinitely wide in the z direction. If the top plate moves with a constant velocity U in the x-direction show that for incompressible flow

$$\frac{du}{dx} = 0; \qquad \frac{dp}{dx} = \mu \frac{d^2u}{dy^2}$$

and

$$\rho C_p u \frac{\partial T}{\partial x} = k \frac{\partial^2 T}{\partial y^2} + \mu \left(\frac{du}{dy}\right)^2$$

Hence, by integrating between $y = 0$ and $y = s$, show that if both plates are adiabatic

$$\dot{m} \frac{d}{dx}\left(C_p \bar{T} + \frac{p}{\rho}\right) = U\tau_s$$

where the mass flow rate per unit width $\dot{m} = \int_0^s \rho u \, dy$, $\bar{T}$ is the bulk temperature, that is

$$\bar{T} = \frac{1}{\dot{m}} \int_0^s \rho u T \, dy, \quad \text{and} \quad \tau_s = \mu \left(\frac{du}{dy}\right)_s$$

6-2 For the above system show that if $dp/dx = 0$

$$C_p(\bar{T} - \bar{T}_I)/U^2 = 2(x/s)^2(Ux/\nu)^{-1}$$

where $\bar{T}_I$ is the bulk temperature at $x = 0$.

6-3 The cross-head guide of a diesel engine consists of a flat slider 3 feet long moving over a flat stationary guide. Using the result of the previous problem, estimate the exit temperature of the lubricating oil if it enters the guide at 140°F when the slider has a velocity of 500 ft/min and the clearance is 0·010 inches.

Ans: 145·3°F

6-4 If, in the above problem, the guide is maintained at a constant temperature of 140°F, calculate the temperature of the slider assuming it to be adiabatic. [Hint: consider Eqn (6-12).]

Ans: 143°F

6-5 By considering the balance of pressure and viscous forces in fully developed steady flow in a circular pipe of radius r_0, with its centre-line along the z-axis, show from first principles that the wall shear stress, τ_s, is related to the axial pressure gradient, dp/dz, by

$$\tau_s = -\frac{1}{2}r_0 \frac{dp}{dz}$$

If V_z is the axial velocity component and μ the viscosity, is the above relationship valid for: (a)

$$V_z = -\frac{r_0^2}{4\mu}\frac{dp}{dz}\left[1 - \left(\frac{r}{r_0}\right)^2\right]?$$

(b) developing flow? (c) turbulent flow?

6-6 By considering a heat balance on an elemental pipe length, dz, and diameter d show that if the heat transfer coefficient is constant and the pipe surface is isothermal then for incompressible steady fully developed flow

$$\dot{m}C_p \frac{d\bar{T}}{dz} = \pi \, dq_s$$

where $\dot{m}$ is the mass flow rate, C_p the specific heat, and q_s the heat flux through the pipe wall. Using the above relation show that

$$\frac{\bar{T} - T_s}{\bar{T}_0 - T_s} = \exp\left(-\lambda \frac{z}{d}\right)$$

where $\lambda \equiv 4N_d/(PR_d)$, $\bar{T}$ being the bulk temperature at a distance z, $\bar{T}_0$ being the bulk temperature at $z = 0$, and T_s the surface temperature of the pipe. Other symbols have their usual meaning.

6-7 In a heat exchanger, oil enters with a bulk temperature of 80°C and leaves at 60°C. The oil cooler consists of a continuous coiled tube through which the oil is passed and the inside surface is maintained at a constant temperature of 10°C. The tube bore is 0·01 m and the bulk velocity of the oil is 1 m/s. Assuming a constant Nusselt number of $N_d = 3·66$, and using the result of the previous question, calculate the length of tube required. For the oil assume the following properties are constant: $C_p = 1·92$ kJ/kg K, $k = 0·143$ W/m K, $v = 231 \times 10^{-6}$ m²/s, $\alpha = 0·084 \times 10^{-6}$ m²/s.

Ans: 27·3 m

6-8 Air at atmospheric pressure with a free stream temperature of 60°C and a velocity of 100 m/s flows over a plate with a uniform temperature of 100°C. Evaluating the fluid properties at the mean film temperature (80°C) use the results of Section 6-4 to calculate the heat flux through the plate surface at a distance of 0·01 m from the leading edge if the recovery factor is evaluated from Eqn (6-68). What is the error if frictional heating is ignored (that is, the recovery factor is taken as zero)?

Ans: 6·99 kW/m²; 11·6%

6-9 Liquid ammonia with a free stream temperature of $-30°C$ and velocity of 5 ft/s flows over a plate with a constant temperature of $-10°C$. If the transition from laminar to turbulent flow occurs when the local Reynolds number, $\mathbf{R}_x$, is 3×10^5 calculate (a) the heat transfer coefficient at the transition point, and (b) the heat loss from one side of the plate up to the transition point, given the plate is 1 foot wide. Evaluate the fluid properties at the mean film temperature ($-20°C$).

Ans: 298 Btu/ft² h °R; 5,280 Btu/h

6-10 For an incompressible constant velocity fluid flowing over an isothermal plate, solve the integral boundary layer equations to find values for the thermal and velocity boundary layer thicknesses and the local Nusselt number assuming that

$$u/U_\infty = y/\delta$$

and
$$(T - T_s)/(T_\infty - T_s) = \tfrac{3}{2}(y/\delta_T) - \tfrac{1}{2}(y/\delta_T)^3$$

where all symbols have their usual meanings. Discuss the validity of the assumed profiles, and compare the answers obtained with the exact solutions of Section 6-4.

6-11 A tank 0·5 m long, 0·2 m wide, and 0·15 m high contains water at 40°C. If the top and bottom of the tank are completely insulated, and the walls are at a constant 30°C, calculate the heat loss from the tank using Eqn (6-94).

Ans: 1·08 kW

6-12 A thin oil-filled electric radiator, 6 feet long and 10 inches high, is situated in a living-room with an air temperature of 80°F. If the outside temperature of the radiator is controlled at a constant 140°F, calculate the electric power supplied assuming radiation is negligible.

Ans: 149 W

6-13 For steady laminar flow, the boundary layer equations can be written as

$$\frac{\partial}{\partial x}(\rho u) + \frac{\partial}{\partial y}(\rho v) = 0$$

$$\rho\left(u\frac{\partial u}{\partial x} + v\frac{\partial u}{\partial y}\right) = -\frac{dp}{dx} + \frac{\partial}{\partial y}\left(\mu\frac{\partial u}{\partial y}\right)$$

$$\rho C_p\left(u\frac{\partial T}{\partial x} + v\frac{\partial T}{\partial y}\right) = \frac{\partial}{\partial y}\left(k\frac{\partial T}{\partial y}\right) + \mu\left(\frac{\partial u}{\partial y}\right)^2$$

where all symbols have their usual meaning. By assuming that ρ, μ, and k are constant and that the temperature depends solely on the velocity component, u, i.e. $T = T(u)$, show that (a)

$$C_p\frac{dT}{du}\left\{\frac{\mathbf{P}-1}{\mathbf{P}}\frac{\partial \tau}{\partial y} - \frac{dp}{dx}\right\} = \left\{1 + \frac{C_p}{\mathbf{P}}\frac{d^2T}{dy^2}\right\}\tau\frac{\partial u}{\partial y}$$

(b) state the three conditions that are necessary for (a) to be valid; (c) hence show under these conditions that the adiabatic wall temperature for flow past a flat plate with a free stream velocity U_∞ and temperature T_∞ is

$$T_{\text{ad, s}} = T_\infty + \tfrac{1}{2}U_\infty^2/C_p$$

(d) from the above show that the Reynolds analogy is valid such that

$$\mathbf{N}_x = \tfrac{1}{2}\mathbf{R}_x C_f$$

6-14 Repeat the above question for the case of turbulent flow where μ is replaced by $\mu + A_\tau$, k by $k + C_p A_q$, and $\mathbf{P}_t = A_\tau / A_q = 1$.

6-15 A thin aerofoil section 6 inches high by 9 inches long is placed in the middle of a wind tunnel where the air is at atmospheric pressure and 80°F. When the air velocity is 120 ft/s, a force balance records a total drag of 0·3 lbf on the aerofoil. Assuming that the drag is due solely to skin friction, use Eqn (6-117) to estimate the mean heat transfer coefficient for the aerofoil at these conditions.

Ans: 117 Btu/ft^2 h °R

6-16 Liquid methylchloride with a free stream velocity of 5 m/s, and temperature of -40°C, flows over a flat plate maintained at -20°C. Calculate the local heat transfer coefficient at a distance of 1 m from the leading edge of the plate. Use Eqn (6-148) and evaluate the fluid properties at the mean film temperature.

Ans: 4·78 kW/m^2 K

6-17 0·96 kg/s of air flows along a pipe 0·1 m bore. At station A, a distance of 100 diameters from the inlet, the air has a static pressure of 1·7 bars and a bulk temperature of 112°C; a manometer connected to the pipe at station A and a point 10 diameters downstream registers a head loss of 58 mm water. Using

$$\mathbf{N}_d = \tfrac{1}{2} \mathbf{R}_d \mathbf{C}_f \mathbf{P}^{1/3}$$

calculate the heat transfer coefficient, h, at station A. For the air take $C_p = 1·01$ kJ/(kg K), $k = 0·0315$ W/m K, $\mu = 22·2 \times 10^{-6}$ kg/m s, and the density of water as 1,000 kg/m^3.

Ans: 224 W/m^2 K

6-18 Repeat Problem 6-17 using the Prandtl–Taylor analogy where

$$\mathbf{N}_d = \frac{\tfrac{1}{2} \mathbf{R}_d \mathbf{C}_f \mathbf{P}}{1 + (\mathbf{P} - 1) u_l / \overline{U}}$$

and $u_l / \overline{U} = 2·44 \, \mathbf{R}_d^{-1/8}$

If a heater is used to keep the pipe wall at 150°C, what power/unit length of pipe must be supplied at station A?

Ans: 206 W/m^2 K; 2·46 kW/m

6-19 Water with a bulk velocity of 30 m/s and a mean bulk temperature of 20°C flows through a tube of 0·15 m bore. Calculate the mean heat transfer coefficient given that

$$\mathbf{N}_d = \tfrac{1}{2} \mathbf{R}_d \mathbf{C}_f \mathbf{P}^{1/3}$$

where for large Reynolds numbers

$$\mathbf{C}_f^{-1/2} = 4·07 \log_{10} (2 \mathbf{R}_d \mathbf{C}_f^{1/2}) - 1·82$$

For the water take $C_p = 4·183$ kJ/kg K, $k = 0·598$ W/m K, $\mu = 0·001$ kg/m s, $v = 1·006 \times 10^{-6}$ m^2/s.

Ans: 20·1 kW/m^2 K

6-20 Recalculate Problem 6-12, for the case of a radiator 30 inches high, using Eqn (6-160).

Ans: 306 W

6-21 A steel plate, 1 m square, is placed vertically into a tank of water at 20°C. If the plate has an initial temperature of 90°C what is its initial mean heat transfer coefficient?

Ans: 701 W/m² K

7

Solution of engineering convection problems

7-1   Introductory remarks

In Chapter 6 a number of boundary layer systems have been analysed. The relatively simple problems of laminar flow past a flat plate and the Couette flow between a fixed and a moving plate were solved exactly. For more practical problems, and for turbulent flow, exact solutions are not feasible. The engineer must turn to approximate solutions of simplified equations, numerical solutions of more detailed equations, or the semi-empirical formulae produced from the correlation of experimental results. In this chapter we shall present useful design formulae for the case of forced and free convection in a number of common engineering systems. We shall also present a powerful numerical method that is capable of predicting boundary layer flows for a wide range of conditions—*the Spalding–Patankar method.*

With respect to the design formulae, it was shown in Section 5-3 that for constant property fluids we normally expect correlations of the form

$$\text{forced convection:} \quad \mathbf{N} = \mathbf{N}(\mathbf{R}, \mathbf{P})$$
$$\text{free convection:} \quad \mathbf{N} = \mathbf{N}(\mathbf{G}, \mathbf{P})$$

In practice, however, owing to property variations through the boundary layer it is necessary either to evaluate the properties at a particular reference temperature or to include the effect of temperature differences in the correlation. There are several reference temperatures in common use; for internal flows the bulk temperature, $\bar{T}$, defined in Eqn (6-17), is often used, while for external flows it is convenient to use an arithmetic mean T_m, based on the body surface and free stream temperatures where

$$T_m = \tfrac{1}{2}(T_s + T_\infty) \tag{7-1}$$

For high-velocity flow, Eckert[1]† proposed a reference temperature T^*, where

$$T^* = T_\infty + 0.5(T_s - T_\infty) + 0.22(T_{ad,s} - T_\infty) \tag{7-2}$$

and the adiabatic wall temperature, $T_{ad,s}$, is defined by Eqn (6-67). The appropriate reference temperature will be quoted in the following formulae.

Finally, it should be pointed out that heat transfer experiments are often subject to large errors, and a scatter of 20 per cent is not uncommon. The

† Full details of references cited are given at the end of the chapter.

empirical relations are usually obtained from the correlation of a large number of data, and while errors are reduced by this means the resulting equations should always be treated as approximate.

7-2   Engineering formulae for free and forced convection

7-2-1   Forced convection over external surfaces

For the case of constant surface temperature and zero pressure gradient the relations derived in Section 6-4 can be used if properties are evaluated at the mean film temperatures T_m for low-speed flow or T^* for high-speed flow.

Laminar flow

The local Nusselt number, defined by Eqn (6-72a) as

$$\mathbf{N}_x \equiv q_s x/[k(T_s - T_{ad,s})]$$

can be calculated from

$$\mathbf{N}_x = a_1(\mathbf{P})\mathbf{R}_x^{1/2} \qquad (7\text{-}3)$$

where
$$
\begin{aligned}
a_1(\mathbf{P}) &= 0{\cdot}332\mathbf{P}^{1/3} & 0{\cdot}6 < \mathbf{P} < 10 \\
&= 0{\cdot}564\mathbf{P}^{1/2} & \mathbf{P} \to 0 \\
&= 0{\cdot}339\mathbf{P}^{1/3} & \mathbf{P} \to \infty
\end{aligned}
$$

The mean Nusselt number, $\overline{\mathbf{N}}_l$, defined by Eqn (6-72b), can be calculated from

$$\overline{\mathbf{N}}_l = 2a_1(\mathbf{P})\mathbf{R}_l^{1/2} \qquad (7\text{-}4)$$

For low-speed flow, that is, $\mathbf{M} < 0{\cdot}3$,

$$T_{ad,s} \approx T_\infty$$

Turbulent flow

For moderate Prandtl numbers, Eqn (6-148) has been simplified by Johnson and Rubesin[2] to

$$\mathbf{N}_x = 0{\cdot}0292\mathbf{P}^{1/3}\mathbf{R}_x^{0{\cdot}8} \qquad (7\text{-}5)$$

However, account must be taken of the initial laminar region, thus,

$$\overline{\mathbf{N}}_l = 0{\cdot}332\mathbf{P}^{1/3} \int_0^{x_{crit}} \frac{1}{x}\left(\frac{U_\infty x}{\nu}\right) dx + 0{\cdot}0292\mathbf{P}^{1/3} \int_{x_{crit}}^l \frac{1}{x}\left(\frac{U_\infty x}{\nu}\right)^{0{\cdot}8} dx$$

where, as illustrated in Fig. 7-1, x_{crit} is the distance from the leading edge to the point at which transition occurs (assuming instantaneous transition). Thus

$$\overline{\mathbf{N}}_l = 0{\cdot}0365\mathbf{P}^{1/3}\left[\mathbf{R}_l^{0{\cdot}8} - \mathbf{R}_{crit}^{0{\cdot}8} + 18{\cdot}2\mathbf{R}_{crit}^{0{\cdot}5}\right] \qquad (7\text{-}6)$$

where $\mathbf{R}_{crit}$ is the critical Reynolds number.

Example 1. Air at 14·7 lbf/in², 70°F, with a velocity of 134 ft/s flows over a flat plate maintained at 390°F. If the plate is 3 ft long calculate the heat loss per foot width assuming (a) a fully turbulent boundary layer and (b) that transition occurs at $\mathbf{R}_{crit} = 4 \times 10^5$. Take $\mu = 14·6 \times 10^{-6}$ lb/ft s, $k = 0·0179$ Btu/ft h °R and $C_p = 0·241$ Btu/lb °R.

$$T_m = \tfrac{1}{2}(390 + 70) = 230°F = 690°R$$

$$\rho = \frac{14·7 \times 144}{53·3 \times 690} = 0·0576 \text{ lb/ft}^3$$

$$\mathbf{R}_l = \frac{\rho U_\infty l}{\mu} = \frac{0·0576 \times 134 \times 3}{14·6 \times 10^{-6}} = 1·59 \times 10^6$$

$$\mathbf{P} = \frac{\mu C_p}{k} = \frac{14·6 \times 10^{-6} \times 0·241}{0·0179/3,600} = 0·71$$

(a) $\overline{\mathbf{N}}_l = 0·0365\mathbf{P}^{1/3}\mathbf{R}_l^{0·8}$

$$= 0·0365 \times 0·71^{1/3} \times (1·59 \times 10^6)^{0·8} = 2,980$$

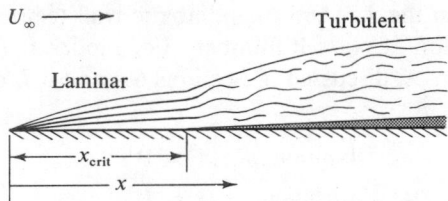

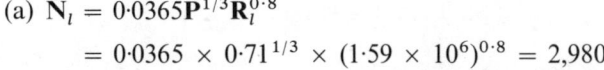

Fig. 7-1   Forced flow over a flat plate

The mean heat transfer coefficient, $\bar{h}$, is given by

$$\bar{h} = \overline{\mathbf{N}}_l \frac{k}{l} = 2,980 \times \frac{0·0179}{3} = 17·8 \text{ Btu/ft}^2 \text{ h °R}$$

Total heat loss per side for a width of 1 ft is

$$\dot{Q} = \bar{h}l(T_s - T_\infty) = 17·8 \times 3 \times (390 - 70)$$

$$= 17,100 \text{ Btu/h per side}$$

(b) $\mathbf{R}_{crit} = 4 \times 10^5$, $\mathbf{R}_{crit}^{0·8} = 30,320$, $\mathbf{R}_{crit}^{1/2} = 633$

$$\overline{\mathbf{N}}_l = 0·0365\mathbf{P}^{1/3}[\mathbf{R}_l^{0·8} - \mathbf{R}_{crit}^{0·8} + 18·2\,\mathbf{R}_{crit}^{0·5}]$$

$$= 0·0365\mathbf{P}^{1/3}[\mathbf{R}_l^{0·8} - 19,800]$$

$$= 0·0365 \times 0·71^{1/3}[(1·59 \times 10^6)^{0·8} - 19,800]$$

$$= 2,340$$

$$\bar{h} = 2{,}340 \times \frac{0{\cdot}0179}{3} = 14 \text{ Btu/ft}^2 \text{ h } ^{\circ}\text{R}$$

$$\dot{Q} = 14 \times 3 \times (390 - 70) = 13{,}400 \text{ Btu/h per side}$$

that is, there is a 27·6 per cent discrepancy if the laminar boundary layer is ignored.

Arbitrary boundary conditions

The above formulae are valid for constant surface temperatures in a zero pressure gradient flow. Heat transfer from arbitrary shaped bodies in laminar flow has been considered by Eckert[3], and the effect of arbitrary temperatures has been calculated by Levy[4]. For turbulent flow, the effect of arbitrary temperature or flux distributions has been calculated by Reynolds *et al.*[5] Free stream turbulence and pressure gradients can have a significant effect on heat transfer, as has been shown by Kestin[6], Junkham and Serovy[7], and Smith and Kuethe[8].

High-speed flow

When the Mach number of the flow is greater than 0·3 the fluid properties should be based on the T^*, and the adiabatic wall temperature should be used in the definition of Nusselt number. For moderate Prandtl numbers, the recovery factor, $\mathscr{R}$, discussed in Sections 6-4 and 6-7, can be calculated from

$$\text{laminar:} \quad \mathscr{R} = \mathbf{P}^{1/2} \tag{7-7a}$$

$$\text{turbulent:} \quad \mathscr{R} = \mathbf{P}^{1/3} \tag{7-7b}$$

Using Eqn (6-67) for the adiabatic wall temperature, where

$$T_{\text{ad, s}} = T_{\infty} + \tfrac{1}{2}\mathscr{R}U_{\infty}^2/C_{\text{p}}$$

Eqns (7-3)–(7-6) can be used for compressible flow. Details of the effect of compressibility are given by Eckert[1] and Lin[9], for turbulent flow, when $10^7 < \mathbf{R}_x < 10^9$, Eqn (7-5) should be replaced by

$$\mathbf{N}_x = 0{\cdot}185\mathbf{P}^{1/3}\mathbf{R}_x(\log_{10}\mathbf{R}_x)^{-2{\cdot}584} \tag{7-8}$$

7-2-2 Heat transfer inside tubes

In engineering applications the flow of fluids in ducts is particularly important for the effective design of heat exchangers, as described in Chapter 10, and, for example, in the calculation of heat losses from distribution lines. It has already been stated that the variation of fluid properties, particularly in oils, has a significant effect on the heat transfer rates and these variations must be allowed for. Liquid metals, where $\mathbf{P} \ll 1$, are correlated separately.

Laminar flow

It was shown in Section 6-3 that for a constant property system the Nusselt number, N_d, defined as $N_d \equiv q_s \, d/[k(T_s - \bar{T})]$, where d is the pipe diameter, was given for a constant heat flux as $N_d = 4 \cdot 36$ and for a constant surface temperature as $N_d = 3 \cdot 66$. In systems where fluids with variable properties flow in situations in which buoyancy forces exist, more elaborate formulae are necessary.

For the case of a *vertical pipe* with a constant surface temperature, T_s, Martinelli and Boelter[10] obtained the result

$$N_d = 1 \cdot 75 F_1 \left\{ \frac{\pi}{4} R_d P \left(\frac{d}{l} \right) \pm 0 \cdot 0722 \left(\frac{d}{l} G_d P \right)^n F_2 \right\}^{1/3} \tag{7-9}$$

It should be noted that the fluid properties in the free convection term, that is G_d and P, are evaluated at the surface temperature, T_s (further, G_d is based on the initial temperature difference, $T_s - \bar{T}_1$) while all other properties are evaluated at the average bulk temperature, $\bar{T}_a$, where $\bar{T}_a = \frac{1}{2}(\bar{T}_1 + \bar{T}_2)$, $\bar{T}_1$ and $\bar{T}_2$ being the bulk temperatures at two stations a distance l apart. The functions F_1 and F_2 are given in Table 7-1, the function F_2 allowing the buoyancy term to vanish as the fluid temperature approaches that of the wall. The index n was calculated as $0 \cdot 75$, but in comparison with data for oils, the index of $n = 0 \cdot 84$ gave better agreement. The buoyancy term adopts the positive sign for the case of cooled fluids flowing downwards or heated fluids upwards (that is, when forced and free convection effects assist each other), and the negative sign applies for the converse cases. The experimental data are correlated by Eqn (7-9) with a maximum deviation of 25 per cent.

Table 7-1 The functions F_1 and F_2 for laminar flow in vertical pipes

$\dfrac{\bar{T}_2 - \bar{T}_1}{T_s - \bar{T}_a}$	0	0·10	0·30	0·50	1·00	1·50	1·80	1·90	1·95	2·00
F_1	1	0·997	0·990	0·978	0·912	0·770	0·610	0·573	0·445	0
F_2	1	0·952	0·869	0·787	0·588	0·403	0·272	0·212	0·164	0

For natural and forced convection in steam-heated *horizontal pipes* with a constant wall temperature, Eubank and Proctor[11] obtained the following empirical expression for petroleum oils with $R_d < 2{,}100$

$$N_d = 1 \cdot 75 \left(\frac{\bar{\mu}}{\mu_s} \right)^{0 \cdot 14} \left\{ \frac{\pi}{4} R_d P \left(\frac{d}{l} \right) + 0 \cdot 04 \left(\frac{d}{l} G_d P \right)^{0 \cdot 75} \right\}^{1/3} \tag{7-10}$$

The properties in this expression are evaluated at the same condition as those in Eqn (7-9), although Eqn (7-10) includes an effect due to the ratio of the bulk to the wall viscosity. This expression correlates data for oils with Prandtl numbers from 140 to 15,200 and for $3.3 \times 10^5 < G_d P < 8.6 \times 10^8$ with a maximum deviation of 30 per cent.

For small values of d and $(T_s - \bar{T}_a)$, where buoyancy effects are negligible, Sieder and Tate[12] obtained the following approximate correlation

$$N_d = 1.86 \left(\frac{\bar{\mu}}{\mu_s}\right)^{0.14} \left\{ R_d P \frac{d}{l} \right\}^{1/3} \tag{7-11}$$

where all properties, except μ_s, are evaluated at the average bulk temperature condition.

Example 2. 4,000 kg of oil flows through a 0·01 m diameter tube 2 m long in 1 hour. If the pipe wall is maintained at 80°C and the oil enters at 20°C calculate the mean heat transfer coefficient using Eqn (7-10). The volume expansion coefficient, β, may be taken as $0.8 \times 10^{-3} \text{ K}^{-1}$, and the other properties are given below.

T (°C)	C_p (kJ/kg K)	k (W/m K)	v (m²/s)	ρ (kg/m³)
20	1·85	0·144	0·000890	894
40	1·92	0·143	0·000235	882
60	2·00	0·142	0·000082	870
80	2·08	0·141	0·000039	851

First estimate. As the outlet bulk temperature is not known, assume the average bulk temperature to be 20°C.

$$P = \frac{v\rho C_p}{k} = \frac{890 \times 10^{-6} \times 894 \times 1.85 \times 10^3}{0.144} = 10,200$$

The mass flow rate, $\dot{m}$, is given by

$$\dot{m} = \frac{4,000}{3,600} = 1.67 \text{ kg/s}$$

$$\bar{U} = \frac{\dot{m}}{\rho \pi d^2/4} = \frac{1.67}{894 \times \pi \times 0.01^2/4} = 23.7 \text{ m/s}$$

$$R_d = \frac{\bar{U}d}{v} = \frac{23.7 \times 0.01}{890 \times 10^{-6}} = 266$$

$$\left(\frac{\bar{\mu}}{\mu_s}\right)^{0.14} = \left(\frac{0.000890 \times 894}{0.000039 \times 851}\right)^{0.14} = 1.6$$

$$\frac{d}{l} = \frac{0.01}{2} = 0.005$$

For Eqn (7-10) it is necessary to evaluate the free convection terms at the pipe surface temperature, that is

$$\mathbf{G_d P} = [d^3 g (\beta/\nu)_s (T_s - \bar{T}_1)][\nu/\alpha]_s$$

$$= [0.01^3 \times 9.81 \times \left(\frac{0.8 \times 10^{-3}}{0.039 \times 10^{-3}}\right) \times (80 - 20)]$$

$$\times \left[\frac{0.039 \times 10^{-3}}{0.141/(851 \times 2.08 \times 10^3)}\right]$$

$$= 5.91$$

Thus, from Eqn (7-10),

$$\mathbf{N_d} = 1.75 \times 1.6$$

$$\times \left\{\frac{\pi}{4} \times 266 \times 10,200 \times 0.005 + 0.04 \times (0.005 \times 5.91)^{3/4}\right\}^{1/3}$$

$$= 61.6$$

$$\bar{h} = \mathbf{\bar N_d} \frac{k}{d} = 61.6 \frac{0.144}{0.01} = 888 \text{ W/m}^2 \text{ K}$$

Total heat supplied per second, $\dot{Q}$, is given by

$$\dot{Q} = 888 \times \pi \times 0.01 \times 2 \times (80 - 20) = 3,350 \text{ W}$$

also

$$\dot{Q} = \dot{m} C_p (\bar{T}_2 - \bar{T}_1)$$

hence

$$\bar{T}_2 = 20 + \frac{3,350}{1.67 \times 1.85 \times 10^3} = 21.1°\text{C}$$

As $\bar{T}_2 \approx \bar{T}_1$ it is not necessary to repeat the calculation using the new average bulk temperature.

For a summary of laminar flow with flux boundary conditions, or convection in the transition region ($2,100 < \mathbf{R_d} < 10,000$), the reader is referred to the work of McAdams[13].

Turbulent flow

In Chapter 6 the Reynolds analogy was used to determine Nusselt numbers for turbulent flow with a Prandtl number of unity, and this was then modified by the Prandtl–Taylor analogy to take account of the viscous sublayer for non-unity Prandtl numbers. For the case of a uniform heat flux, Boelter et al.[14] were able to integrate the boundary layer equation assuming a three-layer model. In practice the resulting expression is found inconvenient to use, and the empirical correlations are generally used. A particularly simple expression is that due to Dittus and Boelter[15] where

$$N_d = 0.023 \, R_d^{0.8} \, P^{0.4} \tag{7-12a}$$

where all properties are evaluated at the average bulk temperature.

This expression is convenient to use and reasonably accurate for small temperature differences ($T_s - \bar{T} < 6°C$ (11°F) for liquids or 56°C (101°F) for gases).

For viscous fluids such as oil, or for large temperature differences, the viscosity effect has been allowed for by Sieder and Tate[12] who provided the result

$$N_d = 0.027 R_d^{0.8} P^{1/3} (\bar{\mu}/\mu_s)^{0.14} \tag{7-12b}$$

All properties, except μ_s, are evaluated at the average bulk temperature, and Eqn (7-12b) is applicable for $0.7 < P < 16,700$.

The above turbulent flow correlations are only valid when entrance effects are negligible (that is, $l/d > 60$). For the effect of the l/d ratio the reader is referred to McAdams[13]. For the case of turbulent flow in a *concentric annulus*, Eqns (7-12a) and (7-12b) can still be employed if the *equivalent diameter* d_e is used, where

$$d_e = 4 \left[\frac{\frac{1}{4}\pi(d_2^2 - d_1^2)}{\pi(d_2 - d_1)} \right] = d_2 - d_1$$

d_2 and d_1 being the outer and inner diameters, respectively. The results can be generalized to an arbitrary shaped tube by defining the equivalent diameter as

$$d_e \equiv 4 \times \text{flow area} \div \text{'wetted' perimeters}$$

Caution must be used in applying such results unless experimental or theoretical corroboration is available.

Example 3. An oil cooler contains a bank of 100 tubes, $\frac{1}{2}$ in bore, through which 2 ft^3 of cooling water flows each second. Assuming that the oil–water temperature difference is small, calculate the heat transfer coefficient inside the cooling tubes. For water take $k = 0.345$ Btu/ft h °R, $\mu = 672 \times 10^{-6}$ lb/ft s, $C_p = 1$ Btu/lb °R, $\rho = 62.4$ lb/ft^3.

$$\text{Total flow area} = \frac{100 \times \pi \times \frac{1}{2}^2}{4 \times 144} = 0.136 \text{ ft}^2$$

$$\bar{U} = \frac{2}{0.136} = 14.7 \text{ ft/s}$$

$$R_d = \frac{62.4 \times 14.7 \times \frac{1}{2}}{12 \times 672 \times 10^{-6}} = 5.68 \times 10^4$$

As the flow is turbulent, and as the temperature difference is small, Eqn (7-12a) can be used:

$$P = \frac{672 \times 10^{-6} \times 1}{0.345/3,600} = 7.01$$

hence

$$N_d = 0.023 \times (5.68 \times 10^4)^{0.8} \times (7.01)^{0.4} = 319$$

$$h = N_d \frac{k}{d} = 319 \times \frac{0.345}{\frac{1}{2}/12} = 2,640 \text{ Btu/ft}^2 \text{ h } ^\circ R$$

For *liquid metals*—used extensively in nuclear power plants—with a Prandtl number less than 0·1, Martinelli's[16] analogy has been simplified by Lyon[17] to give

constant wall flux: $\qquad N_d = 7 + 0.025 (R_d P)^{0.8}$ $\qquad$ (7-13a)

constant wall temperature: $\quad N_d = 5 + 0.025 (R_d P)^{0.8}$ $\qquad$ (7-13b)

which approximates Martinelli's equation within 10 per cent. An empirical relationship by Lubarsky and Kaufman[18] correlates most liquid metals, except mercury, by

$$N_d = 0.625 (R_d P)^{0.4} \qquad (7-14)$$

Equations (7-13) and (7-14) are for turbulent flow and are based on bulk temperatures, but Eqn (7-14) is preferred for design purposes.

Example 4. Compare the Nusselt numbers calculated by Eqns (7-13b) and (7-14) for the case of liquid sodium flowing at 5 m/s through a tube 0·01 m bore. For sodium

$$v = 3.22 \times 10^{-7} \text{ m}^2/\text{s}, \qquad P = 0.005$$

$$R_d = \frac{5 \times 0.01}{3.22 \times 10^{-7}} = 1.55 \times 10^5$$

Equation (7-13b): $\quad N_d = 5 + 0.025 \times (1.55 \times 10^5 \times 0.005)^{0.8} = 10.2$

Equation (7-14): $\quad N_d = 0.625 \times (1.55 \times 10^5 \times 0.005)^{0.4} = 8.95$

For *mass transfer in tubes* a strong analogy exists between the Nusselt number and the Sherwood number. The equivalent mass transfer result to Eqn (7-12a) was found by Gilliland and Sherwood[19] who compiled data on vaporization of different liquids into air at approximately the same temperature. The correlation is given by

$$Sh = 0.023 R_d^{0.8} Sc^{0.4} \qquad (7-15)$$

where the Sherwood and Schmidt numbers, **Sh** and **Sc**, are defined by Eqns (5-102) and (5-100) respectively. The Reynolds number is based on the velocity of the air relative to the wall, and both $\mathbf{R}_d$ and $\mathbf{Sc}$ are based on the properties of the mixture. The range of validity is $2 \times 10^3 < \mathbf{R}_d < 35 \times 10^3$, $0.6 < \mathbf{Sc} < 2.5$. An example of the application of Eqn (7-15) for the condensation of water through air is given in Chapter 9.

7-2-3   Heat transfer outside tubes

The main difference between flow over a cylinder and flow over a flat plate is that a point of separation exists on a cylinder after which a pattern of eddies develops giving rise to a turbulent wake. A large number of data are available for the mean Nusselt number (that is, the average over the whole surface) of single heated cylinders. Hilpert[20], whose data cover a wide range of Reynolds numbers, found a correlation for air of the form

$$\mathbf{N}_d = C\mathbf{R}_d^n \qquad (7\text{-}16)$$

The 'constants' C and n vary with Reynolds number, and are given in Table 7-2. The properties are evaluated at T_m for external flows, or for internal flow at the *mean film temperature*, $\bar{T}_m$, given by

$$\bar{T}_m = \tfrac{1}{2}(T_s + \bar{T})$$

Table 7-2   Values of C and n for forced convection over a single cylinder using Eqn (7-16)

$\mathbf{R}_d$	1–4	4–40	40–4,000	4,000–40,000	40,000–250,000
C	0.891	0.821	0.615	0.174	0.0239
n	0.330	0.385	0.446	0.618	0.805

The effect of free stream turbulence intensity (that is, the root-mean-square of the fluctuating velocity components) is to increase the heat transfer. The report of Comings *et al.*[21] shows that if the turbulence intensity is from 2 to 26 per cent of the free stream velocity, the average Nusselt number is increased by up to 31 per cent.

For liquids McAdams[13] quotes the results of several authors with a recommended relation for single cylinders:

$$\bar{\mathbf{N}}_d = [0.35 + 0.56\, \mathbf{R}_d^{0.52}]\, \mathbf{P}^{0.3} \qquad (7\text{-}17)$$

for a range $50 < \mathbf{R}_d < 10,000$.

Example 5. A hot wire anemometer consists of a platinum wire 0.001 in diameter, 0.15 in long, held between two stainless steel supports. The anemometer is placed in an airstream at atmospheric pressure (that is, 14.7 lbf/in²) with a temperature of 70°F, and the wire is electrically heated to a constant temperature of 500°F. Find the relationship between the air velocity and the electric current for a range of velocities from 200 ft/s to 1,000 ft/s. For air at the mean film temperature take $\mu = 15.4 \times 10^{-6}$ lb/ft s, $k = 0.0189$ Btu/ft h °R, $C_p = 0.242$ Btu/lb °R, and the gas constant, $R = 53.3$ ft lbf/lb °R. The resistivity of platinum at 500°F is 8.71 μΩ in.

$$T_m = \tfrac{1}{2}(500 + 70) = 285°F = 745 °R$$

$$\rho = \frac{14.7 \times 144}{53.3 \times 745} = 0.0533 \text{ lb/ft}^3$$

$$\mathbf{R_d} = \frac{\rho U_\infty d}{\mu} = \frac{0.0533 \times 0.001}{12 \times 15.4 \times 10^{-6}} U_\infty = 0.289 U_\infty$$

The Reynolds number range is between 58 and 289, and so from Eqn (7-16) and Table 7-2

$$\mathbf{N_d} = 0.615 \mathbf{R_d^{0.446}}$$

and
$$\bar{h} = \mathbf{N_d}\frac{k}{d} = 0.615 \times (0.289 U_\infty)^{0.446} \times \frac{0.0189}{0.001/12}$$

$$= 80.2 U_\infty^{0.446} \text{ Btu/ft}^2 \text{ h °R}$$

Heat loss from the wire, $\dot{Q}$, for a wire of area A, is given by

$$\dot{Q} = \bar{h}A(T_s - T_\infty) = 80.2 U_\infty^{0.446} \times \pi \times \frac{0.001}{12} \times \frac{0.15}{12} \times (500 - 70)$$

$$= 0.113 U_\infty^{0.446} \text{ Btu/h} = 0.0331 U_\infty^{0.446} \text{ W}$$

(N.B. 1 Btu/s = 1,055 W.)

Also
$$\dot{Q} = I^2 R_e \text{ W}$$

where I is the current in amps and R_e the wire resistance in ohms.

$$R_e = 8.71 \times 10^{-6} \frac{0.15}{\pi \times 0.001^2/4} = 1.66 \Omega$$

$$U_\infty = \left(\frac{\dot{Q}}{0.0331}\right)^{2.24} = \left(\frac{1.66 I^2}{0.0331}\right)^{2.24}$$

hence,
$$U_\infty = 6430 \, I^{4.48} \text{ ft/s}$$

Therefore, knowing the current the velocity can be calculated. It should be noted that the above calculation takes no account of conduction from the wire to the supports, nor of radiation from the wire to the surroundings.

For the case of banks of tubes, which are commonly used in heat exchangers, the Nusselt number depends on the tube configuration, the number of rows in the bank, and the type of baffles used. The reader is referred to reference (13) for a more complete discussion of forced convection in tube banks.

7-2-4 Stagnation point heating

For the case of external flow over a blunt nosed body there will be an impingement region, before the boundary layer commences, where the heat transfer rates may be exceptionally high. Stagnation heating occurs on space re-entry vehicles or on the leading edge of missiles, aircraft, turbine blades, and many other examples of engineering importance. For the case of subsonic flow the situation is illustrated in Fig. 7-2(a), and for supersonic flow a detached shock wave stands away from the body surface, as shown in Fig. 7-2(b).

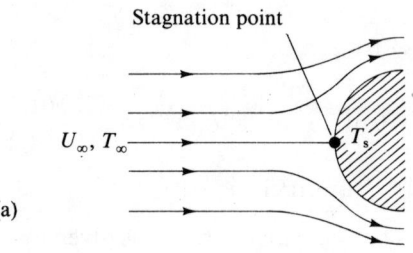

(a)

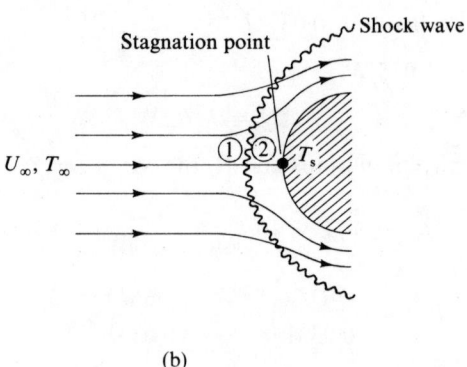

(b)

Fig. 7-2   (a) Stagnation heating in subsonic flow; (b) stagnation heating in supersonic flow

The following relationships have been recommended by van Driest [see reference (9)]:

Incompressible flow $\begin{cases} \text{cylinder:} & \mathbf{N_d} = 1 \cdot 14\mathbf{R_d^{0 \cdot 5}P^{0 \cdot 4}} \\ \text{sphere:} & \mathbf{N_d} = 1 \cdot 32\mathbf{R_d^{0 \cdot 5}P^{0 \cdot 4}} \end{cases}$ $\qquad$ (7-18a)

$\qquad\qquad\qquad\qquad\qquad\qquad\qquad$ (7-18b)

$$\text{Supersonic flow} \begin{cases} \text{cylinder:} & \mathbf{N_d} = 0.95\mathbf{R_d^{0.5}P^{0.4}}(\rho_\infty/\rho_0)^{0.25} & \text{(7-19a)} \\ \text{sphere:} & \mathbf{N_d} = 1.28\mathbf{R_d^{0.5}P^{0.4}}(\rho_\infty/\rho_0)^{0.25} & \text{(7-19b)} \end{cases}$$

For incompressible flow, properties are evaluated at the mean film temperature given by T_m in Eqn (7-1), and the motivating temperature difference for heat transfer is $T_s - T_\infty$. For supersonic flow, properties are evaluated at the stagnation state after the shock wave, and ρ_∞/ρ_0 is the ratio of the gas density in the free stream to the stagnation density behind the shock wave. The motivating temperature difference for heat transfer is $T_0 - T_s$, where T_0 is the stagnation temperature after the shock wave.

Near the stagnation point, normal shock relationships, which are well tabulated for air[22, 23] can be used. If $\mathbf{M_1}$ and $\mathbf{M_2}$ are the Mach numbers before and after the shock, respectively, then

$$\frac{\rho_1}{\rho_2} = \frac{1 + \frac{1}{2}(\gamma - 1)\mathbf{M_1^2}}{\frac{1}{2}(\gamma + 1)\mathbf{M_1^2}} \tag{7-20a}$$

$$\frac{\rho_0}{\rho_2} = [1 + \tfrac{1}{2}(\gamma - 1)\mathbf{M_2^2}]^{1/(\gamma - 1)} \tag{7-20b}$$

where
$$\mathbf{M_2^2} = \frac{1 + \frac{1}{2}(\gamma - 1)\mathbf{M_1^2}}{\gamma\mathbf{M_1^2} - \frac{1}{2}(\gamma - 1)} \tag{7-20c}$$

and where $\gamma \; (= C_p/C_v)$ is the ratio of the specific heats. Equations (7-20a)–(7-20c) are tabulated for air, thus (ρ_∞/ρ_0) can be found as $\rho_\infty = \rho_1$. The stagnation temperature after the shock, T_0, can be calculated from

$$\frac{T_0}{T_1} = 1 + \tfrac{1}{2}(\gamma - 1)\mathbf{M_1^2} \tag{7-20d}$$

Thus, as $T_1 = T_\infty$, T_0 can be found from Eqn (7-20d). Under these conditions the motivating temperature difference for heat transfer is $T_0 - T_s$.

Example 6. A supersonic aircraft travels at a Mach number of 2·2 at an altitude of 10,000 m where the atmospheric conditions are 3×10^4 N/m² (0·3 bars) and $-44°C$. Treating the leading edge of the wing as a cylindrical surface of radius 75 mm, calculate the amount of heat removed to maintain the leading edge at 100°C. For the atmosphere take $\gamma = 1·403$, the gas constant $R = 287$ J/kg K, $\mathbf{P} = 0·71$, $C_p = 1·03$ kJ/kg K, and $\mu = 17·1 \times 10^{-6}$ $(T/273)^{0·75}$ kg/m s, where T is the absolute temperature.

The speed of sound, a_∞ is given by

$$a_\infty = \sqrt{(\gamma R T_\infty)} = \sqrt{(1·403 \times 287 \times 229)} = 309 \text{ m/s}$$

and
$$U_\infty = a_\infty \mathbf{M_1} = 309 \times 2·2 = 681 \text{ m/s}$$

All properties are evaluated at the stagnation state after the shock wave, and it is therefore necessary to evaluate $\mathbf{M_2}$.

From Eqn (7-20c):

$$\mathbf{M}_2^2 = \frac{1 + \frac{1}{2}(1 \cdot 403 - 1) \times 2 \cdot 2^2}{1 \cdot 403 \times 2 \cdot 2^2 - \frac{1}{2}(1 \cdot 403 - 1)} = 0 \cdot 3$$

From Eqn (7-20a):

$$\frac{\rho_2}{\rho_1} = \frac{\frac{1}{2}(1 \cdot 403 + 1) \times 2 \cdot 2^2}{1 + \frac{1}{2}(1 \cdot 403 - 1) \times 2 \cdot 2^2} = 2 \cdot 94$$

From Eqn (7-20b):

$$\frac{\rho_0}{\rho_2} = [1 + \frac{1}{2}(1 \cdot 403 - 1) \times 0 \cdot 3]^{1/(1 \cdot 403 - 1)} = 1 \cdot 16$$

$$\rho_1 = \rho_\infty = \frac{p_\infty}{RT_\infty} = \frac{3 \times 10^4}{287 \times 229} = 0 \cdot 456 \text{ kg/m}^3$$

$$\rho_0 = \rho_1 \frac{\rho_0}{\rho_2} \frac{\rho_2}{\rho_1} = 0 \cdot 456 \times 1 \cdot 16 \times 2 \cdot 94 = 1 \cdot 56 \text{ kg/m}^3$$

and $\quad \dfrac{\rho_\infty}{\rho_0} = \dfrac{\rho_1}{\rho_0} = \dfrac{0 \cdot 456}{1 \cdot 56} = 0 \cdot 292$

From Eqn (7-20d)

$$T_0 = 229 \times [1 + \frac{1}{2}(1 \cdot 403 - 1) \times 2 \cdot 2^2] = 451 \text{ K}$$

$$\mu_0 = 17 \cdot 1 \times \left(\frac{451}{273}\right)^{0 \cdot 75} = 24 \cdot 9 \times 10^{-6} \text{ kg/m s}$$

$$\mathbf{R}_d = \frac{\rho_0 U_\infty d}{\mu_0} = \frac{1 \cdot 56 \times 681 \times 0 \cdot 075 \times 2}{24 \cdot 9 \times 10^{-6}} = 6 \cdot 4 \times 10^6$$

From Eqn (7-19a):

$$\mathbf{N}_d = 1 \cdot 95 \times (6 \cdot 4 \times 10^6)^{0 \cdot 5} \times (0 \cdot 71)^{0 \cdot 4} = 4{,}310 \cdot$$

$$k = \frac{\mu_0 C_p}{\mathbf{P}} = \frac{24 \cdot 9 \times 10^{-6} \times 1 \cdot 03}{0 \cdot 71} = 3 \cdot 62 \times 10^{-5} \text{ kW/m K}$$

and $\quad \bar{h} = \mathbf{N}_d \dfrac{k}{d} = 4{,}310 \times \dfrac{3 \cdot 62 \times 10^{-5}}{0 \cdot 15} = 1 \cdot 04 \text{ kW/m}^2 \text{ K}$

Heat flux removal required to cool the wing to 100°C (373 K), q, is

$$q = h(T_0 - T_s) = 1 \cdot 04 \times (451 - 373) = 81 \cdot 2 \text{ kW/m}^2$$

7-2-5 Free convection over external surfaces

For free convection, all fluid properties other than β are evaluated at the mean film temperature, T_m, defined by Eqn (7-1). β is evaluated at the free stream temperature.

Vertical surfaces

For vertical plates and cylinders, the characteristic dimension for the Nusselt and Grashof numbers is the vertical height. The data of Weise[24] and Saunders[25] are shown plotted in Fig. 7-3 for short vertical plates in air.

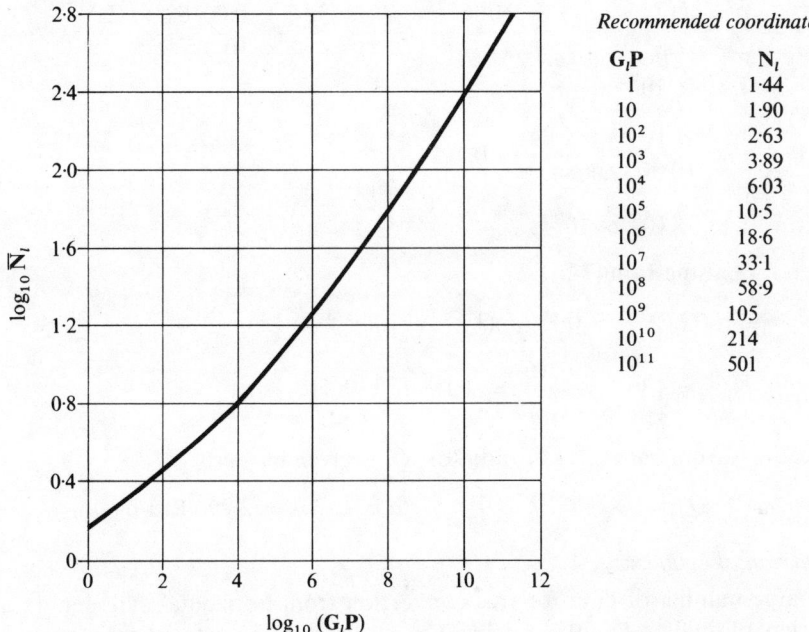

G_lP	N_l
Recommended coordinates	
1	1·44
10	1·90
10^2	2·63
10^3	3·89
10^4	6·03
10^5	10·5
10^6	18·6
10^7	33·1
10^8	58·9
10^9	105
10^{10}	214
10^{11}	501

Fig. 7-3 Free convection from vertical plates in air (after McAdams[13])

McAdams[13] recommends that a correlation of the form

$$\overline{N}_l = C(G_lP)^n \qquad (7\text{-}21)$$

be used, where

for laminar flow $(10^4 < G_lP < 10^9)$: $\quad C = 0\cdot59, n = \frac{1}{4}$

for turbulent flow $(10^9 < G_lP < 10^{12})$: $\quad C = 0\cdot13, n = \frac{1}{3}$

It should be noted that when $n = \frac{1}{3}$, $\overline{h}$ is independent of l. For $G_lP < 10^4$ it is recommended that Fig. 7-3 be used.

Example 7. Calculate the convective heat loss from a radiator 4 ft wide and 2 ft high maintained at a temperature of 200°F in a room at 14·7 lbf/in² and 70°F. Treat the radiator as a vertical plate, and for air take $\mu = 13\cdot2 \times 10^{-6}$ lb/ft s, $k = 0\cdot016$ Btu/ft h °R, and $C_p^1 = 0\cdot24$ Btu/lb °R.

$$T_m = \tfrac{1}{2}(200 + 70) = 135°F = 595 \; °R$$

$$\rho = \frac{p}{RT_m} = \frac{14\cdot7 \times 144}{53\cdot3 \times 595} = 0\cdot0667 \; \text{lb/ft}^3$$

$$\beta = 1/T_\infty = 1/530 = 0\cdot00189 \; °R^{-1}$$

$$G_l = \frac{\rho^2 l^3 g\beta(T_s - T_\infty)}{\mu^2} = \frac{0\cdot0667^2 \times 2^3 \times 32\cdot2 \times 0\cdot00189 \times (200 - 70)}{(13\cdot2 \times 10^{-6})^2}$$

$$= 1\cdot61 \times 10^9$$

$$P = \frac{13\cdot2 \times 10^{-6} \times 0\cdot24}{0\cdot016/3{,}600} = 0\cdot712$$

$$PG_l = 1\cdot15 \times 10^9 > 10^9$$

therefore, using Eqn (7-21),

$$\overline{N}_l = 0\cdot13(G_lP)^{1/3} = 0\cdot13 \times (1\cdot15 \times 10^9)^{1/3} = 136$$

$$\bar{h} = \overline{N}_l\frac{k}{l} = 136 \times \frac{0\cdot016}{2} = 1\cdot1 \; \text{Btu/ft}^2 \; h \; °R$$

Heat loss from *both* sides of radiator, $\dot{Q}$, is given by

$$\dot{Q} = 2 \times 4 \times 2 \times 1\cdot1 \times (200 - 70) = 2{,}290 \; \text{Btu/h}$$

Horizontal cylinders

A large number of data for free convection from horizontal cylinders in a variety of fluids, using the outside diameter as the characteristic dimension, have been correlated by McAdams[13] by

$$\overline{N}_d = C(G_dP)^n \tag{7-22}$$

The correlation is shown in Fig. 7-4, and

for laminar flow $(10^4 < G_dP < 10^9)$: $C = 0\cdot53, n = \tfrac{1}{4}$
for turbulent flow $(10^9 < G_dP < 10^{12})$: $C = 0\cdot13, n = \tfrac{1}{3}$

For $G_dP < 10^4$ it is recommended that Fig. 7-4 be used.

Horizontal plates

Fishenden and Saunders[26] correlated a large number of data for horizontal square plates in air using the length, l, as a characteristic dimension. Equation (7-21) can be used for horizontal plates if the constants C and n are evaluated according to the conditions stated below:

Laminar flow: upper surface heated, or lower surface cooled $(10^5 < G_l P < 2 \times 10^7)$ $\Big\}$ $C = 0.54, n = \frac{1}{4}$

Turbulent flow: upper surface heated, or lower surface cooled $(2 \times 10^7 < G_l P < 3 \times 10^{10})$ $\Big\}$ $C = 0.14, n = \frac{1}{3}$

Laminar flow: upper surface cooled, or lower surface heated $(3 \times 10^5 < G_l P < 3 \times 10^{10})$ $\Big\}$ $C = 0.27, n = \frac{1}{4}$

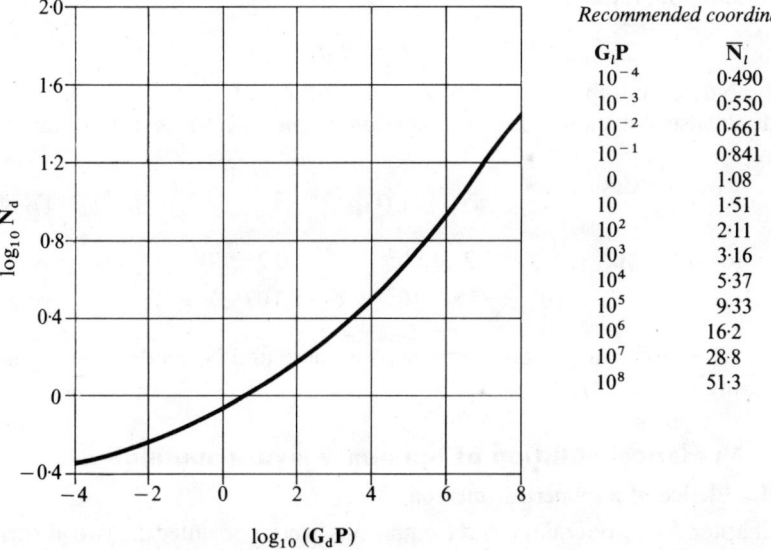

$G_l P$	$\overline{N}_l$
10^{-4}	0·490
10^{-3}	0·550
10^{-2}	0·661
10^{-1}	0·841
0	1·08
10	1·51
10^2	2·11
10^3	3·16
10^4	5·37
10^5	9·33
10^6	16·2
10^7	28·8
10^8	51·3

Recommended coordinates

Fig. 7-4 Free convection from horizontal cylinders in gases and liquids (after MacAdams[13])

7-2-6 Free convection in enclosed spaces

It is convenient to base the Nusselt and Grashof numbers on the distance, s, between the enclosing surfaces, and the temperature difference between the surfaces. Thus,

$$\overline{N}_s = \frac{q_s s}{k(T_{s1} - T_{s2})}$$

and

$$G_s = \frac{s^3 g \beta (T_{s1} - T_{s2})}{v^2}$$

where T_{s1} and T_{s2} are the interior surface temperatures.

For the case of *vertical enclosed air spaces*, Jakob[27] correlated a large number of data, with a length to gap ratio of $3 \cdot 1 < l/s < 42 \cdot 2$ by

$$\overline{N}_s = C(l/s)^{-1/9}(G_s P)^n \tag{7-23}$$

where for

$$2 \cdot 1 \times 10^3 < G_s < 2 \times 10^4: \quad C = 0 \cdot 20, n = \tfrac{1}{4}$$
$$2 \cdot 1 \times 10^5 < G_s < 1 \cdot 1 \times 10^7: \quad C = 0 \cdot 071, n = \tfrac{1}{3}$$

and for $l/s < 17 \cdot 5$ Eqn (7-21) should be used as the flow tends to that of a single plate.

For $G_s < 2 \times 10^3$, free convection is suppressed and heat is transferred by conduction, that is,

$$q_s = k(T_{s1} - T_{s2})/s$$

or the Nusselt number, $\overline{N}_s$, tends to unity and Eqn (7-23) is invalid.

Jakob also correlated data for *horizontal enclosed air spaces* using the form

$$\overline{N}_s = C(G_s P)^n \tag{7-24}$$

where for $10^4 < G_s < 3 \cdot 2 \times 10^5: \quad C = 0 \cdot 21, n = \tfrac{1}{4}$
$$3 \cdot 2 \times 10^5 < G_s < 10^7: \quad C = 0 \cdot 075, n = \tfrac{1}{3}$$

For $G_s < 10^3$ conduction effects predominate and $\overline{N}_s$ tends to unity, invalidating Eqn (7-24).

7-3   Numerical solution of boundary layer equations

7-3-1   Choice of a numerical method

In Chapter 5 the boundary layer equations were presented in two forms: the partial differential equations and the integral equations. Either class of equation can be solved numerically by a variety of methods, and each method has advantages and disadvantages. However, there are a number of criteria that can be used to facilitate selection, and among the most important requirements are that the method should:

(i) give predictions that are in accord with experimental results or exact solutions, if these are available;

(ii) be generally applicable to a wide range of conditions with the minimum of modification;

(iii) be capable of extension, in the case of turbulent flow, to include more advanced turbulence models when these become available;

(iv) be economical of computer storage space and time;

(v) be capable of being readily assimilated and programmed by the user.

Until the wide availability of high-speed digital computers, the integral equations were found to be more convenient to solve. However, it is necessary with these equations to postulate velocity and temperature distributions and to assume relationships between the integral quantities and the wall fluxes (that is, heat flux and shear stress). Examples of this for laminar flow (which is less troublesome owing to the fact that the wall gradient can be used to estimate the flux) and turbulent flow (where empirical relationships are assumed) have been presented in Chapter 6. The results for relatively simple problems have been satisfactory, but as problems become more complex it is necessary to include additional parameters in the profiles and to introduce auxiliary relationships. Examples of such integral methods can be found in references (28)–(30). There are two important limitations of such techniques; one physical and one mathematical. First, every time a new phenomenon is included in the calculation procedure, new empirical relationships—which require experimental corroboration—are necessary. Second, as the number of parameters increase so does the computation time. One integral method that has a relatively large applicability, and whose auxiliary relations are based on the mixing length hypothesis, is that of Kutateladze and Leont'ev[31]. It has been tested for a range of heat transfer conditions with both impermeable and permeable walls (where transpiration cooling is employed) and, for many problems, has produced acceptable results.

Turning to the partial differential boundary layer equations, it is apparent that these equations give a more detailed description of the physical situation than the integral equations. However, with the differential equations it is necessary to relate the shear stress and heat flux to the velocity and temperature throughout the boundary layer and not just at the wall, as is done with the integral equations. For laminar flow there is no great problem, but for turbulent flow some hypotheses are necessary to provide the necessary relationships. Many such hypotheses employ, explicitly or implicitly, the *effective viscosity* model, where an effective viscosity, μ_{eff}, is used for laminar or turbulent flow such that

$$\tau = \mu_{\text{eff}} \frac{\partial u}{\partial y} \tag{7-25}$$

the bar having been omitted from time-average values. It can be seen by reference to Eqn (6-96) that

$$\mu_{\text{eff}} = \mu + A_\tau \tag{7-26}$$

For laminar flow $A_\tau = 0$, but for turbulent flow if the model is to be used for quantitative predictions the function A_τ must be calculated from semi-empirical relationships. A useful, if limited, hypothesis which can be used to provide these relationships is Prandtl's mixing length, which is used in Eqn (6-97). It is apparent that under these conditions

$$\mu_{\text{eff}} = \mu + \rho l^2 \left| \frac{\partial u}{\partial y} \right| \qquad (7\text{-}27)$$

The distribution of mixing length, l, must be assumed for any particular problem, but certain basic distributions which are common to many situations will be discussed later. This model has the virtue of simplicity and was the first to be incorporated in the numerical method of Patankar and Spalding.[32] An alternative approach is to solve additional equations for the turbulence (for example, the turbulence energy and its length scale) and to make assumptions about the relationship between the turbulent shear stress and other properties of turbulence. Bradshaw et al.[33] have produced a numerical procedure for the fluid dynamics problem in which the turbulence energy equation is solved together with the momentum equation. A later version of the method is also capable of calculating heat transfer in boundary layers.[34] The Spalding–Patankar method has also been extended to solve additional turbulence equations (a two-equation method has been developed), but the effective viscosity model is still employed, and the basic solution procedure remains unchanged. Both of the above numerical methods are solved by *marching procedures* which consist of solving the differential equation in the cross-stream direction at one particular station and then moving downstream to the next station until the whole boundary layer has been calculated. Marching procedures applied to parabolic problems have been demonstrated in Section 4-3, and further examples of their application to boundary layer problems can be found in references (35)–(38).

Another method of solving the partial differential equations is *cross-stream integration* where, at a given station, the equations are transformed by a finite difference approximation into ordinary differential equations which are integrated numerically by an iterative procedure. Examples of these are given in references (39)–(42). There is also the *parametric-integral* method where the partial differential equations are multiplied by weighting functions of the independent or dependent variables and integrated across the boundary layer to give a set of ordinary differential equations. These equations are presented in a matrix form with the differential coefficients as the unknown vector which is then expressed in terms of the known quantities by matrix inversion. The resulting differential equations are solved by standard techniques. Examples of this method are given in references (43)–(47).

There is not, nor is there likely to be, a numerical method that is the 'best' for every one of the five criteria stated at the start of this section. Indeed, some requirements are mutually exclusive, and so choice must finally be left to the user. With the current trend of increasing speed and storage of electronic computers, and the necessity for the designer and research worker to be able to predict results more accurately (and with the minimum of experimental corroboration), it would appear that the differential methods show the greatest potential.

In writing this chapter the authors consider that the presentation of one, very general, method in depth is more useful to the reader than a superficial discussion of several methods. The numerical procedure selected, the Spalding–Patankar method,[32] has been developed over many years and is capable of including the advanced models of turbulence that are currently being tested by research workers. The choice of this method is not intended to reflect upon the other techniques, mentioned and unmentioned, developed by eminent workers in the field of thermo-fluids. The reader is to be encouraged to consult the references given in this section, and to review continually the literature devoted to boundary layer prediction methods.

From a relatively detailed presentation of the numerical method, the interested reader should be capable of writing a program for himself related to a specific problem. Alternatively, he can turn to the references given if he requires a fuller exposition. However, before commencing with the Spalding–Patankar method it is illuminating to consider a finite difference formulation of a relatively simple equation: the laminar momentum equation. From this example it is possible to give a continuity to the ensuing numerical procedure, which would otherwise appear disjointed.

7-3-2   A finite difference solution of the incompressible, laminar momentum equation

Schlichting[48] includes an example of a finite difference method which should facilitate the understanding of the Spalding–Patankar method presented in the remaining sections of this chapter. The equations to be solved are (5-76) and (5-77), written below as

$$u \frac{\partial u}{\partial x} + v \frac{\partial u}{\partial y} = -\frac{1}{\rho} \frac{dp}{dx} + v \frac{\partial^2 u}{\partial y^2} \tag{7-28}$$

$$\frac{\partial u}{\partial x} + \frac{\partial v}{\partial y} = 0 \tag{7-29}$$

with the boundary conditions $u = v = 0$ at $y = 0$; $u = U_\infty$ at $y = \delta$, where δ is the defined edge of the boundary layer. The finite difference grid is illustrated in Fig. 7-5, and Δx and Δy, the streamwise and cross-stream steps, respectively, are taken to be constant. The finite difference method employed is *fully implicit*, as described in Chapter 4, and so the step length can be chosen for accuracy, and not for stability, requirements. A central difference formula is used for the y derivatives and a backward difference formula is used for the x derivative. Using Taylor's series we write

$$u_{i-1,j} = u_{i,j} - \Delta x \left(\frac{\partial u}{\partial x}\right)_{i,j} + \frac{(\Delta x)^2}{2!} \left(\frac{\partial^2 u}{\partial x^2}\right)_{i,j} - o(\Delta x)^3$$

$$u_{i-2,j} = u_{i,j} - 2\Delta x \left(\frac{\partial u}{\partial x}\right)_{i,j} + \frac{(2\Delta x)^2}{2!} \left(\frac{\partial^2 u}{\partial x^2}\right)_{i,j} - o(\Delta x)^3$$

where o means 'order of' as defined in Chapter 4.

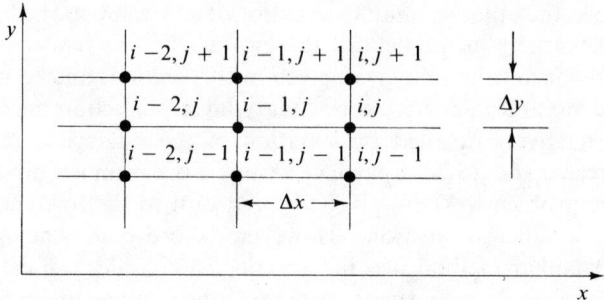

Fig. 7-5 Finite difference mesh for a laminar boundary layer

Therefore,

$$\left(\frac{\partial u}{\partial x}\right)_{i,j} = \frac{3u_{i,j} - 4u_{i-1,j} + u_{i-2,j}}{2\Delta x} - o(\Delta x)^2$$

$$u_{i,j} = 2u_{i-1,j} - u_{i-2,j} + o(\Delta x)^2$$

A similar relationship for $v_{i,j}$ can be obtained, such that

$$v_{i,j} = 2v_{i-1,j} - v_{i-2,j} + o(\Delta x)^2$$

Similarly,

$$u_{i,j+1} = u_{i,j} + \Delta y \left(\frac{\partial u}{\partial y}\right)_{i,j} + \frac{(\Delta y)^2}{2!} \left(\frac{\partial^2 u}{\partial y^2}\right)_{i,j} + o(\Delta y)^3$$

$$u_{i,j-1} = u_{i,j} - \Delta y \left(\frac{\partial u}{\partial y}\right)_{i,j} + \frac{(\Delta y)^2}{2!} \left(\frac{\partial^2 u}{\partial y^2}\right)_{i,j} - o(\Delta y)^3$$

whence $\quad \left(\frac{\partial u}{\partial y}\right)_{i,j} = \frac{u_{i,j+1} - u_{i,j-1}}{2\Delta y} + o(\Delta y)^2$

and $\quad \left(\frac{\partial^2 u}{\partial y^2}\right)_{i,j} = \frac{u_{i,j+1} - 2u_{i,j} + u_{i,j-1}}{(\Delta y)^2} + o(\Delta y)^2$

We can substitute these finite difference approximations into the nonlinear equation, Eqn (7-28), to produce the linear system

$$A_j u_{i,j-1} + B_j u_{i,j} + C_j u_{i,j+1} = F_j \qquad (7\text{-}30)$$

where

$$A_j = -\frac{\Delta x}{2\Delta y}(2v_{i-1,j} - v_{i-2,j}) - v\frac{\Delta x}{(\Delta y)^2}$$

$$B_j = \tfrac{3}{2}(2u_{i-1,j} - u_{i-2,j}) + 2v\frac{\Delta x}{(\Delta y)^2}$$

$$C_j = \frac{\Delta x}{2\Delta y}(2v_{i-1,j} - v_{i-2,j}) - v\frac{\Delta x}{(\Delta y)^2}$$

$$F_j = \tfrac{1}{2}(2u_{i-1,j} - u_{i-2,j})(4u_{i-1,j} - u_{i-2,j}) - \frac{\Delta x}{\rho}\left(\frac{dp}{dx}\right)_i$$

For the case of N grid points across the boundary layer, there will be $N-2$ linear algebraic equations for $u_{i,j}$ $(j = 2, 3, \ldots, N-1)$. As the coefficients A_j, B_j, C_j, and F_j contain only known quantities (that is, previously calculated or initial values of u and v and the prescribed value of dp/dx), the unknown values of $u_{i,j}$ can be found from the resulting matrix equation. As the coefficients form a band matrix, the equations are readily solved by a recurrence relationship, discussed in Chapter 4, together with the boundary conditions $u_{i,1} = 0$, $u_{i,N} = U_\infty$. Having calculated the $u_{i,j}$, the $v_{i,j}$ are found from the continuity equation where

$$v = -\int_0^y \frac{\partial u}{\partial x} \, dy \qquad (7\text{-}31)$$

The quadrature can be performed by trapezoidal integration where the integrand is approximated by a series of straight line segments, the area under each being that of a trapezium. Hence,

$$v_{i,j} = v_{i,j-1} - \frac{\Delta y}{2}\left[\left(\frac{\partial u}{\partial x}\right)_{i,j-1} + \left(\frac{\partial u}{\partial x}\right)_{i,j}\right] \qquad (7\text{-}32)$$

where $v_{i,1} = 0$. Having established values for the velocity at a given downstream station, the wall shear stress, τ_s, can be found from the differential of an interpolation polynomial such that

$$\tau_s = \mu\left(\frac{\partial u}{\partial y}\right)_s = \frac{\mu}{6\Delta y}(18u_{i,2} - 9u_{i,3} + 2u_{i,4}) + o(\Delta y)^3 \qquad (7\text{-}33)$$

The above technique has one important defect; the coordinate system (cartesian or polar) will not, in general, conform to the shape of the boundary layer. As the boundary layer grows, more grid points must be added or be carried from the outset of the problem. The above technique involves testing for the outer edge of the boundary layer, at every downstream step, by checking the change in the value of $u_{i,j}$ at each cross-stream step; the outer edge being defined by the point where $|(\partial u/\partial y)_{i,j}|$ is less than a prescribed tolerance. Such a method is uneconomic of computer storage and time. A better technique is to employ a coordinate system that grows with the boundary layer.

Instead of using y as the cross-stream coordinate, we could use the normalized distance y/δ such that $y/\delta = 1$ corresponds to the defined edge of the boundary layer. In this way the finite difference grid for $0 \leqslant y/\delta \leqslant 1$ would always fit the boundary layer region, and the grid points selected initially, at $x = 0$, say, would lie within the region of interest for all values of x.

Another improvement to the conventional coordinate systems is the choice of the stream function, ψ, as the independent cross-stream variable. Such a change of coordinates, known as the *von Mises transformation*, removes the cross-stream velocity, v, from the boundary layer equations producing a simpler form. If this stream function is then normalized, such that the boundary layer is always confined between the values of zero and unity in the cross-stream direction, we have produced a simpler set of equations that can be bounded by a grid that grows with the boundary layer. The resulting *streamline coordinate system* is discussed in more detail below.

7-3-3 A streamline coordinate system

For the case of steady, compressible flow in a cartesian or an axisymmetric polar system, Eqn (5-84a) can be written

$$\frac{\partial}{\partial x}(\rho u r) + \frac{\partial}{\partial y}(\rho v r) = 0 \tag{7-34}$$

For both cartesian and polar coordinates x and y are the streamwise and cross-stream distances, respectively, and u and v are the streamwise and cross-stream velocity components, respectively. For cartesian coordinates, $r = 1$, and for polar coordinates, r is the radius from the axis of symmetry. The boundary layer momentum equation for compressible flow in cartesian coordinates, Eqn (5-84b), can be written (neglecting body forces) as

$$\rho\left(u\frac{\partial u}{\partial x} + v\frac{\partial u}{\partial y}\right) = -\frac{dp}{dx} + \frac{1}{r}\frac{\partial}{\partial y}(r\tau) \tag{7-35}$$

Similarly, for a polar boundary layer system, Eqns (5-17) can be truncated to the above form. For radial flow, as in Fig. 7-6(a), u, v, x, and y are equivalent to v_r, v_z, r and z, respectively, and for laminar flow

$$\frac{\partial}{\partial y}(r\tau) = \frac{\partial}{\partial z}\left(r\mu\frac{\partial v_r}{\partial z}\right)$$

For axial flow, as shown in Fig. 7-6(b), u, v, x, and y are equivalent to v_z, v_r, z, and r respectively, and for laminar flow

$$\frac{\partial}{\partial y}(r\tau) = \frac{\partial}{\partial r}\left(r\mu \frac{\partial v_z}{\partial r}\right)$$

In both the above cases the second derivative in the streamwise direction is ignored in comparison with that for the cross-stream direction. For cartesian or polar coordinates, the shear stress will be generalized for turbulent flow by the use of Eqn (7-25), where μ is replaced by μ_{eff}.

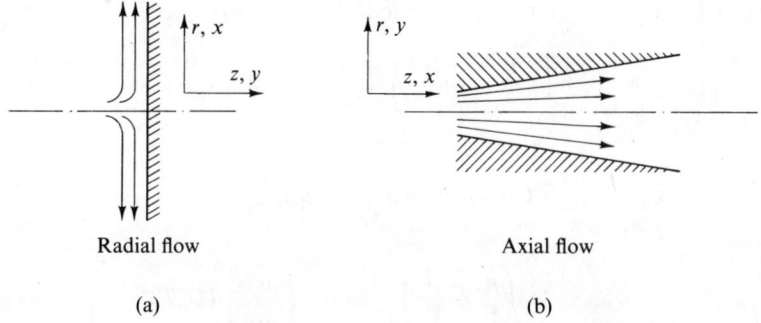

Radial flow   Axial flow

(a)   (b)

Fig. 7-6   Axisymmetric polar systems

It is now convenient to introduce the stream function, ψ, where

$$\rho u r = \frac{\partial \psi}{\partial y} \quad \text{and} \quad \rho v r = -\frac{\partial \psi}{\partial x} \tag{7-36}$$

and, on substitution into Eqn (7-34), it is apparent that the stream function satisfies the continuity requirement. The above orthogonal coordinate system (x, y) can be transformed to a streamline system (ζ, ψ) using the von Mises transformation where

$$\zeta \equiv x$$

and

$$\frac{\partial u}{\partial x} = \frac{\partial u}{\partial \psi}\frac{\partial \psi}{\partial x} + \frac{\partial u}{\partial \zeta}\frac{\partial \zeta}{\partial x} = -\rho v r \frac{\partial u}{\partial \psi} + \frac{\partial u}{\partial \zeta}$$

$$\frac{\partial u}{\partial y} = \frac{\partial u}{\partial \psi}\frac{\partial \psi}{\partial y} + \frac{\partial u}{\partial \zeta}\frac{\partial \zeta}{\partial y} = \rho u r \frac{\partial u}{\partial \psi}$$

$$\frac{dp}{dx} = \frac{dp}{d\zeta} \quad \left(\text{ignoring } \frac{\partial p}{\partial \psi}\right)$$

$$\frac{1}{r}\frac{\partial}{\partial y}(r\tau) = \rho u \frac{\partial}{\partial \psi}(r\tau)$$

Equation (7-35) can therefore be expressed as

$$\frac{\partial u}{\partial \zeta} = \frac{\partial}{\partial \psi}(r\tau) - \frac{1}{\rho u}\frac{dp}{d\zeta}$$

or, using Eqn (7-25),

$$\frac{\partial u}{\partial \zeta} = \frac{\partial}{\partial \psi}\left(\rho u r^2 \mu_{\text{eff}}\frac{\partial u}{\partial \psi}\right) - \frac{1}{\rho u}\frac{dp}{d\zeta} \tag{7-37}$$

Before expressing the energy equation in the von Mises coordinates it is convenient to express the equation as the conservation of stagnation enthalpy, $\tilde{h}\,(\equiv C_p T + \frac{1}{2}u^2)$, as was done in Eqn (6-110) where

$$\rho\left(u\frac{\partial \tilde{h}}{\partial x} + v\frac{\partial \tilde{h}}{\partial y}\right) = \frac{\partial}{\partial y}(-q + u\tau) \tag{7-38}$$

$$\tau = (\mu + A_\tau)\frac{\partial u}{\partial y} = \mu_{\text{eff}}\frac{\partial u}{\partial y} \tag{7-39}$$

$$q = -C_p\left(\frac{\mu}{\mathbf{P}} + \frac{A_\tau}{\mathbf{P_t}}\right)\frac{\partial T}{\partial y} = -\frac{\mu_{\text{eff}}}{\mathbf{P_{eff}}}\frac{\partial}{\partial y}(C_p T) \tag{7-40}$$

$\mathbf{P}$, $\mathbf{P_t}$, and $\mathbf{P_{eff}}$ being the laminar, turbulent and 'effective' Prandtl numbers, respectively. Equation (7-38) can be transformed in a similar way to Eqn (7-35) such that

$$\frac{\partial \tilde{h}}{\partial \zeta} = -\frac{\partial}{\partial \psi}[(q - u\tau)r]$$

or

$$\frac{\partial \tilde{h}}{\partial \zeta} = \frac{\partial}{\partial \psi}\left\{\rho u r^2\left[\frac{\mu_{\text{eff}}}{\mathbf{P_{eff}}}\frac{\partial \tilde{h}}{\partial \psi} + \left(\mu_{\text{eff}} - \frac{\mu_{\text{eff}}}{\mathbf{P_{eff}}}\right)\frac{\partial}{\partial \psi}(\tfrac{1}{2}u^2)\right]\right\} \tag{7-41}$$

where, as before, r is the radius from the axis of symmetry and is equal to unity for plane flows.

Finally, the mass diffusion equation, (5-96), can be expressed as

$$\rho\left(u\frac{\partial c_j^*}{\partial x} + v\frac{\partial c_j^*}{\partial y}\right) = -\frac{\partial \dot{m}_j'}{\partial y} + R_j \tag{7-42}$$

where [by analogy with Eqns (5-93) and (5-100)] the mass flux of constituent j is given by

$$\dot{m}_j' = -\left(\frac{\mu_{\text{eff}}}{\mathbf{Sc_{eff}}}\right)\frac{\partial c_j^*}{\partial y} \tag{7-43}$$

Sc_{eff} being the 'effective' Schmidt number and c_j^* the mass fraction of constituent j, and R_j its rate of creation. Generalizing Eqn (7-43) for axisymmetric systems and transforming to the von Mises coordinates

$$\frac{\partial c_j^*}{\partial \zeta} = -\frac{\partial}{\partial \psi}(\dot{m}_j' r) + \frac{R_j}{\rho u}$$

or
$$\frac{\partial c_j^*}{\partial \zeta} = \frac{\partial}{\partial \psi}\left(\rho u r^2 \frac{\mu_{eff}}{Sc_{eff}} \frac{\partial c_j^*}{\partial \psi}\right) + \frac{R_j}{\rho u} \tag{7-44}$$

Equations (7-37), (7-41), and (7-44) form a system of parabolic partial differential equations, expressed in the von Mises coordinates, that are capable of being solved by the same numerical method.

Having expressed the equations in a simplified form, it is now convenient to normalize the cross-stream variable, ψ, such that its value lies between 0 and 1. By so doing the finite difference grid that will be employed in the solution of the boundary layer equations will automatically grow or contract to fit the defined boundary layer. We therefore introduce a dimensionless stream function, η, where

$$\eta \equiv (\psi - \psi_I)/(\psi_E - \psi_I) \tag{7-45}$$

ψ_I and ψ_E being the value of the stream function at the interior and exterior boundaries, respectively. In general, ψ_I and ψ_E will be functions of ζ although, for the case of an impermeable wall, the stream function will be constant on that boundary. If $\dot{m}_I'$ and $\dot{m}_E'$ are the mass fluxes across the I and E boundaries (which may be due to entrainment of fluid at the outer edge of a boundary layer, or to mass transfer through a permeable wall), then from Eqn (7-36)

$$\left.\begin{array}{l} \dfrac{d\psi_I}{d\zeta} = -(\rho v r)_I = -r_I \dot{m}_I' \\[2mm] \dfrac{d\psi_E}{d\zeta} = -(\rho v r)_E = -r_E \dot{m}_E' \end{array}\right\} \tag{7-46}$$

and

Noting that $\partial \psi/\partial \zeta$ is zero, Eqns (7-37), (7-41), and (7-44) can all be expressed in (ζ, η) coordinates in the form

$$\frac{\partial \Phi}{\partial \zeta} + (a + b\eta)\frac{\partial \Phi}{\partial \eta} = \frac{\partial}{\partial \eta}\left(c\frac{\partial \Phi}{\partial \eta}\right) + d \tag{7-47a}$$

where Φ is a dummy variable which may be u, $\tilde{h}$, or c_j^*. The coefficients are defined as

$$a \equiv \frac{r_I \dot{m}_I'}{\psi_E - \psi_I} \tag{7-47b}$$

$$b \equiv \frac{r_E \dot{m}_E' - r_I \dot{m}_I'}{\psi_E - \psi_I} \tag{7-47c}$$

$$c \equiv r^2 \rho u \frac{(\mu_{eff}/\sigma_{eff})}{(\psi_E - \psi_I)^2} \tag{7-47d}$$

The values of σ_{eff} and d, the latter being known as the *source term*, are given in Table 7-3 for the appropriate value of Φ.

Table 7-3

Dependent variable, Φ	σ_{eff}	d
u	1	$-\dfrac{1}{\rho u}\dfrac{dp}{d\zeta}$
$\tilde{h}$	$\mathbf{P}_{eff}$	$\dfrac{\partial}{\partial \eta}\left[\dfrac{r^2 \rho u}{(\psi_E - \psi_I)^2}\left(\mu_{eff} - \dfrac{\mu_{eff}}{\mathbf{P}_{eff}}\right)\dfrac{\partial}{\partial \eta}\left(\tfrac{1}{2}u^2\right)\right]$
c_j^*	$\mathbf{Sc}_{eff}$	$\dfrac{R_j}{\rho u}$

Equation (7-47a) is valid for any two-dimensional plane or axisymmetric flow system, and the boundary layer will always be located between $\eta = 0$ and $\eta = 1$, as illustrated in Fig. 7-7. Having expressed the basic partial differential equations in a convenient form it is now necessary to convert them to finite difference equations in order to obtain numerical solutions.

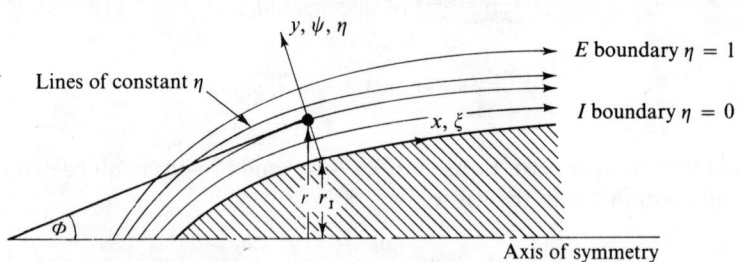

Fig. 7-7 Streamline coordinate system

7-3-4 Finite difference equations for the interior region

In order to formulate the finite difference equations, it is convenient to consider a grid consisting of a finite number of nodes, an example of which is illustrated in Fig. 7-8. At any streamwise location (ζ = constant), the grid lines divide the region between $\eta = 0$ and $\eta = 1$ into N strips. It should be noted that $\eta_1 = \eta_2 = 0$ and $\eta_{N+2} = \eta_{N+3} = 1$. The strip between η_1 and η_3 is referred to as the I boundary region; the strip between η_{N+1} and η_{N+3} is the E boundary region; and the $N - 2$ strips between η_3 and η_{N+1} are referred to as the interior region.

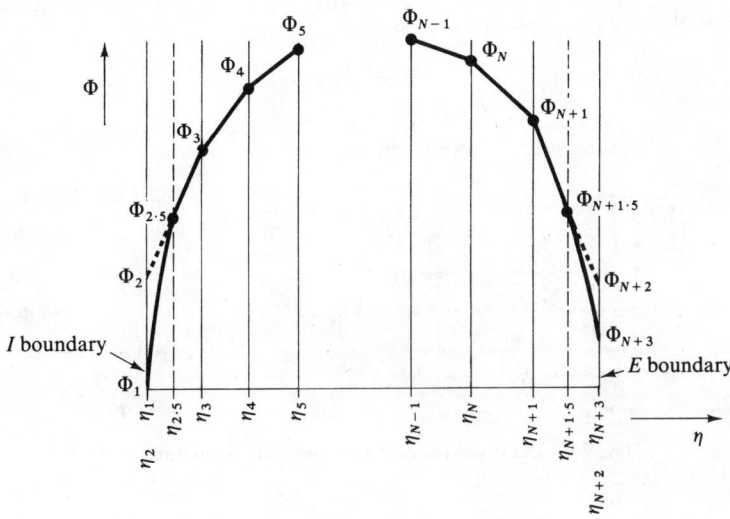

Fig. 7-8 Finite difference grid at one streamwise location

It is necessary to pay particular attention to the wall regions, where the gradient of velocity, temperature, and so on are usually much larger than in the interior region. In heat transfer, where the heat flux through a wall is either specified or required, it is important to determine the wall gradients as accurately as possible. One way of doing this is to have a large number of grid points near the wall, which is uneconomic of computer storage and time; another method is to postulate a distribution of velocity, temperature, and so on, close to the wall. The latter technique will be used in the method described here, and its implementation is effected by the use of *slip values*, which will be discussed more fully later. Suffice it to say at this juncture that Φ_2 and Φ_{N+2} are the slip values that correspond to assumed profile variations.

Instead of forming the finite difference equations from Taylor series expansions, as was done in Section 7-3-2, a micro-integral is evaluated over a control surface shown in Fig. 7-9. This ensures that the conservation equations will be satisfied over any part of the boundary layer. The micro-integral over the element $(i, j + \frac{1}{2})$, $(i - 1, j + \frac{1}{2})$, $(i - 1, j - \frac{1}{2})$, $(i, j - \frac{1}{2})$, shown in Fig. 7-9, where $\eta_{j+1/2} = \frac{1}{2}(\eta_j + \eta_{j+1})$, etc., is performed using the following assumptions:

(i) Φ varies linearly† between adjacent cross-stream steps, that is, $\eta = \eta_{i,j+1/2}$:

$$\Phi = \tfrac{1}{2}(\Phi_{i,j} + \Phi_{i,j+1})$$

(ii) Φ varies in a stepwise manner between adjacent streamwise steps, that is $\zeta_{i-1} < \zeta \leqslant \zeta_i$:

$$\Phi = \Phi_{i,j}.$$

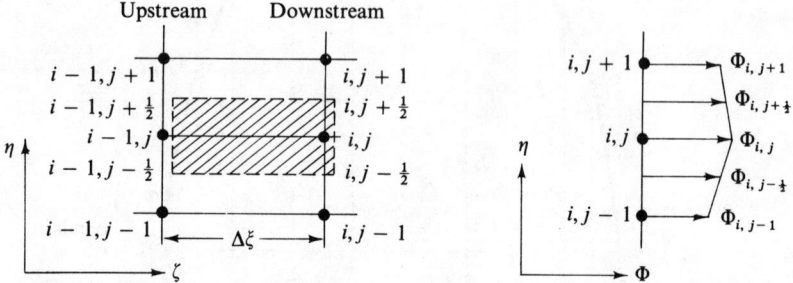

Fig. 7-9 Details of the control volume for an interior point

We now rewrite Eqn (7-47a) in the micro-integral form:

$$\int_{i-1}^{i} \int_{j-1/2}^{j+1/2} \left\{ \frac{\partial \Phi}{\partial \zeta} + (a + b\eta)\frac{\partial \Phi}{\partial \eta} \right\} d\eta \, d\zeta$$

$$= \int_{i-1}^{i} \int_{j-1/2}^{j+1/2} \left\{ \frac{\partial}{\partial \eta}\left(c\frac{\partial \Phi}{\partial \eta}\right) + d \right\} d\eta \, d\zeta \quad (7\text{-}48)$$

Using the above assumptions

$$\int_{i-1}^{i} \int_{j-1/2}^{j+1/2} \left\{ \frac{\partial \Phi}{\partial \zeta} + (a + b\eta)\frac{\partial \Phi}{\partial \eta} \right\} d\eta \, d\zeta$$

$$= \int_{j-1/2}^{j+1/2} [\Phi]_{i-1}^{i} \, d\eta + \left\{ [(a + b\eta)\Phi]_{j-1/2}^{j+1/2} - b \int_{j-1/2}^{j+1/2} \Phi \, d\eta \right\}_i \Delta\zeta$$

and

$$\int_{i-1}^{i} \int_{j-1/2}^{j+1/2} \frac{\partial}{\partial \eta}\left(c\frac{\partial \Phi}{\partial \eta}\right) d\eta \, d\zeta = \left\{ \left[c\frac{\partial \Phi}{\partial \eta}\right]_{j-1/2}^{j+1/2} \right\}_i \Delta\zeta$$

† This assumption is reasonable when the flux due to mass transfer into the boundary layer, due for example to fluid entrained from the free stream, is small compared with the diffusive flux. If this is not the case, a *high-lateral-flux modification* is used, as described in detail in reference (32).

where a, b, and c are assumed to be constant between streamwise steps. The above expressions can all be evaluated by using the linearized expression for Φ, where

$$\Phi_i(\eta) = \begin{cases} \Phi_{i,j} + (\Phi_{i,j+1} - \Phi_{i,j})(\eta - \eta_j)/(\eta_{j+1} - \eta_j) & \text{if } \eta_j \leqslant \eta \leqslant \eta_{j+1/2} \\ \Phi_{i,j} + (\Phi_{i,j-1} - \Phi_{i,j})(\eta - \eta_j)/(\eta_{j-1} - \eta_j) & \text{if } \eta_j \geqslant \eta \geqslant \eta_{j-1/2} \end{cases}$$

Owing to its complexity, the remaining term—the source term, d—requires special consideration, and a detailed discussion of this term and its numerical treatment is given in reference (32). The source term can be evaluated by using the fact that

$$(\psi_E - \psi_I) \int_{j-1/2}^{j+1/2} d \, d\eta = \int_{j-1/2}^{j+1/2} d. \rho u r \, dy \tag{7-49a}$$

$$= -\frac{dp}{dx} \int_{j-1/2}^{j+1/2} r \, dy \qquad \text{if } \Phi \equiv u \tag{7-49b}$$

$$= \left[r \left(\mu_{\text{eff}} - \frac{\mu_{\text{eff}}}{\mathbf{P}_{\text{eff}}} \right) u \frac{\partial u}{\partial y} \right]_{j-1/2}^{j+1/2} \qquad \text{if } \Phi \equiv \tilde{h} \tag{7-49c}$$

$$= \int_{j-1/2}^{j+1/2} Rr \, dy \qquad \text{if } \Phi \equiv c^* \tag{7-49d}$$

where the subscript j, referring to the constituent j, has been omitted from R and c^* to avoid confusion. If we now define

$$S_{i,j} \equiv \frac{1}{\Delta\zeta} \int_i^{i+1} (\psi_E - \psi_I) \int_{j-1/2}^{j+1/2} d \, d\eta \, d\zeta$$

we can then approximate the source term by

$$S_{i,j} \approx -\left[\left(\frac{dp}{dx} \right) r_j \frac{y_{j+1} - y_{j-1}}{2} \right]_{i-1} \qquad \text{if } \Phi \equiv u$$

$$S_{i,j} \approx \left[r_{j+1/2} \left(\mu_{\text{eff}} - \frac{\mu_{\text{eff}}}{\mathbf{P}_{\text{eff}}} \right)_{j+1/2} u_{j+1/2} \frac{u_{j+1} - u_j}{y_{j+1} - y_j} \right]_{i-1}$$

$$- \left[r_{j-1/2} \left(\mu_{\text{eff}} - \frac{\mu_{\text{eff}}}{\mathbf{P}_{\text{eff}}} \right)_{j-1/2} u_{j-1/2} \frac{u_j - u_{j-1}}{y_j - y_{j-1}} \right]_{i-1} \qquad \text{if } \Phi \equiv \tilde{h}$$

$$S_{i,j} \approx [\tfrac{1}{2} R_j r_j (y_{j+1} - y_{j-1})]_{i-1} \qquad \text{if } \Phi \equiv c^*$$

The source term is evaluated at the upstream station to avoid nonlinearities, but for problems where the source term changes significantly in the streamwise direction it may be necessary to modify the above relationships to include downstream effects.

Having linearized the source term, and assuming a linear variation of Φ between adjacent grid points, Eqn (7-48) can be multiplied by $2(\psi_E - \psi_1)/\Delta\zeta$ and written in finite difference form as

$$\left(\frac{\psi_E - \psi_1}{4\Delta\zeta}\right)\{(\eta_j - \eta_{j-1})(\Phi_{i,j-1} - \Phi_{i-1,j-1})$$

$$+ 3(\eta_{j+1} - \eta_{j-1})(\Phi_{i,j} - \Phi_{i-1,j}) + (\eta_{j+1} - \eta_j)(\Phi_{i,j+1} - \Phi_{i-1,j+1})\}$$

$$+ (\psi_E - \psi_1)\{(a + b\eta_{j+1/2})(\Phi_{i,j} + \Phi_{i,j+1})$$

$$- (a + b\eta_{j-1/2})(\Phi_{i,j} + \Phi_{i,j-1})\}$$

$$- \frac{(\psi_E - \psi_1)b}{4}\{(\eta_j - \eta_{j-1})\Phi_{i,j-1}$$

$$+ 3(\eta_{j+1} - \eta_{j-1})\Phi_{i,j} + (\eta_j - \eta_{j-1})\Phi_{i,j-1}\}$$

$$= 2(\psi_E - \psi_1)\left\{c_{j+1/2}\frac{\Phi_{i,j+1} - \Phi_{i,j}}{\eta_{j+1} - \eta_j} - c_{j-1/2}\frac{\Phi_{i,j} - \Phi_{i,j-1}}{\eta_j - \eta_{j-1}}\right\} + 2S_{i,j}$$

$$(7\text{-}50)$$

It should be noted that $(\psi_E - \psi_1)$, a, b, and c are evaluated at the upstream conditions (that is, at the $i - 1$ station) where their values are known. From the definition of c in Eqn (7-47d), and with the use of Eqns (7-36) and (7-45), it can be seen that

$$(\psi_E - \psi_1)\frac{c_{j+1/2}}{\eta_{j+1} - \eta_j} = \left\{\left(r^2\rho u\frac{\mu_{eff}}{\sigma_{eff}}\right)_{j+1/2}\frac{1}{(\psi_E - \psi_1)(\eta_{j+1} - \eta_j)}\right\}_{i-1}$$

$$= \left\{\left(r\frac{\mu_{eff}}{\sigma_{eff}}\right)_{j+1/2}\frac{1}{y_{j+1} - y_j}\right\}_{i-1}$$

Using the above expression for $c_{j+1/2}$, and a similar one for $c_{j-1/2}$, Eqn (7-50) can be written as

$$\Phi_{i,j} = A_j\Phi_{i,j+1} + B_j\Phi_{i,j-1} + C_j \qquad (7\text{-}51)$$

where $\qquad A_j = A_j'/D_j, \qquad B_j = B_j'/D_j, \qquad C_j = C_j'/D_j \qquad (7\text{-}52a)$

and

$$A_j' = \left\{\frac{2}{y_{j+1} - y_j}\left(r\frac{\mu_{eff}}{\sigma_{eff}}\right)_{j+1/2}\right\}_{i-1} - \{(\psi_E - \psi_1)(a + b\eta_{j+1/2})\}_{i-1}$$

$$- \left\{\frac{\psi_E - \psi_1}{4}\left(\frac{1}{\Delta\zeta} - b\right)(\eta_{j+1} - \eta_j)\right\}_{i-1} \qquad (7\text{-}52b)$$

$$B'_j = \left\{ \frac{2}{y_j - y_{j-1}} \left(r \frac{\mu_{\text{eff}}}{\sigma_{\text{eff}}} \right)_{j-1/2} \right\}_{i-1} + \{(\psi_E - \psi_I)(a + b\eta_{j-1/2})\}_{i-1}$$

$$- \left\{ \frac{\psi_E - \psi_I}{4} \left(\frac{1}{\Delta\zeta} - b \right)(\eta_j - \eta_{j-1}) \right\}_{i-1} \quad (7\text{-}52c)$$

$$C'_j = \left\{ \frac{\psi_E - \psi_I}{4\Delta\zeta} [3\Phi_j(\eta_{j+1} - \eta_{j-1}) \right.$$

$$\left. + \Phi_{j+1}(\eta_{j+1} - \eta_j) + \Phi_{j-1}(\eta_j - \eta_{j-1})] \right\}_{i-1} + 2S_{i,j} \quad (7\text{-}52d)$$

$$D_j = A'_j + B'_j + (\eta_{j+1} - \eta_{j-1})[(\psi_E - \psi_I)/\Delta\zeta]_{i-1} \quad (7\text{-}52e)$$

It should be remembered that the coefficients of Eqn (7-51) are only valid for an interior point; the boundary regions must be treated separately, as discussed below.

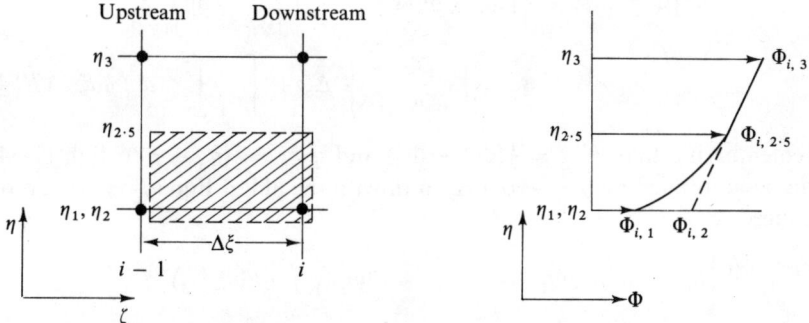

Fig. 7-10 Details of the control volume for a boundary point

7-3-5 The finite difference equations for the boundary regions

The control surface for an I boundary point is shown in Fig. 7-10, and a micro-integral equation over the region $(i, 2{\cdot}5)$, $(i - 1, 2{\cdot}5)$, $(i - 1, 1)$, $(i, 1)$ is used to express the relationship between $\Phi_{i,1}$, $\Phi_{i,2}$, and $\Phi_{i,3}$. The line between $\Phi_{i,2}$, the 'slip value', and $\Phi_{i,3}$ is linear such that $\Phi_{i,2{\cdot}5} = \frac{1}{2}(\Phi_{i,2} + \Phi_{i,3})$. The line between $\Phi_{i,1}$ and $\Phi_{i,2{\cdot}5}$ will depend upon the boundary conditions (a solid or permeable wall, a symmetry axis, or the free stream edge of the boundary layer) and upon the dependent variable (velocity, enthalpy, or mass flux). Thus, while the interior region is relatively simple to treat, the boundary region needs special attention. The principal difficulty arises in the finite difference representation of the $\partial/\partial\eta$ gradient at the I boundary (it should be noted that this section also applies to the E boundary if the appropriate subscripts are changed). In order to permit flexibility for a wide range of boundary conditions, it is proposed that the cross stream derivative at the I boundary should be approximated in the following manner

$$\left[(\psi_E - \psi_I)c\frac{\partial \Phi}{\partial \eta}\right]_{i,2} = \left[-r\frac{\mu_{eff}}{\sigma_{eff}}\frac{\partial \Phi}{\partial y}\right]_{i,2}$$

$$= \Gamma_I[\tfrac{1}{2}(\Phi_{i,2} + \Phi_{i,3}) - \Phi_{i,1} + \delta\Phi_I] \quad (7\text{-}53)$$

where Γ_I and $\delta\Phi_I$ depend on which dependent variable and boundary condition is used (more details are given in Section 7-3-6).

Applying the micro-integral to the element of Fig. 7-10 we arrive at an equation analogous to Eqn (7-48) such that

$$\int_{i-1}^{i}\int_{2}^{2.5}\left\{\frac{\partial \Phi}{\partial \eta} + (a + b\eta)\frac{\partial \Phi}{\partial \eta}\right\} d\eta \, d\zeta = \int_{i-1}^{i}\int_{2}^{2.5}\left\{\frac{\partial}{\partial \eta}\left(c\frac{\partial \Phi}{\partial \eta}\right) + d\right\} d\eta \, d\zeta$$

Using the linearizing assumptions this can be simplified to

$$\int_{2}^{2.5}[\Phi]_{i-1}^{i}\, d\eta + \left\{[(a + b\eta)\Phi]_{2}^{2.5} - b\int_{2}^{2.5}\Phi \, d\eta\right\}_i \Delta\zeta$$

$$= \left\{\left[c\frac{\partial \Phi}{\partial \eta}\right]_{2}^{2.5}\right\}_i \Delta\zeta + \int_{i-1}^{i}\int_{2}^{2.5} d \, d\eta \, d\zeta \quad (7\text{-}54)$$

Remembering that $\Phi_{2.5} = \tfrac{1}{2}(\Phi_2 + \Phi_3)$ and $\eta_{2.5} = \tfrac{1}{2}(\eta_3 - \eta_2)$, Eqn (7-54) can be multiplied by $(\psi_E - \psi_I)/\Delta\zeta$ and, with the aid of Eqn (7-53), it can be reduced to

$$\frac{\psi_E - \psi_I}{8\Delta\zeta}\left\{(\eta_3 - \eta_2)[(3\Phi_{i,2} + \Phi_{i,3}) - (3\Phi_{i-1,2} + \Phi_{i-1,3})]\right.$$

$$+ (\psi_E - \psi_I)\left\{(a + b\eta_{2.5})\frac{\Phi_{i,2} + \Phi_{i,3}}{2} - a\Phi_{i,1}\right\}$$

$$-\frac{\psi_E - \psi_I}{8}\{b(\eta_3 - \eta_2)(3\Phi_{i,2} + \Phi_{i,3})\}$$

$$= \left\{(\psi_E - \psi_I)c_{2.5}\frac{\Phi_{i,3} - \Phi_{i,2}}{\eta_3 - \eta_2} - \Gamma_I[\tfrac{1}{2}(\Phi_{i,2} + \Phi_{i,3}) - \Phi_{i,1} + \delta\Phi_I]\right\} + S_{i,I}$$

$$(7\text{-}55)$$

where $S_{i,I}$ is the average value of $S_{i,j}$ between η_2 and $\eta_{2.5}$.

Using the relationship

$$(\psi_E - \psi_I)\frac{c_{2.5}}{\eta_3 - \eta_2} = \left\{\left(r\frac{\mu_{eff}}{\sigma_{eff}}\right)_{2.5}\frac{1}{y_3 - y_2}\right\}_{i-1}$$

it is possible to express Eqn (7-55) in the same form as Eqn (7-51). However, the value of the coefficients will depend on whether the boundary condition specifies $\Phi_{i,1}$ (a function value) or a flux (a derivative boundary condition).

In the former case, *if* $\Phi_{i,1}$ *is known*, then

$$\Phi_{i,2} = A_2\Phi_{i,3} + B_2\Phi_{i,1} + C_2 \tag{7-56a}$$

and a comparison between Eqns (7-55) and (7-56a) reveals that

$$A_2 = A_2'/D_2, \qquad B_2 = B_2'/D_2, \qquad C_2 = C_2'/D_2 \tag{7-57a}$$

where

$$A_2' = \left\{ \frac{2}{y_3 - y_2}\left(r\frac{\mu_{\text{eff}}}{\sigma_{\text{eff}}}\right)_{2\cdot5}\right\}_{i-1} - \left\{(a + b\eta_{2\cdot5})(\psi_{\text{E}} - \psi_1)\right\}_{i-1}$$
$$- \Gamma_{\text{I}} - \left\{\frac{\psi_{\text{E}} - \psi_1}{4}\left(\frac{1}{\Delta\zeta} - b\right)(\eta_3 - \eta_2)\right\}_{i-1} \tag{7-57b}$$

$$B_2' = 2\{\Gamma_{\text{I}} + a(\psi_{\text{E}} - \psi_1)\}_{i-1} \tag{7-57c}$$

$$C_2' = \left\{\frac{\psi_{\text{E}} - \psi_1}{4\Delta\zeta}(3\Phi_2 + \Phi_3)(\eta_3 - \eta_2)\right\}_{i-1} - 2\Gamma_{\text{I}}\,\delta\dot{\Phi}_{\text{I}} + 2S_{i,\text{I}} \tag{7-57d}$$

$$D_2' = A_2' + B_2' + (\eta_3 - \eta_2)[(\psi_{\text{E}} - \psi_1)/\Delta\zeta]_{i-1} \tag{7-57e}$$

Similar arguments can be applied to the E boundary where

$$\Phi_{i,N+2} = A_{N+2}\Phi_{i,N+3} + B_{N+2}\Phi_{i,N+1} + C_{N+2} \tag{7-56b}$$

It can easily be verified that

$$A_{N+2} = A_{N+2}'/D_{N+2}, \qquad B_{N+2} = B_{N+2}'/D_{N+2}, \qquad C_{N+2} = C_{N+2}'/D_{N+2} \tag{7-57f}$$

where

$$A_{N+2}' = 2\{\Gamma_{\text{E}} - (a + b)(\psi_{\text{E}} - \psi_1)\}_{i-1} \tag{7-57g}$$

$$B_{N+2}' = \left\{ \frac{2}{y_{N+2} - y_{N+1}}\left(r\frac{\mu_{\text{eff}}}{\sigma_{\text{eff}}}\right)_{N+1\cdot5}\right\}_{i-1} + \{(a + b\eta_{N+1\cdot5})(\psi_{\text{E}} - \psi_1)\}_{i-1}$$
$$- \Gamma_{\text{E}} - \left\{\frac{\psi_{\text{E}} - \psi_1}{4}\left(\frac{1}{\Delta\zeta} - b\right)(\eta_{N+2} - \eta_{N+1})\right\}_{i-1} \tag{7-57h}$$

$$C_{N+2}' = \left\{\frac{\psi_{\text{E}} - \psi_1}{4\Delta\zeta}(3\Phi_{N+2} + \Phi_{N+1})(\eta_{N+2} - \eta_{N+1})\right\}_{i-1}$$
$$- 2\Gamma_{\text{E}}\,\delta\Phi_{\text{E}} + 2S_{i,\text{E}} \tag{7-57i}$$

$$D_{N+2}' = A_{N+2}' + B_{N+2}' + (\eta_{N+2} - \eta_{N+1})[(\psi_{\text{E}} - \psi_1)/\Delta\zeta]_{i-1} \tag{7-57j}$$

and $S_{i,\text{E}}$ is the average value of $S_{i,j}$ between $\eta_{N+1\cdot5}$ and η_{N+2}.

When $\Phi_{i,1}$ is not known, but the *total flux crossing the* I *boundary is prescribed* (for example, when the heat flux is specified), then Eqn (7-55) must be modified. For the general case where $\dot{m}_1' \neq 0$, the total flux crossing the

I boundary (J_1) will consist of two components: the flux due to mass transfer, $J_m = \dot{m}_1' \Phi_{i,1}$, and the flux due to the diffusion gradient, $J_d = -(\mu_{\text{eff}}/\sigma_{\text{eff}})$ $(\partial \Phi / \partial y)$—that is, $r_1 J_d = \Gamma_1[\Phi_{i,1} - \tfrac{1}{2}(\Phi_{i,2} + \Phi_{i,3}) - \delta \Phi_1]$.

Thus, referring to Eqn (7-55),

$$r_1 J_1 = a(\psi_E - \psi_1)\Phi_{i,1} + \Gamma_1[\Phi_{i,1} - \tfrac{1}{2}(\Phi_{i,2} + \Phi_{i,3}) - \delta \Phi_1]$$

and if J_1 is known, $\Phi_{i,1}$ can be eliminated such that

$$\Phi_{i,2} = A_2 \Phi_{i,3} + C_2 \tag{7-58a}$$

where
$$A_2 = A_2'/D_2, \qquad C_2 = C_2'/D_2 \tag{7-59a}$$

and

$$A_2' = \left\{ \frac{2}{y_3 - y_2} \left(r \frac{\mu_{\text{eff}}}{\sigma_{\text{eff}}} \right)_{2\cdot5} \right\}_{i-1} - \{(a + b\eta_{2\cdot5})(\psi_E - \psi_1)\}_{i-1}$$
$$- \left\{ \frac{\psi_E - \psi_1}{4} \left(\frac{1}{\Delta \zeta} - b \right)(\eta_3 - \eta_2) \right\}_{i-1} \tag{7-59b}$$

$$C_2' = \left\{ \frac{\psi_E - \psi_1}{4\Delta \zeta}(3\Phi_2 + \Phi_3)(\eta_3 - \eta_2) \right\}_{i-1} + 2r_1 J_1 + 2S_{i,1} \tag{7-59c}$$

$$D_2 = A_2' + (\eta_3 - \eta_2)[(\psi_E - \psi_1)/\Delta \zeta]_{i-1} + 2[a(\psi_E - \psi_1)]_{i-1} \tag{7-59d}$$

If the total flux crossing the E boundary (J_E) is specified, then similar arguments can be used to give

$$\Phi_{i,N+2} = B_{N+2} \Phi_{i,N+1} + C_{N+2} \tag{7-58b}$$

where
$$B_{N+2} = B_{N+2}'/D_{N+2}, \qquad C_{N+2} = C_{N+2}'/D_{N+2} \tag{7-59e}$$

and

$$B_{N+2}' = \left\{ \frac{2}{y_{N+2} - y_{N+1}} \left(r \frac{\mu_{\text{eff}}}{\sigma_{\text{eff}}} \right)_{N+1\cdot5} \right\}_{i-1} + \{(a + b\eta_{N+1\cdot5})(\psi_E - \psi_1)\}_{i-1}$$
$$- \left\{ \frac{\psi_E - \psi_1}{4} \left(\frac{1}{\Delta \zeta} - b \right)(\eta_{N+2} - \eta_{N+1}) \right\}_{i-1} \tag{7-59f}$$

$$C_{N+2}' = \left\{ \frac{\psi_E - \psi_1}{4\Delta \zeta}(3\Phi_{N+2} + \Phi_{N+1})(\eta_{N+2} - \eta_{N+1}) \right\}_{i-1} - 2r_E J_E + 2S_{i,E} \tag{7-59g}$$

$$D_{N+2} = B_{N+2}' + (\eta_{N+2} - \eta_{N+1})[(\psi_E - \psi_1)/\Delta \zeta]_{i-1}$$
$$- 2[(a + b)(\psi_E - \psi_1)]_{i-1} \tag{7-59h}$$

We now have a complete set of finite difference equations, but it is necessary to define the parameters Γ_1 and $\delta \Phi_1$ which control the flux on the I boundary. For the case of a *symmetry axis*, where the cross-stream gradient and the diffusive flux are zero, $J_d = 0$. In this case, Γ_I is put equal to zero and $\Phi_{i,1}$ is put equal to $\frac{1}{2}(\Phi_{i,2} + \Phi_{i,3})$. For the case of a *free boundary*, J_d is again zero and so Γ_I is put equal to zero; however, $\Phi_{i,1}$ is usually specified and hence $\Phi_{i,2}$ is calculated from Eqn (7-56a). Similar results apply for a free boundary or symmetry axis on the E boundary. This leaves the case of the *wall boundary*, which will be discussed in the next section.

7-3-6 Calculation of wall fluxes

In order to solve the finite difference equations when a wall is present on one or both of the boundaries (the traditional boundary layer problem) it is necessary to define the parameters Γ_I and $\delta\Phi_I$, which requires a knowledge of the conditions near the wall. It is the object of this section to present a few examples and recommendations: for more details the reader can refer to reference (32) or he can devise his own relationships.

A considerable simplification that can be made in the wall regions, where the 2·5 grid is close enough to the wall to make streamwise convection negligible, is that the partial differential equations can be made into ordinary differential equations.

The momentum equation

For convenience we can rewrite Eqn (7-35) as

$$\rho\left(u\frac{\partial u}{\partial x} + v\frac{\partial u}{\partial y}\right) = -\frac{dp}{dx} + \frac{1}{r}\frac{\partial}{\partial y}(r\tau) \tag{7-60}$$

Near an impermeable wall, u and v tend to zero as y tends to zero. Near a stationary permeable wall, where fluid effuses with, say, a velocity v_1, u tends to zero but v tends to v_1 as y tends to zero. From continuity considerations, as $u \rightarrow 0$,

$$\frac{\partial}{\partial x}(\rho ur) = -\frac{\partial}{\partial y}(\rho vr) \rightarrow 0$$

which implies that ρvr is constant and equals $(\rho vr)_1$ which, in turn, is equal to $r_1\dot{m}_1'$. If the region close to the wall is thin enough to ignore radial variations, we can simplify Eqn (7-60) to

$$\dot{m}_1'\frac{du}{dy} = -\frac{dp}{dx} + \frac{d\tau}{dy} \tag{7-61}$$

which can be integrated between 0 and y to give

$$\dot{m}_1'u = -\frac{dp}{dx}y + (\tau - \tau_1) \tag{7-62}$$

Equation (7-62) is sometimes referred to as a Couette flow equation, as inertial terms are neglected. It is convenient to treat the region from the wall (where $\eta = 0$) to the grid (at $\eta = 2 \cdot 5$) as a Couette flow region, and if we make assumptions about the relationship between τ and u, Eqn (7-62) can be solved. For the case of laminar flow we can say that

$$\tau = \mu \frac{du}{dy} = \tau_1 + \frac{dp}{dx} y + \dot{m}'_1 u \qquad (7\text{-}63)$$

As the wall shear stress is obviously a function of the local velocity, the pressure gradient, and the mass transfer rate through the wall, we define a number of groups

$$Su_1 \equiv \tau_1/(\rho u^2)_{2 \cdot 5}, \qquad Re_1 \equiv (uy/v)_{2 \cdot 5}$$

$$P_1 \equiv \left(\frac{dp}{dx} \frac{y}{\rho u^2} \right)_{2 \cdot 5}, \qquad M_1 \equiv \dot{m}'_1/(\rho u)_{2 \cdot 5}$$

where the subscript $2 \cdot 5$ refers to the grid at $\eta = \eta_{2 \cdot 5}$. Equation (7-63) can be integrated between $y = 0$ and $y = y_{2 \cdot 5}$ to give

$$Su_1 = \frac{M_1 \{1 - [\exp (M_1 Re_1) - 1 - M_1 Re_1] P_1/(M_1^2 Re_1)\}}{\exp (M_1 Re_1) - 1} \qquad (7\text{-}64)$$

For the case of an impermeable wall, where $M_1 = 0$, this reduces to

$$Su_1 = Re_1^{-1} - \tfrac{1}{2} P_1 \qquad (7\text{-}65)$$

It is now convenient to relate the parameters Γ_1 and $\delta\Phi_1$ to the dimensionless shear stress, Su_1. In Section 7-3-5 we saw that

$$J_d = -\frac{\mu_{\text{eff}}}{\sigma_{\text{eff}}} \frac{\partial\Phi}{\partial y} = \frac{\Gamma_1}{r_1} [\Phi_{i,1} - \tfrac{1}{2}(\Phi_{i,2} + \Phi_{i,3}) - \delta\Phi_1] \qquad (7\text{-}66)$$

For $\Phi = u$, we have $\Phi_{i,1} = 0$ (for a stationary wall) and $\tfrac{1}{2}(\Phi_{i,2} + \Phi_{i,3}) = u_{2 \cdot 5}$. As $u_{2 \cdot 5}$ typifies the gradient at the wall (that is, if $u_{2 \cdot 5} = 0$, $(\partial u/\partial y)_1 = 0$), we assign $\delta\Phi_1 = 0$; that is, as $\tau \sim \partial\Phi/\partial y$, $\delta\Phi_1$ is zero.

Thus Eqn (7-66) becomes

$$-J_d = \tau_1 = \left(\mu_{\text{eff}} \frac{\partial u}{\partial y} \right)_1 = \frac{\Gamma_1}{r_1} u_{2 \cdot 5} \qquad (7\text{-}67)$$

and from the definition of Su_1

$$\tau_1 = (\rho u^2)_{2 \cdot 5} \, Su_1$$

Therefore, for the *velocity boundary layer*, where $\Phi \equiv u$, $\delta\Phi_1 = 0$, and $\Gamma_1 = (\rho u)_{2 \cdot 5} Su_1 r_1$, Su_1 can be calculated from Eqns (7-64) or (7-65).

For turbulent flow the above equations will be valid if the $2 \cdot 5$ grid lies within the viscous sublayer, that is $yu^*/v < 5$, where $u^* = (\tau_1/\rho)^{1/2}$. This can be ensured by judicious spacing of the cross-stream grid positions such that the distance between adjacent grids increases with distance from the wall. An alternative procedure is to use a coarse grid spacing with the 2·5 grid located in the fully turbulent region ($u^*y/v > 70$). For the case of zero pressure gradient and zero mass transfer ($P_1 = M_1 = 0$) the law of the wall, Eqn (6-129), can be expressed as

$$Su_1 = \kappa/\ln (E^*Re_1 Su_1^{1/2}) \tag{7-68}$$

where $\kappa = 0.4$ and $E^* = 9.025$. Corresponding analytical expressions for Su_1 are given in reference (32) for the case when $P_1 \neq 0$ or $M_1 \neq 0$. However, as E^* is usually a function of wall roughness, pressure gradient, and mass transfer rate, the resulting expressions for Su_1 are of limited validity. For situations where F_1 and M_1 are negligible, Eqn (7-68), which can be solved by direct iteration, can be used: in general it is more convenient to ensure that the 2·5 grid is in the viscous sublayer and use Eqn (7-64) or (7-65).

The energy equation

If the convection and source terms are neglected close to the wall, the energy boundary layer equation—by similar arguments to those outlined above—can be reduced to

$$q = q_1 + \dot{m}_1'(\tilde{h}_1 - \tilde{h}) + u\tau \tag{7-69}$$

For the case of laminar flow over an impermeable wall, Eqn (7-69) can be written as

$$-k\frac{dT}{dy} = q_1 + \mu u \frac{du}{dy} \tag{7-70}$$

If property changes close to the wall are neglected, Eqn (7-70) can be integrated between 0 and y to give

$$q_1 y C_p/k = C_p(T_1 - T) - \tfrac{1}{2}Pu^2$$
$$= \bar{\bar{h}}_1 - \bar{\bar{h}} \tag{7-71}$$

where
$$\bar{\bar{h}} \equiv C_p T + \tfrac{1}{2}Pu^2 \quad \text{and} \quad \bar{\bar{h}}_1 = C_p T_1 \tag{7-72}$$

Thus for heat transfer in laminar flow the motivating potential is $\bar{\bar{h}}_1 - \bar{\bar{h}}$, which can be compared with Eqn (6-12). A similar argument can be applied to fully turbulent flow where Eqn (7-71) is still approximately valid if $\mathbf{P}_t$ is used instead of $\mathbf{P}$ in Eqn (7-72).

By analogy with the definition of Su_1 we define Sh_1 such that

$$Sh_1 \equiv q_1/[\rho u(\bar{\bar{h}}_1 - \bar{\bar{h}})]_{2.5} \tag{7-73}$$

From Eqn (7-66)

$$J_d = -\frac{\mu_{\text{eff}}}{\sigma_{\text{eff}}}\frac{\partial \Phi}{\partial y} = q_1 = \frac{\Gamma_1}{r_1}(\bar{\bar{h}}_1 - \bar{\bar{h}})_{2.5} \tag{7-74}$$

But from Eqn (7-72)

$$\bar{h}_1 - \bar{h}_{2.5} = \tilde{h}_1 - \tilde{h}_{2.5} - \tfrac{1}{2}(\mathbf{P} - 1)u_{2.5}^2$$

where $\mathbf{P}$ is replaced by $\mathbf{P}_t$ for flow in the fully turbulent region. Hence comparing Eqns (7-66) and (7-74) we see that

$$\delta\Phi_1 = \tfrac{1}{2}(\mathbf{P} - 1)u_{2.5}^2 \quad \text{and} \quad \Gamma_1 = (\rho u)_{2.5}Sh_1r_1$$

It should be noted that as q is not proportional to $\partial\Phi/\partial y$, $\delta\Phi_1 \neq 0$.

For laminar flow over an impermeable wall, Eqn (7-71) reveals that

$$Sh_1 = (\mathbf{P}Re_1)^{-1} \tag{7-75}$$

and for a permeable wall the solution of Eqn (7-69) yields

$$Sh_1 = M_1/[\exp(\mathbf{P}M_1Re_1) - 1] \tag{7-76}$$

For turbulent Couette flow an expression of the form

$$Sh_1 = Su_1/[\mathbf{P}_t(1 + PrSu_1^{1/2})] \tag{7-77}$$

can be derived, where Pr is a function of $\mathbf{P}/\mathbf{P}_t$. It is recommended, however, that turbulent calculations should make use of Eqns (7-75) and (7-76) which are valid for the viscous sublayer.

The diffusion equation

If the source term is negligibly small near the wall, the laminar Couette flow arguments can be used to show that for an impermeable wall

$$Sc_1 = (\mathbf{Sc}Re_1)^{-1} \tag{7-78}$$

where $\qquad Sc_1 \equiv \dot{m}'_{j,1}/[\rho u(c^*_{j,1} - c^*_j)]_{2.5} \tag{7-79}$

Similarly, for a permeable wall, Eqn (7-76) can be used for Sc_1 if the Prandtl number $\mathbf{P}$ is replaced by the Schmidt number $\mathbf{Sc}$. As $\dot{m}'_j \propto \partial\Phi/\partial y$,

$$\delta\Phi_1 = 0 \quad \text{and} \quad \Gamma_1 = (\rho u)_{2.5}Sc_{j,1}r_1.$$

Concluding remarks

Γ_1 and $\delta\Phi_1$ can be regarded as parameters that provide compatibility between the finite difference equations and the fluxes at the inner boundary walls. The relationships derived are also valid for the E boundary if the subscripts I and 2·5 are replaced by E and $N + 1·5$. In the numerical procedure used for solving the boundary layer equations, it is only necessary to specify the boundary conditions, and the dependent variable that is being solved, and the appropriate Γ_1 and $\delta\Phi_1$, or Γ_E and $\delta\Phi_E$, can be calculated and used in the finite difference equations for the boundary regions.

For cases not included in the above examples, the reader is recommended to apply the same technique to develop values for $\delta\Phi_1$ and Γ_1; that is, to derive the appropriate Couette flow equation and to solve this for laminar flow. For turbulent flow, it is recommended that the 2·5 grid (or the $N + 1·5$ grid) should be located in the viscous sublayer and the appropriate laminar relationships used.

Having described how $\delta\Phi$ and Γ can be calculated for a wall boundary, it is now convenient to return to a free stream boundary where fluid is entrained into the boundary layer. It is the object of the next section to describe a method by which the amount of fluid entrained into the boundary layer, and the subsequent boundary layer growth, can be calculated.

7-3-7 Calculation of the entrainment of fluid at a free stream boundary

It has already been stated that the normalized streamline coordinates permit the boundary layer equations to be framed in such a way that the finite difference grid always contains the relevant part of the boundary layer. For the case of a wall or symmetry axis the edge of the boundary layer is specified; but for the case of a free stream boundary it is necessary to estimate the amount of fluid entrained into the boundary layer and the subsequent growth of the stream function ($\psi_E - \psi_1$).

A better understanding of the problem involved can be gained if Eqn (7-47) is written out in full, for the case of $\Phi = u$, where

$$\frac{\partial u}{\delta\zeta} + \frac{r_1\dot{m}_1' + \eta(r_E\dot{m}_E' - r_1\dot{m}_1')}{\psi_E - \psi_1}\frac{\partial u}{\partial\eta} = \frac{\partial}{\partial\eta}\left\{\frac{r^2\rho u\mu_{\text{eff}}}{(\psi_E - \psi_1)^2}\frac{\partial u}{\partial\eta}\right\} - \frac{1}{\rho u}\frac{dp}{d\zeta} \quad (7\text{-}80)$$

Consider the case of the I boundary being at the edge of the free stream where

$$-\frac{1}{\rho u}\frac{dp}{d\zeta} = \frac{\partial u}{\partial\zeta}$$

that is, the pressure gradient is prescribed from the free stream velocity. Just inside the boundary layer [where $\partial u/\partial\zeta \approx -(1/\rho u)(dp/d\zeta)$] Eqn (7-80) can be simplified to

$$r_1\dot{m}_1' = \lim_{\eta\to\eta_I}\left\{\frac{\partial}{\partial\eta}\left[\frac{r^2\rho u\mu_{\text{eff}}}{\psi_E - \psi_1}\frac{\partial u}{\partial\eta}\right]\bigg/\frac{\partial u}{\partial\eta}\right\}$$

or

$$r_1\dot{m}_1' = \lim_{y\to y_I}\left\{\frac{\partial}{\partial y}\left[r\mu_{\text{eff}}\frac{\partial u}{\partial y}\right]\bigg/\frac{\partial u}{\partial y}\right\}$$

$$= \lim_{y\to y_I}\left\{\frac{\partial}{\partial y}(r\mu_{\text{eff}}) + r\mu_{\text{eff}}\frac{\partial^2 u}{\partial y^2}\bigg/\frac{\partial u}{\partial y}\right\} \quad (7\text{-}81)$$

The recommendation is that the second term in Eqn (7-81) be omitted, and the subsequent expression is then written in finite difference form as

$$r_I \dot{m}_I' = \frac{(r\mu_{\text{eff}})_{2\cdot5}}{\frac{1}{2}(y_3 - y_2)} \tag{7-82}$$

For the case of a fluid with a Prandtl number much smaller than unity, where the thermal boundary layer is much thicker than the velocity boundary layer, it is suggested that μ_{eff} be replaced by $\mu_{\text{eff}}/\mathbf{P}_{\text{eff}}$ in Eqn (7-82). For a more detailed discussion of this problem the reader is referred to reference (32).

Having calculated $r_I \dot{m}_I'$ (and, if appropriate, $r_E \dot{m}_E'$) the change in stream function $(\psi_E - \psi_I)$ can be found by using Eqns (7-46), where

$$\frac{d}{d\zeta}(\psi_E - \psi_I) = r_I \dot{m}_I' - r_E \dot{m}_E'$$

or, in finite difference notation,

$$(\psi_E - \psi_I)_i = (\psi_E - \psi_I)_{i-1} + (r_I \dot{m}_I' - r_E \dot{m}_E')_{i-1} \Delta\zeta \tag{7-83}$$

As pointed out in Section 7-3-4, the value of $(\psi_E - \psi_I)$ used in the formation of the coefficients for the finite difference equations [Eqns (7-52) and also Eqns (7-57) and (7-59)] is based upon the previously calculated step—that is, for the i-step $(\psi_E - \psi_I)_{i-1}$ is used—and is subsequently updated by means of Eqn (7-83).

The streamwise step length, $\Delta\zeta$, may be chosen to be constant; alternatively, it can be selected to be a fraction of the boundary layer thickness. However, when the growth of the boundary layer thickness is slow it is convenient to control the step by comparing the amount of fluid entrained with that already in the layer; for example,

$$\Delta\zeta = \text{constant} \times (\psi_E - \psi_I)/(r_I \dot{m}_I' - r_E \dot{m}_E')$$

where a value of $0\cdot05$ for the constant has been shown to be satisfactory for many problems.

Having calculated the entrainment of fluid into the boundary layer, we are able to control the finite difference grid. However, there are still terms in the coefficients of the finite difference equations that have yet to be calculated, and amongst these is the cross-stream distance, y. From the stream-line coordinate system it is necessary to be able to return to conventional coordinates, and we now turn our attention to this problem.

7-3-8   Calculation of the cross-stream distance

In order to calculate the cross-stream distance, y, from the streamline co-ordinates use is made of Eqns (7-36) and (7-45) where, for a fixed downstream location ($\zeta = x = $ constant), we can write

$$r\, dy = (\psi_E - \psi_1)(\rho u)^{-1}\, d\eta \qquad (7\text{-}84)$$

In general, referring to Fig. 7-7, r and y are related by

$$r \approx r_1 + y \cos \phi \qquad (7\text{-}85)$$

where, for convenience, $\tan \phi$ is taken as the tangent of the body surface on the I boundary (for a boundary layer, where $y \ll r$, ϕ does not change appreciably with y). For the case of radial flow (Fig. 7-6(a)), $\phi = \pi/2$ and r_1 is the radius at the appropriate ζ station. For pipe flow, $\phi = 0$ and $r_1 = 0$; while for plane flow, $\phi = \pi/2$ and $r_1 = 1$.

From Eqns (7-84) and (7-85) we can write

$$y = 2I/[r_1 + (r_1^2 + 2I \cos \phi)^{1/2}] \qquad (7\text{-}86)$$

where $\qquad I \equiv (\psi_E - \psi_1)\displaystyle\int_0^\eta (\rho u)^{-1}\, d\eta = \int_0^y r\, dy \qquad (7\text{-}87)$

For the interior regions of the finite difference grid at the i station, the integral can be approximated numerically, using the values of $(\rho u)_j$ [the subscript i is not necessary in this section as all values are taken at the i station] which have been calculated, by

$$I_{j+1} - I_j = 2(\psi_E - \psi_1)\frac{\eta_{j+1} - \eta_j}{(\rho u)_j + (\rho u)_{j+1}} \qquad (7\text{-}88)$$

However, for the boundary regions, where the velocity distribution may be highly nonlinear, as can be seen illustrated in Fig. 7-10, it is necessary to include a parameter, Ψ_1 say, to allow for nonlinearities in the region between η_2 and $\eta_{2.5}$. Similarly, Ψ_E is used for the region between $\eta_{N+1.5}$ and η_{N+2}. For the I boundary we have

$$I_1 = 0 \qquad (7\text{-}89)$$

$$I_{2.5} = \eta_{2.5}(\psi_E - \psi_1)/[\Psi_1(\rho u)_{2.5}] \qquad (7\text{-}90)$$

$$I_3 = I_{2.5} + 2(\eta_3 - \eta_{2.5})/[(\rho u)_{2.5} + (\rho u)_3] \qquad (7\text{-}91)$$

and I_4 and subsequent values up to I_{N+1} are found from Eqn (7-88). Similar relationships can be obtained for the E boundary, where $\eta_{N+2} = \eta_{N+3} = 1$.

It is now necessary to establish Ψ_1 and as

$$\eta_{2.5} = \frac{1}{\psi_E - \psi_1}\int_0^{2.5} \rho u r\, dy \quad \text{and} \quad I_{2.5} = \int_0^{2.5} r\, dy$$

Eqn (7-90) implies that

$$\Psi_1 = \int_0^{2.5} \rho u r\, dy / [(\rho u)_{2.5}\int_0^{2.5} r\, dy] \qquad (7\text{-}92)$$

For the case of a *free stream boundary or an axis of symmetry*, Eqn (7-92) can be evaluated by assuming a quadratic form for ρu, where

$$\rho u = (\rho u)_1 + [(\rho u)_{2.5} - (\rho u)_1](y/y_{2.5})^2$$

hence $\quad \Psi_1 = \dfrac{(r_1/r_{2.5})[5(\rho u)_1/(\rho u)_{2.5} + 1] + 3[(\rho u)_1/(\rho u)_{2.5} + 1]}{6[(r_1/r_{2.5}) + 1]}$ (7-93)

For *plane flows*, or where $r_1/r_{2.5} = 1$, Eqn (7-93) reduces to

$$\Psi_1 = \tfrac{1}{3} + \tfrac{2}{3}(\rho u)_1/(\rho u)_{2.5}$$ (7-94)

For *wall boundaries* the wall flux relations derived in Section 7-3-6 can be used. For *laminar flow*, where Eqn (7-64) is valid, it can be shown that

$$\Psi_1 = \frac{Su_1 Re_1 \left[\exp (M_1 Re_1) - 1 - M_1 Re_1\right]}{(M_1 Re_1)^2}$$

$$+ \frac{P_1 Re_1 \left[\exp (M_1 Re_1) - 1 - M_1 Re_1 - \tfrac{1}{2}(M_1 Re_1)^2\right]}{(M_1 Re_1)^3}$$ (7-95)

which, for small M_1 reduces to

$$\Psi_1 \approx Su_1 Re_1 \left[\frac{1}{2!} + \frac{M_1 Re_1}{3!}\right] + P_1 Re_1 \left[\frac{1}{3!} + \frac{M_1 Re_1}{4!}\right]$$ (7-96)

While for *turbulent flow*, where Eqn (7-68) is valid, it follows that

$$\Psi \approx 1 - Su_1^{1/2}/\kappa$$ (7-97)

Having evaluated the appropriate Ψ_1 (and Ψ_E for the E boundary) $I_{2.5}$ and the remaining I_j can be calculated, and the cross-stream distances, y_j, can be found. It is obviously necessary to recalculate the y_j at every streamwise step, for although the values of η_j are constant the values of y_j will, in general, change at each i station.

It should be pointed out that in order to obtain the correct y derivative at the 2·5 grid, a 'slip value' of y, y_2, must be used where

$$\left(\frac{\partial \Phi}{\partial y}\right)_{2.5} = \frac{\Phi_3 - \Phi_2}{y_3 - y_2}$$

Using Eqn (7-84),

$$\left(\frac{\partial \Phi}{\partial y}\right)_{2.5} = \left[\left(\frac{\partial \Phi}{\partial \eta}\right)\left(\frac{\partial \eta}{\partial y}\right)\right]_{2.5} = \frac{\Phi_3 - \Phi_2}{\eta_3 - \eta_2}\left[\frac{\rho u r}{\psi_E - \psi_1}\right]_{2.5}$$

Hence $\quad y_3 - y_2 = (\eta_3 - \eta_2)[(\psi_E - \psi_1)(\rho u r)_{2.5}^{-1}]$

from which y_2, and similarly y_{N+2}, can be calculated.

We have now reached a stage where for laminar, external flow (where $\mu_{eff} = \mu$, $P_{eff} = P$, etc., and $dp/d\zeta$ is prescribed), all of the coefficients of the finite difference equations could be calculated and the equations themselves could be solved by the recurrence relationships given in Section 7-3-11. However, for the case of turbulent flow the effective transport properties have yet to be discussed, and this will be done in Section 7-3-9. The calculation of the pressure gradient for internal flows will be described in Section 7-3-10.

7-3-9   The effective transport properties for turbulent flow

It is the object of this section to suggest ways in which μ_{eff} and σ_{eff} can be calculated for turbulent flow. One of the advantages of the Spalding–Patankar method is that it provides a computational vehicle on which a wide variety of turbulence hypotheses can be tested by the user, if he so requires. Hypotheses that have been tried to date include the simple Prandtl mixing length model discussed in Chapter 6, the van Driest hypothesis[49], where the mixing length decays exponentially close to a wall, and a synthesis of the ideas of Kolmogorov[50] and Rotta[51]. The proposals of the last two authors have been developed to produce a two-equation model of turbulence where two dependent variables (the kinematic energy of turbulence, and the product of this and a local length scale) are solved numerically from the generic equation, Eqn (7-47), simultaneously with the other Φ variables.

While multi-equation models of turbulence provide a more complete picture of turbulent processes, and also prove to be more universal compared with mixing length hypotheses, they do this at the expense of added complexity and increased computer time and storage. The mixing length model has been used to give satisfactory prediction for a large variety of problems and, despite its shortcomings, it is proposed to devote the remainder of this section to simple models.

Prandtl's mixing length hypothesis implies that

$$\mu_{\text{eff}} = \mu + A_\tau = \mu + \rho l^2 \left| \frac{\partial u}{\partial y} \right| \qquad (7\text{-}98)$$

and Escudier[52] recommends that near a wall (on the I boundary)

$$0 < y \leqslant \lambda y_l/\kappa: \quad l = \kappa y \qquad (7\text{-}99\text{a})$$

$$\lambda y_l/\kappa < y: \quad l = \lambda y_l \qquad (7\text{-}99\text{b})$$

where λ, y_l, and κ are empirical constants that are selected at the outset of the calculation procedure. y_l is taken to be a representative distance (for example, the boundary layer thickness or duct diameter), while λ is found to lie between 0·07 and 0·12. The value of κ, traditionally 0·4 as given in Eqn (7-68), ranges from 0·35 to 0·6. For the case of free jets or wakes, where no walls are present, Eqn (7-99b) can be used where y_l is taken as a characteristic thickness of the jet or wake. Once the three mixing length constants have been chosen to give the best fit with available experimental data for a particular problem, they can be used to calculate similar problems.

A more plausible model of the mixing length distribution near a wall is that due to van Driest[49]. In his model it is postulated that near a wall

$$l = \kappa y \left\{ 1 - \exp\left[-y \frac{\sqrt{(\tau\rho)}}{\mu A^+} \right] \right\} \tag{7-100}$$

where A^+ is a constant (typically $A^+ = 26$). Most of the turbulent calculations in reference (32) make use of this hypothesis, and special wall-flux relations have been developed from Eqn (7-100). However, a large variety of problems can be solved satisfactorily using Eqn (7-99a), which is merely an asymptote of Eqn (7-100), and added complexity is not necessary if the answers obtained are acceptable.

Assumptions have also to be made for the values of σ_t for turbulent flow— that is, $\mathbf{P}_t$ and $\mathbf{Sc}_t$—and a discussion on the variation of $\mathbf{P}_t$ across a boundary layer can be found in the review by Kestin and Richardson[53]. For flows where a wall is present, constant values of σ_t between 0·8 and 1·0 generally provide satisfactory predictions, while for jets and wakes values between 0·5 and 0·7 appear to be more appropriate. [54, 55, 56] Again the value should be selected to give the best agreement with available experimental data. The effective transport property for the energy equation can then be expressed as

$$\frac{\mu_{\text{eff}}}{\mathbf{P}_{\text{eff}}} = \frac{\mu}{\mathbf{P}} + \frac{A_\tau}{\mathbf{P}_t}$$

and similarly for the diffusion equation.

7-3-10   Calculation of the pressure gradient

For the case of a free stream on, say, the I boundary the pressure gradient is prescribed by the condition at the edge of the boundary layer where

$$-\left(\frac{1}{\rho u} \frac{dp}{d\zeta} \right)_{\text{I}} = \left(\frac{du}{d\zeta} \right)_{\text{I}}$$

In forming the finite difference equations it is assumed that the u term varies in a stepwise fashion in the ζ direction; that is, for $\zeta_{i-1} < \zeta < \zeta_i$:

$$u(\zeta, \eta) = u_{i,j}$$

However, nonlinear terms such as $(\rho u)^{-1}$ are evaluated at the $i - 1$ station to avoid iteration. For consistency of the pressure gradient with the rest of the finite difference equations it is therefore necessary that

$$-\left(\frac{dp}{d\zeta} \right)_i = (\rho u)_{i-1,\text{I}} \frac{u_{i,\text{I}} - u_{i-1,\text{I}}}{\Delta\zeta} \tag{7-101}$$

A similar expression holds if the free stream is on the E boundary.

While for external flows the pressure gradient is given explicitly, for internal flows it is an implicit unknown. However, for internal flows there is an additional source of information: the continuity equation can be integrated across the duct width to enable the mass flow rate to be obtained in terms of the velocity. As, in general, the mass flow rate and duct width are specified, the extra information enables the pressure gradient to be calculated, and the following techniques can be used.

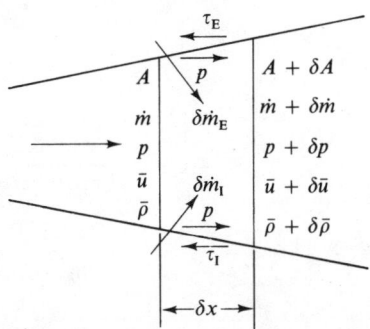

Fig. 7-11   Control volume for the flow through a duct

For the case of a varying area duct with mass transfer through the walls, as illustrated in Fig. 7-11, the one-dimensional continuity and momentum equations can be expressed as

$$\dot{m} = \bar{\rho} A \bar{u}$$

or
$$\frac{d\dot{m}}{\dot{m}} = \frac{d\bar{\rho}}{\bar{\rho}} + \frac{dA}{A} + \frac{d\bar{u}}{\bar{u}} \qquad (7\text{-}102)$$

and

$$(pA + p\,\delta A) - (p + \delta p)(A + \delta A) - F_\tau\,\delta x = (\dot{m} + \delta\dot{m})(\bar{u} + \delta\bar{u}) - \dot{m}\bar{u}$$

or
$$A\,dp + F_\tau\,dx + \bar{u}\,d\dot{m} + \dot{m}\,d\bar{u} = 0 \qquad (7\text{-}103)$$

where (in this section only) A is the duct area, $\dot{m}$ the mass flow rate, and $\bar{u}$ and $\bar{\rho}$ are mean values of the velocity and density, respectively. F_τ is the frictional force per unit length of duct—for example, for a radial diffuser (see Fig. 7-12(a)): $F_\tau = 2\pi r(\tau_I + \tau_E)$; for an axial diffuser (see Fig. 7-12(b)): $F_\tau = 2\pi(\tau_I r_I + \tau_E r_E)$. In order to satisfy the momentum equation, $\bar{u}$ is defined as

$$\bar{u} \equiv \int_I^E r\rho u^2\,dy \Big/ \int_I^E r\rho u\,dy$$

Combining Eqns (7-102) and (7-103) it follows that

$$\frac{dp}{dx} = -\frac{F_\tau}{A} - \frac{2\bar{u}}{A}\frac{d\dot{m}}{dx} + \frac{\dot{m}\bar{u}}{A}\left(\frac{1}{A}\frac{dA}{dx} + \frac{1}{\bar{\rho}}\frac{d\bar{\rho}}{dx}\right) \tag{7-104}$$

which can be used to calculate the pressure gradient at any streamwise location. F_τ can be found from the wall flux relations, and $d\dot{m}/dx$ can be calculated if $\dot{m}_I'$ and $\dot{m}_E'$ are known. The advantage of the one-dimensional equation [Eqn (7-104)] is that iteration is not necessary to solve for the

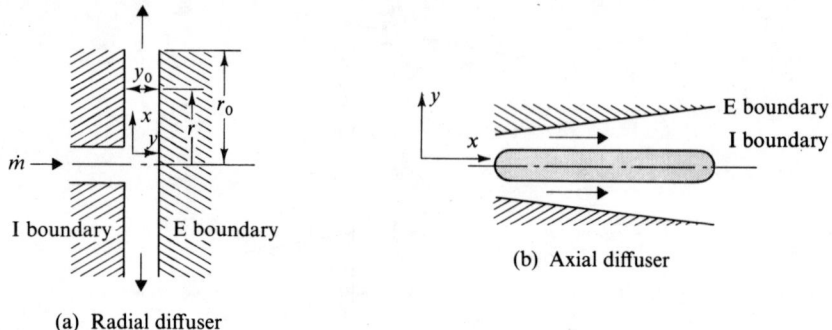

(a) Radial diffuser

(b) Axial diffuser

Fig. 7-12   Internal flow in diffusers

pressure gradient, but any errors will result in a discrepancy between the prescribed duct area A and the calculated duct area A' (that is, A' is the area corresponding to the values of y derived from Section 7-3-8). In general, $A \neq A'$, and a way of correcting for the error is to write

$$\frac{dA}{dx} = \frac{A_i - A'_{i-1}}{\Delta x}$$

where the subscripts $i - 1$ and i refer to the upstream and downstream stations. This can be expressed as

$$\frac{dA}{dx} = \frac{A_i - A_{i-1}}{\Delta x} + c'\frac{A_{i-1} - A'_{i-1}}{\Delta x} \tag{7-105}$$

where c' is unity. However, to avoid instability, which could lead to the calculated area oscillating about the true value at each step, the error term can be moderated by making c' less than unity. For many problems a value of 0·1 should be satisfactory, but it may be found necessary to reduce the size of Δx (that is, $\Delta \zeta$) if acceptable accuracy is not achieved.

For the case of compressible flow, $\bar{\rho}$ can be calculated from the perfect gas law where

$$\frac{d\bar{\rho}}{\bar{\rho}} = \frac{dp}{p} - \frac{d\overline{T}}{\overline{T}}$$

$\bar{T}$ being a mean gas temperature. For moderate pressure changes (low Mach numbers), the term dp/p can be neglected, but for the general case Eqn (7-104) becomes

$$\frac{dp}{dx} = \left[-\frac{F_\tau}{A} - \frac{2\bar{u}\,d\dot{m}}{A\,dx} + \frac{\dot{m}\bar{u}\,dA}{A^2\,dx} - \frac{\dot{m}\bar{u}\,d\bar{T}}{A\bar{T}\,dx} \right] \bigg/ \left[1 - \frac{\dot{m}\bar{u}}{A\bar{\rho}} \right] \qquad (7\text{-}106)$$

An alternative approach for calculating the pressure gradient in internal flows, in which the integrated continuity equation and the momentum equation are solved simultaneously by direct iteration, can be found in reference (57). This technique has proved useful for the case of a radial diffuser where one wall is rotating (introducing a swirl velocity component), as well as for the case of flow through pipes with transpiration cooled walls.[58]

Having dealt with the physical inputs to the numerical method it is now convenient to explain how the equations can be solved.

7-3-11 Solution of the finite difference equations

The general finite difference equation for both the interior and exterior regions of the boundary layer can be written as

$$\Phi_{i,j} = A_j\Phi_{i,j+1} + B_j\Phi_{i,j-1} + C_j \quad (j = 2, N + 2) \qquad (7\text{-}107)$$

and the coefficients can be obtained from Eqns (7-52), (7-57), and (7-59). Equation (7-107) can readily be solved by the recurrence relationships described in Chapter 4, but it should be noted that the coefficients A_j and B_j, which comprise a band matrix, and the coefficients C_j, which comprise a vector, change at each downstream location. It is therefore necessary to repeat the recurrence procedure at each i-step, as once-for-all solutions are not possible.

Equation (7-107) can be reduced to

$$\Phi_{i,j} = A1_j\Phi_{i,j+1} + B1_j \qquad (7\text{-}108)$$

where

$$\left.
\begin{aligned}
A1_j &= A_j/(1 - B_j A1_{j-1}) \\
B1_j &= (B_j B1_{j-1} + C_j)/(1 - B_j A1_{j-1}) \\
A1_2 &= A_2 \\
B1_2 &= B_2\Phi_1 + C_2
\end{aligned}
\right\} \qquad (7\text{-}109)$$

and

Having calculated $A1_j$ and $B1_j$ for $j = 2, N + 2$, Eqn (7-108) is then solved by back-substitution from $j = N + 2$ to $j = 2$. For the case of a prescribed flux on the I boundary, where B_2 is zero, $\Phi_{i,1}$ can be found from the wall flux relations of Section 7-3-6.

The reader is now in a position to write a computer program to solve a particular problem, or he can find in reference (32) a general program that will cope with a large number of variables and boundary conditions. Having dealt with the basic technique it is now interesting to discuss a few problems and compare the predictions with the experimental data.

7-3-12   Applications of the numerical method

It is the intention of this section to compare some results calculated using the Spalding–Patankar method with experimental data or, for laminar flow, exact solutions of the boundary layer equations. The examples are by no means comprehensive, but they illustrate a few of the laminar and turbulent, internal and external, flows that have been tackled. Where appropriate, values of the empirical constants used will be specified and the computational details will be given.

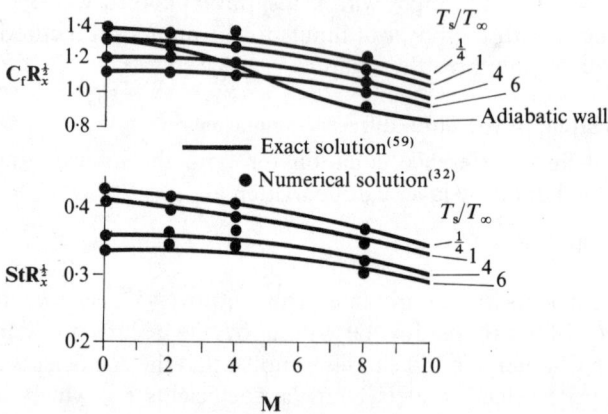

Fig. 7-13   Compressible laminar flow over a flat plate: comparison between the numerical and exact solutions

Exact solutions, evaluated numerically, have been obtained by van Driest[59] for *laminar compressible flow over an isothermal impermeable plate* with a range of plate to free stream temperature ratios, T_s/T_∞. The ratio of specific heats is taken as 1·4, $\mathbf{P} = 0·75$, the density is assumed to be inversely proportional to the absolute temperature, T, and the viscosity variation is given by

$$\mu = \mu_\infty \left(\frac{T}{T_\infty}\right)^{1/2} \frac{1·505}{1 + 0·505(T_\infty/T)}$$

The friction factor, $\mathbf{C_f}$, and the Stanton number, $\mathbf{St}$, where

$$\mathbf{St} = \mathbf{N_x}/\mathbf{R_x}\mathbf{P}$$

have been calculated for the problem by Patankar and Spalding[32] solving Eqns (7-37) and (7-41). A grid with $N = 16$ was used and the boundary was located so that at the outer grid $u = 0.999U_\infty$. Despite the coarse grid used, the variations of $\mathbf{C_f R}_x^{1/2}$ and $\mathbf{St\ R}_x^{1/2}$ with the Mach number, $\mathbf{M}$, which are illustrated in Fig. 7-13, can be seen to agree closely with van Driest's solution.

The *laminar incompressible flow in the entry region of an annular duct* has been solved exactly by Sparrow and Lin[60]. Treating the problem as an internal flow, the Spalding–Patankar method has been used by Morris[58] to solve Eqn (7-37) numerically. A fine grid with $N = 100$ was used, and integration was continued downstream from the duct entrance, where a flat velocity profile was assumed, until a fully developed velocity profile was formed. An example of the dimensionless velocity profiles at different locations for a duct with a radius ratio of $r_E/r_I = 0.8$ is shown in Fig. 7-14. Agreement between the numerical and exact solutions is again very satisfactory.

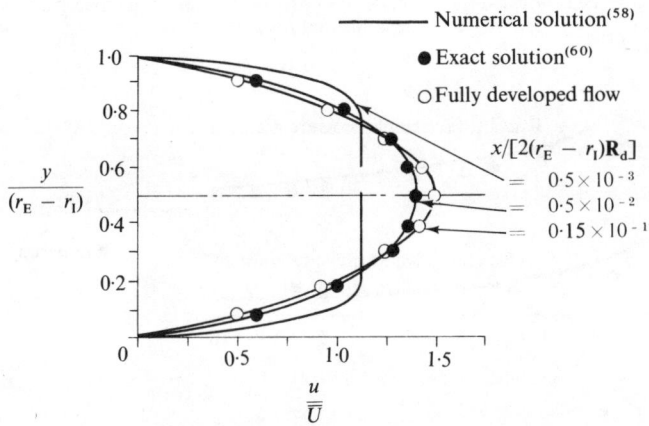

Fig. 7-14 Incompressible laminar flow in the entry region of an annular duct with a radius ratio of 0·8: comparison between the numerical and exact solutions

Having illustrated that the method can be used for external and internal laminar flows we can now turn to turbulent problems. It should be remembered, however, that the numerical method is a computational vehicle that is no better than the physical hypotheses it encompasses. The following examples all feature the mixing length hypothesis, and it is to be expected that more advanced turbulence models will give better predictions.

In order to employ the mixing length model it is necessary to establish values of the empirical constants y_l, λ, and κ in Eqns (7-99). In the calculations conducted by Patankar and Spalding for *turbulent flow over impermeable plane surfaces*, y_l was chosen as the distance from the wall to the point where $u/U_\infty = 0.99$. Values of $\kappa = 0.435$ and $\lambda = 0.09$ gave satisfactory agreement with the experimental data of Coles[61] and the Spalding–Chi correlation[62] for drag on a flat plate, which is shown in Fig. 7-15 as a plot of $\mathbf{C_f}$

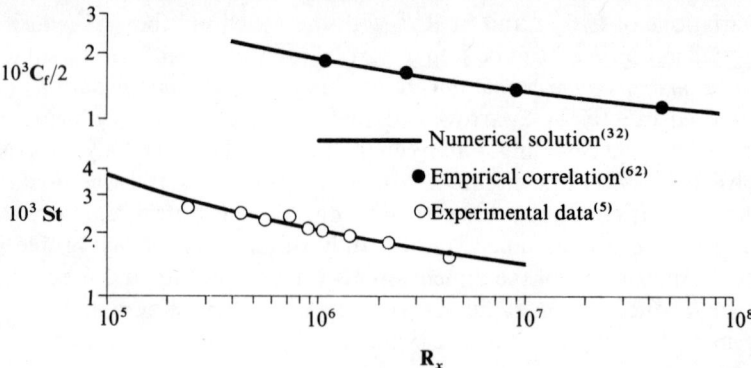

Fig. 7-15   Turbulent flow over a flat plate: comparison between the numerical solutions and experimental results

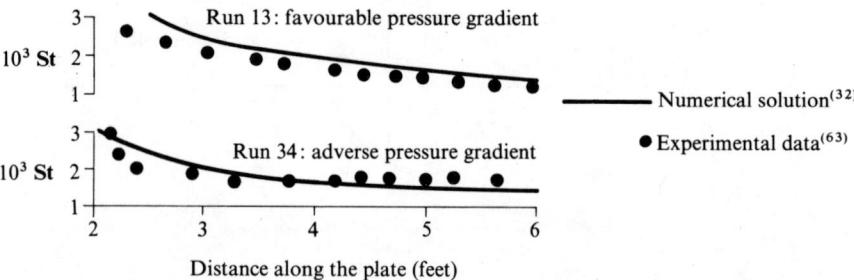

Fig. 7-16   Turbulent flow with a step in wall temperature: comparison between the numerical solutions and experimental results

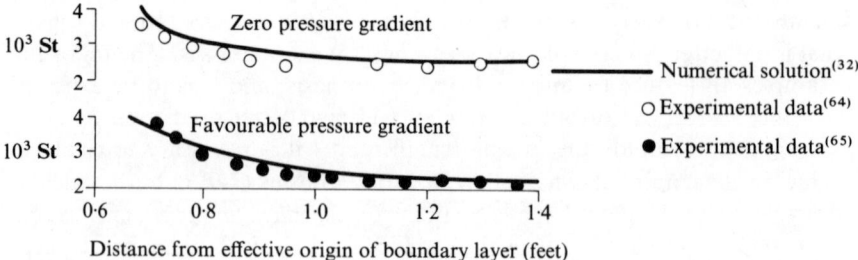

Fig. 7-17   Turbulent flow with a step in heat flux: comparison between the numerical solution and experimental results

against $\mathbf{R}_x$. A constant value of $\mathbf{P}_t = 0.9$ enabled the heat transfer data of Reynolds et al.[5] to be accurately predicted, as illustrated in Fig. 7-15. It should be pointed out that the van Driest hypothesis was used in the above, and subsequent, predictions for the calculation of the wall fluxes. Having established the values of the empirical constants, Patankar and Spalding proceeded to calculate the heat transfer to plane surfaces with and without pressure gradients, as illustrated in Figs 7-16 and 7-17. In these calculations a grid of $N = 16$ was used and the step length, $\Delta\zeta$, was controlled to ensure that the amount of fluid entrained into the boundary layer during any step was 5 per cent of the fluid already in the layer.

Figure 7-16 shows the comparison between the calculated results and the experimental data of Moretti and Kays[63] for plane flow with adverse (positive) and favourable (negative) pressure gradients and a step in wall temperature. It should be pointed out that with an adverse pressure gradient the Stanton number is slightly underestimated, while a favourable gradient causes an overestimate. For large, favourable pressure gradients the turbulent flow can be 'laminarized' (where strong acceleration causes the decay of turbulence with a retransition to laminar flow). In these circumstances a simple, effective viscosity model will be unsuitable unless information can be supplied to indicate laminarization. The same comment holds with flows where flux boundary conditions are used, but in the examples of Fig. 7-17 the pressure gradients are not severe, and agreement between the numerical calculations and the data of Hartnett et al.[64] and Macarthy and Hartnett[65] is satisfactory.

We conclude the examples with an internal flow problem of the *turbulent radial diffuser* that has been investigated by Moller[66]. The Spalding–

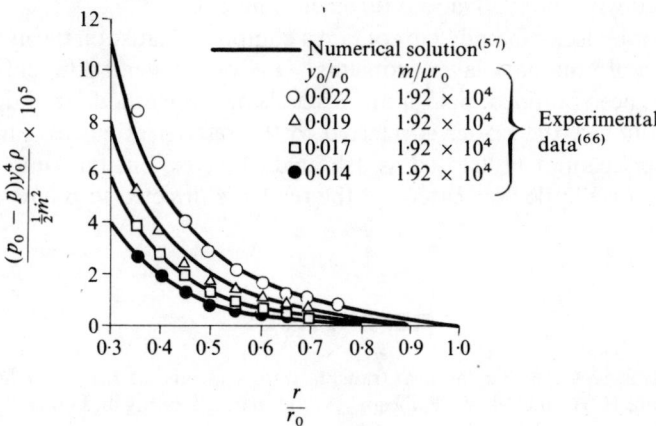

Fig. 7-18 Turbulent flow in a radial diffuser: comparison between the numerical solutions and experimental results

Patankar method has been applied to this system by Bayley and Owen[57] using the Prandtl mixing length and the van Driest hypothesis. The values of the empirical constants used for this geometry (and also applied for the case of rotation of one wall of the diffuser to simulate a turbine rotor system) were $y_l = \frac{1}{2}y_0$, $\kappa = 0.4$, and $\lambda = 0.12$. For the results shown in Fig. 7-18, forty cross-stream steps were used and fifty equal radial steps were employed between $r/r_0 = 0.3$ and 1. The dimensionless pressure coefficient was calculated by the iterative method described in the reference. The underestimate at the small radius ratios is attributed to inlet separation of the incoming fluid as it is turned through $90°$.

7-4   Concluding remarks

A method has been presented that is capable of producing solutions of the differential boundary layer equations for laminar and turbulent flows. An outline of the procedure for formulating and solving the finite difference equations has been presented, and examples of the treatment of a number of boundary conditions have been given. Means of calculating entrainment for external flows and the pressure gradient for internal flows have also been discussed.

A number of comparisons between results calculated by the Spalding–Patankar method and experimental data, or theoretical predictions, have been made to illustrate some of the problems that have been treated. There are numerous other problems featuring permeable walls with transpiration cooling, the flow in combustion chambers, and flows with swirl that can be, and have been, treated by this method. For turbulent flow, however, predictions will only be as good as the hypotheses used; and while the mixing length hypothesis is obviously not universally applicable, the basic procedure can be used with more advanced turbulence models.[67, 68]

This chapter is not intended to contain a complete treatise on the numerical calculation of boundary layer problems. One very general differential procedure has been outlined, but for more details of this method, or of the other methods quoted, the reader is referred to the references given. Finally, if the problem cannot be treated as a boundary layer, but falls into the recirculating or elliptic flow category, the reader is directed to reference (69).

REFERENCES

1. Eckert, E. R. G. Survey on heat transfer at high speeds. *A.R.L. Rept 189*, 1961.
2. Johnson, H. A. and M. W. Rubesin. Aerodynamic heating and convective heat transfer. *Trans. A.S.M.E.*, 1946, **68**, 124.
3. Eckert, E. R. G. Die Berechnung des Wärmeübergangs in der laminaren Grenzschicht umströmter Körper. *VDI-ForschHft*, 1942, **416**.

4. Levy, S. Heat transfer to constant property laminar boundary layer flows with power-function free-stream velocity and wall temperature variation. *J. Aero. Sci.*, 1952, **19**, 341.

5. Reynolds, O., W. M. Kays and S. J. Kline. Heat transfer in the turbulent incompressible boundary layer. *NASA Memos 12-1-58W, 12-2-58W, 12-3-58W, 12-4-58W*, 1958.

6. Kestin, J. The effect of free-stream turbulence on heat transfer rates. *Adv. in Heat Transfer*, 1966, **3**.

7. Junkham, G. H. and G. K. Serovy. Effects of free-stream turbulence and pressure gradient on flat-plate boundary-layer velocity profiles and on heat transfer. *Trans. A.S.M.E., J. Heat Transfer*, 1967, **89** (Series C).

8. Smith, M. C. and A. M. Kuethe. Effects of turbulence on laminar skin friction and heat transfer. *Physics Fluids*, 1966, **9**.

9. Lin, C. C. (ed.). Turbulent flows and heat transfer in *High Speed Aerodynamics and Jet Propulsion*, vol. V. Princeton University Press, 1959.

10. Martinelli, R. C. and L. M. K. Boelter. *Univ. Calif. Publs Engng*, 1942, **5**, 23.

11. Eubank, O. C. and W. S. Proctor. S.M. Theses in Chemical Engineering, M.I.T., 1951.

12. Sieder, E. N. and G. E. Tate. Heat transfer and pressure drop of liquids in tubes. *Ind. Engng Chem.*, 1936, **28**, 1429.

13. McAdams, W. H. *Heat Transmission* (3rd edn). McGraw-Hill, New York, 1954.

14. Boelter, L. M. K., R. C. Martinelli and F. Jonassen. Remarks on the analogy between heat transfer and momentum transfer. *Trans. A.S.M.E.*, 1941, **63**, 447.

15. Dittus, F. W. and L. M. K. Boelter. *Univ. Calif. Publs Engng*, 1930, **2**, 443.

16. Martinelli, R. C. Heat transfer to molten metals. *Trans A.S.M.E.*, 1947, **69**, 947.

17. Lyon, R. N. Liquid metal heat transfer coefficients. *Chem. Engng Prog.*, 1951, **47**, 75.

18. Lubarsky, B. and S. J. Kaufman. Review of experimental investigations of liquid metal heat transfer. *NACA Tech. Note 3336*, 1955.

19. Gilliland, E. R. and T. K. Sherwood. Diffusion of vapours into air streams. *Ind. Engng. Chem.*, 1934, **26**, 516.

20. Hilpert, R. Wärmeabgabe von Geheitzen drahten and Röhren in Luftstrom. *Forsch. Geb. IngWes.*, 1933, **4**, 215.

21. Comings, E. W., J. T. Clapp and J. F. Taylor. Air turbulence and transfer processes. *Ind. Engng. Chem.*, 1948, **40**, 1076.

22. Keenan, J. H. and J. Kaye. *Gas Tables*. John Wiley, London, 1948.

23. Houghton, E. L. and A. E. Brock. *Tables for the Compressible Flow of Dry Air*. Arnold, London, 1961.

24. Weise, R. Wärmeausgang durch freie Konvektion. *Forsch. Geb. IngWes.*, 1935, **6**, 281.

25. Saunders, O. A. Effect of pressure on natural convection in air. *Proc. Roy. Soc. A*, 1936, **157**, 278.

26. Fishenden, M. and O. A. Saunders. *An Introduction to Heat Transfer*. Oxford University Press, London, 1950.

27. Jakob, M. *Heat Transfer*, vol 1. John Wiley, London, 1949.

28. Truckenbrodt, E. Ein Quadraturverfahren zur Berechnung der laminaren und turbulenten Reibungsschicht bei ebener und rotation symmetrischer Strömung. *Ing. Arch.*, 1952, **20**, 211.

29. Spalding, D. B. A unified theory of friction, heat transfer and mass transfer in the turbulent boundary layer and wall jet. *A.R.C. CP 829*, 1965.

30. Head, M. R. Entrainment in the turbulent boundary layer. *A.R.C. R. and M. 3152*, 1960.
31. Kutateladze, S. S. and A. I. Leont'ev. *Turbulent Boundary Layers in Compressible Gases*. Arnold, London, 1964.
32. Patankar, S. V. and D. B. Spalding. *Heat and Mass Transfer in Boundary Layers* (2nd edn). International Textbook Company, 1970.
33. Bradshaw, P., D. H. Ferriss and N. P. Atwell. Calculation of boundary layer development using the turbulent energy equation. *J. Fluid Mech.*, 1967, **28**, 593.
34. Bradshaw, P. and D. H. Ferriss. Calculation of boundary layer development using the turbulent energy equation. Part IV: Heat transfer with small temperature differences. *N.P.L. Aero. Rept 1271*, 1968.
35. Flugge-Lotz, I. The computation of the laminar compressible boundary layer. *Rept R-352-30-7*, Stanford University, 1954.
36. Fussell, D. D. and J. D. Hellams. The numerical solution of boundary layer problems. *A.I.Ch.E. Jl*, 1965, **11**, 733.
37. Glushko, G. S. Turbulent boundary layer on plane plates incompressible fluid. *Izv. Akad. Nauk. Mech.*, 1965 (No. 4), 13.
38. Priskonov, V. M. A standard programme for the solution of boundary layer problems. *NASA TTF-300, TT65-50138*, 1966, 74.
39. Hartree, D. R. and J. R. Womersley. A method for the numerical or mechanical solution of certain types of partial differential equations. *Proc. Roy. Soc. A*, 1937, **161**, 353.
40. Smith, A. M. O. and D. W. Clutter. Machine calculation of compressible laminar boundary layers. *A.I.A.A. Jl*, 1965, **3**, 639.
41. Smith, A. M. O., N. A. Jaffe and R. C. Lind. Study of a general method of solution of the incompressible turbulent boundary-layer equations. *Rept RRRE-4*, Bureau of Naval Weapons, Fluid Mechanics and Flight Dynamics Branch, 1965.
42. Mellor, G. L. Turbulent boundary layers with arbitrary pressure gradients and divergent or convergent cross flows. Princeton Univ., Dept. Aerospace Mech. Sci. Gas Dynamics Lab. Ref. 775, 1966.
43. Pohlhausen, K. Zur naherungsweisen Integration der Differentialgleichungen der laminaren Reibungsschicht, *ZAMM*, 1921, **1**, 252.
44. Pallone, A. Non-similar solutions of the compressible laminar boundary layer equations with applications to the upstream-transpiration cooling problem. *J. Aero. Sci.*, 1961, **28**, 449.
45. Moses, H. L. The behaviour of turbulent boundary layers in adverse pressure gradients. *Rept No. 73*, M.I.T. Gas Turbine Laboratory, 1964.
46. Libby, P. A. and H. Fox. A moment method for compressible laminar boundary layers and their applications. *Int. J. Heat Mass Transfer*, 1965, **8**, 1451.
47. Patankar, S. V. and D. B. Spalding. A calculation procedure for heat transfer by forced convection through two-dimensional uniform-property turbulent boundary layers on smooth impermeable walls. *Proc. Third Int. Heat Transfer Conf.*, Chicago; *A.I.Ch.E.*, 1966, **II**, 50.
48. Schlichting, H. *Boundary Layer Theory* (6th edn). McGraw-Hill, New York, 1968.
49. van Driest, E. R. On turbulent flow near a wall. *J. Aero. Sci.*, 1956, **23**, 1007.
50. Kolmogorov, A. N. Equations of turbulent motion of an incompressible fluid. *Izv. Akad. Nauk SSSR* ser. phys. VI, 1942 (No. 1–2), 56, English transl: *Rept ON/6*, Imperial College, London, Mech. Engng Dept, 1968.

51. Rotta, J. C. Heat transfer and temperature distribution in turbulent boundary layers at supersonic and hypersonic flow. *Rept 65A07*, Aerodynamische Versuchsanstatt, 1965.

52. Escudier, M. P. The distribution of the mixing length in turbulent flows near walls. *Rept TWF/TN/1*, Imperial College, London, Mech. Engng Dept, 1965.

53. Kestin, J. and P. D. Richardson. Heat transfer across turbulent incompressible boundary layers. *Int. J. Heat Mass Transfer*, 1963, **6**, 147.

54. Hinze, J. O. and B. G. van der Hegge Zijnen. Transfer of heat and matter in the turbulent mixing zone of an axially symmetric jet. *Appl. Scient. Res.*, 1949, sec. **AA1**, 435.

55. Forstall, W. and A. H. Shapiro. Momentum and mass transfer in coaxial gas jets. *J. appl. Mech.*, 1950, **17**, 399.

56. Abramovich, G. N. *The Theory of Turbulent Jets*. M.I.T. Press, 1963.

57. Bayley, F. J. and J. M. Owen. The flow between a rotating and a stationary disc. *Aero. Quart.*, 1969, **XX**, 333.

58. Bayley, F. J., W. D. Morris, J. M. Owen and A. B. Turner. Boundary layer prediction methods applied to cooling problems in the gas turbine. *A.R.C. C.P. 1164*, 1971.

59. van Driest, E. R. Investigation of laminar boundary layers in compressible fluids using the Crocco method. *NACA Tech. Note 2597*, 1952.

60. Sparrow, E. M. and S. H. Lin. Developing laminar flow and pressure drop in the entrance region of an annular duct. *A.S.M.E.* Paper No. 64FE-1, 1964.

61. Coles, D. E. The turbulent boundary layer in a compressible fluid. *Rand Corp. Rept R-403-pr*, 1962.

62. Spalding, D. B. and S. W. Chi. The drag of a compressible turbulent boundary layer on a smooth flat plate with and without heat transfer. *J. Fluid Mech.*, 1964, **18**, 117.

63. Moretti, P. M. and W. M. Kays. Heat transfer through an incompressible turbulent boundary layer with varying free-stream velocity and varying surface temperature. *Rept No. PG-1*, Stanford Univ., Mech. Engng Dept, 1964.

64. Hartnett, J. P., R. C. Birkebak and E. R. G. Eckert. Velocity distributions, temperature distributions, effectiveness and heat transfer in cooling of a surface with a pressure gradient. *International Developments in Heat Transfer*, Pt IV, 682 (published by A.S.M.E.), 1961.

65. Macarthy T. F. and J. P. Hartnett. Turbulentnye pogranischnye Sloi s Prodol'nym Gradientom Davleniya i Teploobmenom. *Vsesoyuznoe Soveshehanie po Teplo-i-Massoobmeny*, Minsk, 1964.

66. Moller, P. S. Radial flow without swirl between parallel discs. *Aero. Quart*, 1963, **XIV**, 163.

67. Rodi, W. and D. B. Spalding. A two-parameter model of turbulence and its application to free jets. *Rept BL/TN/B/12*, Imperial College, London, Mech. Engng Dept, 1969.

68. Ng, K. H. and D. B. Spalding. Predictions of two-dimensional boundary layers on smooth walls with a two-equation model of turbulence. *Rept BL/TN/A/25*, Imperial College, London, Mech. Engng Dept, 1970.

69. Gosman, A. D., W. M. Pun, A. K. Runchal, D. B. Spalding and M. W. Wolfshtein. *Heat and Mass Transfer in Recirculating Flow*. Academic Press, London, 1969.

PROBLEMS

Note: where not stated, fluid properties should be obtained from the Appendix.

7-1 Liquid sulphur dioxide at 120°F flows at 10 ft/s over a flat plate maintained at 50°F. If transition from laminar to turbulent flow occurs at $\mathbf{R}_x = 3 \times 10^5$, calculate the local and mean heat transfer coefficients at a distance of 1 ft from the leading edge.
Ans: 859; 1,000 Btu/ft^2 h °R

7-2 Air at 0·2 MN/m^2 and 500°C flows at 300 m/s over a flat surface maintained at 20°C. Calculate the local heat transfer coefficient 2 m from the leading edge using Eqn (7-8). Compare the answer with that obtained from Eqn (7-5). (N.B. T^* should be used for property evaluation.)
Ans: 0·532, 0·492 kW/m^2 K

7-3 5,000 imperial gal/h of engine oil flows through a tube of 2 inches bore and 10 feet length with a surface temperature of 100°C. If the tube is vertical and the oil enters at the bottom with a bulk temperature of 20°C, calculate the mean heat transfer coefficient.
Ans: 27·5 Btu/ft^2 h °R

7-4 Repeat Problem 7-3 for the case of a horizontal tube including buoyancy effects.
Ans: 43·2 Btu/ft^2 h °R

7-5 6 m^3/s of cooling water pass through the tubes of a condenser. If there are 1,000 tubes of 0·05 m bore and 10 m length, calculate the mean heat transfer coefficient and temperature rise according to Eqn (7-12a). Take the initial bulk temperature of the cooling water as 10°C, and assume the inner surface of the tube is at the temperature of the condensing steam, 39°C. (Hint: temperature rise can be calculated from result of Problem 6-6.)
Ans: 7·46 kW/m^2 K, 10·9°C

7-6 Repeat Problem 7-5 using Eqn (7-12b) basing the properties on (a) the initial bulk temperature and (b) the mean bulk temperature.
Ans: (a) 8·26 kW/m^2 K; 11·7°C; (b) 8·95 kW/m^2 K; 11·8°C

7-7 Engine oil flows at 23·3 ft/s through the rectangular passages of a heat exchanger. The oil enters the passages, which are 4 inches high, 6 inches wide, and 6 feet long, at 176°F. Calculate the mean heat transfer coefficient if the passage surface is maintained at 140°F.
Ans: 118 Btu/ft^2 h °R

7-8 10 kg/s of sodium–potassium eutectic flows through a 0·15 m diameter tube. If the tube is maintained at 1,000 K and at a given location the liquid metal has a temperature of 800 K, calculate the heat flux at this location using Eqn (7-14).
Ans: 552 kW/m^2

7-9 Repeat Problem 7-8 where the tube surface instead of being maintained at 1,000 K is cooled by blowing air over the outside of the tube. The air, which flows normal to the tube with a velocity of 10 m/s, is at 5 atmospheres and 80°C. Neglect the tube thickness and evaluate the gas properties at the mean film temperature. What is the average tube surface temperature?

Ans: 50 kW/m^2; 509°C

7-10   Calculate the local heat transfer coefficient at the stagnation point on the tube of the above question.

Ans: 117 W/m^2 K

7-11   A sphere of 1 inch diameter is placed in a wind tunnel where **M** = 3. Calculate the heat transfer coefficient at the stagnation point if the static pressure and temperature of the undisturbed air is 20 lbf/in^2 and 80°F, and the sphere is maintained at 100°F. Assume that γ = 1·403.

Ans: 681 Btu/ft^2 h °R

7-12   Using Eqn (7-21) for free convection from a vertical plate of height *l*, show that for air the following formulae are valid:

$$\text{laminar flow:} \quad h = 0\cdot31 \, (T_s - T_\infty)^{1/4} \, l^{-1/4}$$
$$\text{turbulent flow:} \quad h = 0\cdot23 \, (T_s - T_\infty)^{1/3}$$

where the temperature is in °F, *l* in feet, and *h* in Btu/ft^2 h °R. For air at atmospheric conditions and 80°F take ρ = 0·0735 lb/ft^3, μ = 1·24 × 10^{-5} lb/ft s, C_p = 0·24 Btu/lb °R, and k = 0·0152 Btu/ft h °R.

7-13   A glass window, 4 feet high by 8 feet wide and $\frac{1}{4}$ inch thick, is located in the outside wall of a room with an interior temperature of 70°F. Using the information in Problem 7-12 calculate the heat loss through the window when the outside temperature is 30°F. Take the conductivity of glass as 0·44 Btu/ft h °R.

Ans: 380 Btu/h

7-14   Using Eqn (7-21) for free convection from a horizontal surface, show that for air with the three cases given:

(a) $h = 1\cdot41 \, (T_s - T_\infty)^{1/4} l^{-1/4}$
(b) $h = 1\cdot71 \, (T_s - T_\infty)^{1/3}$
(c) $h = 0\cdot71 \, (T_s - T_\infty)^{1/4} l^{-1/4}$

where the temperature difference is in °C, the length *l* in metres, and *h* is in W/m^2 K. For air at atmospheric conditions and 20°C take **P** = 0·71, v = 1·47 × 10^{-5} m^2/s, k = 0·0254 W/m K.

7-15   Using the above results calculate the heat loss through the ceiling of a square room of 3 m side with an air temperature of 20°C. The ceiling is 20 mm thick and has a conductivity of 0·28 W/m K, and the air space above the ceiling is at 5°C. Recalculate the problem if 50 mm of fibreglass (k = 0·04 W/m K) is put on top of the ceiling.

Ans: 198 W; 63 W

7-16   Using the arguments made in Section 7-3-3, verify that Eqn (7-47) is valid for the transfer of momentum, energy, and mass in two-dimensional laminar and turbulent boundary layers.

7-17 Verify that Eqns (7-51) and (7-52) are a finite difference representation of Eqn (7-48).

7-18 In the turbulent Couette-flow region close to a wall, if $\mu_{\text{eff}} = \rho K^2 y^2 |(\partial u/\partial y)|$ show that Eqn (7-60) can be reduced to

$$\frac{du_+}{dy_+} = \frac{1}{K} \frac{(1 + m_+ u_+ + p_+ y_+)^{1/2}}{y_+}$$

where

$$u_+ = u(\tau_I/\rho)^{-1/2}, \qquad\qquad y_+ = y(\tau_I \rho)^{1/2}/\mu$$
$$p_+ = \mu(dp/dx)(\tau_I \rho)^{-1/2}, \qquad m_+ = \dot{m}_I'(\tau_I \rho)^{-1/2}$$

State all your assumptions.

7-19 Show from the result of the previous question that if

(a) $p_+ = m_+ = 0; u_+ = \dfrac{1}{K} \ln (E^* y_+)$

(b) $p_+ = 0; u_+ = \dfrac{1}{K} \ln (E^* y_+) + \dfrac{m_+}{4K^2} \{\ln (E^* y_+)\}^2$

(c) $m_+ = 0;$

$$u_+ = \frac{1}{K} \left\{ [2(1 + p_+ y_+)^{1/2} - 1] + \ln \left[\frac{4 E^* y_+}{2 + p_+ y_+ + 2(1 + p_+ y_+)^{1/2}} \right] \right\}$$

where E^* is an integration constant. Express the above solutions in terms of Su_1, R_1, P_1, and M_1.

7-20 In the Couette-flow region show that for laminar flow

$$\frac{du_+}{dy_+} = 1 + m_+ u_+ + p_+ y_+$$

and hence

$$u_+ = \frac{\exp (m_+ y_+) - 1}{m_+} + \frac{p_+}{m_+^2} [\exp (m_+ y_+) - 1 - m_+ y_+]$$

where the symbols are defined in Problem 7-18. Use the above result to verify Eqn (7-64).

7-21 Using the notation of Problem 7-20, show that for laminar conditions in the Couette-flow region

$$\frac{d\Phi_+}{dy_+} = P(1 + m_+ \Phi_+) + \tfrac{1}{2}(1 - P)W \frac{d}{dy_+} (u_+^2)$$

where

$$\Phi_+ \equiv (\tilde{h}_I - \tilde{h})\tau_I^{1/2}\rho^{1/2}/q_I \quad \text{and} \quad W \equiv -\tau_I^{3/2}\rho^{-1/2}/q_I$$

Hence show that

$$\Phi_+ = \frac{\exp (Pm_+ y_+) - 1}{m_+} + \tfrac{1}{2}(1 - P)Wu_+^2$$

Use this result to verify Eqn (7-76).

8

Radiation

In earlier chapters we have discussed the methods by which heat is transferred through solid and fluid substances, in the latter case by motions which can often be readily observed. It is an everyday experience, however, that thermal energy can be transmitted between bodies without an intervening carrier, and indeed, life on this planet is supported by energy from the sun, 93 million miles distant. The means by which energy is transmitted between bodies not in tangible contact is referred to as *radiation*, a general term covering the whole range of manifestations of energy, from radio waves, through visible light to which our eyes react, to X-rays, γ-rays, and cosmic rays, which are the concern of the nuclear physicist. These are all phenomena of similar form and differ only in their placings in the wavelength or frequency scale of the spectrum of *electromagnetic radiation* shown in Fig. 8-1. All bodies continuously emit electromagnetic radiation, but the predominant

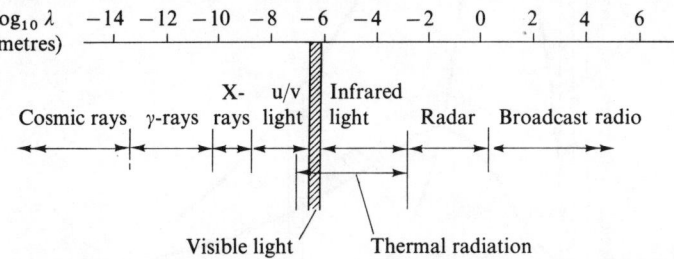

Fig. 8-1 Thermal radiation and the electromagnetic spectrum

form of the emitted energy depends upon many factors, including the nature of the body and the form of any external excitation to which it may be subject. Thus, certain elements quite spontaneously emit X-rays, while others do so if excited by atomic bombardment. Electrical conductors emit radio waves if excited by an alternating current. All bodies emit visible light if raised to a sufficiently high temperature and, indeed, emit energy near to light in the electromagnetic spectrum in *quantities that are dependent principally upon the temperature of the body*. It is this manifestation of electromagnetic radiation, referred to as *thermal radiation*, with which we are concerned in this chapter.

It was the detailed study of thermal radiation and its temperature-dependence which led to much of the basis of modern particle physics. Max Planck in 1900 conceived the idea of the quantum of energy—a then revolutionary, but now basic, concept in physics—to explain his semi-empirical equation for the distribution of intensity of radiation with wavelength in the electromagnetic spectrum:

$$q_{b\lambda} = \frac{c_1 \lambda^{-5}}{\exp(c_2/\lambda T) - 1} \tag{8-1}$$

In this equation $q_{b\lambda}$† is the *monochromatic emissive power* corresponding to the wavelength of emission, λ, such that the area under the curve of $q_{b\lambda}$ against λ is the total rate of emission per unit of emitter surface, q_b, as shown in Fig. 8-2. c_1 and c_2 are universal constants involving Planck's constant and the quantum of energy, while T is the absolute temperature at which emission occurs.

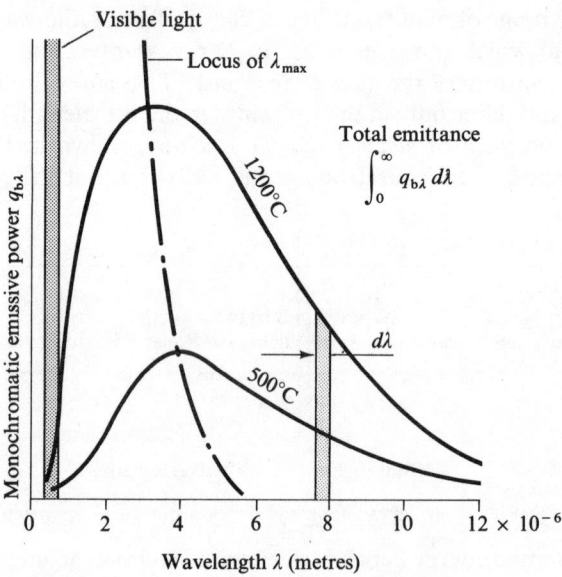

Fig. 8-2 Variation of emissive power with wavelength

The form of Planck's radiation equation, as demonstrated in Fig. 8-2, exhibits a clear peak of emissive power at a wavelength which is strongly temperature-dependent, and conforms to *Wien's Displacement Law* which states that the wavelength of maximum emission is inversely linearly proportional to the absolute temperature. At temperatures of the order of 500°C significant emission occurs only in the infrared region, at wavelengths just too large to be visible to the human eye. As the temperature is raised the energy distribution moves towards the lower wavelength region, and at

† The symbol q is adopted for heat flux throughout this text; however, the symbol W is often used in studies of radiation.

about 700°C in a darkened room a dull redness is usually visible. At 2,000°C objects glow brilliantly ('white heat'), while at the effective surface temperature of the sun (6,000°C) the emission peak occurs in the centre of the visible light waveband. This ranges approximately from 4×10^{-7} to 7.5×10^{-7} m, and for practical purposes the effective waveband of thermal radiation may be taken as 10^{-7} to 10^{-4} m. The variation of energy distribution with temperature is of considerable practical utility; glass, for example, can be made transparent to low wavelength radiation, like sunlight, and opaque to higher wavelength emission from low temperature sources, giving a net heating effect in greenhouses or domestic living rooms; equally, surface finishes can be made to reflect radiant energy from high-temperature sources, and to transmit strongly at low temperature, to produce a net radiant cooling effect. To appreciate the detailed working of these ideas, however, we must first postulate the basic laws governing the interchange between bodies of thermal radiation.

8-1 Kirchhoff's law

Consider a small body of surface area, A_1 in a large radiating enclosure, and suppose that energy falls at the rate q_b on unit surface of that body. Of this energy, generally, a fraction, α, will be absorbed, a fraction, ρ, will be reflected, and the remainder, τ, may be transmitted through the body. In practice, most solid bodies are not significantly transparent to thermal radiation, and even glass will transmit energy only over a narrow band of wavelength. For practical purposes τ may thus be regarded as zero, and since absorption of thermal energy takes place in only a few atomic layers of a solid body, we may assume that at the surface

$$\alpha + \rho = 1 \qquad (8\text{-}2)$$

Thus the energy absorbed by the small body under consideration is $\alpha_1 A_1 q_b$, in which α_1 is the *absorptivity* of the body (ρ is its *reflectivity*). When thermal equilibrium is attained, the rate at which the body absorbs heat must exactly balance its rate of emission, say q_1 per unit surface. Thus we have

$$A_1 q_1 = \alpha_1 A_1 q_b \qquad (8\text{-}3)$$

Equally, for a second body, represented by suffix 2 and subjected also to radiation q_b per unit surface from the surrounding enclosure, we obtain

$$A_2 q_2 = \alpha_2 A_2 q_b \qquad (8\text{-}4)$$

By considering a generality of bodies the simple result is obtained that

$$q_b = \frac{q_1}{\alpha_1} = \frac{q_2}{\alpha_2} = \cdots = \frac{q}{\alpha} \qquad (8\text{-}5)$$

Now, the greatest value that α can have is unity, for which case a body absorbs all the energy incident upon it, and equally radiates from itself the greatest possible amount for the given conditions. Such a body is known as a *perfect emitter* or, more commonly, as a *black body* because of its usual visual appearance.

The conditions of the hypothetical experiment considered in the above derivation, and thus the implicit assumptions, must be closely borne in mind. First, the bodies were supposed to be enclosed in a radiating surrounding surface, which will behave like a perfect emitter since, although its absorptivity may be less than unity if the enclosed bodies are small, reflected energy will have infinite opportunities for final absorption by the originally emitting surface. Indeed, a laboratory black body may be made from an enclosure with a very small orifice for the passage of incident or emitted radiation, as shown in Fig. 8-3. Thus the radiation incident upon the small bodies considered above is as from a perfect emitter, and conveniently represented as q_b.

Fig. 8-3 Laboratory 'black' body

The ratio of the energy emitted by the enclosed small bodies to that emitted by the surrounding perfect emitter is known as the emissivity, ε. Thus, we have

$$q = \varepsilon q_b \qquad (8\text{-}6)$$

and comparison with Eqn (8-5) shows that

$$\varepsilon = \alpha \qquad (8\text{-}7)$$

Before enlarging on this result as a universal law, we must consider the second important implication in the argument leading to Eqn (8-5). This arises from the inevitable assumption of the attainment of thermal equilibrium in the system comprised by the small bodies and the enclosure. For this isolated system thermal equilibrium must correspond to equality of temperature, so that for each body in the system the incident distribution of energy through the radiation spectrum coincides with the emitting distribution in accord with Planck's law. Thus, the conclusion that the emissivity and absorptivity of a body are equal is true only for total radiations of identical wavelength distribution, and, to take account of this important reservation, the corresponding statement, known as *Kirchhoff's law*, is usually stated in terms of radiation at a single wavelength, or *monochromatic radiation*: 'The monochromatic emissivity of a body is equal to the monochromatic absorptivity at the same wavelength.'

8-2   The Stefan–Boltzmann law

Electromagnetically interacting systems may be analysed by techniques similar to those employed for the more conventional (in engineering, at least) chemically or mechanically interacting thermodynamic systems. By these analytical techniques, which are beyond the scope of this book and irrelevant to its purpose, it may be shown that for a *perfect emitter* or *black body* the *total emitted radiation is proportional to the fourth power of its absolute surface temperature*.

The constant of proportionality in this relation is known as the *Stefan–Boltzmann* constant after two nineteenth-century physicists, contemporaries of Planck. In particular, Stefan was the first to show experimentally the fourth-power dependence of thermal radiation on absolute temperature. From his experiments and the later theoretical work we may write, for a perfect emitter,

$$q_b = \sigma T^4 \qquad (8\text{-}8)$$

where the Stefan–Boltzmann constant, σ, is

$$\sigma = 56 \cdot 7 \times 10^{-12} \text{ kW/m}^2 \text{ K}^4$$
$$= 0 \cdot 171 \times 10^{-8} \text{ Btu/ft}^2 \text{ }^\circ\text{R}^4$$
$$= 1 \cdot 00 \times 10^{-8} \text{ Chu/ft}^2 \text{ K}^4$$

It must be clearly appreciated that q_b is the total emission from unit area of a black body, usually known as the *emittance*, *radiant flux density*, or *total hemispherical intensity*, since it is the total radiation passing through a hemisphere enclosing a differential area of the body, divided by that area. Equation (8-8) enables q_b to be easily determined, but the principal difficulty in dealing with the radiation interchange between black bodies lies in determining the fraction of the emittance from each that is incident upon the other. This problem is considered in the next section.

8-3   Radiation interchange between black bodies

Since, by definition of a black body or perfect emitter, all the radiation incident upon it is absorbed ($\alpha = 1$), calculation of the net interchange between surfaces of this type reduces to a problem in solid geometry which, with the irregular shapes commonly occurring in engineering practice, can often be of considerable complexity.

Consider first the radiation from the elementary area dA of Fig. 8-4, which will, of course, be wholly intercepted by the hemisphere of radius r, say, in whose base dA lies. If dA_1 is a corresponding elemental area on the normal from dA to the hemisphere, the solid angle it subtends at dA, $d\omega_1$, is by definition dA_1/r^2. (Thus, the solid angle subtended by the complete hemisphere is $2\pi r^2/r^2$; that is, 2π.) If the flow of energy through dA_1 from dA is $d\dot{Q}_1$ then the so-called *intensity of radiation* is defined by the equation

$$d\dot{Q}_1 = i_n \, d\omega_1 \, dA \tag{8-9}$$

In this case the suffix n reminds us that the receiving area, dA_1, is normal to the radiation from the emitting area dA, but note that it is the latter that is incorporated in the definition of the intensity; the role of the receiving area is contained in the solid angle $d\omega_1$.

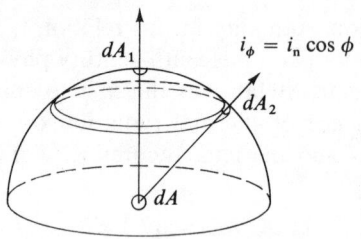

Fig. 8-4   Radiation from an elementary surface

A second elemental receiving area, say dA_2, lying at angle ϕ with the normal from dA will intercept radiant energy as given by

$$d\dot{Q}_2 = i_\phi \, d\omega_2 \, dA \tag{8-10}$$

in which $d\omega_2 = dA_2/r^2$ and i_ϕ is given by *Lambert's cosine law* as

$$i_\phi = i_n \cos \phi \tag{8-11}$$

If the surface of the hemisphere is divided into strips subtending the angle ϕ with the normal from dA (as shown in Fig. 8-4), then each will also subtend the solid angle at dA given by

$$d\omega = \frac{2\pi r \sin \phi \, r \, d\phi}{r^2}$$

$$= 2\pi \sin \phi \, d\phi \tag{8-12}$$

The radiation through each will be given by Eqns (8-10), (8-11), and (8-12) as

$$d\dot{Q}_\phi = i_n \cos \phi \, 2\pi \sin \phi \, d\phi \, dA$$

and the total radiation through the hemisphere will be

$$\dot{Q} = \int_0^{\pi/2} (i_n 2\pi \, dA) \sin \phi \cos \phi \, d\phi$$

$$= i_n \pi \, dA \qquad (8\text{-}13)$$

Since $\dot{Q}$ will also be given by the Stefan–Boltzmann equation [Eqn (8-8)] as

$$\dot{Q} = \sigma T^4 \, dA$$

it follows that
$$i_n = \frac{\sigma T_4}{\pi} \qquad (8\text{-}14)$$

Let us now consider the general case of two perfect emitters or black bodies separated by a non-absorbing space, and between which there is a net interchange of thermal radiation. Their geometries are shown in Fig. 8-5, and suffixes 1 and 2 refer separately to each body.

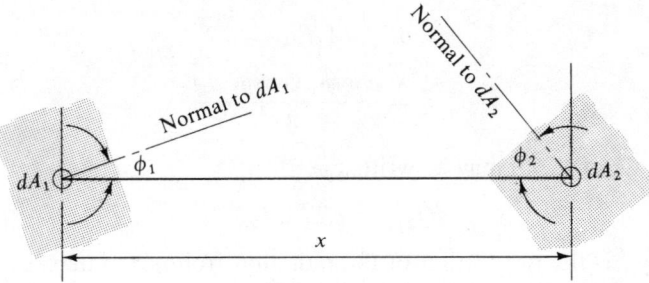

Fig. 8-5 Radiation between two black surfaces

The small area dA_2 on the second body subtends a solid angle $d\omega_1$ at dA_1 on the first where

$$d\omega_1 = \frac{dA_2 \cos \phi_2}{x^2} \qquad (8\text{-}15)$$

If the normal intensity of radiation from body 1 is i_{n_1}, the radiation incident on dA_2 is, by definition,

$$d\dot{Q}_{1.2} = i_{n_1} \cos \phi_1 \, d\omega_1 \, dA_1$$

$$= \frac{i_{n_1} \cos \phi_1 \cos \phi_2 \, dA_1 \, dA_2}{x^2} \qquad (8\text{-}16a)$$

Similarly, the radiation from dA_2 to dA_1 is given by

$$d\dot{Q}_{2.1} = \frac{i_{n_2} \cos \phi_2 \cos \phi_1 \, dA_2 \, dA_1}{x^2} \qquad (8\text{-}16b)$$

The net radiant interchange, if the bodies are respectively at temperatures T_1 and T_2, is thus given by Eqns (8-16) as

$$\dot{Q} = \dot{Q}_{1.2} - \dot{Q}_{2.1} = \int \int \frac{\sigma \cos \phi_1 \cos \phi_2 \, (T_1^4 - T_2^4)}{\pi x^2} \, dA_1 \, dA_2 \quad (8\text{-}17)$$

This double integration over the complete surfaces of the two thermally radiating bodies is normally a very tedious process, but one which may conveniently be performed by finite element methods using digital computers. Photographic techniques can also be used, and by methods such as these geometrical data for a range of commonly used combinations of surfaces have been made available. It is usual to represent this geometrical information as a *view factor*, often given the symbol F. This is defined as the fraction of the radiation from one of the two radiating surfaces which is intercepted by the other. Thus, Eqn (8-16a) may be written in terms of the view factor as

$$d\dot{Q}_{1.2} = \frac{i_{n_1} \cos \phi_1 \cos \phi_2 \, dA_1 \, dA_2}{x^2}$$

$$= \sigma T_1^4 \, dA_1 F_{1.2} \quad (8\text{-}18)$$

in which

$$F_{1.2} = \frac{\cos \phi_1 \cos \phi_2 \, dA_2}{\pi x^2} \quad (8\text{-}19a)$$

Equally, Eqn (8-16b) may be written as

$$d\dot{Q}_{2.1} = \sigma T_2^4 \, dA_2 F_{2.1} \quad (8\text{-}19b)$$

where $F_{2.1}$ is the proportion of the radiation from dA_2 intercepted by the area dA_1. Since there can be no net interchange of thermal radiation when $T_1 = T_2$, it follows that for the elementary areas

$$dA_1 F_{1.2} = dA_2 F_{2.1} \quad (8\text{-}20)$$

Thus the net radiant interchange is, in terms of the view factors,

$$\begin{aligned} \dot{Q} = \dot{Q}_{1.2} - \dot{Q}_{2.1} &= A_1 F_{1.2} \sigma (T_1^4 - T_2^4) \\ &= A_2 F_{2.1} \sigma (T_1^4 - T_2^4) \end{aligned} \quad (8\text{-}21)$$

Some values of the view factor for parallel and perpendicular rectangular plates, which were originally calculated by Hottel[1][†], are given in Fig. 8-6.

8-3-1 Re-radiating black surfaces

Hottel, who was responsible for much of the available data on thermal radiation of engineering interest, gives data in reference (1) for other pairs of surfaces and also for the commonly occurring situation in which radiation

† Full details of references cited are given at the end of the chapter.

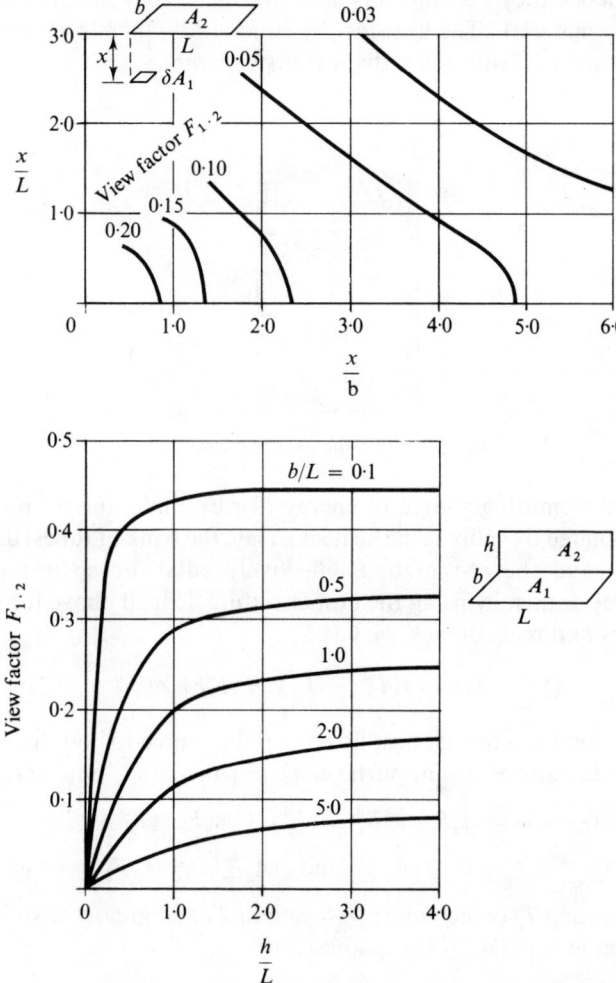

Fig. 8-6 View factors $F_{1.2}$ for parallel and orthogonal rectangles (after Hottel[1])

passes between two surfaces, not only directly but also via a third *re-radiating surface.*

In a boiler, for example, a strongly radiating bed of burning fuel will usually be exchanging heat with rows of tubes containing the evaporating fluid. These two surfaces, however, will largely be enclosed by refractory walls, and these can contribute to the net heat transfer process—as shown, for example, in Fig. 8-7—even though they may essentially be adiabatic in that the algebraic sum of the heat flows will be zero. Strictly, heat will be conducted to the surroundings from the rear of the enclosing surfaces which, since in practice they become very hot, are made of a heat-resisting, low-conductivity material. The heat lost by conduction, therefore, is invariably negligible compared with the radiant transfer rates.

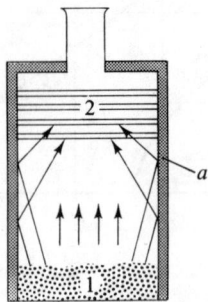

Fig. 8-7   Re-radiation in an enclosure

Thus, if the emitting source of energy (for example, the burning bed of fuel) is designated by suffix 1, the heat sink (say, the bank of tubes) designated by suffix 2, and the re-radiating, effectively adiabatic enclosing surface designated by a, then by using the nomenclature defined above for radiating black bodies in direct contact, we have

$$\dot{Q}_1 = A_1 F_{1.2}\sigma(T_1^4 - T_2^4) + A_1 F_{1.a}\sigma(T_1^4 - T_a^4) \qquad (8\text{-}22)$$

In this equation $\dot{Q}_1$ is the net heat flow from the source to both the heat sink, 2, and the adiabatic enclosing surface, a. For the latter, however,

$$\dot{Q}_a = 0 = A_a F_{a.1}\sigma(T_a^4 - T_1^4) + A_a F_{a.2}(T_a^4 - T_2^4) \qquad (8\text{-}23)$$

But $\qquad\qquad A_a F_{a.1} = A_1 F_{1.a} \quad \text{and} \quad A_a F_{a.2} = A_2 F_{2.a}$

Thus, eliminating T_a (which must be less than T_1 and greater than T_2 so that the first term in Eqn (8-23) is negative) gives

$$\dot{Q}_1 = \left[A_1 F_{1.2} + \frac{1}{(1/A_1 F_{1.a}) + (1/A_2 F_{2.a})} \right] \sigma(T_1^4 - T_2^4) \qquad (8\text{-}24)$$

In this way an appreciable augmentation of the net radiation from 1 to 2 is obtained, as shown by Eqn (8-24). Note that in its derivation it was implicitly assumed that a single view factor sufficed for each of the three pairs of surfaces in the above example. In practice, this may not apply if, for example, there is a complex geometric arrangement of tubes in a boiler so that $F_{2.a}$ may vary from row to row, or even from tube to tube. In such a case, however, the enclosing surface may be subdivided into the appropriate number of sections of equal view factor, and the separate sections considered as units in the whole system. Note that for a surface, like the source surface

1 considered above—which is totally enclosed by the other radiating surfaces—the combined view factors must add up to unity since all the emitting
radiation is incident upon one or other of the enclosures. Thus, in the example above, in which the adiabatic surface is now subdivided into sections
$a', a'', a''', \ldots$, we have

$$F_{1.2} + F_{1.a'} + F_{1.a''} + F_{1.a'''} + \cdots = 1 \qquad (8\text{-}25)$$

This equation, together with the requirement that

$$\dot{Q}_{a'} = \dot{Q}_{a''} = \dot{Q}_{a'''} = \cdots = 0$$

for the adiabatic surface, enables the heat flow $\dot{Q}_{1.2}$ to be obtained, although
the analysis is often tedious. Fortunately, Eqn (8-24) can be made to relate
to most practical re-radiation problems with sufficient accuracy, especially
when account is taken of the often gross approximations necessary to cope
with the consequences of real surfaces being imperfect emitters, as discussed
in the next section.

8-4   Imperfect emitters—non-black surfaces

In the previous section the exchange of thermal radiation between 'perfect'
emitters was discussed, that is, those surfaces which, following the argument
in the development of Kirchhoff's law, had the maximum possible emissive
power, corresponding to an absorptivity, α, of unity. Earlier we saw also
that the distribution of energy emitted throughout the electromagnetic spectrum of wavelength was a function of temperature, with a clearly defined
maximum at a single wavelength.

In practice it is found, however, that no surface behaves exactly as an ideal
emitter, or black body. Invariably the emittance at a particular wavelength
is less than predicted for the ideal surface, and the ratio $q_\lambda/q_{b\lambda}$ is known as
the *monochromatic emissivity*, ε_λ. As a consequence of Kirchhoff's law, as
stated in Section 8-1 the monochromatic emissivity of a surface, ε_λ, is equal
to the absorptivity at the same wavelength, α_λ, and the law was precisely
stated for the monochromatic case for reasons which will now become clear.

Although the distribution of energy emitted through the spectrum of
wavelength from a real surface, q_λ, usually follows the general trend predicted for a black body, there are often irregularities in the distribution, as
shown, for example, in Fig. 8-8, so that the monochromatic emissivity varies
with wavelength. Surfaces which behave in this way are known as *selective
emitters*, and clearly the total emissivity—the ratio of the areas under the distribution curves for the real surface and the corresponding ideal surface—will
vary with temperature. Thus, at one temperature the majority of thermal radiation from a surface may lie, for example, in a wavelength range in which the
emissivity is high; conversely, at a lower temperature, where the majority of the
energy falls in the higher wavelength region in which the monochromatic

emissivities may be low, the total emissivity will correspondingly be low. Consider such a body at a temperature T_1, at which its total emissivity is ε_1. By definition, the total emissive power (per unit surface) will be

$$q_1 = \varepsilon_1 \sigma T_1^4$$

If at the same time incident upon it is radiation from a second body at temperature T_2, then the first body will absorb power at the rate q_1', given by

$$q_1' = \alpha_1 \sigma T_2^4$$

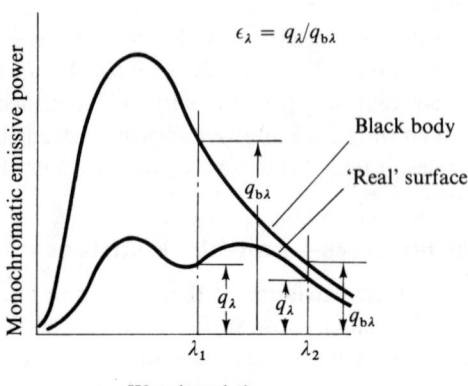

Fig. 8-8 Monochromatic emissive power of a 'real' (non-ideal) emitter

α_1 is the absorptivity of body 1, but its value will be that corresponding to the distribution of radiant energy with wavelength at the temperature of the second body, T_2, and if the first body is a selective emitter (and therefore a selective absorber) then α_1 and ε_1 will not have the same values as might wrongly be inferred from Kirchhoff's law. However, no paradox is involved, since, if bodies 1 and 2 were at the same temperature, a single energy-wavelength distribution would apply to body 1 (and, of course, another to body 2) so that the total or average emissivity would equal the absorptivity and

$$q_1' - q_1 = \varepsilon_1 \sigma T_1^4 - \alpha_1 \sigma T_1^4 = 0$$

In practical radiation problems we are normally dealing with radiation between selective emitters at different temperatures for which ε and α do not normally have the same value. Thus, to the often complex geometrical problems of the ideal, black body, in real cases of radiation interchange must be added the problems of the selective emitter with its wavelength-dependent emissivity and absorptivity. There is also the further complication that in practice these parameters are found also to be dependent upon the angles of emission and incidence of the radiant energy, so that the exact determination of radiation between practical surfaces is an almost impossibly complex task. Sometimes in engineering practice use is made of the selective properties of radiating surfaces, and earlier in this chapter attention was drawn to the use of finishes with a high total emissivity for low-temperature energy-wavelength distributions, and a low emissivity and thus absorptivity for the predominantly short-wave radiation associated with high-temperature radiation. Thus, such surfaces attain a lower equilibrium temperature when subjected to, say, solar radiation than would a non-selective emitter under the same conditions. In engineering practice, however—particularly where the higher operating temperatures associated with the combustion of fuels occur and where the role of radiant heat transfer is frequently at its most critical—extensive use is made of refractories, and metallic surfaces are usually heavily oxidized. It is fortunately the case that such surfaces are not strongly selective emitters, and with adequate accuracy it can be assumed that the monochromatic emissivity is independent of wavelength, so that absorptivity, α, for incident radiation at a temperature different from the emitting temperature can be taken to be equal to the total or average emissivity, ε. Bodies which satisfy this assumption are known as *grey bodies*.

8-5 Radiation interchange between 'grey bodies'

We have here to deal with bodies for which, as explained above, we can assume that emissivity and absorptivity are equal, but less than unity. Thus, of the radiation incident upon a surface, a proportion α is absorbed, and the remainder reflected and re-radiated with the wavelength distribution corresponding to the characteristics of the reflecting surface. It will be assumed, as explained in Section 8-1, that the solid bodies dealt with are opaque to thermal radiation, and that transmissivity, τ, is zero. Then Eqn (8-2) applies, so that

$$\alpha + \rho = 1 \qquad\qquad (8\text{-}2)$$

in which ρ is the reflectivity.

8-5-1 Interchange between small 'grey' bodies

Suppose that two *grey* bodies, represented by suffices 1 and 2, have emissivities ε_1, ε_2. Suppose also that the bodies are very small compared with their distance apart. Then under this condition it may be assumed that of the radiation unabsorbed, and thus reflected, at each surface, a negligible proportion returns again to the original emitting body. Radiation not absorbed on the first incidence is, therefore, considered to move off into space, taking no further part in the energy interchange. It is implicit in this assumption that the surfaces are *diffuse* reflectors; that is, that a beam of unabsorbed incident radiation is reflected uniformly over the solid angle 2π, as shown in Fig. 8-9(a).

Some surfaces, particularly highly reflective ones, reflect specularly; that is, an incident beam is reflected also as a beam, with the angles of incidence and reflection simply related. Under particular geometrical conditions with this type of surface it is conceivable that all the reflected energy can be returned to the original emitting body, even though both are small, with the opportunity for re-absorption and re-reflection. Alternatively, and more usually, all the reflected energy will disappear into space, as in Fig. 8-9(b), and our assumption above is unnecessary. In practice, however, thermally radiating surfaces are usually *diffuse* reflectors.

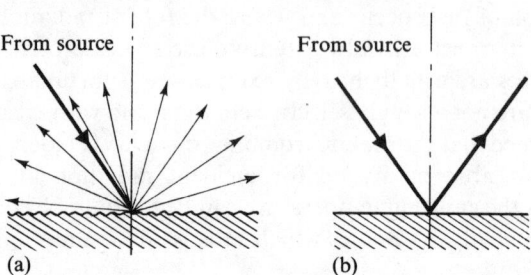

Fig. 8-9   (a) Diffuse reflection; (b) specular reflection

Returning to the two diffusely reflecting, small grey bodies, the energy emitted by body 1 will be $A_1\varepsilon_1\sigma T_1^4$, of which $F_{1.2}A_1\varepsilon_1\sigma T_1^4$ is incident on the second body, and of which $\alpha_2 F_{1.2}A_1\varepsilon_1\sigma T_1^4$ is absorbed. Since bodies 1 and 2 are *grey*, we have $\alpha_2 = \varepsilon_2$ and the energy transfer from 1 to 2 is

$$\dot{Q}_{1.2} = \varepsilon_1\varepsilon_2 A_1 F_{1.2}\sigma T_1^4$$

Similarly, the reverse heat flow $\dot{Q}_{2.1}$ is obtained, and the net radiant interchange between the two bodies is

$$\dot{Q} = \dot{Q}_{1.2} - \dot{Q}_{2.1} = \varepsilon_1\varepsilon_2 AF\sigma(T_1^4 - T_2^4) \tag{8-26}$$

In this equation $AF = A_1 F_{1.2} = A_2 F_{2.1}$, where F is the geometric view factor derived in the consideration of radiant energy exchange between black bodies. Thus the equivalent emissivity, $\bar{\varepsilon}$, of this system of two small grey bodies is

$$\bar{\varepsilon} = \varepsilon_1\varepsilon_2 \tag{8-27}$$

In practice, of course, a proportion of the reflected radiation from each is returned to the other body and re-radiated so that the true equivalent emissivity is greater than given by Eqn (8-27). This may be regarded as giving the least possible value.

8-5-2 Interchange between large parallel grey plates

At the other extreme, consider the exchange of radiation between two very large grey surfaces a small distance apart, so that *all* reflected radiation is returned to the emitter. For this situation the view factor will be unity and the effective areas of the surfaces the same. Using the reflectivity ρ from Eqn (8-2), so that $\rho = 1 - \alpha = 1 - \varepsilon$ for the grey surfaces, we have for the radiation emitted by the first surface per unit area (as in Fig. 8-10) $\varepsilon_1 \sigma T_1^4$, of which $\alpha_2 \varepsilon_1 \sigma T_1^4$ or $\varepsilon_2 \varepsilon_1 \sigma T_1^4$ is absorbed by the second surface on the first incidence. The remainder is reflected; that is $\rho_2 \varepsilon_1 \sigma T_1^4$, of which $\alpha_1 \rho_2 \varepsilon_1 \sigma T_1^4$ or $\rho_2 \varepsilon_1^2 \sigma T_1^4$ is absorbed by the original surface, 1. The remainder is reflected; that is $\rho_1 \rho_2 \varepsilon_1 \sigma T_1^4$, of which $\alpha_2 \rho_1 \rho_2 \varepsilon_1 \sigma T_1^4$ or $\rho_1 \rho_2 \varepsilon_2 \varepsilon_1 T_1^4$ is absorbed by surface 2 on the second incidence. Continuing the argument shows that on the third incidence of the twice-reflected original emission from 1 or 2 the energy absorbed is $(\rho_1 \rho_2)^2 \, \varepsilon_1 \varepsilon_2 \sigma T_1^4$, and so on, so that in general, on the $(n + 1)$th incidence the energy absorbed by surface 2 is $(\rho_1 \rho_2)^n \varepsilon_1 \varepsilon_2 \sigma T_1^4$.

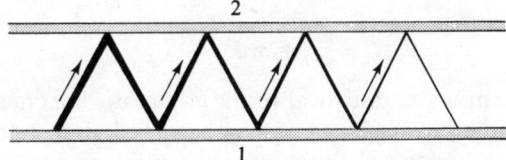

Fig. 8-10 Radiation between two large 'grey' plates

A similar reasoning applies equally for the radiation from surface 2, so that for the net exchange of energy we have, for an area A,

$$\dot{Q} = \dot{Q}_{1.2} - \dot{Q}_{2.1} = \varepsilon_1 \varepsilon_2 A \sigma (T_1^4 - T_2^4)\{1 + \rho_1 \rho_2 + (\rho_1 \rho_2)^2 + \cdots \}$$

(8-28)

ρ_1 and ρ_2 are, by definition, each less than unity, so that the series in brackets is convergent to the sum $(1 - \rho_1 \rho_2)^{-1}$. Thus Eqn (8-28) may be written

$$\dot{Q} = \frac{\varepsilon_1 \varepsilon_2}{1 - \rho_1 \rho_2} A \sigma (T_1^4 - T_2^4)$$

$$= \frac{\varepsilon_1 \varepsilon_2}{1 - (1 - \varepsilon_1)(1 - \varepsilon_2)} A \sigma (T_1^4 - T_2^4)$$

$$= \frac{1}{(1/\varepsilon_1) + (1/\varepsilon_2) - 1} A \sigma (T_1^4 - T_2^4)$$

(8-29)

Thus, the equivalent emissivity, $\bar{\varepsilon}$, for two large grey surfaces a small distance apart is given by

$$\bar{\varepsilon} = \frac{1}{(1/\varepsilon_1) + (1/\varepsilon_2) - 1}$$

(8-30)

8-5-3 Interchange between a small grey body and a grey enclosure

A third extreme of grey body geometry is provided by the case of the small grey body in an enveloping grey enclosure. In this situation the enclosure behaves as a *black* body through the opportunities for ultimate absorption offered by repeated reflection from its walls, as was shown in connection with Fig. 8-3. Thus, if the emission from the small body, 1, is $\varepsilon_1 A_1 \sigma T_1^4$, effectively all of this will ultimately be absorbed by the enclosure. Of the emission from the enclosure $\varepsilon_2 A_2 \sigma T_2^4$, the fraction corresponding to the view factor, $F_{2.1}$, will be incident upon the small body 1, of which $\alpha_1 \varepsilon_2 F_{2.1} A_2 \sigma T_2^4$ or $\varepsilon_1 \varepsilon_2 F_{2.1} A_2 \sigma T_2^4$ will be absorbed. Thus the net exchange of energy is

$$\dot{Q} = \dot{Q}_{1.2} - \dot{Q}_{2.1} = \varepsilon_1 A_1 \sigma T_1^4 - \varepsilon_1 \varepsilon_2 A_2 F_{2.1} \sigma T_2^4 \qquad (8\text{-}31)$$

If the net energy transfer is to be zero when $T_1 = T_2$, then from Eqn (8-31) it is seen that

$$A_1 = A_2 \varepsilon_2 F_{2.1} \qquad (8\text{-}32)$$

so that
$$\dot{Q} = \varepsilon_1 A_1 \sigma (T_1^4 - T_2^4) \qquad (8\text{-}33)$$

This is apparently a paradoxical result in that ε_2, the emissivity (and absorptivity) of the enclosing surface, may be varied almost at will from that associated with a highly polished metal to a blackened refractory. On the other hand, A_1, A_2, and $F_{1.2}$ are uniquely related by the geometrical procedures leading to Eqn (8-17).

The apparent paradox lies in the assumption that the enclosure absorbs the emission from the small body like a perfect emitter. In fact there will be a number of reflections before the radiation from 1 is absorbed, and the greater the number of reflections—that is, the lower is ε_2—the higher the probability that the reflected radiation will be re-incident on the small body. The effect of the further incidences is to increase the proportion of the original radiation intercepted by the small enclosed body—that is, to increase the effective $F_{2.1}$ in Eqn (8-31). Thus, the inverse relationship between ε_2 and $F_{2.1}$ required to satisfy Eqn (8-32) is completely conceivable. Indeed, it is the interrelation between the geometric view factors and the emissivities and absorptivities of non-black surfaces which makes the exact computation of real problems in radiation heat transfer so complex (and significantly more so if the grey body approximation is not allowed, so that account has to be taken of the selective emission and absorption properties of real surfaces).

8-5-4 The generalized emissivity and geometric factor, $\mathcal{F}$

Hottel[1] defines a factor, $\mathcal{F}$, representing the combined emissivities of a grey body and the effective geometric view factor, allowing for multiple reflections. Thus, for the first case considered here of two small well-separated surfaces, writing Eqn (8-26) with Hottel's factor would give

$$A_1 \mathscr{F}_{1.2} = \varepsilon_1 \varepsilon_2 A_1 F_{1.2} \tag{8-34}$$

For the close, large grey plates, Eqn (8-29) becomes

$$A_1 \mathscr{F}_{1.2} = \frac{1}{(1/\varepsilon_1) + (1/\varepsilon_2) - 1} A_1 F_{1.2} \tag{8-35}$$

(with $F_{1.2} = 1$, of course). For the small enclosed surface considered above,

$$A_1 \mathscr{F}_{1.2} = \varepsilon_1 A_1 F_{1.2} \tag{8-36}$$

(again, with $F_{1.2} = 1$).

Most radiating solid surfaces in engineering practice have emissivities in the range 0·7–0·9. For two surfaces with these values of $\varepsilon_1 = 0·7$, $\varepsilon_2 = 0·9$, Eqns (8-34) and (8-35), which can be considered to represent, respectively, opposite extreme geometric situations, give

$$\mathscr{F}_{1.2} = 0·63 F_{1.2} \quad \text{and} \quad \mathscr{F}_{1.2} = 0·65 F_{1.2}$$

a difference of barely 3 per cent. Where one of the surfaces is so large as to almost completely enclose the other and to behave like a black body, the simple form of Eqn (8-36) may be used (to which, of course, Eqns (8-34) and (8-35) reduce if the emissivity of the second surface is put equal to unity). Only in rare cases is additional complexity of analysis worthwhile.

8-6   Radiation from non-luminous gases

In considering heat transfer by radiation up to this point it has been implicitly assumed that the spaces between the radiating surfaces have been empty. In most practical cases this will not be the situation, and usually the spaces will be filled by a gas. Many common gases are very largely transparent to thermal radiation, so that their effects upon the radiant heat transfer process are often negligibly small. This is not universally true, however, and in particular the polyatomic gas molecules can, in certain circumstances, absorb and emit thermal energy at significant rates. Water vapour, for example, at a pressure of 1 atm in a duct 1 ft in diameter would emit at 100°C about 30 per cent of the energy from a black body as also would sulphur dioxide; carbon dioxide at the same temperature would emit at about 15 per cent of the rate from an ideal emitter. The diatomic carbon monoxide, on the other hand, at the same conditions, would emit less than one-fiftieth or 2 per cent of the black body radiation. In most practical problems it is the effects of the polyatomic molecules which must therefore be taken into account, and particularly the commonly occurring products of combustion in this category, CO_2 and H_2O.

For solid bodies the emission and absorption of radiation is almost entirely a surface effect, as has been seen, so that temperature gradients within the material are irrelevant. This is not so in gases, for even those which have a significant role to play in thermal radiation are still largely transparent to the incident emission and do not reflect the unabsorbed energy as do solids of low absorptivity. It is thus to be expected that the thickness of a layer of gas will be important in determining its radiation characteristics, for this parameter will determine the number of molecules that can play a part in the interaction of a gas with the energy flux entering or leaving its bounding surface. The characteristic dimension of a layer of gas which determines its radiation characteristics is known as its *beam length*. If we consider a hemispherical shell of gas, of thickness dr, then the emissive power from it will be $(dq/dr)\,dr$, or in terms of ε, the emissivity of the gas element, and q_b, the black body emissive power, $q_b(d\varepsilon/dr)\,dr$.

From Eqn (8-10) the emission from an element dA_2 of gas on the hemispherical surface of radius r, subtending a solid angle $d\omega$ on a small element dA_1 at the centre of the hemisphere (which similarly subtends a solid angle $d\omega_1$ at dA_2) will be

$$d\dot{Q}_{2.1} = i_n \cos\phi\,d\omega_1\,dA_2$$

$$= i_n \cos\phi\,d\omega\,dA_1 \qquad (8\text{-}37a)$$

since
$$d\omega_1 = \frac{dA_1}{r^2} \quad \text{and} \quad d\omega = \frac{dA_2}{r^2}$$

ϕ is the angle which the radius to dA_2 makes with the normal to dA_1. Equation (8-13) shows that the emissive power is π times the normal intensity so that Eqn (8-37a) may be written as

$$d\dot{Q}_{2.1} = \frac{q_b}{\pi}\cos\phi\,d\omega\,dA_1\left(\frac{d\varepsilon}{dr}\right)dr \qquad (8\text{-}37b)$$

The emission from a complete gas volume to a finite area at its base is thus given by

$$\dot{Q} = \frac{q_b}{\pi}\int_{A_1}\int_0^R\int_\omega \cos\phi\left(\frac{d\varepsilon}{dr}\right)d\omega\,dr\,dA_1 \qquad (8\text{-}38)$$

The three stages of this integration are: first, over the enveloping solid angle, which may be performed once and for all for a given enclosing shape; second, to the appropriate extreme length dimension (radius) of the appropriate shape; and third, over the base area.

The form for $d\varepsilon/dr$ in Eqn (8-38) must be assumed and the simple molecular concept—that the absorption and emissivity of the gas depends upon the number of molecules per unit bounding surface—would require that $d\varepsilon/dr$

were constant. Now in Eqn (8-37b) the volume of the emitting element of gas at dA_2 is

$$dV = r^2 \, d\omega \, dr$$

and the solid angle subtended at dA_2 by dA_1 is

$$d\omega_1 = \frac{dA_1 \cos \phi}{r^2}$$

Thus, if we define a 'volumetric intensity' of emission from the elementary gas volume $i_v = d\dot{Q}_{2.1}/d\omega_1 \, dV$, then

$$i_v = \frac{q_b}{\pi} \frac{d\varepsilon}{dr}$$

and to an enclosing sphere subtending solid angle 4π, the emission will be E_v per unit volume of radiating element of gas, where

$$E_v = 4\pi i_v = 4q_b \frac{d\varepsilon}{dr} \tag{8-39}$$

For a finite volume V of gas, with bounding walls of area A, an equivalent emissivity, ε_g, is defined so that the emission $\dot{Q}$ is given by

$$\dot{Q} = Aq_b\varepsilon_g \tag{8-40}$$

However, if $d\varepsilon/dr$ in Eqn (8-36) can be considered to be constant over the characteristic dimension (*beam length*, L) of the gas volume, then Eqn (8-39) may be written

$$\dot{Q} = 4Vq_b \frac{d\varepsilon}{dr} = 4Vq_b \frac{\varepsilon_g}{L} \tag{8-41}$$

Equating Eqns (8-40) and (8-41) shows that

$$L = \frac{4V}{A} \tag{8-42}$$

Thus, the hypothesis that the role of the molecules in a radiating gas is a linear function of their number per unit bounding surface yields the simple result that the characterizing beam length is equal to the equivalent diameter ('hydraulic mean depth'), which is well-known in fluid dynamics to characterize passages of non-circular cross-section in turbulent flow (see Section 6-8). In practice, this hypothesis oversimplifies the situation, and ignores the self-absorption properties of real gases in which molecules absorb radiation from other molecules in the total volume. Further, in using Eqn (8-40) to derive the expression for the beam length L [Eqn (8-42)], the whole of the bounding surface, A, of the volume was assumed to be receiving radiation uniformly from the gas.

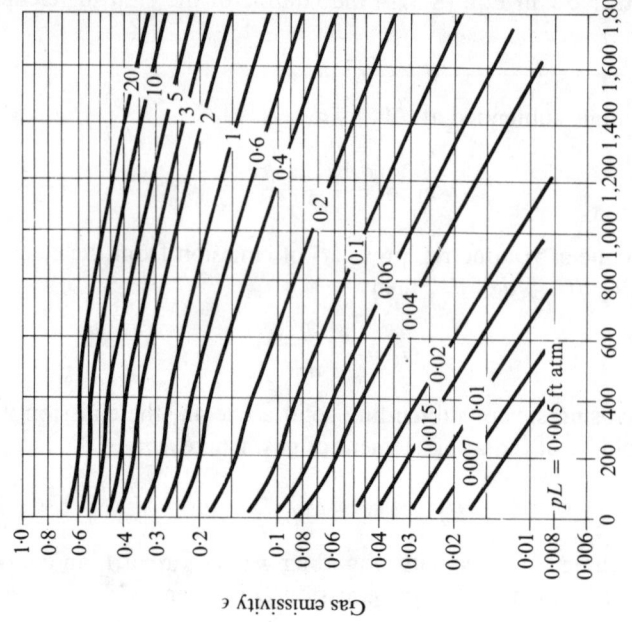

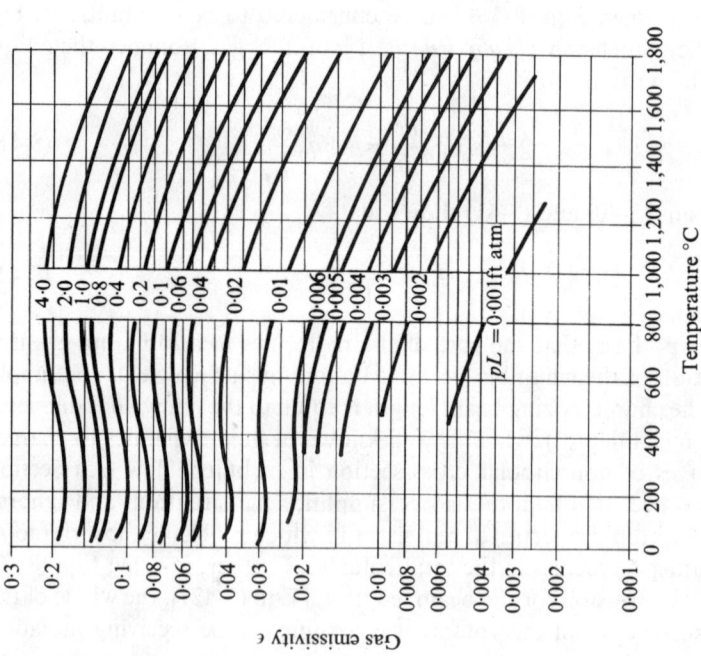

Fig. 8-11 (a) Emissivity of carbon dioxide (after Hottel[1]); (b) emissivity of water vapour (after Hottel[1])

The first of these approximations is sufficiently close to reality when the number of molecules involved in the radiation process is relatively small. This implies that the beam length, L, is small, and equally that the gas pressure is low. Indeed, these two parameters are so similar in their effect that at low total pressures at least, gas radiation at a given temperature is correlated against the product

(partial pressure of radiating constituent) $\times$ (beam length) $= p \cdot L$

If self-absorption cannot be neglected, then the term $d\varepsilon/dr$ in Eqn (8-38) must be appropriately modified.

Also, if the simple geometric assumption in the derivation of the equation for beam length [Eqn (8-42)] is not acceptable, then the consequent complexities will require the integration of Eqn (8-38), usually most conveniently by finite element procedures. In many practical problems, however, the simple analysis is adequate and Table 8-1 lists the beam lengths for some simple shapes obtained from Eqn (8-42) compared with the exact values recommended by Hottel[1] for 'typical' values of the product pL.

Table 8-1

Shape	Characterizing dimension	Beam length, Eqn (8-42)	Beam length, Hottel[1]
Sphere	Diameter, D	$0.66D$	$0.60D$
Infinite cylinder	Diameter, D	D	$0.90D$
Cylinder, height = diameter	Diameter, D	$0.66D$	$0.60D$
Space between tubes with centres on equilateral triangles, clearance = diameter	Diameter, D	$3.45D$	$2.8D$

8-6-1 Determination of emissivities and absorptivities for non-luminous gases

Empirical data on the emissivities of radiating non-luminous gases have been obtained by Hottel[1] and are correlated, as explained in the previous section, against the product

(partial pressure of constituent, p) $\times$ (beam length, L)

and for a wide range of temperatures. Hottel's data for two of the most commonly occurring radiating gases, carbon dioxide and water vapour, are given in Figs 8-11(a) and 8-11(b).

From the known composition of a gas mixture the partial pressures of the radiating constituents are determined from the elementary gas laws.

With the beam length determined as previously described, the emissivity of each gas is determined at its temperature, T_g, say, for the corresponding value of pL. Although the emissivities of the separate constituents of a gas mixture are additive, there is mutual absorption so that if, for example, in a mixture of water vapour and carbon dioxide, the emissivities of each were respectively determined as ε_{H_2O} and ε_{CO_2} for the appropriate conditions, the total emissivity of the mixture, ε_g, is given by

$$\varepsilon_g = \varepsilon_{H_2O} + \varepsilon_{CO_2} - \Delta\varepsilon$$

where $\Delta\varepsilon$ is the correction required to account for the absorption of radiation from the CO_2 in the H_2O, and vice versa. The correction $\Delta\varepsilon$ is small but significant, and is shown in Fig. 8-12. Further correction of the data is required for total pressures of the gas mixture different from 1 atm, and it is here that the greatest uncertainty enters the calculation of radiation from gases. There is need for much more data than are currently available, particularly since it is known that emissivities at a total pressure of only 5 atm can be more than double those at 1 atm for nominally similar values of the usual correlating parameters.

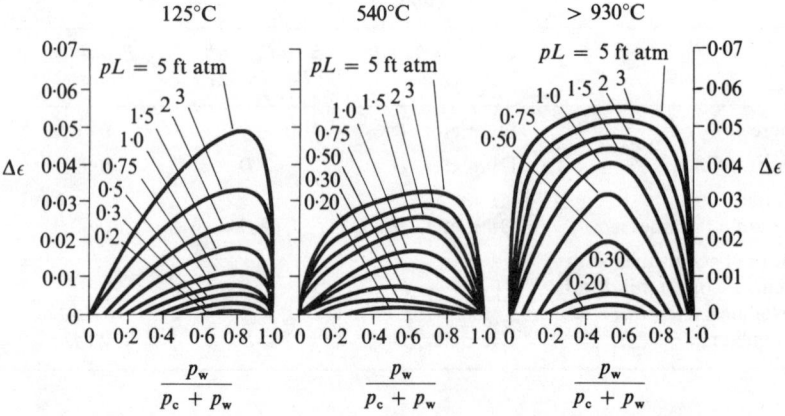

Fig. 8-12 Correction for mutual absorptivity of water and carbon dioxide

The calculation of the absorptivities of gas mixtures follows closely the procedure described above for the emissivities. Figures 8-11(a) and 8-11(b) show, however, that neither CO_2 nor H_2O meets the assumptions of 'grey' bodies for the emissivities, and thus the absorptivities are strongly temperature-dependent. Thus, having determined the emissivity of a mixture at its temperature, the absorptivities of the constituents must be determined as the emissivities at the temperature of the surface which is radiating to the gas, T_s say. However, as might be expected, in view of the discussion in the previous section on the role of the molecular density in gas radiation, the temperature of the gas itself is found to influence its absorptivity to emission with the wavelength distribution corresponding to the bounding surface temperature, T_s. Hottel[1] suggests that, to allow for this effect, the absorptivity of each constituent should be determined at the emitting surface temperature, T_s, and at $(pL)(T_s/T_g)$ instead of at pL. The value of ε obtained from Fig. 8-11 should then be multiplied by $(T_g/T_s)^{0.65}$ in the case of carbon dioxide, and by $(T_g/T_s)^{0.45}$ for water vapour to yield the corresponding absorptivities, α_{CO_2} and α_{H_2O} with a final mutual absorptivity correction, as for ε.

Having thus determined the radiation properties of the gas, the rate of interchange of energy with the surroundings can be determined. If these are *black*, the heat flow from the gas, $\dot{Q}_{gs}$, is given by

$$\dot{Q}_{gs} = \varepsilon_g \sigma T_g^4 A_s$$

where A_s is the area of the surrounding surfaces. From the surroundings to the gas the radiation is $\sigma T_s^4 A_s$, of which the proportion α_g is absorbed, so that for $\dot{Q}_{sg}$ we have

$$\dot{Q}_{sg} = \alpha_g \sigma T_s^4 A_s$$

and the net interchange is

$$\dot{Q} = \dot{Q}_{gs} - \dot{Q}_{sg} = \sigma A_s(\varepsilon_g T_g^4 - \alpha_g T_s^4) \tag{8-43}$$

If the surroundings are *grey*, then energy radiated from the gas is partially reabsorbed. By the same argument used to consider the net interchange between two large parallel plates, it may be shown that if the emissivity, and therefore the absorptivity, of the grey surroundings is ε, then the net heat flow is given by

$$\dot{Q} = \frac{\varepsilon \sigma A_s}{1 - (1 - \alpha_g)(1 - \varepsilon)} [\varepsilon_g T_g^4 - \alpha_g T_s^4] \tag{8-44}$$

Thus, if the absorptivity of the gas is high, thereby reabsorbing most of the original radiation reflected from the surroundings, the net heat flow is reduced to almost ε times the interchange with black surroundings, as given by Eqn (8-43). However, if the absorptivity of the gas, α_g, is small so that the radiation reflected from the surrounding walls passes through the gas, with further opportunities to be re-incident on the walls, the overall heat exchange is little affected. This follows since

$$\lim_{\alpha \to 0} \frac{\varepsilon}{1 - (1 - \alpha_g)(1 - \varepsilon)} = 1 \cdot 0$$

To illustrate the procedures described above for calculating radiant heat transfer from non-luminous gases, consider the following example:

Example 1 A gas turbine combustion chamber is 1 ft in diameter and the walls are maintained at 500°C. The products of combustion are at 1,000°C and a pressure of 1 atm and contain 12 per cent by volume each of carbon dioxide and water vapour. Determine the net radiant heat transfer per unit surface.

Emissivity of CO_2: L (Table 8-1) $= 0.9D = 0.9$ ft
$p_{CO_2} = 0.12$ atm. Therefore $p_{CO_2} = 0.108$ ft atm.
At 1,273 K, from Fig. 11(a), $\varepsilon_{CO_2} =$ **0.071**

Emissivity of H_2O: $p_{H_2O}L$ is as for $CO_2 = 0.108$ ft atm.
At 1,237 K, from Fig. 11(b), $\varepsilon_{H_2O} =$ **0.047**

(Hottel[1] suggests that the emissivity of water vapour should be corrected for the absolute level of its partial pressure, but states that the correction is only approximate, and is not here considered a worthwhile additional complication.)

From Fig. 8-12 the mutual absorptivity correction for $p_{CO_2}L + p_{H_2O}L = 0.216$, and $p_{H_2O}/(p_{CO_2} + p_{H_2O}) = 0.5$ is $\Delta\varepsilon = 0.003$.

$$\varepsilon_g = \varepsilon_{CO_2} + \varepsilon_{H_2O} - \Delta\varepsilon$$
$$= 0.071 + 0.047 - 0.003 = \mathbf{0.115}$$

Absorptivity of CO_2: $p_{CO_2}L(T_s/T_g) = 0.108 \times \dfrac{773}{1,273} = 0.0655$

At $T_{s'} = 773$ K, $\varepsilon_{CO_2} = 0.065$

$$\alpha_{CO_2} = 0.065 \times \left(\frac{1,273}{773}\right)^{0.65} = \mathbf{0.090}$$

Absorptivity of H_2O: $p_{H_2O}(T_s/T_g) = 0.0655$ (as for CO_2)

At $T_s = 773$ K, $\varepsilon_{H_2O} = 0.058$

$$\alpha_{H_2O} = 0.058 \times \left(\frac{1,273}{773}\right)^{0.45} = \mathbf{0.073}$$

The mutual absorptivity correction at 773 K is negligible according to Fig. 8-12 at the appropriate conditions. Therefore,

$$\alpha_g = 0.090 + 0.073 - 0 = \mathbf{0.163}$$

If the combustion chamber walls can be considered as *black surfaces*, then the heat flow from the gas to the walls, $\dot{Q}_{gs}$ is given by

$$\dot{Q}_{gs} = \varepsilon_g \sigma T_g^4$$

and from the walls to the gas is

$$\dot{Q}_{sg} = \sigma T_s^4 \alpha_g$$

so that the net interchange is

$$\dot{Q} = \dot{Q}_{gs} - \dot{Q}_{sg} = \sigma(\varepsilon_g T_g^4 - \alpha_g T_s^4)$$

$$= 1{\cdot}0\left[0{\cdot}115\left(\frac{1{,}273}{100}\right)^4 - 0{\cdot}163\left(\frac{773}{100}\right)^4\right]$$

$$= \textbf{2,430 Chu/ft}^2\textbf{ h}$$

If the combustion chamber walls are *grey* with an emissivity of 0·8, then from Eqn (8-4) we have

$$\dot{Q} = \frac{0{\cdot}8 \times 1{\cdot}0}{1 - (1 - 0{\cdot}145)(1 - 0{\cdot}8)}\left[0{\cdot}115\left(\frac{1{,}273}{100}\right)^4 - 0{\cdot}163\left(\frac{773}{100}\right)^4\right]$$

$$= 0{\cdot}965 \times 2{,}430$$

$$= \textbf{2,350 Chu/ft}^2\textbf{ h}$$

Thus, in this typical case the difference between assuming the combustion chamber walls black or grey makes less than 4 per cent difference in heat flow.

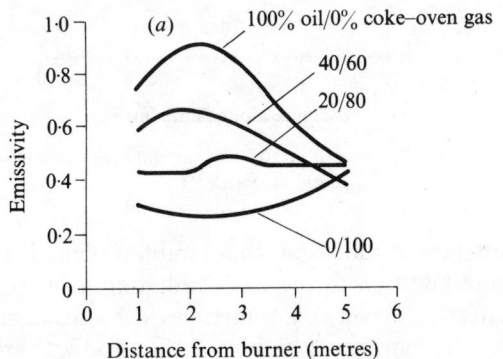

Fig. 8-13 Effect of carburation and luminosity upon flame emissivity (after Béer and Howarth[2])

8-7   Radiation from luminous flames

The polyatomic gases which are principally responsible for the thermal radiation studied in the previous section are formed usually from the combustion of hydrocarbon fuels. An intermediate stage in the combustion process, however, is often the production of clouds of carbon particles which, at the high temperatures associated with combustion, radiate intensely in and near the wavelengths of visible light in the electromagnetic spectrum. It is this visible radiation which is commonly referred to as the 'flame'.

The effect of the presence of a luminous flame in the combustion process is shown graphically by the data of Fig. 8-13.[2] This shows the emissivities measured in a furnace in which the ratio of carbon to hydrogen in the fuel was varied by changing the ratio of oil to coke-oven gas. Despite the almost identical emissivities of the products of combustion at the combustion chamber outlet, the intensely luminous oil flame (curve 'a') shows a peak emissivity approaching that of an ideal emitter. The coke-oven gas, typically of gaseous fuels, shows emissivities in the range associated with non-luminous combustion, so that the total radiation in this case is barely one-quarter of that associated with oil. This difference in performance is, of course, of great significance in the design of combustion chambers.

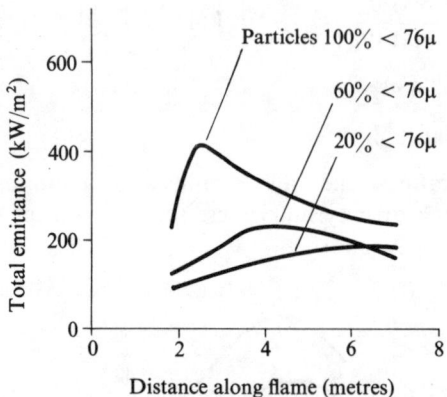

Fig. 8-14   Effect of particle diameter ($\mu = 10^{-6}$ m) on emissivity of a coal flame (after Béer and Howarth[2])

The exact estimation of radiation from luminous flames still poses severe difficulties, since in addition to the basic radiating properties of the combustion product itself—carbon and, sometimes, ash—account must be taken of the geometric form and distribution of the product through the total radiating volume. Thus, fine particles may be expected to radiate differently from coarser products of combustion, and this is found experimentally to be the case. If the size of the particles is very much less than the wavelength of the thermal radiation, the cloud behaves very much like the gases and solid surfaces studied earlier in this chapter, with the absorptivity and emissivity equal for given distributions of wavelength. Clouds of larger particles, on the other hand, behave quite differently in their tendency to *scatter* as well as *absorb* radiation. Thus, quite different radiation properties are found from identical fuels burning at nominally similar combustion temperatures, as may be seen for example in Fig. 8-14.[2] This shows the effective emissivity of flames associated with the combustion of pulverized anthracite of varying initial fineness. Although each curve tends towards the same ultimate value associated with the common products of combustion, the peak emissivity, and correspondingly the total radiation along the flame length, is almost twice as great for the finer products of combustion. Several additional degrees of freedom are therefore added to the already complex problem of radiant heat transfer when luminous flames are present. Although there has been much basic research into the problem of quantifying flame radiation [see, for example, references (1) and (2)], empiricism is still essential in determining the emissivity of flames, and, further, the experimental difficulties of obtaining reliable data are severe. These arise from the problems of measuring both the radiation heat transfer (and separating it from the other effects), and of defining a mean emitting temperature in a chemically reacting and inevitably unsteady flame, in which severe striations of temperature occur.

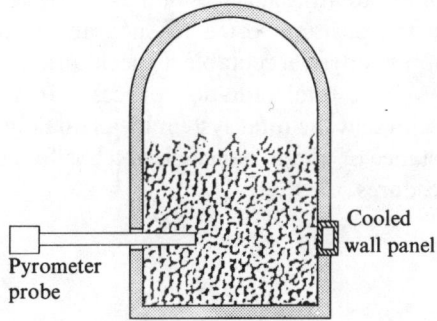

Fig. 8-15   Arrangement for measuring flame emissivity in a furnace

To determine the effective emissivity of a flame, ε_f, a standard technique uses the arrangement shown diagrammatically in Fig. 8-15. First, the emission from the flame within the enclosure—which (as has been seen in Section 8-1) behaves as a black body—is determined with a small pyrometer or heat flux meter. If this radiating quantity is $\dot{Q}_1$, then we have

$$\dot{Q}_1 = \varepsilon_f \sigma T_f^4 + (1 - \alpha_f)\sigma T_B^4 \qquad (8\text{-}45)$$

where ε_f and T_f are the unknown effective flame emissivity and temperature, and σT_B^4 is the emission from the adiabatic black enclosure, of which α_f is absorbed by the flame and the remainder passes through to the pyrometer. If the radiation from the furnace wall is now removed by sighting the pyrometer on to the flame with a cooled section of wall as background, and if the emission from the flame alone is determined, with all other conditions unchanged, then

$$\dot{Q}_2 = \varepsilon_f \sigma T_f^4 \qquad (8\text{-}46)$$

Thus,

$$\dot{Q}_1 - \dot{Q}_2 = (1 - \alpha_f)\sigma T_B^4$$

If in the original experiment it is assumed that the furnace wall attains the effective flame temperature (that is, $T_f = T_B$), then equally for this equilibrium situation Kirchhoff's law applies and $\alpha_f = \varepsilon_f$ so that

$$\varepsilon_f = 1 - \frac{\dot{Q}_1 - \dot{Q}_2}{\sigma T_B^4} \tag{8-47}$$

8-8 Electrical network analogy for radiating systems

In Section 8-5, radiation between two non-ideal emitters was considered, and with the grey body assumption it was shown that results acceptably accurate for most engineering purposes could be obtained by approximating real, but complex, situations by one of the extreme geometric conditions. The interposition of re-radiating surfaces or a gas between the grey radiating bodies complicates the position, often to such an extent that the simple assumptions remain no longer acceptable if the additional components contribute significantly to the total radiation process. In these circumstances it is convenient to represent the total system by an analogous electric circuit, the equivalent resistance or conductance of which is found by standard network solution procedures.

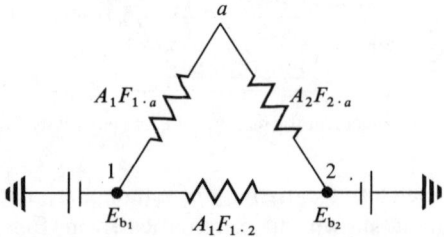

Fig. 8-16   Electric circuit analogue for re-radiating enclosure

To illustrate first the principle with a simple example, consider the furnace represented in Fig. 8-7 in which the bed of burning fuel exchanged heat with the sink of evaporator tubes both directly and via the surrounding refractory surface, a. The equivalent circuit for this system of black bodies is shown in Fig. 8-16. At the furnace 1 a potential is applied, E_{b_1} say, which corresponds to the black body emittance at temperature, T_1, that is $\sigma_1 T_1^4$. Similar potential sources exist at the evaporator tubes 2 and at the adiabatic enclosing surface a. Between each of these components the electrical conductances correspond to the emitting area–view factor products as shown, remembering that Eqns (8-20) and (8-25) apply, so that in this case

$$A_a F_{a.2} = A_2 F_{2.a}$$

and

$$F_{1.a} = 1 - F_{1.2}$$

Writing R for resistance—the reciprocal of conductance—we have, for 1 to 2 direct,

$$R_{1.2} = \frac{1}{A_1 F_{1.2}}$$

For 1 to 2 via a, we have

$$R_{1.a.2} = R_{1.a} + R_{a.2} = \frac{1}{A_1 F_{1.a}} + \frac{1}{A_2 F_{2.a}}$$

For the combined circuits in parallel, the resistance is R_0, say—the reciprocal of conductance C_0—so that

$$C_0 = \frac{1}{R_0} = \frac{1}{R_{1.2}} + \frac{1}{R_{1.a.2}}$$

$$= A_1 F_{1.2} + \frac{1}{(1/A_1 F_{1.a}) + (1/A_2 F_{2.a})} \quad (8\text{-}48)$$

The heat flow, analogous to the current flow, is the product of conductance C_0, and potential difference $E_{b_1} - E_{b_2}$, which results exactly in Eqn (8-24) as already derived in Section 8-3:

$$\dot{Q}_1 = \left[A_1 F_{1.2} + \frac{1}{(1/A_1 F_{1.a}) + (1/A_2 F_{2.a})} \right] \sigma(T_1^4 - T_2^4) \quad (8\text{-}24)$$

Suppose now that the surfaces in the furnace are grey, with emissivities and therefore absorptivities ε with the appropriate suffix. At each of these surfaces the total radiation per unit receiving surface inwards from the other surfaces in the system is q_{in} say, and the radiation leaving, q_{out}, is given by

$$q_{out} = \varepsilon E_b + \rho q_{in} \quad (8\text{-}49)$$

where E_b is black body emittance, defined above as the analogue of electrical potential, and ρ is the reflectivity of the surface given by

$$\rho = 1 - \alpha = 1 - \varepsilon$$

The net heat flow from unit area of the surface considered, q, is given by

$$q = q_{out} - q_{in} \quad (8\text{-}50)$$

Eliminating q_{in} between Eqns (8-49) and (8-50) gives the net outflow of radiant energy from a surface of area A as

$$\dot{Q} = qA = \frac{\varepsilon}{1 - \varepsilon} A(E_b - q_{out}) \quad (8\text{-}51)$$

Equation (8-51) gives the net outflow of heat from a grey surface as a function of its emissivity, and the difference between its black body potential E_b and a hypothetical potential corresponding to the total outward flux of energy,

q_{out}, which is conveniently alternatively represented as E_0. The emissivity–area function can be regarded as a notional conductance, so that the furnace network of our example, when the surfaces are grey, becomes as in Fig. 8-17. The nodal potential values of the black body situation of Fig. 8-16 are replaced by the appropriate E_0 values with the conductances to the black body emittance potentials corresponding to the emissivity–area functions of Eqn (8-51).

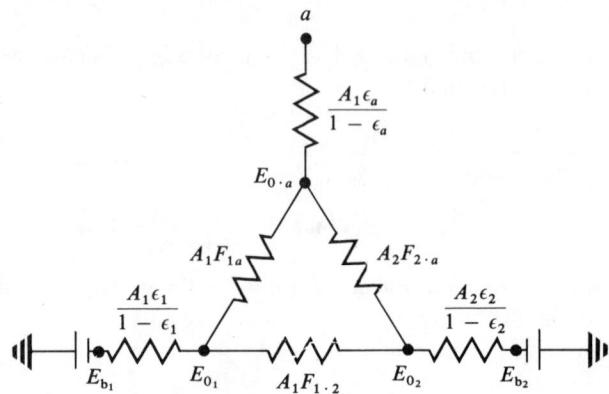

Fig. 8-17 Electric circuit analogue for re-radiating enclosure with grey surfaces

Determining the overall resistance (reciprocal of conductance) from the heat source 1 to the sink 2, and using Hottel's notation $\mathscr{F}$ for the equivalent emissivity–view factor combination as defined in Section 8-5, gives

$$\frac{1}{A_1 \mathscr{F}_{1.2}} = \frac{1}{A_1}\frac{1-\varepsilon_1}{\varepsilon_1} + \frac{1}{C_0} + \frac{1}{A_2}\frac{1-\varepsilon_2}{\varepsilon_2}$$

in which C_0 is the previously determined conductance of the black body network, in this case from E_{0_1} to E_{0_2} as in Eqn (8-48). With some rearrangement we finally obtain, for the overall geometric factor $\mathscr{F}_{1.2}$,

$$\frac{1}{\mathscr{F}_{1.2}} = \left(\frac{1}{\varepsilon_1} - 1\right) + \frac{A_1}{A_2}\left(\frac{1}{\varepsilon_2} - 1\right) + \cfrac{1}{F_{1.2} + \cfrac{1}{[1/A_{1.a}] + [(A_1/A_2)(1/F_{2.a})]}}$$

$$(8\text{-}52)$$

Observe that in this equation the emissivity of the re-radiating surface does not appear, and since this is adiabatic, with no current or heat flow, its potential E_{0_a} is set entirely by the relative magnitudes of the conductances in the $1a2$ circuit.

Note also that if there is no re-radiating surface (that is, $F_{1.a} = F_{2.a} = 0$) and surface 2 encloses surface 1 so that ε_2 and $F_{1.2}$ become effectively unity,

Eqn (8-52) reduces to Eqn (8-36). Equation (8-52) also reduces to Eqn (8-30) for the case when $F_{1.2} = F_{2.1} = 1$ as for large parallel grey plates, so that the derivation of this equation by the analogous circuit method is seen to yield a more general relation for radiant heat transfer between grey bodies.

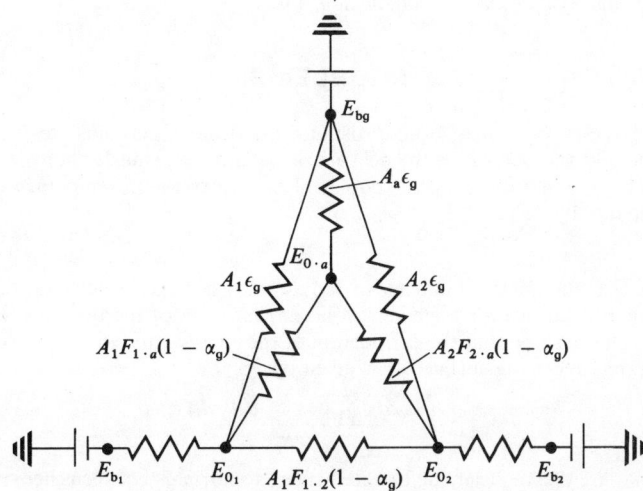

Fig. 8-18   Equivalent electrical network for a grey furnace filled with a grey gas

A further step towards complete generality is obtained by allowing a radiating and absorbing gas to fill the spaces between the heat source, heat sink, and adiabatic re-radiating surfaces. If the gas behaved strictly as described in Section 8-6, with emissivities and absorptivities strongly temperature-dependent, the overall heat transfer problem would be almost impossibly complex. However, the gas will have a mean temperature related to the other temperatures in the system, and if it is assumed to be grey with emissivity and absorptivity, $\varepsilon_g = \alpha_g$, evaluated at this mean temperature, relatively little error is in practice involved. Noting that radiant energy to a gas which is not absorbed is transmitted through it, the equivalent circuit for the present example is modified from the previous case of the grey furnace without the gas, first by the conductances between the grey heat source, sink, and re-radiating surfaces being reduced in the ratio of the transmissivity $(1 - \alpha_g)$. Further, there will be radiant heat transfer between the gas and each of the component surfaces with conductances $A\varepsilon_g$. The gas potential will be E_{bg} (that is σT_g^4), and no grey body modification is required for this potential since the gas, unlike the solid surfaces considered in the derivation of the potential equation [Eqn (8-51)] has no reflective component. The complete equivalent circuit for this general case thus becomes as shown in Fig. 8-18.

REFERENCES

1. Hottel, H. C. Radiant-heat transmission. Chapter 4 of *Heat Transmission* (ed. W. H. McAdams) (3rd edn). McGraw-Hill, 1954.
2. Béer, J. M. and C. R. Howarth. Radiation from flames in furnaces. *Twelfth International Symposium on Combustion*, 1968.

PROBLEMS

8-1 Two large black plates a small distance apart are maintained respectively at 1,000°C and 500°C. Determine the net rate of radiant heat transfer between them in kW/m^2 and show that allowing the cooler plate to increase in temperature to 700° changes the rate by 24%.

Ans: 128·6 kW/m^2

8-2 Show that the normal intensity of radiation, defined as the emittance per unit area of radiating surface through unit solid angle, is $1/\pi$ of the total hemispherical emittance. Hence show that the proportion of the energy emitted from a surface A_1 incident upon a receiving surface A_2 is given by

$$\int_{A_1} \int_{A_2} \frac{\cos \phi_1 \cos \phi_2 \; dA_1 \; dA_2}{\pi x^2}$$

where ϕ_1, ϕ_2 are the angles made by the surfaces to normals between them which are of length x.

8-3 Radiation escapes through a hole 20 mm in diameter from a laboratory black body. It is incident upon a black plate parallel to and coaxial with the hole, 0·2 m in diameter and 0·3 m away. Determine the fraction of the emission from the hole which is intercepted by the plate, as a first approximation by regarding both areas as small compared with the distance between them.

Ans: 0·111

8-4 Obtain a more accurate estimate of the view factor for the conditions of Problem 8-3 by approximating the receiving plate by a spherical cap of radius 0·3 m (so that $\phi_2 = 0$) and of the same projected area. The emitting surface may be assumed small as before.

Ans: 0·100

8-5 A bed of burning coal in a furnace radiates as a plane rectangular black surface, 3 m by 2 m, at 1,500°C, to an opaque bank of black tubes of the same projected area. These are at a surface temperature of 300°C and at such a distance from the fire bed that the view factor is 0·5. Determine the net radiant heat flow to the tube bank and show that enclosing the furnace with adiabatic vertical black walls increases the heat flow by 50%.

Ans: 1,660 kW

8-6 The two parallel plates of Problem 8-1 are maintained as before at 1,000° and 500°C, but now must be considered as imperfect, grey emitters with emissivities 0·8 and 0·5 respectively. Determine the net rate of heat exchange under the new conditions.

Ans: 57·2 kW/m^2

8-7 Suppose the plates of Problem 8-6 are moved sufficiently far apart that the view factor becomes 0·7. Determine the limits between which the net radiant interchange would be expected to lie. Which is likely to be nearer the true value?

Ans: 36·0 and 40·0 kW/m^2

8-8 An exhaust duct 5 ft in diameter conveys gas at 1,000°C and 1 atmosphere with the composition 15% CO_2, 10% H_2O, and 75% N_2 by volume. If the duct walls have an emissivity of 0·8 and operate at a temperature of 250°C estimate the net radiant heat transfer with the gas in Chu/ft^2 h. Use the data of Figs. 8-11 and 8-12 as necessary.

Ans: 5,170 Chu/ft^2 h

8-9 It is proposed to measure the temperature of the gases in a furnace from a thermocouple attached to a disc which is exposed to radiation emitted through a small hole in the furnace wall. This hole is 1 inch in diameter and the disc which is coaxial with the hole and 18 inches from it, is 12 inches in diameter and has an emissivity of 0·8. The disc loses heat by radiation and convection to the surroundings at 60°F, the convective heat transfer coefficient being 1·0 Btu/ft^2 h °F. What temperature will the thermocouple record when the furnace is at 3,000°F?

Ans: 97°F

8-10 A furnace wall, which is lagged on its outer face, comprises two parallel plates between which air at 93°C flows at such a rate as to produce a convective heat transfer coefficient of 57 W/m^2 K. Assuming that heat flows from the inner to the outer plate by radiation only, show that the latter will operate at a temperature of 552°C if the inner plate is maintained at 837°C by the furnace gases. The emissivities of the inner and outer plates are respectively 0·9 and 0·6.

Show also that if a plate of emissivity 0·6 is introduced as a radiation shield between the inner and outer walls of the furnace, the temperature of the latter will fall to 213°C if all other conditions remain unchanged.

8-11 In a gas fire the products of combustion radiate directly to the surroundings and also heat by radiation and convection a ceramic backing plate which thus re-radiates heat through the combustion gases also to the surroundings. Determine the rate at which heat is emitted per square foot from this fire and the surface temperature of the ceramic given the following data:

Mean temperature of combustion products	1,040°F
Emissivity of combustion products	0·2
Emissivity of ceramic	0·7
Convective heat transfer to ceramic	3·0 Btu/ft^2 h °F

The backing plate may be assumed perfectly insulated on its rear side.

Ans: 3,947 Btu/ft^2 h, 710°F

8-12 A separately fired steam superheater comprises a bank of tubes above a furnace in which the temperature is 1,570°F and the total pressure atmospheric. The effective emissivity of the flame in the furnace is 0·5 and the proportions of CO_2 and H_2O by volume each 12·5%. Using the data given below and in Figs. 8-11 and 8-12, determine whether the unit is adequate for the specified duty and the maximum heat flux in the tube bank:

Thermal duty:

Required steam outlet temperature   700°F (enthalpy 1,373 Btu/lb)
Steam inlet conditions   Saturated at 200 lbf/in^2 (enthalpy 1198 Btu/lb)
Rate of steam flow   8,000 lb/h
Rate of gas flow   6,000 lb/h

Tube bank geometry:

Gas flows across tubes; 10 rows of tubes, 8 tubes per row, staggered; tube o.d. 4·0 in, i.d. 3·7 in; length, 5 ft; pitching, 8 in transverse, 6 in longitudinal.

Heat transfer rates:

Steam side convection, 30 Btu/ft^2 h °F; gas side convection, 7 Btu/ft^2 h °F; emissivity of tube surfaces, 0·9; mean specific heat of gases, 0·27 Btu/lb °F.

Ans: Adequate; 6,800 Btu/ft^2 h in first two rows

9
Heat transfer with change of phase

9-1 Regimes of boiling heat transfer

It is a thermodynamic axiom that there exists a temperature, called the *saturation temperature*, dependent only upon the prevailing pressure, at which a liquid and its vapour can exist together in equilibrium. It is an everyday observation that if a liquid is heated to this temperature bubbles of vapour form at the heated boundaries of the liquid mass and move to the free surface, producing as they do so an often violent agitation of the liquid. This agitation enhances the convective motion of the liquid and results in the high rates of heat transfer associated with boiling. The heating of an otherwise quiescent mass of liquid is known as *pool boiling*, so that the motion of the fluid is due only to the convective currents of liquid and vapour, and is akin to natural or free convection. Equally, vapour ebullition may be found in a forced motion of the liquid, analogous to forced convection. In either form of boiling heat transfer the rate of vapour formation can rise to a sufficiently high level that the individual bubbles coalesce to a single film of vapour covering the heated surface. This leads to a sharp fall in the heat transfer coefficient and consequently, if there is no reduction in heat flux, a rise in surface temperature with often drastic results. These three modes of boiling heat transfer—pool boiling, forced boiling, and film boiling—are considered separately in this chapter.

9-1-1 Pool boiling

As an initially quiescent mass of liquid is steadily heated, motion will begin through the action of buoyancy forces on the hotter, and therefore lighter, fluid particles near the heated surface. Experimental observations confirm that at low rates of heat transfer the coefficients conform to the predictions of the theory of natural or free convection, as set out in Chapters 6 and 7, with evaporation of the liquid occurring only at the free surface and having no effect upon the heat transfer processes. As the heating rate is increased, however, the temperature of the heated surface rises towards the saturation condition of the liquid mass, and bubbles of vapour form locally. On detaching from the heated surface, through the action of the high vapour–liquid buoyancy forces, these bubbles collapse in the cooler main liquid bulk. Nevertheless, their brief existence affects the heat transfer process since liquid must move into the space vacated by the bubbles on the heated surface, so that the convective motion is augmented compared with the situation with no change of phase. This is known as the *subcooled* or *local* boiling regime.

Further increase in the rate of heating leads to a general rise in temperature in the complete system until finally the whole liquid mass reaches saturation temperature. In this condition the bubbles formed at the heated surface persist throughout their motion to the free surface, indeed often growing through further evaporation or as a result of reducing hydrostatic pressure. Under these conditions the formation of vapour affects the motion of the whole liquid mass, and it is with this *nucleate boiling* or *bulk boiling* regime that the highest heat transfer coefficients are associated. Under favourable conditions heat fluxes in excess of 10^6 Btu/ft^2 h or 3,000 kW/m^2 can be obtained with temperature differences between heated surface and liquid bulk saturation temperature of less than 100°F or 55°C.

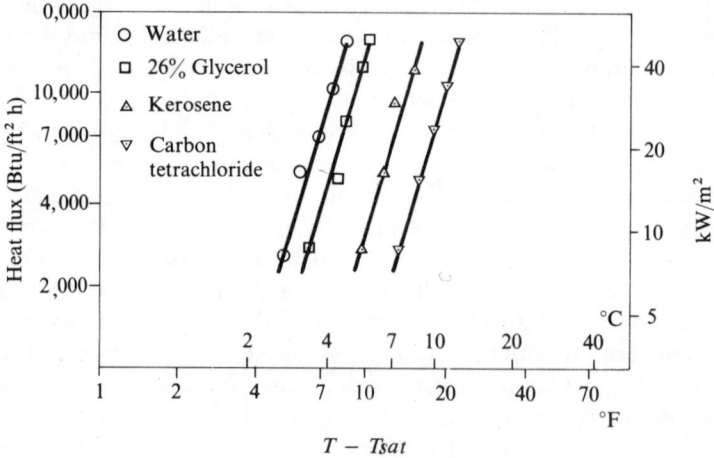

Fig. 9-1   Pool boiling data at atmospheric pressure (after Cryder and Finalborgo[2])

It will be clear that analysis of the full nucleate boiling processes poses severe and, to date, intractable difficulties. The hydrodynamics of the creation and motion of the bubbles themselves are not fully understood. Further, the existence of 'preferred' sites on the heated surface for the initiation of vapour bubbles, which can be observed in even the most primitive boiling experiment, shows that the physical nature of this surface plays an important part in determining the magnitude of pool boiling heat transfer coefficients. Thus, any analysis or attempt at correlating boiling heat transfer data which omits a model of the state of the heated surface is neglecting what, in many situations, may be the controlling influence on the heat transfer process.

Even if the bubble motion were fully understood the consequent movement of the liquid mass would require a highly sophisticated analysis. It would need to deal with an irreversible time-variant motion in three dimensions, and it is this motion which in the end determines the magnitude of the heat transfer rates. To demonstrate this, Tong[1]† calculates from experimental observations that in a typical high heat flux situation only about 2 per cent of the heat flow is accounted for as latent heat of evaporation in the vapour bubbles; thus 98 per cent of the heat flow is due to the consequent convective liquid motion.

Empirical pool boiling heat transfer data, to which recourse must inevitably be made for design purposes, are usually presented as graphs of heat flux against the difference between the temperature of the heated surface, T, and the liquid saturation temperature, T_{sat}, corresponding to the prevailing pressure. Typical data for the nucleate boiling of a range of liquids at atmospheric pressure are shown in Fig. 9-1,[2] while the effect of pressure on pool boiling of water is shown in Fig. 9-2.[3]

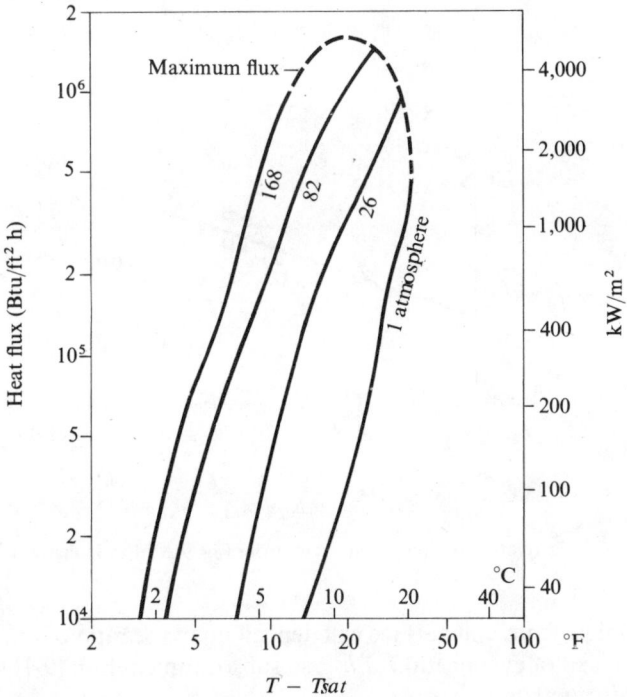

Fig. 9-2 Effect of pressure on pool boiling in water (after Addoms[3])

Attempts to obtain a general correlation of data of this type have had the most (but still limited) success when representing the bubble motion at the instant of departure from the heated surface as a 'bubble' Reynolds number, $\mathbf{R}_b$, defined as

$$\mathbf{R}_b = \frac{d_b G_b}{\mu_L} \tag{9-1}$$

† Full details of references cited are given at the end of the chapter.

where d_b is the bubble diameter at separation, G_b the vapour mass velocity or mass flow per unit cross-sectional area from the surface, and μ_L the viscosity of the liquid.

The bubble diameter at departure is a function of the separating buoyancy forces, and thus of the difference in density between vapour and liquid, $\rho_L - \rho_v$, and of the resisting forces due to the surface tension of the liquid, σ. Fritz[4] derived the relationship between these variables as

$$d_b \propto \sqrt{\frac{2\sigma}{g(\rho_L - \rho_v)}} \qquad (9\text{-}2)$$

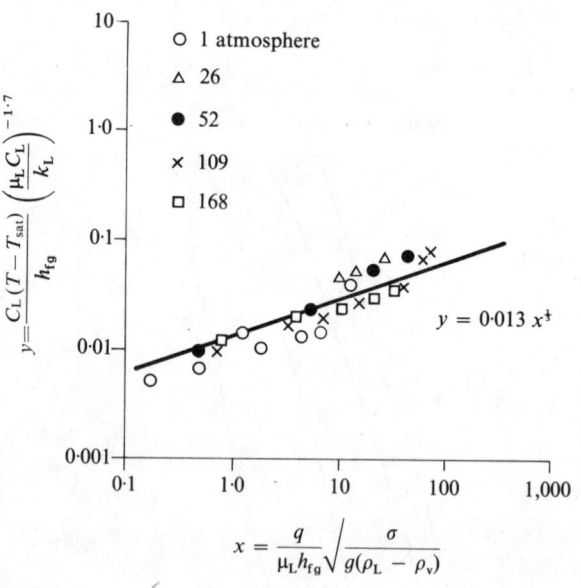

Fig. 9-3 Correlation of pool boiling data from Fig. 9-1 (after Rohsenow[5])

The vapour flow from unit surface will depend upon the ratio of the heat flux to the latent heat of evaporation, $\dot{q}/h_{fg}$, so substituting in Eqn (9-1) gives, for the Reynolds number,

$$\mathbf{R}_b \propto \frac{\dot{q}}{\mu_L h_{fg}} \sqrt{\frac{\sigma}{g(\rho_L - \rho_v)}} \qquad (9\text{-}3)$$

The heat transfer coefficient is represented as a Stanton number, that is,

$$\frac{\dot{q}}{(T - T_{sat})G_b C_L} = \frac{h_{fg}}{(T - T_{sat})C_L} \qquad (9\text{-}4)$$

in which C_L is the specific heat of the liquid. The remaining liquid properties can be represented as a conventional Prandtl number, $\mu_L C_L/k_L$, and Rohsenow, in this way was able to correlate pool boiling data for a number of liquid–surface combinations.[5] Figure 9-3 shows the effectiveness of the correlation for water over a range of pressures from 1 to 168 atm, and the mean line through the experimental points from reference (3) corresponds to

$$\frac{C_L(T - T_{sat})}{h_{fg}} = C_{sf} \left\{ \frac{\dot{q}}{\mu_L h_{fg}} \sqrt{\frac{\sigma}{g(\rho_L - \rho_v)}} \right\}^{1/3} \left(\frac{\mu_L C_L}{k_L}\right)^{1.7} \qquad (9\text{-}5)$$

in which the proportionality constant C_{sf} has the value 0·013 for the water–platinum surface used in Addoms's experiments[3] shown in Fig. 9-3. For other liquid–surface combinations, C_{sf} ranges between 0·003 and 0·015, and its value is clearly dependent upon many factors which are only partially understood. Certainly the roughness or smoothness of the heated surface, as has been earlier implied, will be important since it is believed that minute crevasses form the sites on which the first vapour bubbles form. If these are not present and the nucleating radii conform to the predictions of classical thermodynamics,[6] much greater superheat temperature differences $(T - T_{sat})$ are necessary than are observed to be the case with surfaces of conventional 'engineering' finish. Equally, the presence of dissolved gases in the heated liquid can affect bubble formation by providing, as nucleation sites, bubbles of gas driven out of solution by the heating process. The roles of such effects as these are still not fully understood, but heat transfer coefficients in pool boiling are usually sufficiently high not to be controlling in most engineering situations, and a large margin of uncertainty is rarely critical.

9-1-2 Boiling in forced flow

By the same method as the lower rates of heating produce heat transfer coefficients in pool boiling in accord with the predictions of simple natural convection, so does gradually increased heating in forced boiling begin effectively as simple forced convection. As might be expected, however, the hydrodynamic development of forced boiling flows is more complex than in pool boiling of a quiescent mass of liquid. Figure 9-4 shows how the physical appearance of a forced boiling flow changes as its enthalpy is raised during its passage along the heated surface.

Liquid enters at the lower end of the tube, its forced motion resulting from direct pumping, as in a forced circulation boiler, or from the difference in mean fluid density between the heated tube and an unheated 'downcomer' feeding liquid to the bottom of the tube, as in a natural circulation boiler. After a period of forced convection in which evaporation is zero or negligible, and of an extent depending upon the amount by which the entering liquid is *subcooled* below the saturation temperature, T_{sat}, *local* or *subcooled* boiling

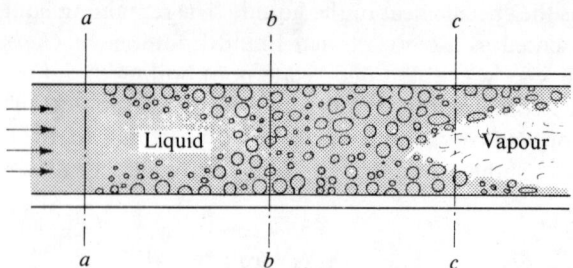

Fig. 9-4   Regimes of forced boiling flow

will begin at a section such as *a–a*. Vapour bubbles will form at the wall but will not carry into the main liquid stream, whose bulk temperature is still below the saturation condition. At a section *b–b* the continuing heating will have raised the bulk temperature to T_{sat} and full nucleate or bulk boiling will have commenced, as in the pool boiling situation. In the forced flow, however, which is necessarily constrained into a limited cross-sectional area, the reduced mean fluid density will have produced a substantial acceleration of the flow. In the streamwise pressure gradient the less dense vapour phase acquires a higher velocity than the liquid and begins to stratify into a separate region, usually in the core of the flow, as shown from section *c–c* in Fig. 9-4. In certain circumstances, particularly at lower saturation pressures where the difference between liquid and vapour densities is greatest, the very high core velocity appears to stabilize the annular film of liquid on the walls, and to suppress nucleation within it so that the heat transfer characteristics revert to those associated with forced convection.

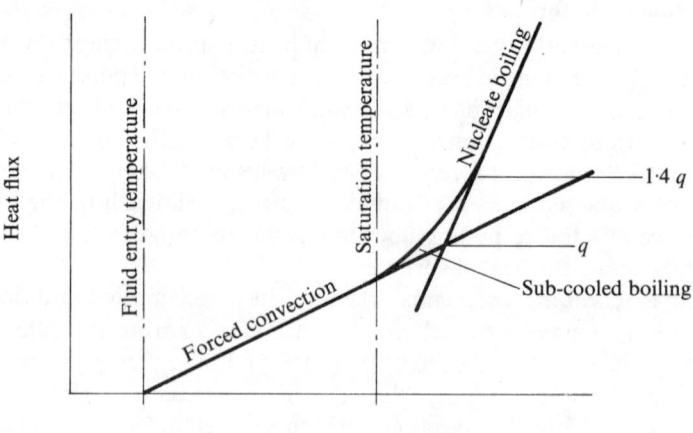

Fig. 9-5   Boiling curve for forced flow

There is not a clearly defined, universal dividing line between the separate regimes of forced flow boiling, as might be expected in a hydrodynamic and thermodynamic process dependent upon so many imperfectly understood variables, although certain broad guidelines may be laid down.

It is assumed that full nucleate forced boiling is little different from pool boiling in terms of the heat flux–temperature difference relationship. Thus, for a given set of forced convection conditions (that is, liquid velocity and physical properties in a duct of specified geometry) the heat flux–wall temperature plot of Fig. 9-5 may be drawn, using one of the correlations from Chapter 7, to represent the region before boiling commences. Superimposed on this can be the corresponding variation of heat flux with wall temperature for full nucleate pool boiling, also as in Fig. 9-5. The placing of this line will be independent of fluid velocity—as is generally found in forced boiling in the nucleate region such as b–c of Fig. 9-4—implying that our initial assumption is not far wrong. Forster and Grief[7] suggest that full nucleate boiling is established where the heat flux is 40 per cent above that corresponding to the flux at the intersection of the convection and nucleate boiling curves. Thus, the intervening region, as shown in Fig. 9-5, may be taken to correspond to the subcooled boiling region (a–b of Fig. 9-4). Theoretical and empirical procedures for developing the heat flux curve in this region have been developed, but in view of the uncertainties of the whole boiling process, such an exercise seems not to be worthwhile at this stage of our knowledge.

The position of the transition between the full nucleate or bulk boiling region and the reversion to the annular flow system, in which nucleation is partially suppressed, depends clearly upon the operating pressure. Thus, at low saturation pressures, where the difference between liquid and vapour densities is greatest, such a transition is likely at a lower dryness fraction χ (mass of vapour/mass of vapour and liquid) than at high pressures. The transition between the nucleate boiling regime and the annular convecting layer regime is demonstrated in Fig. 9-6. This figure shows heat transfer data obtained by Bennett and his co-workers[8] in experiments on the forced boiling of water. The observed heat transfer coefficients are represented in the ordinate as the multiple of the value that would result from forced convection only in the liquid, with no boiling present. The abscissa employs the so-called Martinelli parameter, X_{tt}, which represents the vapour–liquid mass ratio in terms of the dryness fraction, together with the vapour–liquid density ratio and the viscosity ratio in a form which has been found empirically to correlate hydrodynamic data on forced two-phase flows.[9] The results of reference (8) are presented in Fig. 9-6 using the Martinelli parameter in inverse form, so that the left region corresponds to low vapour mass or dryness fractions. In this region the heat transfer coefficients are observed to be principally dependent upon the heat flux, and virtually independent of the total mass rate of fluid flow. At high dryness fractions, however, as at the right of the figure, the heat transfer coefficients become essentially dependent upon the rate of vapour flow, showing the reversion to a forced convection type of heat transfer mechanism.

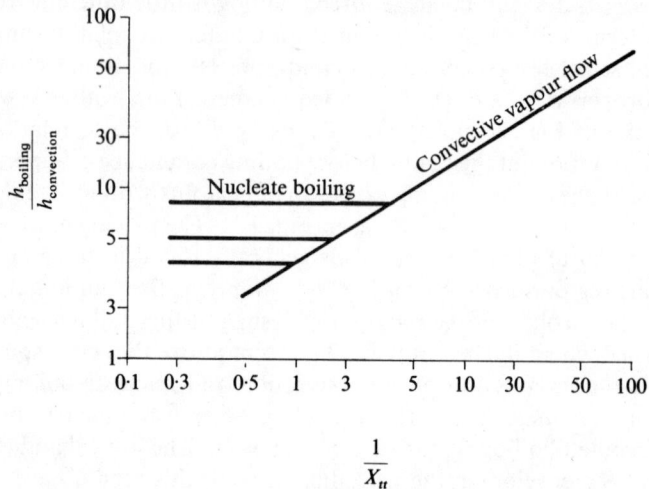

Fig. 9-6   Correlation for heat transfer in forced boiling flow (after Bennett *et al.*[8])

The transition region is clearly dependent upon the operating conditions, even in the limited range of variables shown in reference (8), but observe the effect of pressure if it is supposed that a single value of the Martinelli parameter represents a given heat flux and total mass flow. Thus, taking the mean condition of transition between nucleate boiling and the convective regime to be when $1/X_{tt}$ has the value 2·5, shows that at a saturation pressure of 1 atm (represented approximately as 1 bar $= 10^5$ N/m²)

$$\frac{1}{X_{tt}} = \left(\frac{\chi}{1-\chi}\right)^{0·9}\left(\frac{\rho_L}{\rho_v}\right)^{0·5}\left(\frac{\mu_L}{\mu_v}\right)^{0·1} = 2·5 \qquad (9\text{-}6)$$

that is, in SI units,

$$\left(\frac{\chi}{1-\chi}\right)^{0·9} = 2·5 \times \left(\frac{0·59}{956}\right)^{0·5}\left(\frac{1·3 \times 10^{-5}}{2·8 \times 10^{-4}}\right)^{0·1}$$

whence $\qquad\qquad\qquad \chi = 0·031$

At a pressure of 100 bar, however,

$$2·5 = \left(\frac{\chi}{1-\chi}\right)^{0·9} \times \left(\frac{55·5}{685}\right)^{0·5} \times \left(\frac{3·3 \times 10^{-5}}{1·8 \times 10^{-4}}\right)^{0·1}$$

and $\qquad\qquad\qquad \chi = 0·36$

These values should not be taken to represent absolutely the dryness fraction at which nucleate boiling in forced flow will be suppressed in the annular liquid film, and almost certainly they underestimate the vapour mass fraction up to which nucleate conditions will apply. There is a clear need for further research to determine exactly the important factors that control transition between the different regimes of forced boiling flow and, indeed, of the absolute levels of heat transfer coefficient under such conditions. Just as with pool boiling, however, these are rarely 'controlling' coefficients in engineering practice, and usually the designer's principal task is to ensure that the maximum heat flux is not exceeded, with the possibly disastrous consequences of film boiling, as discussed in the next section.

9-1-3 Film boiling

Figure 1-4 has shown how the heat flux associated with boiling heat transfer varies over a wide range of the driving temperature difference. Although from the more detailed study of boiling mechanisms in this chapter it is clear that Fig. 1-4 is an oversimplification of the processes involved, its salient features are generally valid and, in particular, the clearly defined maximum heat flux is associated with all conditions of boiling heat transfer. The magnitude of this maximum flux, and the prediction of its occurrence, pose the really critical problems of a study of boiling heat transfer.

The existence of a maximum heat flux in boiling is logically simple to conceive, particularly in the light of the role of the nucleation process studied so far in this chapter. While a continuing rise in the heat supply will, for a period, increase the agitation of the liquid mass—and, hence, the convective heat transfer rate—it is clear that the increasingly more numerous nucleation sites must ultimately coalesce to a continuous film of vapour. This will restrict, and ultimately prevent, access of liquid to the heated surface, with a consequent reduction in the heat transfer rate. In processes in which the heat flux is virtually independent of the surface temperature (as with electric resistance heating or nuclear generation of energy), or from a source at a much higher temperature (as in a furnace), the reduction in the rate of heat extraction leads to a violently unstable situation in which rising surface temperature further reduces the rate of heat removal. The often catastrophic consequences of this instability have engendered dramatic technological terms for the transition to *film boiling*, such as *the boiling crisis*, and *burnout*.

Although the logical conception of a film boiling regime, with its reduced rates of heat transfer, is simple enough, the controlling features of the mechanism which determines the exact conditions of its onset are very complex. To the complicated mechanism of the nucleate boiling process must be added the physical conditions of the reactions between the vapour bubbles. In pool boiling these lead in practice to an initial transient situation in which there may be large fluctuations in surface temperature as coalescing bubbles intermittently separate from the heated surface. Finally, however, a stable, continuous film of vapour is formed, known as the *Liedenfrost state*.

In forced boiling there are also the different flow regimes in which film boiling may commence. Thus, in a fully nucleate situation associated with low vapour quantities, as we saw in the last section, transition to film boiling may follow broadly the mechanism of the equivalent pool boiling processes, and be associated with very large heat fluxes. In the annular flow regime, however, possibly with already suppressed nucleation, the transition mechanism will be quite different and will occur at lower heat fluxes. Under this latter condition, necessarily associated with a high vapour core velocity and the consequential convective cooling, the final onset of film boiling will not usually produce as dramatic a discontinuity in heat transfer rate as from full nucleation; for this reason this phenomenon is known sometimes as *slow burnout*.

Although attempts have been made to analyse the mechanism of transition from unrestrained nucleate boiling to stable film boiling, notably by Zuber[10], for the purposes of engineering design recourse must inevitably be made to empirical correlations of experimental data. Many of these exist, of varying degrees of complexity, and attempt to relate the peak heat flux, $\dot{q}_{max}$, to the relevant physical properties of the system. One of the most successful in relating data over a wide range of conditions is due to Kutateladze[11] and is based upon a dimensional analysis of the saturated pool boiling process. It takes the form

$$\dot{q}_{max} = K_f\{h_{fg}(\rho_v)^{1/2}[\sigma g(\rho_L - \rho_v)]^{1/4}\} \tag{9-7a}$$

The constant K_f is found to vary between 0·13 and 0·19 for a range of surfaces. Taking the mean value of 0·16 gives, for the critical heat flux in water at 1 atm in an acceleration field of 1g:

$$\dot{q}_{max} = 0·16\{2,258 \times (0·59)^{1/2} \times [0·06 \times 9·81 \times (956 - 0·59)]^{1/4}\}$$

$$= 1,350 \text{ kW/m}^2$$

In the foot–pound–pound-force system Eqn (9-7a) becomes†

$$\dot{q}_{max} = K_f\{h_{fg}(\rho_v)^{1/2}[\sigma g g_c(\rho_L - \rho_v)]^{1/4}\} \tag{9-7b}$$

$$= 0·16\{971 \times (0·037)^{1/2} \times [0·004 \times 32·2 \times 32·2(59·8 - 0·037)]^{1/4}\}$$

$$= 119 \text{ Btu/ft}^2 \text{ s}$$

$$= 430,000 \text{ Btu/ft}^2 \text{ h}$$

† The unusual juxtaposition of g, the magnitude of the acceleration field in which the boiling process takes place, and g_c, the Newtonian constant in the pound–pound-force system, should be noted. The resulting dimensional consistency of Eqn (9-7b) is shown as

$$\dot{q}_{max} \sim \frac{\text{Btu}}{\text{lb}}\left(\frac{\text{lb}}{\text{ft}^3}\right)^{1/2}\left[\frac{\text{lbf ft}}{\text{ft s}^2}\frac{\text{lb ft}}{\text{lbf s}^2}\frac{\text{lb}}{\text{ft}^3}\right]^{1/4}$$

$$\sim \frac{\text{Btu}}{\text{ft}^2 \text{ s}}$$

At the higher pressure of 100 atm, Kutateladze's criterion predicts that $\dot{q}_{max}$ will have increased to 5,200 kW/m^2 or $1\cdot65 \times 10^6$ Btu/ft^2 h, since the increased vapour density, ρ_v, more than counteracts the reduced surface tension, σ, and the lesser buoyancy forces due to the difference in liquid and vapour densities, $\rho_L - \rho_v$. Observe that the form of the correlation suggests that there will, however, be a stationary point in the correlation between maximum nucleate boiling flux and pressure, for at the critical pressure the buoyancy forces and the surface tension will have reduced to zero. This turning point has been observed in organic fluids and water[3,12] and appears to occur at about one-third of the critical pressure when the heat flux at the inset of film boiling is approximately four times the corresponding value at 1 atm.

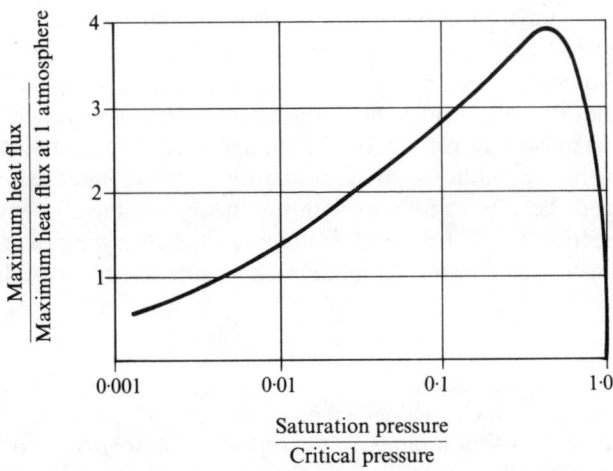

Fig. 9-7 Effect of pressure upon maximum heat flux in pool boiling according to Eqn (9-7a) (after Kutateladze[11])

The predictions of the Kutateladze equation [Eqn (9-7a)] for water are shown in Fig. 9-7 and are seen to be in excellent accord with these observations. Its essentially empirical basis must be borne in mind, however, and the consequent dangers of applying the equation to conditions outside those of the experiments used to quantify it must be emphasized. The observed variations in the constant K represent a margin of uncertainty in predicting $\dot{q}_{max}$ of ±20 per cent for saturated pool boiling conditions. Boiling in a sub-cooled mass of liquid allows a higher maximum heat flux usually to be attained before the onset of film boiling. Any restriction of access to the heated surface, on the other hand, as when the 'pool' is constrained within a tube, leads to a drastic reduction in the maximum achievable heat flux. The correlation proposed in reference (13) for boiling in short tubes of the order of $\frac{1}{2}$ in (12 mm) in diameter suggests that maximum heat fluxes less than one-tenth of the values predicted for unrestricted pool boiling are likely.

In forced boiling flows, as has been implied earlier in this section, the limitation on allowable heat flux is more likely to be set by the hydrodynamic characteristics of the flow containing a substantial fraction of vapour, and here the consequences of exceeding the maximum are less dramatic. The sharp reduction in rate of heat transfer from a very high heat flux occurs if at all in forced flows usually in a region where the average or bulk fluid temperature is less than the saturation value, that is, in the subcooled regions like a–b of Fig. 9-4. Under these conditions maximum heat fluxes are found to increase with mass velocity and with the degree of subcooling and very high values can be achieved which are rarely controlling in an engineering design. The correlations for maximum heat flux in pool boiling may be used to give a conservative estimate for many forced flow systems in which stable film boiling is considered to be a possible situation.

Stable film boiling, as has been implied, is usually to be avoided in practice, although it is sometimes inevitable, particularly in cryogenic systems. There is little information available for the accurate estimate of heat transfer rates under these conditions. By considering the shear interaction between the liquid and the vapour film, as with the theory of film condensation discussed in Section 9-2-1, Bromley[14] obtained the following expression for the heat transfer coefficient in film boiling on a curved surface of diameter d:

$$h = K_c \left[\frac{g\rho_v(\rho_L - \rho_v)h_{fg}k_v^3}{\mu_v d\,\Delta T} \right]^{1/4} \tag{9-8}$$

K_c is a constant for which a value of 0·62 was found to represent satisfactorily the observations for film boiling on the outside of a horizontal tube. For a plane surface Eqn (9-8) is clearly inapplicable since h does not go to zero in this situation. Berenson[15] has suggested that for this case d should be replaced by the bubble diameter, as given by Eqn (9-2), when the constant of proportionality in this equation changes K_c in Eqn (9-8) to 0·43. Clearly, however, estimates of heat transfer rates in film boiling, when radiant heat transfer often becomes significant as well, should be treated with reserve.

9-2   Condensation of a pure vapour

If a vapour at a given pressure is brought into contact with a surface at a temperature less than the corresponding saturation temperature, by however small an amount, condensation to the liquid phase will immediately occur. The resistance to heat transfer by condensation is therefore zero, and the corresponding heat transfer coefficients are infinite. (Condensation, therefore, offers the only truly reversible heat transfer process available to thermodynamics.)

In practice, if condensation is to occur continuously over a surface from which the latent heat is removed by the action of some cooling process on its rear face, then it will be found necessary to maintain a finite, yet usually small, difference in temperature between the surface and the vapour saturation condition. That the resistance to heat transfer coefficient is not infinitely large results from the presence of a *film* of condensate through which the latent heat must be conducted to the cooled surface. Under certain conditions it is possible to break up this film into droplets between which the effectively infinite coefficients of true condensation apply, thus increasing the average coefficient of heat transfer over the whole surface. This situation is known as *dropwise condensation*, in contrast to *film condensation* when the condensate forms as a continuous layer over the cooled surface.

If the vapour is not pure—that is, it is mixed with a different constituent which is not condensable at the prevailing conditions—then before the vapour can come into contact with the cool surface it must diffuse through the non-condensing constituent. The resistance to this diffusion process is always finite and usually high, and to determine its effect upon the overall condensation heat transfer process we must invoke the principles of matter or mass transfer, referred to in Chapters 1 and 5. As will be seen in Section 9-3 the corresponding overall coefficients of heat transfer, when condensation takes place in the presence of a non-condensing constituent, are much reduced compared with the very high values that prevail in a pure vapour.

9-2-1 Film condensation

The basic principles of conduction set out in Chapters 1 and 2 can be simply applied to calculate the resistance of a continuous film of condensate to heat transfer if its thickness is determined from the hydrodynamic conditions applying. Condensation most commonly occurs on stationary surfaces, and any forced motion of the condensate film can only result from viscid interaction between it and the vapour flow over it. This interaction is usually small compared with the gravitational forces acting on the liquid which, in most practical situations, control the condensation process.

As the basic example of the effect of these forces, consider the film of condensate on the vertical plane surface of Fig. 9-8. At a section, distance y from the top of the surface, the film is supposed to have thickness δ, and the boundary condition, since the film begins at the top edge of the plate, will be that $\delta = 0$ where $y = 0$. To determine the functional relation between δ and y consider the equilibrium of the elementary length of film, dy. This will be acted upon by the gravitational forces—which will be resisted by the viscid shear within the liquid film—and, if the relative vapour flow were upward, by the shear at the film boundary. If the vapour flow were downward, this shear would add to the gravitational forces, but we shall suppose (as implied above to be the usual case) that this vapour–liquid interaction is negligible.

Dynamic (that is, acceleration) and surface tension effects in the system are also assumed to be negligible, so that the equation of motion reduces to a balance between gravity and internal liquid shear.

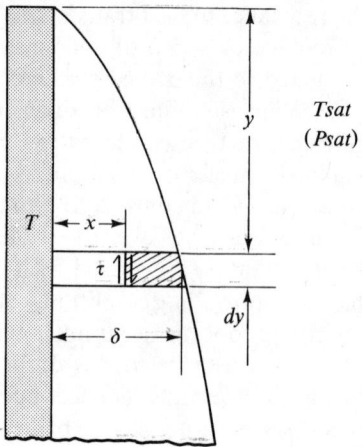

Fig. 9-8   Film of condensate on a plane surface

At a section, distance x from the wall, the shear in the condensate is τ, say, which must resist the gravitational pull on the mass of film between x and δ (that is, for a width of film z in the direction normal to x and y):

$$\tau z \, dy = g(\rho_L - \rho_v)z \, dy(\delta - x) \tag{9-9a}$$

This equation takes account of the buoyancy of the liquid in the vapour by using the density difference, $\rho_L - \rho_v = \rho$, say. Of course, in most practical examples ρ_v is very much less than ρ_L which may normally be used with acceptable accuracy for ρ. The shear stress, τ, can only result from a velocity gradient in the fluid, and the direct proportionality of laminar flow will be assumed, so that

$$\tau = \mu_L \frac{dv}{dx}$$

and Eqn (9-9a) becomes

$$\mu_L \frac{dv}{dx} = g\rho(\delta - x) \tag{9-9b}$$

Integrating this equation with the 'no-slip' condition that $v = 0$ at $x = 0$ gives the distribution of velocity v through the film of condensate as

$$v = \frac{g\rho}{\mu_L}\left(\delta x - \frac{x^2}{2}\right) \tag{9-10}$$

(As an exercise to demonstrate the nature of the terms omitted in this simple derivation, the reader may derive Eqn (9-10) from the general equations of fluid motion in Chapter 5 where the body force, F_y, in Eqns (5-15) is replaced by ρg for this special case of the motion of a liquid film under gravity.)

From Eqn (9-10) the mass rate of flow of condensate past the section y is determined as

$$\dot{m} = \rho_L \int_0^\delta vz \, dx$$

$$= \frac{g\rho^2}{\mu_L} \frac{\delta^3}{3} z \tag{9-11}$$

if ρ_L is assumed equal to ρ, that is, ρ_v is negligible.

Now, over the elementary length of film dy the change in the rate of liquid flow must be related to the rate at which vapour has condensed; that is, to the rate of heat flow through the element of condensate film to the surface which is at a temperature ΔT less than the vapour saturation temperature, T_{sat}. Assuming a linear distribution of temperature through the film gives, for the heat flow rate,

$$d\dot{Q} = k_L z \, dy \frac{\Delta T}{\delta} \tag{9-12a}$$

which must equal the latent heat of the mass $d\dot{m}$ condensed between y and $y + \delta y$; that is,

$$d\dot{Q} = d\dot{m} \, h_{fg} \tag{9-12b}$$

Eliminating $d\dot{Q}$ between these two equations gives, for the change in rate of mass flow,

$$d\dot{m} = \frac{k_L z \, dy}{h_{fg}} \frac{\Delta T}{\delta} \tag{9-13}$$

Differentiating Eqn (9-11) gives

$$d\dot{m} = \frac{g\rho^2 z}{\mu_L} \delta^2 \, d\delta \tag{9-14}$$

so that equating (9-13) and (9-14) leads to

$$\delta^3 \, d\delta = \frac{\mu_L k_L \, \Delta T}{g\rho^2 h_{fg}} dy \tag{9-15}$$

Integrating Eqn (9-15) with the boundary condition that $\delta = 0$ at $y = 0$ gives, for the thickness of the condensate film,

$$\delta = \sqrt[4]{\left(\frac{4\mu_L k_L \, \Delta T \, y}{g\rho^2 h_{fg}} \right)} \tag{9-16}$$

The local heat transfer coefficient at the section y is given by

$$h_y = \frac{d\dot{Q}}{z\,dy\,\Delta T}$$

$$= \frac{k_L}{\delta}$$

$$= \sqrt[4]{\frac{g\rho^2 h_{fg} k_L^3}{4\mu_L\,\Delta T\,y}} \tag{9-17a}$$

Over a finite height, Y, of condensing surface the average coefficient of heat transfer $\bar{h}$, is given by

$$\bar{h} = \frac{1}{Y}\int_0^Y h_y\,dy$$

$$= \tfrac{4}{3}h_y$$

$$= 0.943 \sqrt[4]{\frac{g\rho^2 h_{fg} k_L^3}{\mu_L\,\Delta T\,Y}} \tag{9-1·7b}$$

If the condensing surface is inclined at angle ϕ to the vertical, the gravitational term in Eqns (9-9) becomes $g\cos\phi$ and Eqn (9-17b) is correspondingly modified. For condensation on a horizontal circular tube the gravitational force varies around the perimeter and the heat transfer coefficient is found by integrating between $\phi = 0$ and $\phi = 180°$. For this condition Nusselt[16] showed that the average heat transfer coefficient is given by an equation similar to (9-17b), that is,

$$\bar{h} = 0.73 \sqrt[4]{\frac{g\rho^2 h_{fg} k_L^3}{\mu_L\,\Delta T\,d}} \tag{9-18}$$

where d is the diameter of the tube. Equating Eqns (9-18) and (9-17b) shows that average heat transfer coefficients due to filmwise condensation on a vertical plate of height Y and on a horizontal tube of diameter d will be the same if Y/d is about 2·76. Thus, the perhaps surprising result is obtained that if a tube of greater length/diameter ratio than this value is turned from the horizontal to the vertical position, there is a reduction in the rate of heat transfer by condensation over it.

Equations (9-17a) and (9-18) give a remarkably accurate prediction of filmwise condensation rates, especially in view of the assumptions involved in their derivation. Corrections can be made for the effect of vapour drag on the condensate film, and for the 'sensible' heat transfer required to cool the condensate to the bulk mean value, which is additional to the latent heat only allowed in Eqn (9-12b) in the above derivation. The linear temperature distribution through the film may also be substituted by the true variation which allows for the distortion due to the film velocity. Only very rarely, however, are these corrections to the simple theory significant. Rippling of the condensate film may be a more important effect, and where this occurs it appears to lead to an increase in mean heat transfer coefficient.

The greatest errors in the predictions of the above simple theory result from two important departures from the assumed conditions. First, in deriving Eqn (9-9b), the linear newtonian relationship between shear stress and velocity gradient has been used, which is only valid for *laminar* flow in the condensate film. A Reynolds number for the flow of a film of liquid over a plate of width z may be defined as

$$\mathbf{R}_f = \frac{d_e G}{\mu_L} \tag{9-19a}$$

$$= \frac{1}{\mu_L} \frac{4 z \delta \dot{m}}{z} \frac{\dot{m}}{z \delta}$$

$$= \frac{4}{\mu_L} \left(\frac{\dot{m}}{z} \right) \tag{9-19b}$$

In Eqn (9-19a) d_e is the equivalent diameter of the fluid system defined as

$$d_e = 4 \times \frac{\text{cross-sectional area of fluid flow}}{\text{wetted perimeter}}$$

and G is the mass velocity, that is the mass flow per unit cross-sectional area. In terms of the film Reynolds number $\mathbf{R}_f$, Eqn (9-17b) becomes, from Eqn (9-11)

$$\bar{h} = \frac{4}{3} h_Y$$

$$= \frac{4}{3} \frac{k_L}{\delta}$$

$$= \frac{4}{3} k_L \left(\frac{g \rho^2 z}{\mu_L \dot{m}} \right)^{1/3}$$

$$= \frac{4}{3} \left(\frac{g \rho^2 k_L^3}{\mu_L^2} \frac{4}{4} \frac{\mu_L z}{\dot{m}} \right)^{1/3}$$

which may be written as

$$\bar{h} \left(\frac{\mu_L^2}{g \rho^2 k_L^3} \right)^{1/3} = 1 \cdot 47 \mathbf{R}_f^{-1/3} \tag{9-20}$$

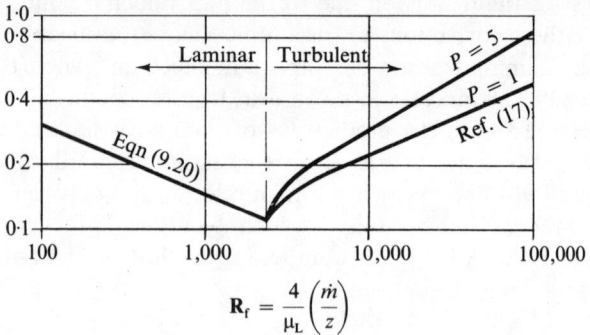

$$\mathbf{R}_f = \frac{4}{\mu_L}\left(\frac{\dot{m}}{z}\right)$$

Fig. 9-9   Heat transfer coefficients in condensation

If experimental observations of condensing heat transfer rates are plotted on this basis, as in Fig. 9-9, Eqn (9-20) is followed satisfactorily until the Reynolds number $\mathbf{R}_f$ reaches a value near 2,000. At higher rates of condensation there is a clear change in the nature of the relationship, due to turbulence in the liquid film, leading to higher heat transfer rates than are predicted by the laminar relationship for shear stress. One of the models for turbulent transport properties described in Chapters 6 and 7 may be used for a detailed analysis of this regime, but Fig. 9-9 shows the predictions of an early semi-empirical analysis by Colburn[17] which is in satisfactory agreement with experimental data and shows, as in forced convection heat transfer, a dependence upon the liquid Prandtl number.

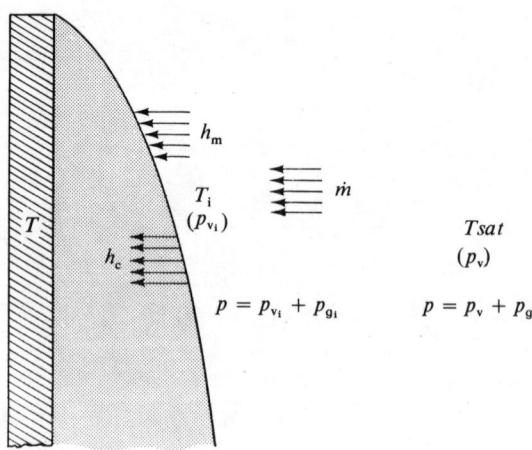

Fig. 9-10   Condensation through a non-condensable gas

9-2-2 Dropwise condensation

The most important departure in practice between the predictions of the simple theory of filmwise condensation in the previous section and experimental observations arises from the breakdown of the continuous film of condensate. This film, as was seen in the introduction to this chapter, presents the only measurable resistance to heat transfer in condensation, so that its breakdown results in sharply increased average coefficients of heat transfer. The continuity of the condensate film can be broken by suitable treatment of the condensing surface, to prevent the liquid 'wetting' it. Under these conditions the condensate coalesces into droplets which rapidly become heavy enough to slide off the condensing surface. Between the droplets bare surface is presented to the vapour over which, as we have seen, the resistance to heat transfer is negligible. While the dropwise mode of condensing heat transfer occurs, coefficients are observed which are, on average, ten times greater than those observed with the equivalent filmwise heat fluxes. Unfortunately it has not so far proved possible to maintain continuously the dropwise mode of condensation. Although certain surface treatments will result in the associated high rates of heat transfer for substantial periods of time, in the end, on metallic surfaces, the promoter is dissipated and there is a return to the filmwise mode. There is also the problem that, although the required surface treatments involve only minute amounts of the appropriate agents (coatings of one or two molecules thickness are often adequate), the very high purities required of the condensate, particularly in modern steam power plant, are significantly contaminated by the continual washing away of the necessary surface coating. In the design of continually operating engineering plant, advantage is not usually taken of the high heat transfer coefficients of dropwise condensation, but since the coefficients of filmwise condensation are often high enough not to be controlling in many designs, the loss is not as significant as the ratio of the coefficients in the two modes suggests.

The mechanism of dropwise condensation is still not fully understood. The above description of the process does not refer to the means by which the condensate migrates from the bare surface to coalesce into large droplets. A recent attempt to analyse the nature of dropwise condensation has been made by Lefevre and Rose[18], while a full empirical study of the process is reported by Hampson and Ozesek[19].

9-3 Condensation in the presence of a non-condensable gas

Consider the condensing system in Fig. 9-10 in which the cooled surface is held at a temperature T as in the previous analysis of condensation of a pure vapour. In the present case the vapour is also at its saturation temperature, T_{sat}, corresponding to its pressure, p_v, but now this is only a *partial pressure*, being less than the total fluid pressure, p, by the partial pressure of the other non-condensable gas, supposed uniformly mixed with the vapour. Thus we have

$$p = p_{sat} + p_g \qquad (9\text{-}21)$$

The Gibbs–Dalton law of partial pressures is assumed to apply, so that each constituent exerts the pressure that it would have if it alone occupied the total volume of the mixture, and has the corresponding internal energy and, thus, enthalpy.

In the equilibrium condition, in which the pressures and temperatures in the system are invariant with time, heat will flow to the wall because of the temperature difference between the gas–vapour mixture and the wall, $T_{sat} - T$, and vapour will continually condense to form the film of liquid shown in Fig. 9-10.

To reach the liquid film, however, the vapour must diffuse through the gas, and as we saw in Chapters 1 and 5 there is a resistance to this transfer of matter governed by principles analogous to those governing the transfer of heat. The potential driving force for this transfer of matter or mass is provided by a difference in concentration of the diffusing species between the main bulk of the mixture and the surface of the condensate film, which is defined by the suffix, i. Thus, at this point the temperature is T_i, the vapour pressure, p_{v_i}, being the saturation value corresponding to this temperature, and the concentration of vapour c_i in moles per unit volume. Despite having saturation conditions in the vapour, it is nevertheless a sufficiently close approximation to assume that the vapour behaves as a perfect gas, so that the concentration can be represented in terms of its partial pressure alone. Thus Eqn (1-9), written in terms of the molar transfer and molar concentrations,

$$\dot{n} = K(c - c_a)$$

becomes, in terms of the partial pressure and the mass transfer rate per unit surface area,

$$\dot{m} = \frac{K}{RT}(p_v - p_{v_i}) \qquad (9\text{-}22a)$$

In the present situation of simultaneous heat and mass transfer the temperature is not constant in the two-component system, but there is a functional dependence between the saturation pressure and temperature, so that in the present notation we have

$$\dot{m} = K_p(p_v - p_{v_i}) \qquad (9\text{-}22b)$$

in which K_p is the mass transfer coefficient defined by Eqn (9-22a) expressed in terms of the difference in partial pressures.

In Chapters 1 and 5 the analogy between the governing equations of mass transfer and heat transfer was demonstrated, and in Chapter 7 the consequential similarity between the correlating empirical equations for turbulent flows were seen. Thus, for heat transfer we had [in Eqn (7-12a)]

$$\mathbf{N} = 0.023\mathbf{R}^{0.8}\mathbf{P}^{0.4} \tag{9-23}$$

while for mass transfer [in Eqn (7-15)],

$$\mathbf{Sh} = 0.023\mathbf{R}^{0.8}\mathbf{Sc}^{0.4} \tag{9-24}$$

where
$$\mathbf{Sh} = K\frac{\bar{p}\,d}{p\,D}$$

In the Sherwood number, which is the dimensionless mass transfer coefficient, K is modified in the analogy by the ratio of the mean partial pressure, $\bar{p}$, of the constituent (in the present case, the non-condensable gas) through which mass transfer is taking place, to the total pressure, p. This modification allows for the phenomenon referred to in Chapter 1, which distinguishes in principle mass transfer from heat transfer, namely, the ability in the former case for matter transport to occur in both directions.

From Eqns (9-23) and (9-24) it follows that if the flow conditions for a mass transfer system and a heat transfer system (as defined by the Reynolds number, $\mathbf{R}$) correspond, and if the physical properties in each case (represented respectively by the Schmidt and Prandtl numbers, $\mathbf{Sc}$ and $\mathbf{P}$) are known, then from a knowledge of the heat transfer coefficient, h, the mass transfer coefficient, K_p, may be obtained; and, of course, vice versa.

Thus, in the present situation of simultaneous heat and mass transfer in a mixture of gas and saturated vapour, we can write, with the nomenclature of Fig. 9-10, the heat balance for unit area of condensing surfaces as

$$\dot{q} = h_c(T_i - T) = h_m(T_{sat} - T_i) + \dot{m}h_{fg} \tag{9-25}$$

In this equation h_c is the true condensation coefficient, given for the complete liquid film by Eqns (9-17a) or (9-18), h_m is the heat transfer coefficient due to the convective motion of the gas–vapour mixture over the condensing surface, h_{fg} is the latent heat of condensation, and $\dot{m}$ is the mass rate of vapour transfer to unit area of the condensate film, given by Eqn (9-22a).

To illustrate the technique necessary for solving Eqn (9-25), and the effect of a non-condensable gas upon heat transfer by condensation, consider the following example.

Example. A tube, 20 mm in diameter, is maintained at a temperature of 95°C and is swept by a steam–air mixture at 99°C and a total pressure of 1 bar (10^5 N/m^2). The mass ratio of air/steam is 0·05 and the convective heat transfer coefficient between the mixture and the tube is 30 W/m^2 K. It may be assumed that the Schmidt and Prandtl numbers for the mixture are the same, and the mass diffusivity, D, is 22 mm^2/s. Find the effective coefficient of heat transfer.

From the Gibbs–Dalton law, the mass ratio of gas to vapour is given by

$$\frac{m_g}{m_v} = \frac{p_g V}{R_g T_m} \frac{1}{V \rho_v}$$

in which V is the total mixture volume, T_m its temperature, p_g the partial pressure of the gas, R_g its characteristic constant, and ρ_v the density of the vapour at its pressure, p_v. This may be approximated by

$$\rho_v = \frac{p_v}{R_v T_m}$$

where R_v is the characteristic constant of the vapour. For steam at about 1 atm, R_v is 454 J/kg K, and for air, $R_g = 287$ J/kg K, so that in the present example

$$0.05 = \frac{p_g}{287} \frac{454}{p_v} = \frac{454}{287} \frac{p_g}{1 - p_g}$$

and

$$p_g = 0.03, \qquad p_v = 0.97 \text{ bar}$$

The corresponding vapour saturation temperature, T_{sat}, is 98·8°C. The convective heat transfer coefficient is given as 30 W/m² K, thus the corresponding Nusselt number is, with the thermal conductivity of the mixture k, taken as 0·028 W/m K, and using the tube diameter, d, as the characterizing dimension

$$\mathbf{N} = \frac{h_m d}{k} = \frac{30 \times 0.02}{0.028} = 21.4$$

This, as we have seen in Eqns (9-23) and (9-24), is equivalent to the Sherwood number, the non-dimensional coefficient of vapour diffusivity, corrected for the corresponding 'outward' diffusion of gas. Thus we have

$$\mathbf{Sh} = \mathbf{N} = K \frac{\bar{p}}{p} \frac{d}{D}$$

so that

$$K = \mathbf{N} \frac{D}{d} \frac{p}{\bar{p}}$$

$$= 21.4 \frac{2.2 \times 10^{-5}}{0.02} \frac{p}{\bar{p}} \left[\frac{m^2}{s} \frac{1}{m} \right]$$

$$= 0.0235 \frac{p}{\bar{p}} \text{ m/s}$$

For the rate of mass transfer of vapour from the mainstream to the condensate surface we have, from Eqn (9-22a),

$$\dot{m} = \frac{K}{R_v T_m} (p_v - p_{v_i})$$

$$= \frac{0 \cdot 0235}{454 \times 371} \frac{p}{\bar{p}} (p_v - p_{v_i}) 10^5$$

in which $T = 371$ K $= 98°C$ is used as the average vapour temperature. Thus, in Eqn (9-22b) we have

$$K_p = 0 \cdot 014 \frac{p}{\bar{p}}$$

if the difference in vapour pressure providing the driving potential for mass transfer is measured in bars.

Alternatively, in the pound–foot–hour–Fahrenheit system, the Nusselt number remains, of course, 21·4, the mass diffusivity is 0·85 ft²/h, and the tube diameter is 0·787 in, so that

$$K = 21 \cdot 4 \frac{0 \cdot 85}{0 \cdot 787/12} \frac{p}{\bar{p}} \left[\frac{\text{ft}^2}{\text{h}} \frac{1}{\text{ft}} \right]$$

$$= 277 \frac{p}{\bar{p}} \text{ ft/h}$$

and for the mass rate of diffusion, with $R = 84 \cdot 5$ ft lbf/lb °R, $T = 668°$R, and the vapour pressure difference (in lbf/ft²), we obtain

$$\dot{m} = \frac{277}{84 \cdot 5 \times 668} \frac{p}{\bar{p}} (p_v - p_{v_i}) \text{ lb/ft}^2 \text{ h}$$

and

$$K_p = 0 \cdot 049 \frac{p}{\bar{p}}$$

Before solving Eqn (9-25) the heat transfer coefficient through the condensate film must be determined from Eqn (9-18) for this case of the circular tube:

$$h_c = 0 \cdot 73 \sqrt[4]{\frac{g \rho^2 h_{fg} k_L^3}{\mu \, \Delta T \, d}}$$

In the first instance, assume that the vapour has free access to the film of condensate, as if no air were present, and the temperature difference through the film is then

$$\Delta T = T_{\text{sat}} - T$$

$$= 98 \cdot 8 - 95 = 3 \cdot 8 \text{ K}$$

Thus, in SI units, for h_c we obtain

$$h_c = 0.73 \sqrt[4]{\left[\frac{9.81 \times (962)^2 \times (2,259 \times 10^3) \times (0.68)^3}{(2.9 \times 10^{-4}) \times 3.8 \times 0.02}\right]}$$

$$\times \left[\frac{m}{s^2}\left(\frac{kg}{m^3}\right)^2 \frac{N\,m}{kg}\left(\frac{N\,m}{m\,s\,K}\right)^3 \frac{m}{N\,s}\frac{1}{K}\frac{1}{m}\right]^{1/4}$$

$$= 17 \text{ kW/m}^2 \text{ K}$$

In lb–ft–h–°F units we obtain

$$h_c = 0.73 \sqrt[4]{\left[\frac{32.2 \times (3,600)^2 \times (59.9)^2 \times 972 \times (0.39)^3}{1.67 \times 10^{-9} g_c \times 6.84 \times (0.787/12)}\right]}$$

$$\times \left[\frac{ft}{h^2}\left(\frac{lb}{ft^3}\right)^2 \frac{Btu}{lb}\left(\frac{Btu}{ft\,h\,°R}\right)^3 \frac{ft^2}{lbf\,h}\frac{lbf\,h^2}{lb\,ft}\frac{1}{°R}\frac{1}{ft}\right]^{1/4}$$

$$= 3,000 \text{ Btu/ft}^2 \text{ h °R}$$

(Note that $g_c = 4.18 \times 10^8$ lb ft/lbf h².)

Equation (9-25) may now be written as

$$\dot{q} = h_c(T_i - T) = h_m(T_m - T_i) + K_p(p_v - p_{v_i})h_{fg}$$

which in SI units becomes

$$\dot{q} = 17\left(\frac{T_i - 95}{3.8}\right)^{-1/4}(T_i - 95)$$

$$= 0.03(99 - T_i) + 0.014\frac{p}{p}(0.97 - p_{v_i})2,259$$

Note that h_c, compared with the value calculated above, is corrected for the changed temperature difference through the film. Observe also that the mixture temperature, 99°C, is used in the convection term since in this example there is a slight superheat, 0.2°C.

Solution of this equation requires a trial and error procedure, since T_i and p_{v_i} are related in a complex way. It is thus necessary to assume a value of T_i, the temperature at the vapour-condensate interface, to extract p_{v_i} from the steam tables and then to repeat the process until the left- and right-hand sides of the heat flux equation balance.

First, assume that $T_i = 96.8°C$. Interpolating in the steam tables gives p_{v_i} as 0.902 bar. Thus, at the interface the partial pressure of the air is

$$p_{g_i} = p - p_{v_i}$$

$$= 0.098 \text{ bar}$$

The mean pressure, $\bar{p}$, is evaluated as the logarithmic mean, that is

$$\bar{p} = \frac{0.098 - 0.03}{\ln \dfrac{0.098}{0.03}} = 0.0574 \text{ bar}$$

(Compare the arithmetic mean value 0.064 bar.) So we have for Eqn (9-25)

$$\dot{q} = 17 \left(\frac{1.8}{3.8}\right)^{-1/4} 1.8 = 0.03 \times 2.2 + 0.014 \frac{1}{0.0574} (0.97 - 0.902) 2{,}259$$

that is, $\qquad 36.9 = 0.066 + 37.2$

$$= 37.3 \text{ kW/m}^2$$

The agreement between the two sides of this equation is satisfactory, and taking the heat flux as the mean, 37.1 kW/m^2, shows that the overall coefficient of heat transfer from the vapour–gas mixture at a saturation temperature of $98.8°C$ and the surface at $95°C$ is

$$h_0 = \frac{37.1}{98.8 - 95} = 9.75 \text{ kW/m}^2 \text{ K}$$

This compares with $h_c = 17.0 \text{ kW/m}^2$ K calculated in the absence of the air, showing that only 5 per cent by weight of non-condensable gas has reduced the heat transfer coefficient by over 40 per cent.

REFERENCES

1. Tong, L. S. *Boiling Heat Transfer and Two-Phase Flow.* John Wiley, New York, 1966.
2. Cryder, D. S. and A. C. Finalborgo. Heat transmission from metal surfaces to boiling liquids. *Trans. A.I.Chem.E.*, 1937, **33**.
3. Addoms, J. N. Heat transfer at high rates to water boiling outside cylinders. Sc.D. Thesis, Dept of Chemical Engineering, M.I.T., 1948.
4. Fritz, W. Maximum volume of vapour bubbles. *Phys. Z.*, 1935, **36**.
5. Rohsenow, W. M. A method of correlating heat transfer data for surface boiling of liquids. *Trans. A.S.M.E.*, 1952, **74**.
6. Ewing, J. A. *Thermodynamics for Engineers.* Cambridge University Press, London, 1946.
7. Forster, K. and R. Grief. Heat transfer to a boiling liquid. *Trans. A.S.M.E., J. Heat Transfer*, 1959, **81**.
8. Bennett, J., J. Collier, H. Pratt, and J. Thornton. Heat transfer to two-phase gas–liquid systems. *Trans. I.Chem.E.*, 1961, **39**.
9. Martinelli, R. C. and R. W. Lockhart. Proposed correlation of data for isothermal two-phase two component flow in pipes. *Chem. Engng. Progr.*, 1949, **45**.
10. Zuber, N. Hydrodynamic aspects of boiling heat transfer. *U.S.A.E.C. Rept AECU-4439*, 1959.
11. Kutateladze, S. S. Boiling heat transfer. *Int. J. Heat and Mass Transfer*, 1961, **4**.

12. Kazakova, E. Maximum heat transfer to boiling water at high pressures. *Précis* in *Engineers' Digest*, 1951, **12**.

13. Cohen, H. and F. J. Bayley. Heat transfer problems of liquid-cooled gas turbine blades. *Proc. I.Mech.E.*, 1955, **169**.

14. Bromley, L. A. Heat transfer in stable film boiling. *Chem. Engng. Progr.*, 1950, **46**.

15. Berenson, P. J. Film-boiling heat transfer from a horizontal surface. *Trans. A.S.M.E., J. Heat Transfer*, 1961, **83**.

16. Nusselt, W. The filmwise condensation of steam. *Z. Ver. dt. Ing.*, 1916, **60**.

17. Colburn, A. P. The calculation of condensation where a portion of the condensate layer is in turbulent flow. *Trans. A.I.Chem.E.*, 1933, **30**.

18. Lefevre, E. J. and J. Rose. A theory of dropwise condensation of steam. *Proc. A.S.M.E.–I.Mech.E. Heat Transfer Conference*, 1966.

19. Hampson, H. and N. Ozesek. An investigation into the condensation of steam. *Proc. I.Mech.E.*, 1953, **1B**.

PROBLEMS

9-1 Use the correlation of Eqn (9-5) to predict the heat transfer coefficient associated with a surface to saturation temperature difference of $10°C$ in the pool boiling of water at $100°C$ and $1·01$ bar (101 kN/m^2). The proportionality constant C_{sf} should be taken as $0·010$ and data in Table A6 in the Appendix may be used as required. The surface tension and latent heat of evaporation of water at these conditions are $0·058$ N/m and $2,257$ kJ/kg respectively.

Ans: $10·2$ kW/m^2 K

9-2 Show that at a pressure of $39·8$ bar, under conditions otherwise similar to those in Problem 9-1, the heat transfer coefficient rises to $24·1$ kW/m^2 K. Consider the principal physical reasons for this change, as shown by the properties in the correlation. Take the surface tension as $0·034$ N/m and latent heat as $2,406$ kJ/kg.

Show also that a 10% change in the assumed value of C_{sf} produces a 33% change in the predicted coefficient.

9-3 In the evaporator tubes of a boiler working at 560 lbf/in^2 ($479°F$) the heat flux is estimated to be $50,000$ Btu/ft^2 h. The tubes are 2 inches in diameter and the feed water enters with a velocity of 3 ft/s. Use the criterion of Fig. 9-5, the turbulent pipe-flow equation (7-12a) and the boiling heat transfer correlation (9-5) to confirm that the designer's assumed value for the heat transfer coefficient in the boiler tubes, $5,000$ Btu/ft^2 h °F, is conservative. Take data as required from Table A6 in the Appendix with appropriate conversion factors; other physical properties are as in Problem 9-2.

9-4 Up to what heat flux could the designer of Problem 9-3 safely ignore the possibility of film boiling in the tubes? Use the Kutateladze criterion of Eqn (9-7b).

Ans: $1·58 \times 10^6$ Btu/ft^2 h

9-5 Use the Kutateladze criterion to show that the maximum heat flux in pool boiling in water is $5,320$ kW/m^2 and occurs at a pressure of about 83 atmospheres. Steam tables and the data in Table A6 should be used as required.

9-6 Liquid ammonia is evaporated at 30°C and 11·7 bar from the outside of tubes 20 mm in diameter. The surface is maintained at 200°C by steam condensing inside the tubes. Use the Bromley equation (9-8) and data from Tables A5 and A7 (vapour density 9·05 kg/m^3, latent heat 1,146 kJ/kg) to estimate the film boiling heat transfer coefficient.

Ans: 0·25 kW/m^2 K

9-7 Steam at 1 atmosphere (T_{sat} = 100°C, latent heat 2,257 kJ/kg) condenses on a vertical plate 1 m high which is at 98°C. Determine the local heat transfer coefficient at mid-height, the average coefficient over the plate and check that the condensate flow will everywhere be laminar. Use the data of Table A6 as required.

Ans: 8·7, 6·9 kW/m^2 K; $\mathbf{R}_f$ = 88

9-8 In a refrigerator, freon at 108 lbf/in^2 (T_{sat} = 86°F, latent heat 57·6 Btu/lb) condenses on tubes maintained at 80°F by convection to ambient air and which are $\frac{1}{4}$ inch diameter and 2 feet long. Determine the number of tubes required for a heat load of 3,000 Btu/h using the data of Table A5 and the appropriate conversion factors.

Show that if the tubes were arranged vertically, instead of horizontally as in the above case, the surface area would have to be increased by 143%.

Ans: 27 tubes

9-9 The condenser of a large steam power plant comprises a bank of horizontal tubes 1 inch diameter inside which flows circulating water at 50°F and at a velocity to give an average forced convection heat transfer coefficient of 500 Btu/ft^2 h °F. Steam at 1·071 lbf/in^2 (T_{sat} = 104°F, latent heat 1,035 Btu/lb) condenses in the film-wise mode on the outer surfaces. If by the use of a suitable promoter dropwise condensation giving five times the condensing coefficient for the above conditions could be assured, show that a saving of about 16% of heat transfer surface could be made at the design stage. Use the data of Table A6 as required and note that a trial and error solution of the heat balance equation will be necessary.

9-10 The total pressure in an acid–gas condenser at a chemical plant is 19·7 lbf/in^2 of which a partial pressure of 5 lbf/in^2 is contributed by the non-condensable gas. The condenser tube surface is maintained at 206°F by circulating water and the condensing coefficient in the absence of the gas is calculated to be 1,400 Btu/ft^2 h °F. The coefficient of mass transfer of moisture through the gas, K_p, is $0·03p/\bar{p}$ lb/lbf s and the convective heat transfer coefficient due to the gas–vapour flow is 34 Btu/ft^2 h °F. Show that under these conditions the effective condensing coefficient is reduced by about 26%. Use steam tables as required.

10

Heat exchangers

The principles of heat transfer find application most commonly in the design of heat exchangers—units for the exchange of heat between two fluid streams at different levels of temperature. Examples of such units are found at all levels of technological sophistication and size, from domestic radiators and refrigerator condensers, through car radiators and aircraft oil coolers, to the vast units of the chemical process and power generation industries.

Essentially, heat exchangers are arrangements of surfaces separating the fluids between which heat transfer takes place. These surfaces are heated by one fluid and cooled by the other. Since tubes are inherently strong for their size, the commonest arrangement of surface is a matrix of tubes inside which one fluid flows at the higher pressure in the system. The second fluid is then arranged to flow over the outside of the tubes.

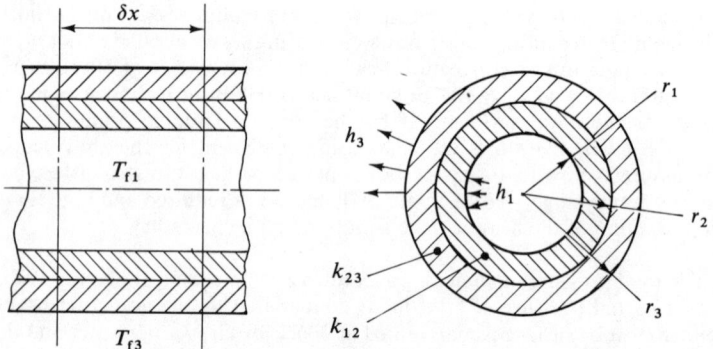

Fig. 10-1   Heat transfer through an element of heat exchanger tube

10-1   Overall heat transfer coefficient

Consider an elementary length δx of such a tube, as shown in Fig. 10-1. The hotter fluid is on the inside and is at a temperature T_{f1} at this section in the complete unit. The cooler fluid is at a temperature T_{f3} outside the tube, and as we saw in Eqn (2-20a) in Chapter 2, the rate of heat transfer $\delta \dot{Q}$ over this elementary length of the unit is given by

$$\delta \dot{Q} = \frac{2\pi(T_{f1} - T_{f3})\,\delta x}{\dfrac{1}{r_1 h_1} + \dfrac{\ln(r_2/r_1)}{k_{12}} + \dfrac{\ln(r_3/r_2)}{k_{23}} + \dfrac{1}{r_3 h_3}} \tag{10-1}$$

In this equation, which was derived for the composite tube with fluid boundary conditions, the coefficients h_1 and h_3 would relate to the convective heat transfer processes on the hot and cold sides of the tube surface. In Eqn (10-1) there would normally be only one logarithmic conduction term relating to the heat flows through the metallic tube wall, although sometimes under severe operating conditions, solid deposits such as carbon from an exhaust gas or dissolved solids from a solution may build up to produce the composite conducting cylinder of Fig. 10-1 (see also Fig. 2-8).

Chapter 2 also introduced the concept of the *overall heat transfer coefficient U*, defined to give the overall rate of heat transfer between two temperature levels through a composite system, as in the present element of heat exchanger tube,

$$\delta\dot{Q} = U\,\delta A(T_{f1} - T_{f3}) \tag{10-2}$$

The element of area, δA, over which the overall coefficient U is presumed to act is arbitrary, and may, for example, be either the inner or outer surface area of the tube. Thus, if the area is taken to be that of the outside of the tube such that $\delta A = 2\pi r_3\,\delta x$ and U is said to be the *overall heat transfer coefficient referred to the tube outer surface* then, a comparison of Eqns (10-1) and (10-2) shows that

$$U = \frac{1}{\dfrac{r_3}{r_1}\dfrac{1}{h_1} + \dfrac{r_3}{k_{12}}\ln\left(\dfrac{r_2}{r_1}\right) + \dfrac{r_3}{k_{23}}\ln\left(\dfrac{r_3}{r_2}\right) + \dfrac{1}{h_3}} \tag{10-3a}$$

Commonly the resistance to heat flow through the metal wall of the tube is quite negligible, so that the term $(r_3/k_{12})\ln(r_3/r_2)$ may be omitted. Equally, the so-called 'fouling' resistances, due to solid deposits comprising the composite conducting system in Fig. 10-1, are often negligible or absent altogether, so that Eqn (10-3a) then reduces to the simpler form

$$\frac{1}{U} = \frac{r_3}{r_1}\frac{1}{h_1} + \frac{1}{h_3} \tag{10-3b}$$

Regarding the heat transfer coefficients as inverse resistances, this equation can be written

$$\text{overall resistance} = \left\{ \begin{array}{l} \text{inside resistance} \\ \text{referred to outside} \\ \text{surface} \end{array} \right\} + \text{outside resistance}$$

The first term on the right-hand side is, as shown by Eqn (10-3b), the inverse of the inside surface heat transfer coefficient scaled to the value corresponding to it acting over the larger outer tube surface. For the case of a thin tube of large radius or the flat plate, Eqn (10-3b) becomes

$$\frac{1}{U} = \frac{1}{h_1} + \frac{1}{h_3} \tag{10-3c}$$

If either of the coefficients h_1 or h_3 is very much greater than the other, then its effect upon the overall coefficient is correspondingly negligible and the lower heat transfer coefficient is said to be *controlling*. Suppose, for example, the heat transfer coefficient within a tube conveying a liquid is, say, $h_1 = 1{,}000$ Btu/ft^2 h °F, and the outside coefficient from the flow of a gas $h_3 = 30$ Btu/ft^2 h °F; then, from Eqn (10-3c),

$$U = \frac{h_3 \times h_1}{h_3 + h_1}$$

$$= \frac{30 \times 1{,}000}{30 + 1{,}000} = 29 \cdot 1 \text{ Btu/ft}^2 \text{ h °F}$$

Doubling h_1 to 2,000 only produces a 2 per cent change in U which is, however, increased by 95 per cent by doubling h_3. Clearly this is the controlling coefficient in which the designer should concentrate his efforts at improvement.

10-1-1   Secondary surfaces

In order to increase the performance of heat exchangers, designers often adopt the technique of adding surface to the side of the system over which the controlling heat transfer coefficient applies. Probably the commonest example of this is found in the motor-car radiator, where the coefficient between the radiator tubes and the engine cooling water is much greater than between the tubes and the atmospheric air. Thus, attached with good thermal contact (usually by soldering or brazing) to the outside of the tubes are metal strips (or *fins*) having a large surface area. Heat is transferred by conduction through the radiator tube and along the fins, which are swept on both sides by only one of the fluids in the system (the atmospheric air in the case of the car radiator) and are known as *secondary surfaces*.

Thus, suppose that to a base (or *primary*) surface, such as the radiator tubes, δA_3, an area of fins or secondary surface of area δA_f is added. It is usual to assume that over both base and fin the same heat transfer coefficient applies, say, h_3. Since the heat has to be conducted along the fin before it can be transferred to the fluid at temperature T_{f3}, there must be a fall in metal temperature outwards from the base surface. (If the hotter fluid flows over the fins, there will conversely be a rise in temperature outwards from the base or primary surface, with a corresponding reduction in potential for heat transfer.)

As was seen in Chapter 3, we can define a *fin efficiency*, ϕ, as the actual rate of heat transfer from the fin divided by the heat transfer rate if all the fin were at the base temperature; that is, $(\bar{T} - T_{f3})/(T_b - T_{f3})$, where $\bar{T}$ is, by definition of ϕ, the mean fin temperature. The rate of heat transfer between the fluid within the tubes (represented by suffix 1) and the outside fluid is

$$\delta\dot{Q} = h_1 \, \delta A_1 (T_{f1} - T_b) = h_3 [\delta A_3 (T_b - T_{f3}) + \delta A_f (\overline{T} - T_{f3})]$$

$$= h_3 \, \delta A_3 [1 + \phi(\delta A_f / \delta A_3)](T_b - T_{f3})$$

This equation assumes, as is usually the case, that the reductions in temperature through the tube wall are negligible, so that the base temperature, both inside and outside, is T_b.

In terms of an overall heat transfer coefficient U_1 referred to the inner base surface δA_1, we have

$$\delta\dot{Q} = U_1 \, \delta A_1 (T_{f1} - T_{f3})$$

whence
$$\frac{1}{U_1} = \frac{1}{h_1} + \frac{\delta A_1}{\delta A_3} \frac{1}{h_3 [1 + \phi(\delta A_f / \delta A_3)]} \tag{10-4}$$

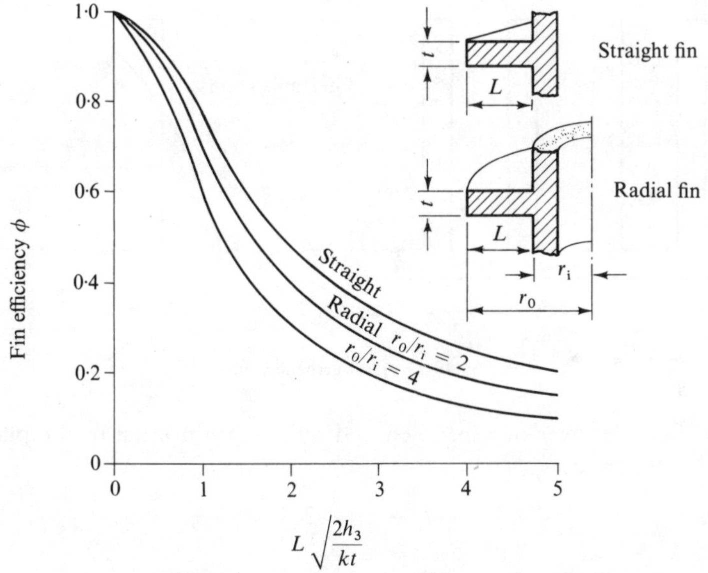

Fig. 10-2 Efficiencies of several fin geometries (after Gardner[1])

Values of the fin efficiency, ϕ, obtained by Gardner[1]†, were given in Fig. 3-7 for a number of fin geometries, and further examples for some common fin arrangements are also shown in Fig. 10-2. The abscissa of the graph is the parameter

$$L\sqrt{\frac{2h_3}{kt}}$$

† Full details of references cited are given at the end of the chapter.

in which L is the length or height of the fin, t its thickness, k the thermal conductivity of the metal, and h_3 the heat transfer coefficient over the fin surface.

In the example of the preceding section, suppose that straight fins $\frac{1}{16}$ in thick and $\frac{1}{2}$ in high are added on the gas side at a pitching of $\frac{3}{16}$ in, as shown in Fig. 10-3. If the fins are constructed of mild steel ($k = 26$ Btu/h ft °F) the efficiency correlating parameter has the value, with $h_3 = 30$ as before,

$$\frac{1}{24} \sqrt{\frac{2 \times 30}{26 \times \frac{1}{12} \times \frac{1}{16}}} = 0{\cdot}88$$

The efficiency from Fig. 10-2 is thus 0·8. The ratio of the fin area for heat transfer to base area, $\delta A_f / \delta A_3$, is obtained from the geometry of the system.

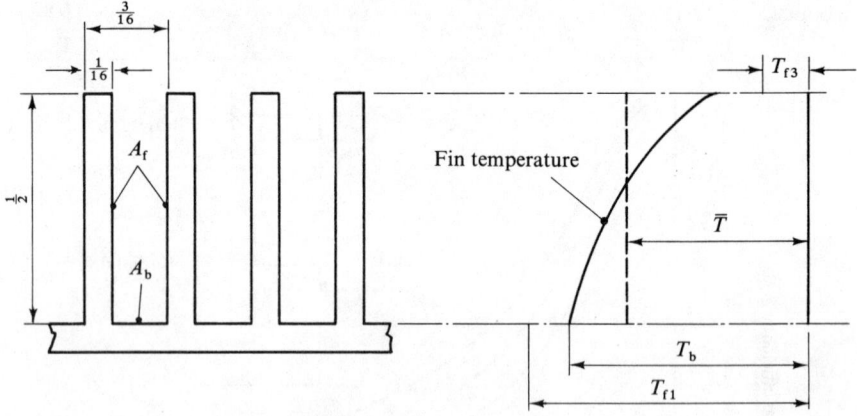

Fig. 10-3 Arrangement of fins

In this example, over one fin pitch and unit length normal to the plane of Fig. 10-3, we have

$$\delta A_f = 2L + 1t = 2 \times \tfrac{1}{2} + \tfrac{1}{16}$$
$$\delta A_3 = p - t = \tfrac{2}{16}$$

that is,
$$\frac{\delta A_f}{\delta A_3} = \frac{17}{2} = 8{\cdot}5$$

The inner area, δA_1, is unchanged and corresponds to the element of one pitch length chosen to evaluate these ratios—$\frac{3}{16}$ in. Thus, for the overall coefficient U_1 we have

$$\frac{1}{U_1} = \frac{1}{1{,}000} + \frac{\frac{3}{16}}{\frac{2}{16}} \frac{1}{30(1 + 0{\cdot}8 \times 8{\cdot}5)}$$

whence $U_1 = 135$, which compares with $U = 29 \cdot 1$ Btu/ft^2 h °F before the addition of the fins.

Attention must be drawn to the approximations of this procedure. First, the heat flows in the heat exchanger wall, particularly near the base of the fin, are complex and strictly need calculation by the numerical techniques of Chapter 4. The assumption that their effect on the overall heat transfer process can be accounted for by the simple linear addition of a fin surface of efficiency ϕ to the base surface δA_3 is clearly an approximation that takes insufficient regard of the complex effects of fin geometry.

Further, in determining the fin efficiencies plotted in Fig. 10-2, heat transfer into the fin tip has been ignored, so that the 'radially' outward temperature gradient at this point is assumed to be zero. Although this is an adequate approximation for most practical applications, it will certainly introduce some error leading to an overestimation of the fin efficiency. Although semi-empirical proposals have been made to reduce this effect, as described in Chapter 3, the usual practice is to use the fin efficiencies of Figs. 3-7 and 10-2, but to ignore the contribution of the tip area to the total fin area. In the example above, $\delta A_f / \delta A_3$ would be reduced to $8 \cdot 0$ (instead of $8 \cdot 5$) and U_1 becomes 129 (instead of 135). This uncertainty of about $4\frac{1}{2}$ per cent is typical and by no means negligible, but while data on the variation of the convective coefficients over the fin and base surfaces—which is known to effect fin efficiencies—are unavailable, the additional complexity of more precise procedures is not considered worthwhile.

10-1-2 Radiation in heat exchangers

As was seen in Chapter 8, radiant heat transfer involves the difference in the fourth power of the absolute temperatures of the exchanging media. Thus, the simple linear temperature dependence appropriate to convection and conduction, which enables the useful concept of an overall heat transfer coefficient to be derived by the elimination of intermediate temperatures, is no longer strictly applicable when a significant amount of radiant interchange occurs. However, an equivalent heat transfer coefficient, h_r, can be defined for radiating systems as

$$d\dot{Q} = h_r \, \delta A (T_1 - T_2)$$

where $\delta \dot{Q}$, the net heat transfer between elementary areas of the systems (as shown in Chapter 8, and with the nomenclature there defined), is given by

$$\delta \dot{Q} = \sigma \, \delta A \mathscr{F}_{1 \cdot 2}(T_1^4 - T_2^4)$$

Thus
$$h_r = \frac{\sigma \mathscr{F}_{1 \cdot 2}(T_1^4 - T_2^4)}{T_1 - T_2} \tag{10-5}$$

With the levels of temperature normally associated with significant radiant heat transfer in engineering systems, substantial variations in the lower temperature, T_2, of Eqn (10-5) have little effect upon the numerator and, hence, the net rate of heat flow. Thus, this coefficient, h_r, behaves approximately like a convective coefficient in linking heat flow linearly with temperature difference for a fixed upper temperature T_1, and at least allows the relative magnitude of convection and radiation on one side of a heat exchanger system to be determined.

10-2 Ideal contraflow heat exchanger

Up to this point consideration of the heat transfer process within heat exchangers has been restricted to elements of surface over which the fluid temperatures could be considered constant. However, the object of a heat exchanger is to change the temperatures of the fluids, and procedures will now be developed by which account may be taken of these variations.

Consider a heat exchanger in which the two fluids move in opposing directions, so that the originally hotter fluid enters the heat exchanger at the end where the originally cooler, but now heated, fluid leaves, as in Fig. 10-4. This is known as the contra- or counterflow heat exchanger.

The heat balance over the element δA for the two fluids requires that

$$d\dot{Q} = \dot{m}_1 C_1 \, \delta T_1 = \dot{m}_2 C_2 \, \delta T_2 \qquad (10\text{-}6)\dagger$$

where $\dot{m}$ and C are, respectively, the rates of mass flow and specific heats of the fluids represented by suffices 1 and 2, and δT the changes in their temperatures over the element. δT_1 and δT_2 must, of course, be of opposite sign.

Consider first the situation in which

$$\dot{m}_1 C_1 = \dot{m}_2 C_2 = \dot{m}C \qquad (10\text{-}7)$$

so that the heat capacities or 'water equivalents' of the two fluid streams are equal. Such conditions occur sometimes in practice, usually where a single fluid leaves the heat exchanger to undergo some process, after which it returns to the other side of the unit for regenerative heat recovery. For this simple case the changes in both fluid temperatures over each element are equal in magnitude, and the fluid temperatures over the full heat transfer surface of the heat exchanger, from end a to end b, can be represented as parallel lines (see Fig. 10-4). If the now constant temperature difference between the fluids is ΔT, we have

$$\Delta T = T_{1a} - T_{2a} = T_{1b} - T_{2b} \qquad (10\text{-}8)$$

Now, in terms of the overall heat transfer coefficient, U, and the heat exchanger surface area A to which it is referred, we have

$$\dot{Q} = UA \, \Delta T \qquad (10\text{-}9)$$

---

† The suffix 'f' used previously to distinguish fluid temperatures from surface temperatures (as in Chapter 2) is now discarded.

and, for the originally hotter fluid,

$$\dot{Q} = \dot{m}C(T_{1a} - T_{1b}) \qquad \text{(10-10a)}$$

and for the colder fluid,

$$\dot{Q} = \dot{m}C(T_{2a} - T_{2b}) \qquad \text{(10-10b)}$$

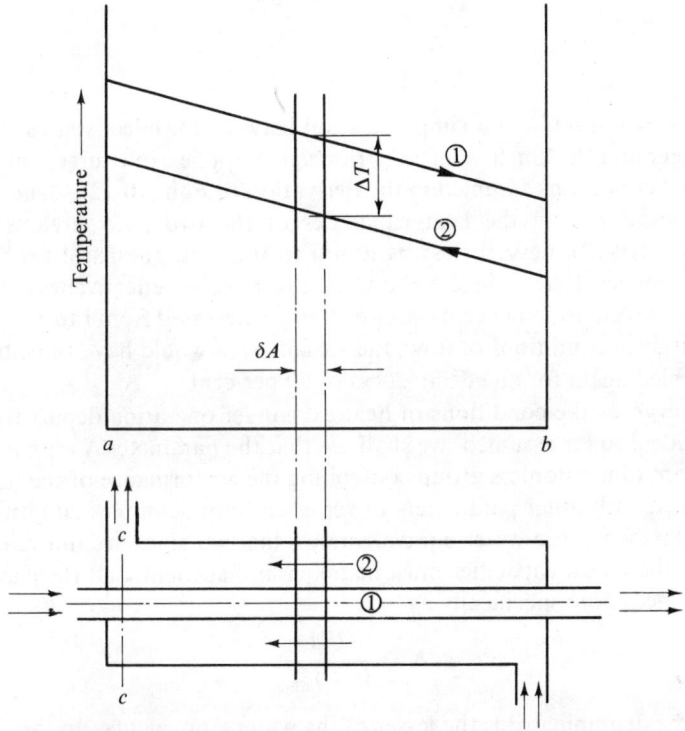

Fig. 10-4   Conditions in an ideal contra- or counterflow heat exchanger

If there were no resistance to heat transfer between the two fluids—that is, if U were infinite—ΔT would become zero, and we would have the thermodynamically perfect situation in which the initially colder fluid left the heat exchanger at the temperature of the entering hot fluid, the highest in the system. This, of course, is not possible in practice, and to measure the performance of a heat exchanger against this ideal a parameter is employed called the *thermal ratio* or *effectiveness*, ε. This is defined as the ratio of the actual rise in temperature of the colder fluid to its maximum possible rise; that is, in the nomenclature of the present example,

$$\varepsilon = \frac{T_{2a} - T_{2b}}{T_{1a} - T_{2b}} \qquad \text{(10-11a)}$$

From Eqns (10-8), (10-9), and (10-10), for the effectiveness we obtain

$$\varepsilon = \frac{T_{2a} - T_{2b}}{(T_{1a} - T_{2a}) + (T_{2a} - T_{2b})}$$

$$= \frac{\dot{Q}/\dot{m}C}{(\dot{Q}/UA) + (\dot{Q}/\dot{m}C)} \tag{10-11b}$$

$$= \frac{\Lambda}{\Lambda + 1} \tag{10-12}$$

where $\Lambda = UA/\dot{m}C$. This simple relation between the effectiveness of a heat exchanger and the dimensionless parameter Λ applies, of course, only to the idealized conditions assumed in the derivation of Eqn (10-12)—true contra-flow operation with the heat capacities of the two fluid streams equal. Equation (10-12), nevertheless, is useful to illustrate the basic problem of heat exchanger design, since it shows that to raise the effectiveness of a unit from 50 per cent to 75 per cent requires Λ to be increased from 1 to 3, implying that for given conditions of flow, the surface area would have to be trebled, and trebled again for an effectiveness of 90 per cent.

However, as the conditions of heat exchanger operation depart from the simple ideal so far assumed, we shall see that the parameter Λ remains as an important dimensionless group controlling the performance of the unit, but is coupled with other parameters to represent more complex situations. In particular, when the water equivalents on the two sides are not equal, the ratio of these heat capacities must be taken into account, and the parameter Λ is defined more specifically as

$$\Lambda = \frac{UA}{(\dot{m}C)_{min}}$$

where the denominator is the lesser of the water equivalents. In this form Λ is commonly called the *number of transfer units* and is given the symbol Ntu, as proposed in the design method of reference (2) and described in Section 10-5-7. Also, of course, when the heat capacities of the two streams are not equal the temperature difference between the fluid streams is no longer constant over the heat transfer surface, and expressions must be obtained for the mean values of ΔT to be applied with the overall coefficient U in Eqn (10-9) for a range of conditions.

10-3   Logarithmic mean temperature difference

For a contraflow heat exchanger, as in the preceding section, but in which $\dot{m}_1 C_1 \neq \dot{m}_2 C_2$, Eqn (10-6) applies and Eqn (10-9) is written in elementary form as

$$d\dot{Q} = U \, dA \, \Delta T \tag{10-9a}$$

in which ΔT now varies over the surface area A, and at a point is given by

$$\Delta T = T_1 - T_2$$

The change in ΔT over the element dA is $d(\Delta T)$ and is given by

$$d(\Delta T) = dT_1 - dT_2 \qquad (10\text{-}13)$$

as illustrated in Fig. 10-5. Also, substituting from the heat balance equation [Eqn (10-6)],

$$d(\Delta T) = -d\dot{Q}\left(\frac{1}{\dot{m}_1 C_1} - \frac{1}{\dot{m}C_2}\right) \qquad (10\text{-}14)$$

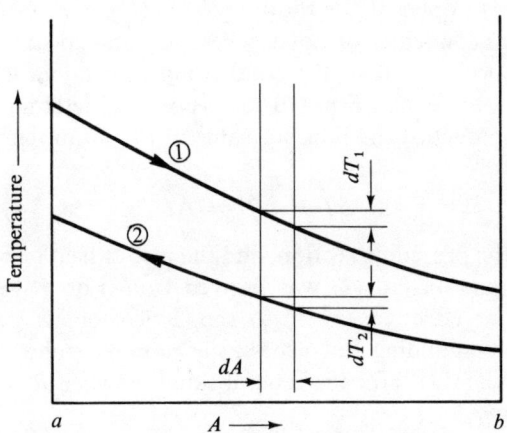

Fig. 10-5 Temperatures in a contraflow heat exchanger

The negative sign ensures that the heat flow is positive in the direction of increasing A, even when $\dot{m}_1 C_1 > \dot{m}_2 C_2$. Integrating Eqn (10-14) over the whole unit between a and b gives

$$\Delta T_b - \Delta T_a = -\dot{Q}\left(\frac{1}{\dot{m}_1 C_1} - \frac{1}{\dot{m}_2 C_2}\right) \qquad (10\text{-}15)$$

Substituting from Eqn (10-14) into (10-9a) for $d\dot{Q}$ gives

$$-\frac{d(\Delta T)}{(1/\dot{m}_1 C_1 - 1/\dot{m}_2 C_2)} = U\, dA\, \Delta T$$

which integrates to

$$-\ln\frac{\Delta T_b}{\Delta T_a} = UA\left(\frac{1}{\dot{m}_1 C_1} - \frac{1}{\dot{m}_2 C_2}\right) \qquad (10\text{-}16)$$

Dividing Eqn (10-15) by (10-16) gives

$$\dot{Q} = UA \left[\frac{\Delta T_b - \Delta T_a}{\ln (\Delta T_b / \Delta T_a)} \right] \qquad (10\text{-}17)$$

The bracketed term in Eqn (10-17) represents an effective temperature difference by which the product of overall heat transfer coefficient, U, and the total surface area in the heat exchanger, A, must be multiplied to yield the total rate of heat flow. It is known, because of its form, as the *logarithmic mean temperature difference*, LMTD. It will be found to be the appropriate expression for the effective temperature difference if the direction of flow of one of the fluids in the example above is reversed to form a *parallel flow heat exchanger*, and it applies if the ratio $\dot{m}_2 C_2 / \dot{m}_1 C_1$ ($= R$, say) becomes zero or infinite as in a condenser or boiler where the latent heat leads to an effectively infinite specific heat of the condensing or evaporating fluid. When $\dot{m}_1 C_1$ and $\dot{m}_2 C_2$ are equal, Eqn (10-17) becomes indeterminate, but as this condition is approached the limiting value of the mean temperature difference becomes

$$\Delta T = \Delta T_a = \Delta T_b$$

as observed in the preceding section, and the rate of heat transfer is given by Eqn (10-9). Equation (10-12) was derived from Eqn (10-9), relating the thermal ratio, or effectiveness, ε, to the dimensionless parameter, $\Lambda = UA/\dot{m}C$. A corresponding but necessarily more complex relation can be derived from Eqn (10-17) for the general situation when $R = \dot{m}_2 C_2 / \dot{m}_1 C_1$ is not equal to unity.

If R is less than unity, so that $\dot{m}_2 C_2 < \dot{m}_1 C_1$, then, from the definition of effectiveness in contraflow,

$$\varepsilon = \frac{T_{2a} - T_{2b}}{T_{1a} - T_{2b}}$$

and

$$T_{1a} = T_{2b} + \frac{T_{2a} - T_{2b}}{\varepsilon} \qquad (10\text{-}18a)$$

that is,

$$\Delta T_a = T_{1a} - T_{2a} = \left(\frac{1}{\varepsilon} - 1 \right)(T_{2a} - T_{2b}) \qquad (10\text{-}18b)$$

Also, for the ratio of the heat capacities, R, we have

$$R = \frac{\dot{m}_2 C_2}{\dot{m}_1 C_1}$$

$$= \frac{T_{1a} - T_{1b}}{T_{2a} - T_{2b}}$$

so that

$$T_{1b} = T_{1a} - R(T_{2a} - T_{2b})$$

Subtracting T_{2b} from each side of this equation and substituting for T_{1a} from Eqn (10-18a) gives

$$\Delta T_b = T_{1b} - T_{2b} = \left(\frac{1}{\varepsilon} - \frac{1}{R}\right)(T_{2a} - T_{2b}) \qquad (10\text{-}19)$$

The rate of heat transfer in the unit is given by the heat balance as

$$\dot{Q} = \dot{m}_2 C_2 (T_{2a} - T_{2b})$$

Equating this expression for $\dot{Q}$ to the logarithmic mean temperature difference relation of Eqn (10-17), and substituting for ΔT_a and ΔT_b from Eqns (10-18b) and (10-19), gives

$$\dot{m}_2 C_2 (T_{2a} - T_{2b}) = UA(T_{2a} - T_{2b})\frac{(1/\varepsilon - 1) - (1/\varepsilon - R)}{\ln\left[(1/\varepsilon - 1)(1/\varepsilon - R)\right]}$$

which, with some rearrangement, and putting $\Lambda = UA/\dot{m}_2 C_2$, becomes

$$\varepsilon = \frac{1 - \exp\left[-\Lambda(1 - R)\right]}{1 - R\exp\left[-\Lambda(1 - R)\right]} \qquad (10\text{-}20)$$

This expression can be generalized for contraflow units to the situation when $\dot{m}_2 C_2$ is greater than $\dot{m}_1 C_1$ by defining Λ and R in terms of the least thermal capacity of the two streams, that is,

$$\Lambda = \frac{UA}{(\dot{m}C)_{min}}; \qquad R = \frac{(\dot{m}C)_{min}}{(\dot{m}C)_{max}}$$

and the effectiveness as

$$\varepsilon = \frac{\dot{Q}}{(\dot{m}C)_{min}(T_{1i} - T_{2i})}$$

where suffix 1 refers to the hotter fluid, 2 to the colder fluid, and 'i' to the inlet condition of each. Note that although, as has been previously observed, Eqn (10-17) continues to define the effective mean temperature even when the direction of one of the streams is reversed to give a parallel flow heat exchanger, the derivation of the effectiveness relation, Eqn (10-20), is no longer applicable. In this latter case, with the temperatures in the unit as in Fig. 10-6, the effectiveness becomes

$$\varepsilon = \frac{T_{2b} - T_{2a}}{T_{1a} - T_{2a}}$$

Proceeding exactly as for the contraflow case gives

$$\Delta T_a = T_{1a} - T_{2a} = \frac{T_{2b} - T_{2a}}{\varepsilon}$$

Similarly, since

$$R = \frac{\dot{m}_2 C_2}{\dot{m}_1 C_1} = \frac{T_{1a} - T_{1b}}{T_{2b} - T_{2a}}$$

we obtain

$$\Delta T_b = T_{1b} - T_{1a} = \left(\frac{1}{\varepsilon} - 1 - R\right)(T_{2b} - T_{2a})$$

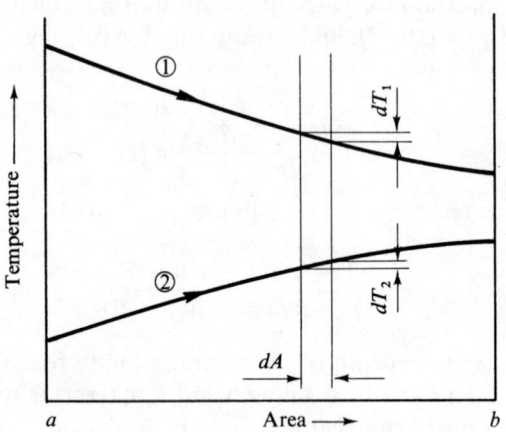

Fig. 10-6 Temperatures in a parallel flow heat exchanger

Equating the expressions for the heat transfer rate, $\dot{Q}$, from the heat balance and Eqn (10-17) gives

$$\dot{Q} = \dot{m}_2 C_2 (T_{2b} - T_{2a}) = UA \frac{1/\varepsilon - (1/\varepsilon - 1 - R)}{\ln\left[(1/\varepsilon)(1/\varepsilon - 1 - R)\right]} \qquad (10\text{-}21)$$

whence

$$\varepsilon = \frac{1 - \exp(-\Lambda)(1 + R)}{1 + R}$$

Observe that for the special case of $R = 1$, Eqn (10-20) becomes, in the limit,

$$\varepsilon = \frac{\Lambda}{\Lambda + 1}$$

which is Eqn (10-12), showing the effectiveness to approach unity as Λ approaches infinity. By contrast, Eqn (10-21) when $R = 1$ becomes

$$\varepsilon = \frac{1 - \exp(-2\Lambda)}{2}$$

showing the effectiveness to approach 50 per cent as Λ becomes very large. Clearly, the parallel flow arrangement is less attractive than the contraflow unit which makes the greatest use of the available temperature differences between the fluids. The general comparison between the performance of these two types of heat exchanger can be seen in Figs. 10-7(a) and 10-7(b) in which effectiveness for a range of the independent variables, Λ and R, is plotted according to Eqns (10-20) and (10-21).

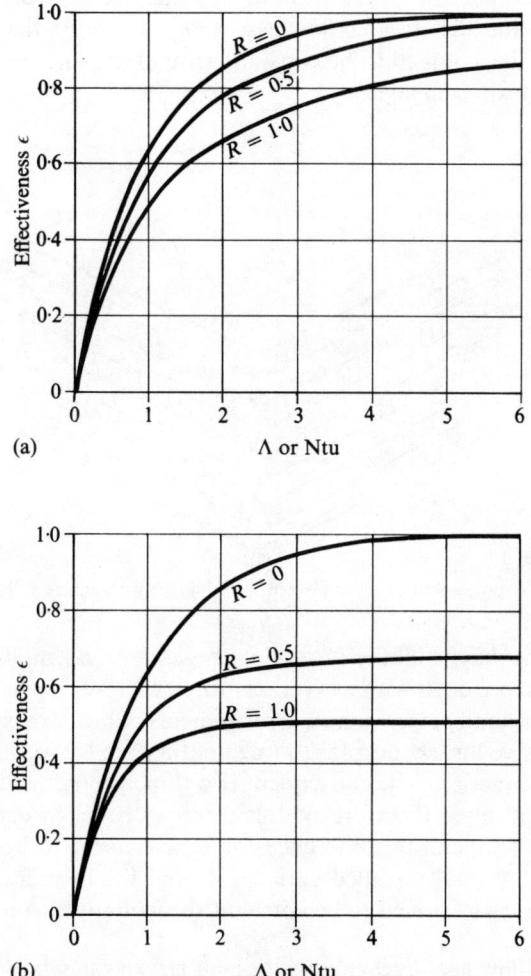

Fig. 10-7   (a) Effectiveness for a contraflow heat exchanger; (b) Effectiveness for a parallel flow heat exchanger

10-4   Cross-flow heat exchangers

Even in the diagrammatic representation of the ideal contraflow heat
exchanger of Fig. 10-4, the close observer will note that to achieve a realistic
means of entry and exit for the fluids at each end requires that one of the
streams on these regions should flow *across* the other. As a consequence, at
a section like c–c the temperature difference between the two fluids will vary
around the heat transfer surface, being greater at the lower side than at the
upper where the cooler leaving fluid will have received more heat in its passage across the tube. On the other hand, at a mid-section where the element
δA is situated and the streams flow essentially coaxially, the temperatures
can reasonably be assumed to be constant around the surface and changing
only *along* the heat exchanger.

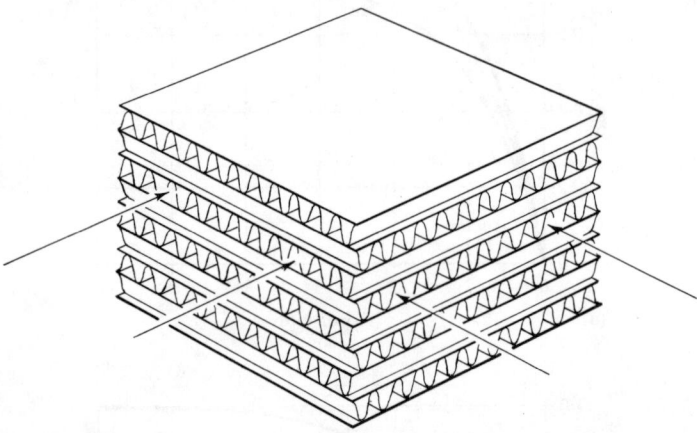

Fig. 10-8   Arrangement of cross-flow heat exchanger with unmixed fluid streams

Although the extent of the cross-flow regions in essentially contraflow
units (such as those dealt with in Sections 10-2 and 10-3) is usually negligible
in its effect on overall performance, arrangements of heat exchanger in which
the streams move largely normal to one another are frequently convenient
in engineering practice. The consequent two-dimensionality of the temperature distribution must therefore be taken into account in determining the
effective mean temperature difference for heat transfer to be used in con-
junction with the total installed surface, A, and the overall coefficient, U,
which is invariably assumed to be constant throughout the unit.

10-4-1   Cross-flow heat exchanger with both streams unmixed

A frequently occurring form of heat exchanger matrix is shown in Fig. 10-8.
The two fluid streams move orthogonally to one another in the alternate
spaces between parallel flat plates. These spaces contain corrugated plates

which act as secondary heat transfer surfaces (the principles of operation of which were discussed in Section 10-1-1) if they are in good thermal contact with the separating flat plates which represent the primary or base heat transfer surfaces.

Fluid entering a passage formed by each corrugation and the adjacent primary surface will experience temperature changes appropriate only to that individual passage, and each fluid stream is therefore said to be *unmixed*. The temperatures within the complete unit can therefore be represented (as shown in Fig. 10-9) in terms of a two-coordinate system formed by viewing the heat exchanger in a direction normal to both flows, and with coordinates measured respectively in the flow directions. As in Fig. 10-9, the x co-ordinate lies in the direction of flow of the hotter fluid, represented as before

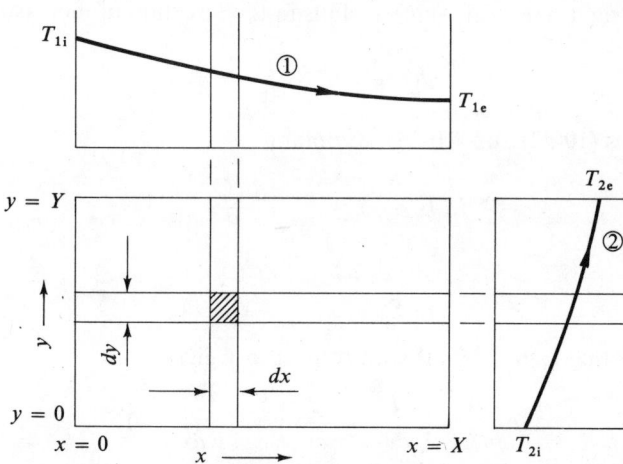

Fig. 10-9 Coordinate system for a cross-flow heat exchanger with both fluids unmixed

by suffix 1; the cooler fluid, suffix 2, flows in the y direction. The inlet and exit boundary conditions in each case are represented, respectively, by suffices 'i' and 'e' and the linear limits of integration are between 0 and X and Y for the two fluids. The overall heat transfer coefficient, U, is assumed constant over the domain, and is referred to the base surface of area XY for a single element of surface, which would, of course, be multiplied by the number of plate-elements in the z direction normal to both flows in designing a practical heat exchanger. The effect upon the effective value of U of the secondary surfaces in a unit like that of Fig. 10-8 would be calculated according to principles set out in Section 10-1-1.

From the definition of U, the rate of heat transfer between the two fluids over the elementary area $dx\,dy$ is given by

$$d\dot{Q} = U(T_1 - T_2)\,dx\,dy \qquad (10\text{-}22)$$

Assuming that the two mass flows over the single plate-element under consideration are distributed uniformly over the respective dimensions X and Y, we shall have

$$dm_1 = \frac{m_1}{Y} \, dy$$

$$dm_2 = \frac{m_2}{X} \, dx$$

so that the heat balances for the two streams are

$$d\dot{Q} = -\frac{m_1 C_1}{Y} \frac{\delta T_1}{\delta x} \, dx \, dy \qquad (10\text{-}23a)$$

(negative right-hand side since T_1 falls in the direction of increasing x), and

$$d\dot{Q} = \frac{m_2 C_2}{X} \frac{\partial T_2}{\partial y} \, dy \, dx \qquad (10\text{-}23b)$$

From Eqns (10-22) and (10-23) we obtain

$$U(T_1 - T_2) = -\frac{m_1 C_1}{Y} \frac{\partial T_1}{\partial x} \qquad (10\text{-}24a)$$

$$-\frac{m_1 C_1}{Y} \frac{\partial T_1}{\partial x} = \frac{m_2 C_2}{X} \frac{\partial T_2}{\partial y} \qquad (10\text{-}24b)$$

Differentiating Eqn (10-24a) with respect to y gives

$$\frac{UY}{m_1 C_1} \left(\frac{\partial T_1}{\partial y} - \frac{\partial T_2}{\partial y} \right) + \frac{\partial^2 T_1}{\partial y \, \partial x} = 0$$

and substituting $\partial T_2 / \partial y$ from Eqn (10-24b) leads to

$$\frac{\partial^2 T_1}{\partial y \, \partial x} + \frac{UY}{m_1 C_1} \frac{\partial T_1}{\partial y} + \frac{UX}{m_2 C_2} \frac{\partial T_1}{\partial x} = 0 \qquad (10\text{-}25a)$$

A similar equation can, of course, be obtained by eliminating T_1 between Eqns (10-22), (10-23), and (10-24). For more general use, Eqn (10-25a) and its equivalent for T_2 can be written in dimensionless form in terms of the parameters

$$\bar{x} = \frac{x}{X}; \qquad \bar{y} = \frac{y}{Y}; \qquad \Lambda_1 = \frac{UA}{m_1 C_1}; \qquad \Lambda_2 = \frac{UA}{m_2 C_2}$$

$$R = \frac{m_2 C_2}{m_1 C_1} = \frac{\Lambda_1}{\Lambda_2}; \qquad \bar{T}_1 = \frac{T_1 - T_{2i}}{T_{1i} - T_{2i}}; \qquad \bar{T}_2 = \frac{T_2 - T_{2i}}{T_{1i} - T_{2i}}$$

Of these parameters, Λ has the form used in Sections 10-2 and 10-3 to represent the thermal performance of the contraflow heat exchanger, but is now specifically related to one of the heat capacities. The dimensionless temperatures, $\bar{T}_1$ and $\bar{T}_2$, take the form associated with the previously defined effectiveness, ε, and if $R \leqslant 1$, an effectiveness of 100 per cent would correspond to $\bar{T}_2$ becoming unity. If $R > 1$, this ultimate in performance would correspond to $\bar{T}_1$ becoming zero.

The dimensionless form of Eqn (10-25a) is found to be

$$\frac{\partial^2 \bar{T}_1}{\partial \bar{y} \, \partial \bar{x}} + \Lambda_1 \frac{\partial \bar{T}_1}{\partial \bar{y}} + \Lambda_2 \frac{\partial \bar{T}_1}{\partial \bar{x}} = 0 \qquad (10\text{-}25b)$$

with the boundary conditions

$$\bar{x} = 0: \qquad \bar{T}_1 = 1$$

$$\bar{y} = 0: \qquad \bar{T}_2 = 0$$

A direct solution cannot be found to Eqns (10-25), which are partial differential equations in the single dependent variable T_1 or $\bar{T}_1$. Numerical solutions can be obtained by finite difference methods similar to those described in Chapter 4, and Nusselt[3] has obtained a series solution. Nusselt's results are presented in terms of the mean outlet temperatures from either side, that is,

$$\bar{T}_{1me} = \int_0^1 \bar{T}_{1e} \, d\bar{y} = \frac{T_{1me} - T_{2i}}{T_{1i} - T_{2i}}$$

$$\bar{T}_{2me} = \int_0^1 \bar{T}_{2e} \, d\bar{x} = \frac{T_{2me} - T_{2i}}{T_{1i} - T_{2i}}$$

Now the rate of heat exchange in the unit is

$$\dot{Q} = \dot{m}_2 C_2 (T_{2me} - T_{2i}) = \dot{m}_1 C_1 (T_{1i} - T_{1me})$$

$$= \dot{m}_2 C_2 \bar{T}_{2me}(T_{1i} - T_{2i}) = \dot{m}_1 C_1 (1 - \bar{T}_{1me})(T_{1i} - T_{2i})$$

Also, an overall effective mean temperature difference in the unit is defined by the equation

$$\dot{Q} = UA \, \Delta T_{\mathrm{m}}$$

so that

$$\overline{\Delta T_{\mathrm{m}}} = \frac{\Delta T_{\mathrm{m}}}{T_{1i} - T_{2i}} = \frac{\bar{T}_{2me}}{\Lambda_2} = \frac{1 - \bar{T}_{1me}}{\Lambda_1} \qquad (10\text{-}26)$$

A logarithmic mean temperature difference ΔT_l may be determined for the present situation, from Eqn (10-17) of Section 10-3 as

$$\Delta T_l = \frac{(T_{1i} - T_{2me}) - (T_{1me} - T_{2i})}{\ln \left[(T_{1i} - T_{2me})/(T_{1me} - T_{2i}) \right]}$$

or, in dimensionless form as

$$\overline{\Delta T_l} = \frac{\Delta T_l}{T_{1i} - T_{2i}} = \frac{(1 - \overline{T}_{2me}) - \overline{T}_{1me}}{\ln\left[(1 - \overline{T}_{2me})/\overline{T}_{me}\right]} \qquad (10\text{-}27)$$

The ratio of the effective mean temperature difference for the cross-flow heat exchanger as here considered, and given by Eqn (10-26), to the mean temperature difference for a contraflow unit giving the same rates of heat flow, and given by Eqn (10-27), may be determined from the solution to Eqn (10-25a). Some values of this ratio, usually given the symbol, F, where

$$F = \frac{\overline{\Delta T}_m}{\overline{\Delta T}_l} \qquad (10\text{-}28)$$

and derived from Nusselt's[3] solution, are given in Table 10-1 below.

Table 10-1   Values of F for unmixed cross-flow heat exchanger

$\overline{\Delta T}_{2ml}$	0	0·1	0·2	0·3	0·4	0·5	0·6	0·7	0·8	0·9	1·0
$R = \frac{1}{2}$	1	0·998	0·994	0·988	0·975	0·955	0·935	0·891	0·824	0·702	0
$= 1$	1	0·996	0·988	0·974	0·948	0·910	0·835	0·710	0·500	0·220	—
$= 2$	1	0·993	0·975	0·935	0·824	—	—	—	—	—	—

Note that the ratio is indeterminate when the effectiveness ($= \overline{T}_{2ml}$ for $R \leqslant 1$) is unity and when the product $R\overline{T}_{2ml}$ exceeds unity so that the fluid of lesser thermal capacity (the hotter in this case) would be cooled to less than the cold fluid inlet temperature.

Table 10-1 shows that this cross-flow heat exchanger is less effective than the equivalent contraflow unit, although to a reduced extent at the points of lower effectiveness. To attain high effectiveness from cross-flow units it is best to design individual units of lower performance and to link them in series contraflow, as will be seen in Section 10-5.

10-4-2   Cross-flow heat exchanger with both streams mixed

Consider a heat exchanger comprising a shell through which fluid 1 flows over a system of tubes laid, as shown in Fig. 10-10, and conveying internally fluid 2 generally across stream 1. If there is a very large number of *passes*, or longitudinal runs of tube, then it may be assumed that the rise in temperature of fluid 2 in each is differentially small and the temperature effectively constant in each pass at, say, T_{2y}. The shell side fluid (fluid 1) flowing over the tubes is not constrained into separate streamwise passages as were the flows in the heat exchanger of the previous section, and the general turbulence of the flow will cause considerable cross-stream mixing. We shall suppose that this is sufficient to produce a uniform cross-stream temperature at each section x, say T_{1x}. This flow is said to be *fully mixed*, and the flow within the tubes behaves as if it were also fully mixed. With the heat exchanger of Fig. 10-10 we have therefore the other extreme of the cross-flow unit, compared with the previous section, in which there was no mixing in either stream.

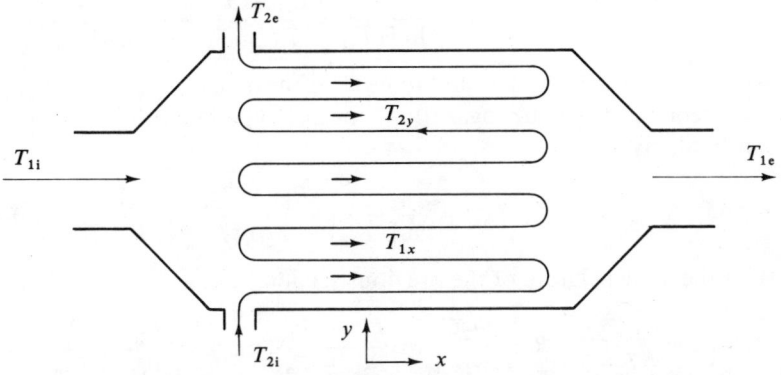

Fig. 10-10   Cross-flow heat exchanger with both streams mixed

Now, suppose that there exists a mean temperature for stream 2, say T_{2m}, which yields the same overall heat transfer rate as with the actual distribution of temperature. Then for the elemental length dx

$$d\dot{Q} = U(T_1 - T_{2m})Y\,dx$$

where Y is the heat transfer surface area per unit length of heat exchanger, or, for a plate heat exchanger, the dimension in the y direction. Then integrating over the full length of unit in the y direction gives

$$\dot{Q} = U \int_0^X (T_1 - T_{2m})Y\,dx = UA(T_{1m} - T_{2m})$$

By an analogous argument for fluid 2, we obtain

$$\dot{Q} = U \int_0^Y (T_{1m} - T_2)X\,dy = UA(T_{1m} - T_{2m})$$

This identical result for $\dot{Q}$ confirms that, irrespective of the variation of temperature in the x or y directions, we can indeed assume that there is a constant average temperature in the other fluid which will give the same rate of heat transfer. Thus, if we consider, first, the cooler fluid to be at a constant mean temperature, T_{2m}, the concept of the logarithmic mean temperature difference applies for the variation in temperature of the hotter fluid, so that for this we may write

$$\Delta T_m = T_{1m} - T_{2m} = \frac{(T_{1i} - T_{2m}) - (T_{1e} - T_{2m})}{\ln [(T_{1i} - T_{2m})/(T_{1e} - T_{2m})]}$$

$$= \frac{T_{1i} - T_{1e}}{\ln [(T_{1i} - T_{2m})/(T_{1e} - T_{2m})]} \qquad (10\text{-}29)$$

Equally, assuming the hotter fluid to be at its mean temperature, and applying the expression for the logarithmic mean temperature difference to the cooler fluid, gives

$$\Delta T_m = T_{1m} - T_{2m} = \frac{T_{2e} - T_{2i}}{\ln [(T_{1m} - T_{2i})/(T_{1m} - T_{2e})]} \qquad (10\text{-}30)$$

With the nomenclature of the previous section, so that

$$F = \frac{\Delta T_m}{\Delta T_l}, \qquad R = \frac{\dot{m}_2 C_2}{\dot{m}_1 C_1} = \frac{T_{1i} - T_{1e}}{T_{2e} - T_{2i}}, \qquad \varepsilon = \frac{T_{2e} - T_{2i}}{T_{1i} - T_{2i}}$$

and putting $(T_{1i} - T_{2m})/(T_{1e} - T_{2m}) = \lambda$ for convenience, we obtain from Eqn (10-29),

$$F = \frac{T_{1i} - T_{1e}}{\ln \lambda} \frac{\ln [(T_{1i} - T_{2e})/(T_{1e} - T_{2i})]}{(T_{1i} - T_{2e}) - (T_{1e} - T_{2i})} \qquad (10\text{-}31)$$

But

$$1 - \varepsilon = \frac{T_{1i} - T_{2e}}{T_{1i} - T_{2i}} \quad \text{and} \quad 1 - \varepsilon R = \frac{T_{1e} - T_{2i}}{T_{1i} - T_{2i}}$$

Also

$$\dot{Q} = \dot{m}_1 C_1 (T_{1i} - T_{1e})$$

$$= \dot{m}_2 C_2 (T_{2e} - T_{2i})$$

so that Eqn (10-31) becomes

$$F = \frac{\ln [(1 - \varepsilon)/(1 - \varepsilon R)]}{\ln \lambda} \frac{\dot{Q}/\dot{m}_1 C_1}{(\dot{Q}/\dot{m}_1 C_1) - (\dot{Q}/\dot{m}_2 C_2)}$$

$$= \frac{R}{R - 1} \frac{\ln [(1 - \varepsilon)/(1 - \varepsilon R)]}{\ln \lambda} \qquad (10\text{-}32)$$

Equation (10-32) gives a relation for F, the ratio of effective mean temperature in the mixed cross-flow heat exchanger to the logarithmic mean temperature difference based upon the prevailing inlet and outlet temperatures, and which would apply for true contraflow. This equation, however, includes the parameter λ which is implicitly determined by the other variables in the system, R and the effectiveness ε. However, equating Eqns (10-29) and (10-30) gives

$$\frac{1}{R} = \frac{T_{2e} - T_{2i}}{T_{1i} - T_{1e}}$$

$$= \frac{\ln\left[(T_{1m} - T_{2i})/(T_{1m} - T_{2e})\right]}{\ln \lambda}$$

Therefore

$$\frac{T_{1m} - T_{2i}}{T_{1m} - T_{2e}} = \lambda^{1/R}$$

Now

$$\frac{\lambda^{1/R}}{\lambda^{1/R} - 1} = \frac{T_{1m} - T_{2i}}{T_{2e} - T_{2i}}$$

and

$$\frac{R\lambda}{\lambda - 1} = \frac{\dot{m}_2 C_2}{\dot{m}_1 C_1} \frac{T_{1i} - T_{2m}}{T_{1i} - T_{1e}} = \frac{T_{1i} - T_{2m}}{T_{2e} - T_{2i}}$$

and from Eqn (10-29)

$$\frac{R}{\ln \lambda} = \frac{\dot{m}_2 C_2}{\dot{m}_1 C_1} \frac{T_{1m} - T_{2m}}{T_{1i} - T_{1e}} = \frac{T_{1m} - T_{2m}}{T_{2e} - T_{2i}}$$

so that

$$\frac{\lambda^{1/R}}{\lambda^{1/R} - 1} + R\left(\frac{\lambda}{\lambda - 1} - \frac{1}{\ln \lambda}\right) = \frac{T_{1m} - T_{2i}}{T_{2e} - T_{2i}} + \frac{T_{1i} - T_{2m}}{T_{2e} - T_{2i}} + \frac{T_{1m} - T_{2m}}{T_{2e} - T_{2i}}$$

$$= \frac{1}{\varepsilon} \qquad (10\text{-}33)$$

Equation (10-33) cannot be solved explicitly for λ, but for given values of the flow ratio, R, corresponding values of λ and ε can be obtained and substituted into Eqn (10-32) to find the effective temperature difference ratio, F. Table 10-2 gives some computed values of F.

Table 10-2 Values of F for a mixed cross-flow heat exchanger

ε	0	0·1	0·2	0·3	0·4	0·5
$R = 1$ F	1·0	0·998	0·996	0·969	0·924	0·798
$R = \frac{1}{2}$ F	1·0	1·0	1·0	0·987	0·967	0·945

Comparison with the values for the unmixed unit analysed in the preceding section shows a significant difference in performance only as the maximum effectiveness is approached. This is found in the unit with $R = 1$ to be about 0·565, and occurs where the parameter $\Lambda = UA/\dot{m}C$ has a value of almost

3, so that the paradoxical situation is found in which an increase in the surface area above this value reduces the effectiveness of the unit. Clearly, it is inefficient to operate a heat exchanger under these conditions, and for high effectiveness it is best to operate cross-flow units of low individual effectiveness in series counterflow, when the ideal contraflow performance can be approached, as will be seen in the next section.

10-5 Thermal design of heat exchangers

There are several ways in which the data specifying the thermal performance of a heat exchanger can be presented, and the nature of the data affects the technique of the design employed. Thus, if the temperatures of the fluids concerned in the overall heat transfer process are known, so that the overall mean temperature difference can be immediately determined according to the principles of the preceding sections, then the design procedure becomes entirely straightforward. For example, consider the simplest possible case of the heat exchanger for a gas turbine plant specified below:

10-5-1 The contraflow unit with $R = 1$

Example 1. A true contraflow heat exchanger for a gas turbine plant is to have a thermal ratio of 80 per cent. The air from the compressor enters the unit at 350°C and receives heat from the turbine exhaust at 700°C. The gas side heat transfer coefficient is 20 Chu/ft^2 h °C and on the air side the coefficient is 30 Chu/ft^2 h °C. Determine the surface area required for a mass flow rate of 10 lb/s on each side. Assume that the specific heat at constant pressure around the cycle is unchanged at 0·25 Chu/lb °C.

In this case the thermal ratio, or effectiveness as defined by Eqn (10-11a), is to be 80 per cent, so

$$0.80 = \frac{T_{2e} - T_{2i}}{T_{1i} - T_{2i}}$$

where suffix 2 refers to the air side, 1 to the gas side, and e and i denote the exit and inlet conditions. Then, with the specified temperatures,

$$0.80 = \frac{T_{2e} - 350}{700 - 350}$$

whence $$T_{2e} = 630°C$$

In this heat exchanger with the mass flow rates and the specific heats on each side equal, that is $\dot{m}_1 C_1 = \dot{m}_2 C_2$, the temperature difference between the fluids is constant through the unit at

$$T_{1i} - T_{2e} = T_{1e} - T_{2i} = 700 - 630 = 420 - 350 = 70°C$$

The overall heat transfer coefficient U is given by Eqn (10-3c) as

$$\frac{1}{U} = \frac{1}{20} + \frac{1}{30}$$

that is,
$$U = \frac{20 \times 30}{20 + 30} = 12 \text{ Chu/ft}^2 \text{ h } °C$$

Thus, in the equations for the overall rate of heat transfer [Eqns (10-10)],

$$\dot{Q} = \dot{m}_1 C_1 (T_{1i} - T_{1e})$$
$$= \dot{m}_2 C_2 (T_{1e} - T_{2i})$$
$$= 10 \times 0.25 \times 280 = 700 \text{ Chu/s}$$

and Eqn (10-9), which gives the rate of heat flow in terms of the overall heat transfer coefficient, and the mean temperature difference, becomes

$$Q = UA \, \Delta T$$

that is,
$$A = \frac{700 \times 3,600}{12 \times 70} = 3,000 \text{ ft}^2$$

We have thus determined the heat transfer surface area required within this unit to meet the specified duty. For this simple and, in practice, unusual case of the contraflow heat exchanger with $R = \dot{m}_1 C_1 / \dot{m}_2 C_2 = 1$ we could, of course, have used Eqn (10-12) directly, which would be $0.80 = \Lambda/(\Lambda + 1)$, that is, $\Lambda = 4 = UA/\dot{m}C$, whence

$$A = \frac{4 \times 10 \times 25 \times 3,600}{12} = 3,000 \text{ ft}^2$$

The designer's task is now to dispose the surface in a practical way; for example, in this case 1,145 tubes 10 ft long and 1 in diameter would give the required surface area. The flow of air and gas through and over these tubes would, of course, determine the heat transfer coefficients in the system according to the correlations given in Chapter 6, as well as the inevitable irreversible viscid pressure that would also in practice be specified. A complete heat exchanger design study almost inevitably becomes a matter of trial and error to obtain an acceptable combination of all the large number of variables involved. In this section, however, consideration is restricted to the thermal consequences of the principles of the earlier sections in this chapter. We shall particularly consider the effects of departures from the simple, ideal contraflow example already examined.

10-5-2 Cross-flow heat exchangers with $R = 1$

Example 2. Suppose the contraflow unit is to be replaced, because of convenience of installation, by a cross-flow unit. Examine the surface area requirements when

(a) both fluids are unmixed;

(b) both fluids are mixed.

(a) Table 10-1 shows that for the case of $R = 1$ and an effectiveness (represented by $\overline{T}_2$ in the table) of 0·8, the ratio of the effective mean temperature difference to the logarithmic mean (in this case the constant) temperature difference corresponding to the inlet and outlet temperatures is 0·5. Thus, ΔT becomes $0·5 \times 70$, so that twice the surface area is required (that is, 6,000 ft^2).

10-5-3 The series contraflow heat exchanger

(b) In Section 10-4-2 it was noted that the effectiveness of the cross-flow unit with both fluids mixed, as given by the equations there derived, reaches a maximum of 0·565. To obtain the required thermal ratio of 0·8 it is necessary to utilize more effectively the temperature differences between the fluids in the complete system, which may conveniently be done by arranging a number of cross-flow units in series contraflow, as shown in Fig. 10-11.

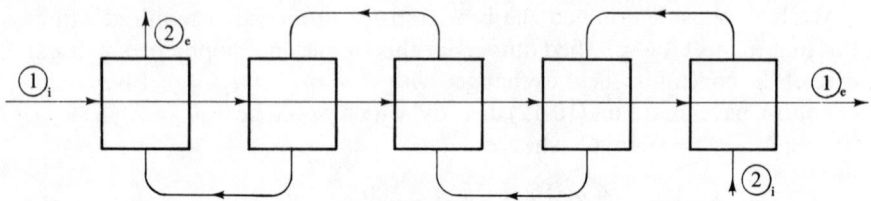

Fig. 10-11 Arrangement of a series contraflow heat exchanger

Suppose in general there are n cross-flow units of identical thermal performance; that is, each of effectiveness ε_s. Then the temperature change of each fluid in each unit will be the same, that is

$$\frac{T_{2e} - T_{2i}}{n} = \frac{T_{1i} - T_{1e}}{n} = DT \text{ (say)}$$

If ΔT is the mean effective temperature difference corresponding to that for the equivalent true contraflow unit, the effectiveness will be, as given by Eqns (10-9) and (10-11b),

$$\varepsilon_s = \frac{DT}{\Delta T + DT} \tag{10-34}$$

Similarly, for the complete unit

$$\varepsilon_0 = \frac{nDT}{\Delta T + nDT}$$

which, on substituting for ΔT from Eqn (10-34), becomes

$$\varepsilon_0 = \frac{n\varepsilon_s}{1 + (n-1)\varepsilon_s} \quad \text{or} \quad \varepsilon_s = \frac{\varepsilon_0}{n - \varepsilon_0(n-1)} \tag{10-35}$$

Thus, for the case of Fig. 10-9, with $n = 5$, and for the present requirement that $\varepsilon_0 = 0.8$, we have

$$\varepsilon_s = \frac{0.8}{5 - 4 \times 0.8} = 0.445$$

Thus, for the required effectiveness of 80 per cent in the complete series-flow unit, each element, if it operated in true contraflow, would need to satisfy Eqn (10-12), as

$$0.445 = \frac{\Lambda}{\Lambda + 1}$$

so that

$$\Lambda = \frac{UA}{\dot{m}C} = 0.802$$

Interpolating in Table 10-2 for a unit effectiveness of 0.445 shows that the ratio of the effective mean temperature difference in the mixed-fluid cross-flow unit to that in the contraflow unit with the same mean fluid outlet temperatures is 0.89. Thus, for each element in the series-flow unit, the required surface area will be given by

$$A = \frac{0.802 \times 10 \times 0.25 \times 3,600}{0.89 \times 12} = 676 \text{ ft}^2$$

and if the complete unit of five elements is to give an effectiveness of 80 per cent, the required installed heat transfer area will need to be $5 \times 676 = 3,380 \text{ ft}^2$; that is, about 13 per cent more than the true contraflow unit.

If, in part (a) of this design example, five elements of an unmixed fluid cross-flow unit had been used, Table 10-1 shows the temperature difference parameter, F, to be 0.932 so that the total installed area in this case is only slightly less than for the mixed flow case at $3,220 \text{ ft}^2$.

10-5-4 The contraflow heat exchanger with $R \neq 1$

Up to this point we have considered only heat exchangers in which the 'water equivalents' of the two fluid streams have been the same, so that $R = 1$. Let us now consider the effects of the temperature asymmetry in the system when this is no longer the case, as in the following example.

Example 3. In the heat exchanger of Example 1, Section 10-5-1, the mass flow rate of the cooler air stream is reduced to 5 lb/s, half the previous value. All other independent variables retain their original values, viz. gas and air side heat transfer coefficients, 20 and 30 Chu/ft^2 h °C respectively; gas flow rate, 10 lb/s; gas and air inlet temperatures, 700 and 350°C; mean specific heats, 0·25 Chu/lb °C. The unit is to continue to transfer 80 per cent of the maximum possible heat flow between the two streams.

In this example, the maximum possible heat flow rate would correspond to the air stream being raised to the gas inlet temperature 700°C, as before, so that the effectiveness of 80 per cent corresponds to the air outlet temperature becoming

$$T_{2e} = 350 + 0·80(700 - 350) = 630°C$$

as before. The corresponding fall in gas temperature is given by the heat balance as

$$\dot{Q} = \dot{m}_1 C_1 (T_{1i} - T_{1e}) = \dot{m}_2 C_2 (T_{2e} - T_{2i})$$

$$= 10 \times 0·25 \times (700 - T_{1e}) = 5 \times 0·25 \times (630 - 350)$$

whence

$$\dot{Q} = 350 \text{ Chu/s} \quad \text{and} \quad T_{1e} = 560°C$$

The appropriate temperature difference for heat transfer is the logarithmic mean defined by Eqn (10-17) in this case as

$$\Delta T_e = \frac{(T_{1i} - T_{2e}) - (T_{1e} - T_{2i})}{\ln\left[(T_{1i} - T_{2e})/(T_{1e} - T_{2i})\right]}$$

$$= \frac{(700 - 630) - (560 - 350)}{\ln(70/210)} = \frac{-140}{-1·099} = 127·4°C$$

This value may be compared with the simple arithmetic mean temperature difference, which is 140°C, that is 10 per cent greater in this example.

The surface area, A, required for heat transfer is determined as before, for the overall heat transfer coefficient $U = 12$ Chu/ft^2 h °C as

$$A = \frac{Q}{U \Delta T_e}$$

$$= \frac{350 \times 3,600}{12 \times 127·4} = 824 \text{ ft}^2$$

10-5-5   Cross-flow heat exchangers with $R \neq 1$

If the true contraflow arrangement is replaced by a cross-flow unit with both fluids unmixed, Table 10-1 shows that with the effectiveness 80 per cent

$(= \bar{T}_{2m})$ and $R = \frac{1}{2}$, the effective mean temperature difference for heat transfer is 0·824 of the logarithmic mean value, so that the required surface area would need to be increased in the ratio 1/0·824 for the same performance, that is to 1,000 ft².

We noted in the earlier example with $R = 1$ that it was not possible to attain the specified effectiveness of 80 per cent, or indeed any value above 0·565, with a cross-flow heat exchanger with mixed fluid streams. In that example a series of cross-flow units were arranged in contraflow to utilize more effectively the temperature differences between the streams and a simple relationship was derived between the effectiveness of the component heat exchangers, ε_s, and that of the complete series flow unit, ε_0, and given by Eqn (10-35).

When the 'water equivalents' or thermal capacities of the two streams are no longer equal (that is, $R \neq 1$), so that the temperature difference between them changes through the complete heat exchanger, the rates of heat transfer in each of the component units correspondingly vary. The analysis of the performance of the units in series is thus more complex, but the overall effectiveness may be obtained as follows.

10-5-6 Series contraflow heat exchangers with $R \neq 1$

Equation (10-16) gives the natural logarithm of the ratio of the temperature differences between the two fluid streams at the opposite ends of a heat exchanger as

$$-\ln \frac{\Delta T_b}{\Delta T_a} = UA \left(\frac{1}{\dot{m}_1 C_1} - \frac{1}{\dot{m}_2 C_2} \right) \tag{10-16}$$

In this case let suffix b refer to the gas outlet end and a to the air outlet of the series contraflow unit. After the gas has traversed one of the n units of equal surface area, the temperature difference between the streams will be ΔT_1, where

$$-\ln \frac{\Delta T_1}{\Delta T_a} = \frac{UA}{n} \left(\frac{1}{\dot{m}_1 C_1} - \frac{1}{\dot{m}_2 C_2} \right) \tag{10-36a}$$

that is,

$$\frac{\Delta T_1}{\Delta T_a} = \left(\frac{\Delta T_b}{\Delta T_a} \right)^{1/n} \tag{10-36b}$$

This assumes, of course, that U is constant through the complete unit, of which the overall effectiveness (ε_0) is, in the nomenclature of Fig. 10-12,

$$\varepsilon_0 = \frac{T_{2e} - T_{2i}}{T_{1i} - T_{2i}} \tag{10-37}$$

Now
$$T_{2e} - T_{2i} = \frac{\dot{Q}}{\dot{m}_2 C_2}$$

where $\dot{Q}$ is the rate of heat exchange in the whole unit. Also,

$$T_{1i} - T_{2i} = T_{1i} - T_{2i_1} + (T_{2e_1} - T_{2i_2}) + (T_{2i_2} - T_{2i_3}) + \cdots$$
$$+ (T_{2i_{n-1}} - T_{2i_n})$$
$$= T_{1i} - T_{2i_1} + \dot{Q}_{n-a}/\dot{m}_2 C_2$$

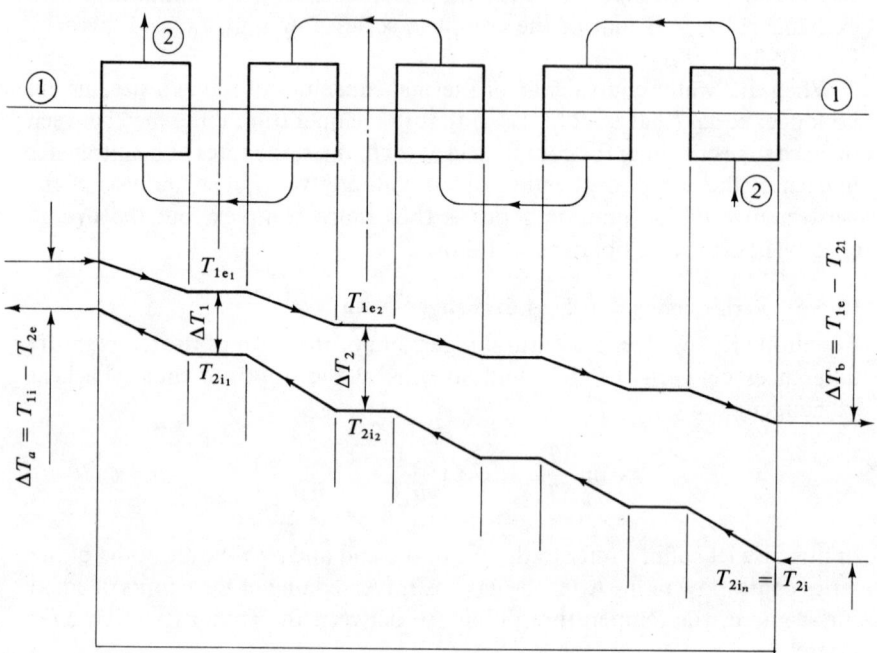

Fig. 10-12 Series contraflow unit with $R \neq 1$

where $\dot{Q}_{n-a}$ is the heat transfer in all the elements except that at end a. Each element has the same effectiveness, ε_s, so that for ε_s at the end a,

$$\varepsilon_s = \frac{T_{2e} - T_{2i_1}}{T_{1i} - T_{2i_1}} = \frac{\dot{Q}_a}{\dot{m}_2 C_2} \frac{1}{T_{1i} - T_{2i_1}}$$

where $\dot{Q}_a$ is the heat transfer in the element at a. Therefore,

$$T_{1i} - T_{2i} = \frac{\dot{Q}_a}{\dot{m}_2 C_2} \frac{1}{\varepsilon_s} + \frac{\dot{Q}_{n-a}}{\dot{m}_2 C_2}$$

and since $\dot{Q} = \dot{Q}_a + \dot{Q}_{n-a}$, we obtain

$$T_{1i} - T_{2i} = \frac{\dot{Q}}{\dot{m}_2 C_2} + \frac{\dot{Q}_a}{\dot{m}_2 C_2}\left(\frac{1}{\varepsilon_s} - 1\right) \qquad (10\text{-}38)$$

Substituting for $T_{1i} - T_{2i}$ in Eqn (10-38) in the expression for ε_0 [Eqn (10-37)] gives

$$\varepsilon_0 = \frac{\dot{Q}}{\dot{m}_2 C_2}\bigg/\left[\frac{\dot{Q}}{\dot{m}_2 C_2} + \frac{\dot{Q}_a}{\dot{m}_2 C_2}\left(\frac{1}{\varepsilon_s} - 1\right)\right]$$

$$= \frac{1}{1 + (Q_a/Q)(1/\varepsilon_s - 1)} \qquad (10\text{-}39)$$

From the definition for U, Eqn (10-9),

$$\frac{\dot{Q}_a}{\dot{Q}} = \frac{A_a \overline{\Delta T_a}}{A \overline{\Delta T}} = \frac{1}{n}\frac{\overline{\Delta T_a}}{\overline{\Delta T}}$$

where $\overline{\Delta T_a}$ and $\overline{\Delta T}$ are the mean effective temperature differences for the element and the complete unit, respectively. Observe that if the temperature difference between the fluids is constant, as when $R = 1$, Eqn (10-39) reduces to Eqn (10-35), previously derived as the relation between ε_0 and ε_s for this special case.

For $\dot{Q}_a/\dot{Q}$ in Eqn (10-39), we may write

$$\frac{\dot{Q}_a}{\dot{Q}} = \frac{U\overline{\Delta T_a}A/n}{U\overline{\Delta T}A} \qquad (10\text{-}40)$$

and substituting the logarithmic mean temperature difference expression from Eqn (10-17) for $\overline{\Delta T_a}$ and $\overline{\Delta T}$ gives

$$\frac{\dot{Q}_a}{\dot{Q}} = \frac{1}{n}\frac{\Delta T_a - \Delta T_1}{\ln(\Delta T_a/\Delta T_1)}\frac{\ln(\Delta T_a/\Delta T_b)}{\Delta T_a - \Delta T_b} \qquad (10\text{-}41)$$

Substituting further for ΔT_1 from Eqn (10-36b) gives

$$\frac{\dot{Q}_a}{\dot{Q}} = \left[\frac{1 - (\Delta T_b/\Delta T_a)^{1/n}}{1 - (\Delta T_b/\Delta T_a)}\right] \qquad (10\text{-}42)$$

The ratio of the temperature differences at the ends of the complete heat exchanger, $\Delta T_b/\Delta T_a$, is now substituted from Eqn (10-16), that is,

$$\frac{\Delta T_b}{\Delta T_a} = \exp\frac{UA}{\dot{m}_2 C_2}(1 - R)$$

$$= \exp n\Lambda_s(1 - R) \qquad (10\text{-}43)$$

where
$$\Lambda_s = \frac{UA/n}{\dot{m}_2 C_2}$$

which is the dimensionless heat exchanger performance parameter for one stage of the complete series flow unit. But from Eqn (10-20) we have for the effectiveness of a stage

$$\varepsilon_s = \frac{1 - \exp(-k)}{1 - R\exp(-k)}$$

where
$$k = \Lambda_s(1 - R)$$

Thus
$$\exp k = \frac{1 - R\varepsilon_s}{1 - \varepsilon_s} \tag{10-44}$$

and
$$\frac{\Delta T_b}{\Delta T_a} = \exp nk \tag{10-45}$$

so that
$$\frac{\dot{Q}_a}{\dot{Q}} = \frac{1 - (1 - R\varepsilon_s)/(1 - \varepsilon_s)}{1 - [(1 - R\varepsilon_s)/(1 - \varepsilon_s)]^n} \tag{10-46}$$

Substituting from this Eqn (10-46) for $\dot{Q}_a/\dot{Q}$ into Eqn (10-39) gives, for ε_0, after some rearrangement,

$$\varepsilon_0 = \frac{[(1 - R\varepsilon_s)/(1 - \varepsilon_s)]^n - 1}{[(1 - R\varepsilon_s)/(1 - \varepsilon_s)]^n - R} \tag{10-47}$$

For the heat exchanger specified in Example 3, Eqn (10-47) shows that if five units are used in the complete heat exchanger (that is, $n = 5$), with $R = \frac{1}{2}$ and $\varepsilon_0 = 80$ per cent, ε_s must be 0·33. If these units each individually operated in contraflow, Eqn (10-20) and Fig. 10-7(a) show that the corresponding value of Λ_s is 0·44 (for the whole heat exchanger, $\Lambda = 5 \times 0·44 = 2·20$, and substitution of this value into Eqn (10-20) gives $\varepsilon_0 = 80$ per cent).

For the cross-flow units used in the series flow arrangement, interpolation in Table 10-1 shows that for $\varepsilon_s = 0·33$ and $R = \frac{1}{2}$, the ratio of the effective mean temperature difference to the logarithmic mean appropriate to counterflow, F, is 0·984 if both fluids are unmixed. Thus, with the overall heat transfer coefficient, $U = 12$ Chu/ft^2 h °C, air and gas mass flows of 5 and 10 lb/s, respectively, with $C = 0·25$ Chu/lb °C, the required surface area per element is given by

$$A_s = \frac{1}{0·984} \times 0·44 \times \frac{5 \times 0·25 \times 3,600}{12}$$

$$= 168 \text{ ft}^2$$

In the complete unit, with an overall effectiveness $\varepsilon_0 = 80$ per cent, the required surface area for heat transfer is $5 \times 168 = 840$ ft^2, which compares with the value of 824 ft^2 required for the true counterflow case. If both fluids are mixed, the effective temperature difference ratio, F, falls only to 0·980, which is negligibly different from the previous example with both fluids unmixed, especially when account is taken of the assumptions in the theoretical analysis of heat exchangers (particularly concerning the constancy of the overall heat transfer coefficient, U, and the extent of mixing in and between the units). Indeed, the examples of heat exchangers analysed in this text, from the ideal counterflow unit to the extremes of cross-flow units with each fluid stream mixed and unmixed, can be made to represent with adequate accuracy most practical designs, particularly if series contraflow arrangements are used to obtain the necessary effectiveness in the latter cases.

10-5-7   The Λ or Ntu method of heat exchanger design

In the examples so far considered in this section the fluid temperatures have been specified, so that the determination of the mean effective temperature difference for heat transfer between the two fluid streams has been straightforward. This is commonly the situation in heat exchanger design, but sometimes it is necessary to determine the performance of a unit in which the surface area and heat transfer coefficients are known, for example, through a change in operating conditions of an already existing design. Because of the hyperbolic form of the mean temperature difference relationship for all but the simple case of the counterflow unit with $R = 1$, solution of the equations for the unknown outlet temperatures of the fluid streams is usually tedious, involving trial and error methods of solution.

However, when the functional relationship between the temperatures in the unit and the heat transfer parameter is plotted in a convenient form—as in Figs. 10-7(a) and 10-7(b)—this difficulty is circumvented, as shown in the following example.

Example 4. Determine the temperature of the air at outlet from the contraflow heat exchanger specified in Example 1 if the gas flow is increased to 10 kg/s at 1,000°C, and the air flow to 7·5 kg/s at 400°C. The corresponding gas and air side heat transfer coefficients are now 200 and 300 W/m^2 K respectively and the specific heat 1,100 J/kg K for each stream. The surface area for heat transfer remains as previously determined, that is, 76·6 m^2.

The overall heat transfer coefficient, U, is given by Eqn (10-3a) as

$$U = \frac{200 \times 300}{200 + 300} = 120 \text{ W/m}^2 \text{ K}$$

The value of Λ or Ntu is determined as

$$\Lambda \text{ (Ntu)} = \frac{UA}{(\dot{m}C)_{\min}} = \frac{120 \times 76 \cdot 6}{7 \cdot 5 \times 1{,}100} = 1 \cdot 114$$

Also,

$$R = \frac{(\dot{m}C)_{\min}}{(\dot{m}C)_{\max}} = \frac{7 \cdot 5 \times 1{,}100}{10 \times 1{,}100} = 0 \cdot 75$$

From Eqn (10-20) or Fig. 10-7(a), the effectiveness with these values of Λ (Ntu) and R is found to be 0·56, so that

$$0 \cdot 56 = \frac{T_{2e} - T_{2i}}{T_{1i} - T_{2i}} = \frac{T_{2e} - 400}{1{,}000 - 400}$$

The air outlet temperature is thus found to be 736°C, and from the heat balance

$$\dot{Q} = (\dot{m}C)_1(T_{1i} - T_{1e}) = (\dot{m}C)_2(T_{2e} - T_{2i})$$

the outlet gas temperature is found to be 748°C and the rate of heat transfer, $\dot{Q}$, to be 2,770 kW.

Kays and London[2] present plots similar to Fig. 10-7 for a whole range of heat exchanger geometries, so that estimates can rapidly be made by the procedure set out here of the performance of every probable arrangement of heat transfer surface. For many practical cases, however, the most appropriate of the three situations considered in this chapter—the contraflow heat exchanger or the cross-flow exchanger with both fluids mixed or unmixed, compounded as necessary to achieve the desired performance—will be found to be a sufficiently close approximation of the real conditions.

10-6   The regenerator or capacitance heat exchanger

In the heat exchangers considered so far in this chapter, the thermal conditions have implicitly been assumed to be invariant with time. Thus, the rates of heat flow have been steady, with the convection (and radiation, if significant) from the hotter fluid continuously equal to the convection to the colder fluid, and both equal to the steady rate of conduction through the separating heat transfer surface. This is the most common situation in heat exchanger practice, and units which operate in this way are known as *recuperators* or *conductance heat exchangers*.

An alternative class of heat exchangers makes use of the thermal capacity of the heat transfer surface, and these are known as *capacitance heat exchangers*, or *regenerators*. In this type of unit all the heat transfer surface (as distinct from one or other side of a tube or plate) is alternatively swept for a predetermined period of time by the hotter fluid, when it absorbs heat, and then by the colder fluid to which the absorbed heat is given up. Although, as will be seen, a capacitance heat exchanger is inevitably less effective than a recuperator with an equivalent surface for heat transfer, the regenerator is found to be a more convenient alternative for certain special applications.

In steelworks, for example, the temperature and general corrosive state of the gases from the blast furnaces and converters make the design of recuperative heat exchangers impractical. A matrix of temperature and corrosion-resistant ceramic material may, however, be conveniently used as a regenerator, through which the hot exhaust gases are passed during the endothermic steel- or iron-making process of one furnace or converter. While this is recharging with ore or iron, the blast air for a second furnace is preheated by blowing through the ceramic matrix. Another area in which regenerators are finding increasing application is as exhaust heat exchangers in small, particularly automotive, gas turbines. In these engines the viscid pressure loss inevitably associated with convective heat transfer, as observed in Chapter 5, extracts a significant penalty in reduced engine performance, so that the overall heat transfer coefficient, U, must be kept low. As a consequence, for an effectiveness high enough to make the heat exchanger a worthwhile additional complication in the gas turbine cycle, the area for heat transfer, A (as seen in this chapter), must be correspondingly high. To obtain a large surface area within the smallest possible volume requires passages of small diameter, and the large number of these required can be very conveniently incorporated in a regenerator matrix. In practice, to cope more conveniently with the continuous fluid flow through a gas turbine, the matrix is arranged as a disc or drum which is rotated so that the elements comprising it move continuously between the hot turbine exhaust and the cooler compressed air.

10-6-1 'Single-blow' operation

Of the two examples of regenerator heat exchangers described briefly above, that for the steelworks conditions operates with one of the fluids blown through the matrix for a substantial period of time—certainly measured in tens of minutes. For these conditions the analysis of the operation may conveniently be considered throughout the period of the 'blow', as follows.

If $\dot{m}$ is the mass rate of flow of fluid past a section at a distance x from the entrance to the matrix, then between x and $x + dx$ the heat transferred from (or to) the matrix in time dt will comprise two components. The first, dQ_1, will be the heat transferred to the mass of gas $\dot{m}\,dt$ in moving between x and $x + dx$, that is,

$$dQ_1 = \dot{m}\,dt\,C\left(\frac{\partial T}{\partial x}\right)_t dx \qquad (10\text{-}48a)$$

The second item, dQ_2, will represent the heat given to the fluid enclosed between x and $x + dx$ as the temperature level changes with time, and is given by

$$dQ_2 = \rho A_c \, dx \, C \left(\frac{\partial T}{\partial t}\right)_x dt \qquad (10\text{-}48b)$$

in which A_c is the cross-sectional area for fluid flow in the matrix, such that if u is the mean velocity

$$\dot{m} = \rho u A_c \qquad (10\text{-}49)$$

It is assumed throughout that the change in density of the fluid is negligible—an acceptable approximation if mean values are used.

The total heat flow to the fluid, $dQ_1 + dQ_2$, must be equal to the heat lost by the matrix material, that is,

$$dQ_1 + dQ_2 = -C_m m'_m \, dx \left(\frac{\partial T_m}{\partial t}\right)_x dt \qquad (10\text{-}50)$$

which is also given by

$$dQ_1 + dQ_2 = hA' \, dx \, (T_m - T) \, dt \qquad (10\text{-}51)$$

In these equations m'_m and A' are, respectively, the mass of material and the heat transfer area per unit length of matrix; C_m is its specific heat, and T_m its temperature; h is the heat transfer coefficient between matrix and fluid.

This analysis of the heat flow in the regenerator makes implicit assumptions concerning the thermal conductivity of the matrix. The representation of the matrix temperature at the section x by a single temperature, T_m, assumes that the transverse temperature gradients are zero and that the thermal conductivity in this direction is strictly infinite. In practice, the walls are usually thin enough for this to be an acceptable approximation. The heat balance equations [Eqns (10-50) and (10-51)] ignore heat conduction in the x direction, implying that the conductivity in this direction is zero. This again is an acceptable approximation, except for short matrices such as occasionally are found in automotive regenerators.

From Eqns (10-48a), (10-49), (10-50), and (10-51) we obtain, for the heat balances,

$$\frac{\dot{m}C}{hA'} \left[\left(\frac{\partial T}{\partial x}\right)_t + \frac{1}{u}\left(\frac{\partial T}{\partial t}\right)_x\right] = T_m - T \qquad (10\text{-}52)$$

and

$$-\frac{m'_m C_m}{hA'} \left(\frac{\partial T_m}{\partial t}\right)_x = T_m - T \qquad (10\text{-}53)$$

These equations can be simplified by representing time t in terms of the volume of fluid which has passed a given section in the matrix in that time. We suppose that when an arbitrary particle has reached the section x then the volume parameter, say v, has the value zero, so that if this happens at time t_0 we have

$$v = \frac{\dot{m}}{\rho}(t - t_0) \tag{10-54a}$$

Note that t_0 contains x since this is the time that the chosen particle will have taken to reach section x from the entrance to the matrix, where $x = 0$; that is,

$$t_0 = x/u \tag{10-54b}$$

Then with this transformation we obtain

$$\left(\frac{\partial T}{\partial x}\right)_v = \left(\frac{\partial T}{\partial x}\right)_t + \left(\frac{\partial T}{\partial t}\right)_x \frac{dt}{dx}$$

$$= \left(\frac{\partial T}{\partial x}\right)_t + \frac{1}{u}\left(\frac{\partial T}{\partial t}\right)_x$$

and

$$\left(\frac{\partial T_\mathrm{m}}{\partial t}\right)_x = \left(\frac{\partial T_\mathrm{m}}{\partial v}\right)_x \left(\frac{\partial v}{\partial t}\right)_x$$

$$= \frac{\dot{m}}{\rho}\left(\frac{\partial T_\mathrm{m}}{\partial v}\right)_x$$

With these substitutions, Eqns (10-52) and (10-53) become

$$\frac{\dot{m}C}{hA'}\left(\frac{\partial T}{\partial x}\right)_v = T_\mathrm{m} - T \tag{10-55}$$

and

$$\frac{m'C_\mathrm{m}}{hA'}\left(\frac{\partial T_\mathrm{m}}{\partial v}\right)_x = T - T_\mathrm{m} \tag{10-56}$$

These equations can be further simplified if they are expressed in terms of the dimensionless parameters

$$\xi = \frac{hA'}{\dot{m}C}x$$

and

$$\eta = \frac{\rho hA'}{\dot{m}\dot{m}'_\mathrm{m}C_\mathrm{m}}v$$

when they become

$$\left(\frac{\partial T}{\partial \xi}\right)_\eta = T_\mathrm{m} - T \tag{10-57}$$

and
$$\left(\frac{\partial T}{\partial \eta}\right)_\zeta = T - T_m \tag{10-58}$$

These equations for the operation of a regenerator were first derived by Hausen[4]. He obtained an analytical solution by assuming the term $(1/u)(\partial T/\partial t)_x$ in Eqn (10-52) to be negligibly small. This is equivalent to neglecting the thermal storage of the fluid (that is, dQ_2). By taking t_0 in Eqn (10-54a) to be negligible compared with the total blow time t, he obtained numerical predictions of regenerator performance for the long blow times typical of steelworks and the process industries in general. As we shall see in the next section, however, many regenerators operate with short blow times for which Hausen's solutions are inadequate. Johnson[5], using one of the earliest digital computers, obtained numerical solutions to Eqns (10-57) and (10-58) of general validity, which are shown in Fig. 10-13. In this figure the gas temperature is represented non-dimensionally in the ordinate as

$$\bar{T} = \frac{T_L - T_{mo}}{T_i - T_{mo}}$$

in which T_{mo} is the initial matrix temperature, T_i the inlet gas temperature, and T_L the gas temperature after it has traversed the full length, L, of the matrix. Thus, the boundary conditions for Eqns (10-57) and (10-58) are:

$$\zeta = 0: \quad T = T_i \quad \text{for all } \eta$$

$$\eta = 0: \quad T = T_{mo} \quad \text{for all } \zeta$$

The abscissa of Fig. 10-13 is the quotient of η and ζ for the full length, L, of the matrix and the total blow time, $t = \rho v/\dot{m}$, where v is the volume of fluid which has passed through the matrix up to this time. This quotient is known as the 'utilization factor', U, and is given by

$$U = \frac{\rho h A'}{\dot{m} m'_m C_m} \frac{\dot{m}}{\rho} t \frac{\dot{m} C}{h A' L}$$

$$= \frac{\dot{m} C}{m'_m L} t$$

$$= \frac{\text{thermal capacity of fluid passed to time } t}{\text{thermal capacity of matrix}}$$

This is a convenient way of representing the fluid temperature changes through the regenerator matrix, and shows the characteristic S-shaped curves of the transient response, with the temperature ultimately asymptotically approaching the final equilibrium situation of the gas inlet temperature after a slow start and an almost linear central region. This form of temperature–utilization factor plot has been found very convenient for obtaining the heat transfer characteristics of matrix materials.[6] By comparing observed temperature–time variations with those predicted, the appropriate value of Λ (Ntu) may be determined, which is equivalent to ξ_L when determined for the whole matrix; that is,

$$\Lambda = \xi_L = \frac{hA'L}{\dot{m}C} = \frac{hA}{\dot{m}C}$$

as used in the earlier sections of this chapter.

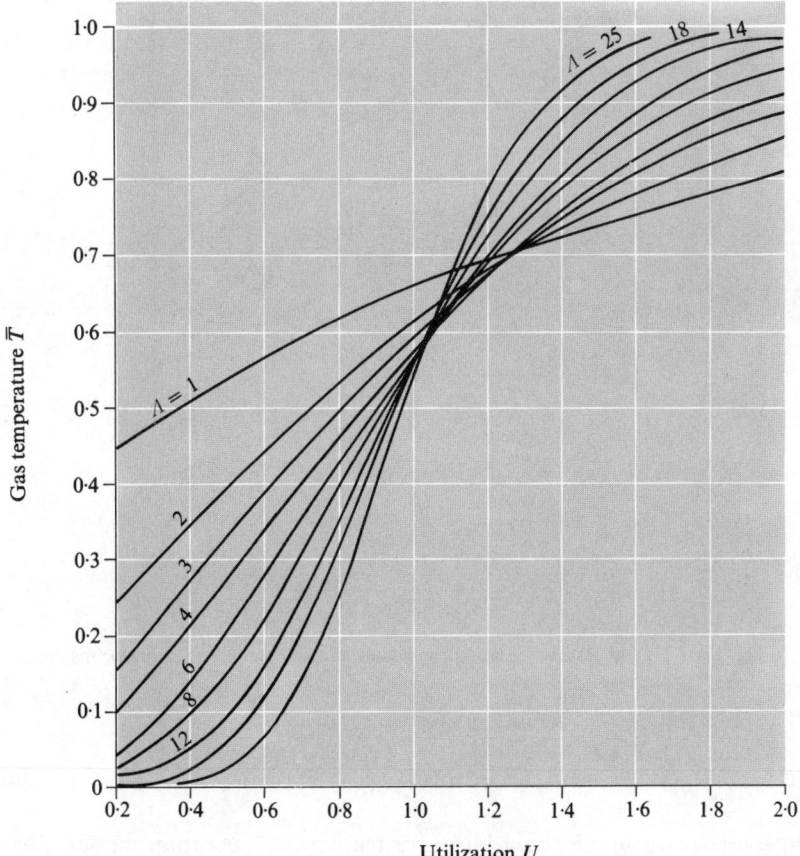

Fig. 10-13 Variation of non-dimensional temperature with time, represented as utilization factor (after Johnson[5])

Observe that, apart from conditions corresponding to Λ less than about 4, each of the curves passes through the point (1·05, 0·59), which provides a useful check upon the accuracy of observed experimental results.

10-6-2 Cyclic operation

If the hot and cold flows through a regenerator matrix are switched at regular intervals, then, after the initial few operations, the matrix temperatures will stabilize to a regular, cyclic variation with time, of the same period as the switching period. Thus, at a section x in the matrix the temperature variations throughout the cycling period, $2t$, will be as shown in Fig. 10-14 if this period is assumed to be equally divided between heating and cooling of the matrix.

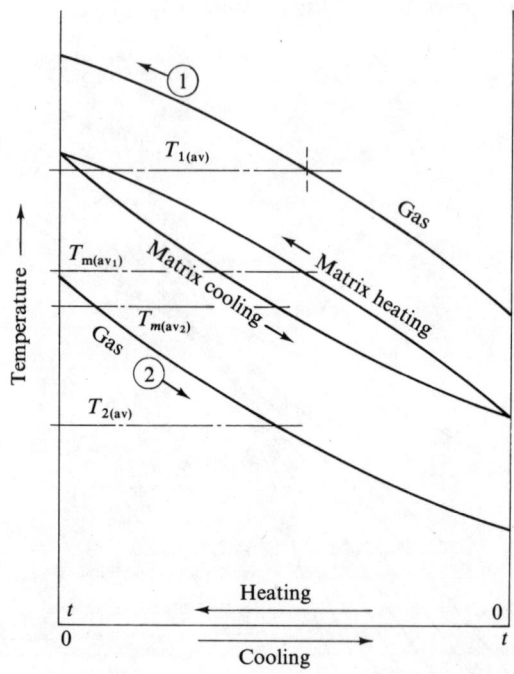

Fig. 10-14   Temperature variations with time at a section in a regenerator matrix

The heat transferred while the hot gas flows will be

$$dQ = h_1 \, dA(T_{1(av)} - T_{m(av_1)})t \qquad (10\text{-}59a)$$

and similarly during the cold blow, for truly cyclic operation the same heat transfer will occur, that is,

$$dQ = h_2 \, dA(T_{m(av_2)} - T_{2(av)})t \qquad (10\text{-}59b)$$

The letters 'av' refer to the average conditions during the time t for the passage of gas 1 (supposed the hotter) and then gas 2, and 'm' refers to the matrix temperature. Combining these equations, and defining the overall coefficient U, as in Eqn (10-36a), as

$$\frac{1}{U} = \frac{1}{h_1} + \frac{1}{h_2}$$

gives

$$d\dot{Q} = U \, dA[(T_{1(av)} - T_{2(av)}) - \Delta T_m] \qquad (10\text{-}60)$$

where

$$\Delta T_m = T_{m(av_1)} - T_{m(av_2)} \qquad (10\text{-}61)$$

which is the fluctuation in temperature of the matrix material. Now, this fluctuation in average temperature for the whole matrix will clearly depend upon the total heat transferred and the thermal capacity of the matrix, so that

$$\Delta T_m = \frac{Q}{m_m C_m}$$

$$= Qt$$

$$= \frac{\dot{m}_2 C_2 (T_{2e} - T_{2i}) t}{m_m C_m}$$

$$= \varepsilon (T_{1i} - T_{2i}) \frac{\dot{m} C_2}{\dot{m}_m C_m}$$

in which ε is the effectiveness of the regenerator if operated in counterflow and $\dot{m}_m$ is the rate at which the mass of matrix material is switched between the hot and cold streams. If this rate is infinite (that is, if the switching time is zero), ΔT_m according to Eqn (10-61) is zero and Eqn (10-60) reduces to the form for the heat transfer rate in a recuperator or simple conductance heat exchanger. From Eqns (10-57) and (10-58), Johnson[5] has calculated the effectiveness of a regenerator operating cyclically in true counterflow with non-zero blowing times, and his results are reproduced as Fig. 10-15. The effectiveness, ε, is plotted against Λ ($= UA/\dot{m}C$) for the case of $\dot{m}_1 C_1 = \dot{m}_2 C_2 = R = 1$ for a range of values of the utilization factor, U, determined for the blow time, t. For a zero blow time, the effectiveness is given by Eqn (10-12) as

$$\varepsilon = \frac{\Lambda}{\Lambda + 1} \quad \text{and} \quad U = \frac{h}{2}$$

if h is the heat transfer coefficient (assumed the same) during the heating and cooling blows. Observe that if $U = 0.1$ the performance of the generator is negligibly different from that of the equivalent recuperator. Cox and Stevens[7] suggest that a mass flow rate of about 1 lb air/s through a rotary regenerator matrix of 1 ft² frontal area is typical for acceptable pressure losses. With a blockage ratio (i.e. throughway area, A_c/frontal area) of 0·25,

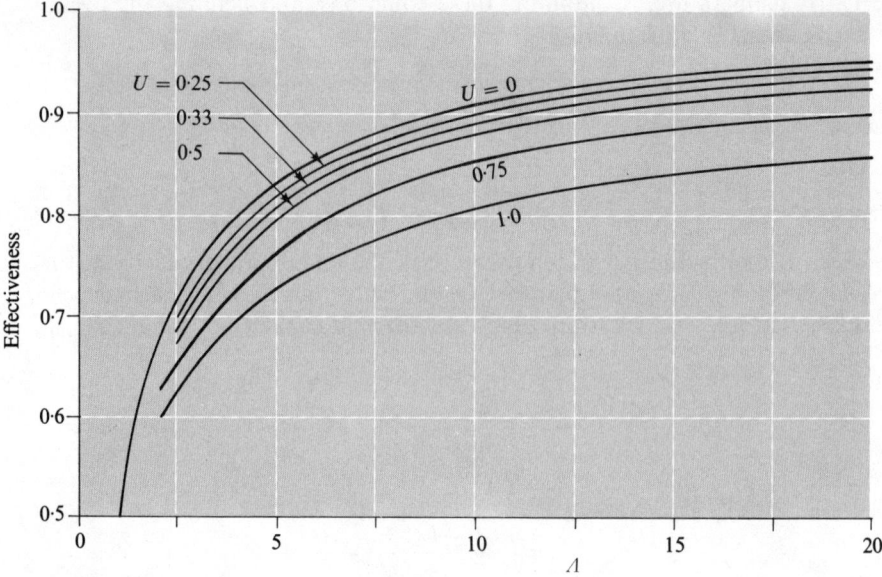

Fig. 10-15 Variation of effectiveness with λ for a cyclically operating regenerator

a steel disc 3 in long would weigh about 60 lb to pass 1 lb air/s through each semicircle in contraflow. If the utilization factor is 0·1, we have

$$0{\cdot}1 = \frac{0{\cdot}24 \times 1{\cdot}0}{60 \times 0{\cdot}12} \times t$$

0·24 and 0·12 are the specific heats of air and steel respectively, so that $t = 3$ seconds, corresponding to rotating the matrix with a cycling period of 6 seconds, that is, at only 10 rev/min. It is thus a simple matter to arrange for a cyclically operating regenerator to have a performance corresponding to the equivalent recuperator, with a negligible loss in performance due to the time-variant conditions. It should be noted, however, that the short flow passages which result from the easily attained small passage diameters (the 3-in long matrix of the example above is typical) can lead to significant errors through the assumption of zero axial conduction, implicit in the derivation of Eqns (10-57) and (10-58). A comprehensive analysis of the effects of axial conduction in heat exchanger matrices will be found in reference (8).

REFERENCES

1. Gardner, K. A. Efficiency of extended surfaces. *Trans. A.S.M.E.*, 1945, **67**.
2. Kays, W. M. and A. L. London. *Compact Heat Exchangers*. McGraw-Hill, New York, 1955.

3. Nusselt, W. *Tech. Mech. Thermodynamik*, 1930, **1**.
4. Hausen, H. On the theory of heat exchange in regenerators. *ZAMM*, 1929, **9**.
5. Johnson, J. E. Regenerator heat exchangers for gas turbines. *A.R.C. R. and M. No. 2630*, 1952.
6. Bayley, F. J. and C. W. Rapley. Heat transfer and pressure loss characteristics of matrices for regenerative heat exchangers. *A.S.M.E. Paper No. 65-HT-35*, 1965.
7. Cox, M. and R. K. P. Stevens. Regenerative heat exchangers for gas turbine power plant. *Proc. I.Mech.E.*, 1950, **163**.
8. Schultz, H. Regenerators with longitudinal heat conduction. Joint I.Mech.E.–A.S.M.E. General Discussion on Heat Transfer, 1951.

PROBLEMS

10-1 In a tubular heat exchanger air flows over the tube bank to cool water flowing within the tubes. If the air side heat transfer coefficient is 100 W/m^2 K and the water side coefficient is 2,000 W/m^2 K, show that doubling the former will reduce the heat transfer surface required by 47·6%, whereas a twofold increase on the water side will save only $2\frac{1}{2}$%.

10-2 If in Problem 10-1 the tubes are 20 mm outside diameter, determine the pitching required of annular fins of 40 mm outside diameter and 3 mm thick to achieve the specified doubling of the air side coefficient. The thermal conductivity of the fin material is 20 W/m K and all other conditions are unchanged. Use Fig. 10-2 to obtain the fin efficiency, and neglect temperature variations through the tube wall and heat flow at the fin tips.

Ans: 16·2 mm

10-3 What fin pitching would be necessary in Problem 10·2 if the fins were perfect conductors of infinite thermal conductivity?

Ans: 27·0 mm

10-4 A plate-fin heat exchanger for a gas turbine is constructed as in Fig. 10-8. Through passages in one direction compressed air flows at a rate giving a heat transfer coefficient of 42 Btu/ft^2 h °F, while exhaust gas moves in cross-flow through the remaining passages to give a gas side heat transfer coefficient of 22 Btu/ft^2 h °F. The plates are 0·3 inch apart and 0·05 inch thick. The corrugations are constructed from sheet 0·03 inch thick and may be treated as straight fins at a pitching of 0·25 inch. Taking the thermal conductivity as 12 Btu/ft h °F, determine the overall coefficient referred to the plate surface, considering carefully the assumptions required.

Ans: 29 Btu/ft^2 h °F

10-5 2·4 kg/s of a fluid having a specific heat 0·8 kJ/kg K enter a contra-flow heat exchanger at 300°C and are heated to 700° by 2 kg/s of a fluid having a specific heat 0·96 kJ/kg K entering the unit at 1,000°. Show that, to heat the cooler fluid to 800°, all other conditions remaining unchanged, would require the surface area for heat transfer to be increased by 87·5%.

10-6 A heat exchanger contains 4,500 1-inch diameter tubes, through which pass 25 lb/s of air to be cooled from 1,000 to 300°F. Water passes in counter flow over the outside of the tubes, rising in temperature from 100 to 185°F. Determine the tube length required for this duty if the water side resistance to heat flow is negligible and the heat transfer coefficient on the air side is as given by Eqn (7-12a) for developed turbulent pipe flow. Physical properties of the air should be evaluated at the average air temperature from the data in Table A8 of the Appendix, using conversion factors as required.

Ans: 6·3 ft

10-7   Show that in a *parallel flow* heat exchanger the overall rate of heat flow between the two fluids is given by Eqn (10-17). Show also that in such a unit, in which the thermal capacity of one stream is twice the other, the greatest achievable effectiveness is 66·7%.

10-8   A contra-flow heat exchanger has a fluid temperature difference at the hot end twice that at the cold end. If there is no change of phase of the fluids concerned and the overall heat transfer coefficient and specific heats of the fluids are not dependent on temperature, what proportion of the heat transfer surface, measured from the hot end, transfers half the total heat flow?

Ans: 0·415

10-9   An exhaust gas calorimeter comprises a pipe $1\frac{1}{2}$ inches bore and $1\frac{7}{8}$ inches outside diameter jacketed by a concentric pipe $2\frac{1}{4}$ inches bore, the unit having an effective length of 18 ft. Exhaust gas enters the inner pipe at 910°F and leaves at 85°. The cooling water flows in parallel with the gas at a rate of 95 lb/min, entering at 50° and leaving at 61·8°F. The heat transfer coefficient on the gas side is 43·3 Btu/ft² h °F. If the coefficient in the water annulus follows the correlation for developed turbulent flow, of the form

$$\mathbf{N} = k\mathbf{R}^{0·8}\mathbf{P}^{0·4}$$

show that these data are consistent with k having the value 0·022. Use physical properties from the Appendix as required.

10-10   Derive the dimensionless differential equation (10-25b) for the temperature of one of the fluids in a cross-flow heat exchanger with unmixed fluid streams.

Such a unit is required to raise the temperature of 1 lb/s of air from 300 to 780°F using 0·5 lb/s of combustion products initially at 1,500°F. If the overall heat transfer coefficient is 30 Btu/ft² h °F and the specific heats of air and combustion products 0·25 Btu/lb °F, determine the heat transfer surface required, using Table 10-1.

Ans: 40 ft²

10-11   Use the procedures of Section 10-4-2 to show that the effectiveness of a cross-flow heat exchanger in which each fluid stream is fully mixed at each cross-section, and of the same thermal capacity, reaches a maximum value of 0·565 when $UA/\dot{m}C$ (in the nomenclature of the text) is about 3·0.

10-12   1,000 kg/min of the product of a chemical plant at 700°C (specific heat 3·6 kJ/kg K) are to be used to heat 1,200 kg/min of incoming feed from 100°C (specific heat 4·2 kJ/kg K). If the installed heat transfer surface is 42 m² and the overall heat transfer coefficient is 1 kW/m² K, compare the feed outlet temperatures with counter and parallel flow arrangements.

Ans: 362°C; 345°C

10-13   A plate-fin heat exchanger, as in Fig. 10-8, has a surface area for heat transfer of 1,760 ft². 10 lb/s of gas with a specific heat of 0·24 Btu/lb °F and at 1,000°F enter the hot side where the heat transfer coefficient is 25 Btu/ft² h °F. On the cold side the heat transfer coefficient is 40 Btu/ft² h °F and 8 lb/s of fluid enter at 200°F with a specific heat of 0·225 Btu/lb °F. Use Table 10-1 to plot effectiveness against $UA/\dot{m}C$ and thus predict the mean exit temperature of the originally cooler stream.

Ans: 840°F

10-14   If the matrix of the heat exchanger in Problem 10-13 is divided into three equal sections which are then arranged in series counter flow, show that the outlet temperature of the initially cooler fluid will now become 886°F.

10-15   A matrix for a regenerative or capacitance heat exchanger is made from wire gauze of specific heat 0·5 kJ/kg K and has a mass of 1·0 kg. The effective surface area for heat transfer is 0·5 m². A 'single blow' test is performed to measure the heat transfer coefficient in the matrix by exposing it, when initially at a uniform temperature of 20°C, suddenly to a gas flow of 0·10 kg/s at 150°C. The following variation of gas exit temperature with time is obtained:

Time (seconds)	2	4	6	8	10
Gas outlet temperature (°C)	38	75	110	130	142

If the specific heat of the gas is 1 kJ/kg K use Fig. 10-13 to evaluate the heat transfer coefficient to which these readings correspond.

Ans: 1·2 kW/m² K

Appendix:

Tables of properties

In the following tables the thermal properties of some common solids, liquids, and gases are presented in SI units. *Table A1* contains a set of conversion factors by which many commonly used quantities can be converted from SI to BS units. In the remaining tables, conversion factors are included at the bottom of each table, where appropriate.

Table A2 has been adapted from data given in Eckert and Drake's *Heat and Mass Transfer*[1]. The density, specific heat, and thermal conductivity for a number of pure metals and alloys of engineering importance are tabulated. Where available, the data for the effect of temperature on thermal conductivity are included. For calculations involving small temperature variations the conductivity may be treated as constant, but for more accurate calculations over a wide temperature range it is necessary to assume a linear, or higher order, variation of conductivity with temperature.

Table A3, also derived from reference (1), presents the density, specific heat, and thermal conductivity of a number of common non-metallic solids. For insulating materials, the conductivity varies with the density of the material as well as with the temperature. It should be noted, however, that the density given is an *apparent density*, that is, the mass per unit volume (including the volume of the voids in a porous material). At low temperatures the voids reduce the thermal conductivity of a porous material, owing to the relatively low conductivity of air; but at high temperatures this effect can be reversed owing to the influence of convection and radiation within the pores.

Table A4 presents values of density, specific heat, dynamic viscosity, thermal conductivity, and Prandtl number as functions of temperature for some liquid metals. This table was obtained from *Thermodynamic and Transport Properties of Fluids* by Mayhew and Rogers[2]. The thermal properties of liquid metals do not vary significantly with pressure. It should be noted that the dynamic viscosity has been multiplied by 10^5; for example, the viscosity of sodium at 600 K is 32×10^{-5} kg/m s or $21\cdot5 \times 10^{-5}$ lb/ft s. The values of the volume expansion coefficient, β, have been computed from the density–temperature values. A smoothing curve (a low-order least squares polynomial) was fitted to the specific volume, v, where $v = \rho^{-1}$, and this curve was numerically differentiated to give $\beta = v^{-1} \, \partial v / \partial T = -\rho^{-1} \, \partial \rho / \partial T$. For additional data on liquid metals the reader is referred to reference (3).

In *Table A5* the density, specific heat at constant pressure, kinematic viscosity, thermal conductivity, and Prandtl number are plotted as functions of temperature for a number of saturated liquids. These data are based on the tables of reference (1) and the value of the volume expansion coefficient, β, is given at one temperature. The thermal properties are not strongly dependent on pressure. For more complete data the reader is directed to references (4) and (5).

Table A6 presents the values of density, specific heat at constant pressure, dynamic viscosity, thermal conductivity, and Prandtl number as functions of temperature for water and steam at its saturation pressure. This table is based on data derived from Haywood's *Thermodynamic Tables in SI units*[6]. It should be noted that the values for saturated water can be used with good accuracy above saturation pressure. Superheated steam can be approximately treated as a perfect gas with a characteristic constant, R, of 461·5 J/kg K. The values of the volume expansion coefficient, β, which were obtained from reference (5), are for atmospheric pressure. Formulae for the transport properties of water and steam, and comprehensive data over a wide range of values, can be found in reference (7).

In *Table A7* the density, specific heat, dynamic viscosity, thermal conductivity, and Prandtl number are presented as functions of temperature for a number of gases; these data were derived from reference (1). For each gas the characteristic constant, R, is given enabling the density to be determined from the perfect gas law as a function of pressure and temperature. The effect of high pressure on the viscosity and conductivity of gases has been calculated from the kinetic theory of gases by Comings *et al.*[8, 9]. The effect of p/p_c and T/T_c (where p_c and T_c, the critical pressure and temperature, are given in Table A7) on the viscosity and conductivity are shown in Figs. A1 and A2. The values μ_1, and k_1, refer to conditions at atmospheric pressure, corresponding to Table A7. For additional data the reader is referred to references (10) and (11).

Table A8 is devoted solely to air, and values of density, specific heat at constant pressure, dynamic viscosity, thermal conductivity, and Prandtl number are given as functions of temperature. The data, valid for dry air at atmospheric pressure, were obtained from reference (2). It should be noted that above 1500 K dissociation occurs, that is, the molecules break down into their constituents, and the thermal properties depend appreciably on pressure. Table A8, supplemented if necessary by Figs. A1 and A2, is only valid if dissociation is negligible. Additional data can be obtained from references (10), (11), and (12).

Table A9 presents the formulae by which the specific heat at constant pressure and the dynamic viscosity can be calculated as a function of absolute temperature at atmospheric pressure. The specific heat can be evaluated by

$$C_p = C_1 + C_2 T^m + C_3 T^n$$

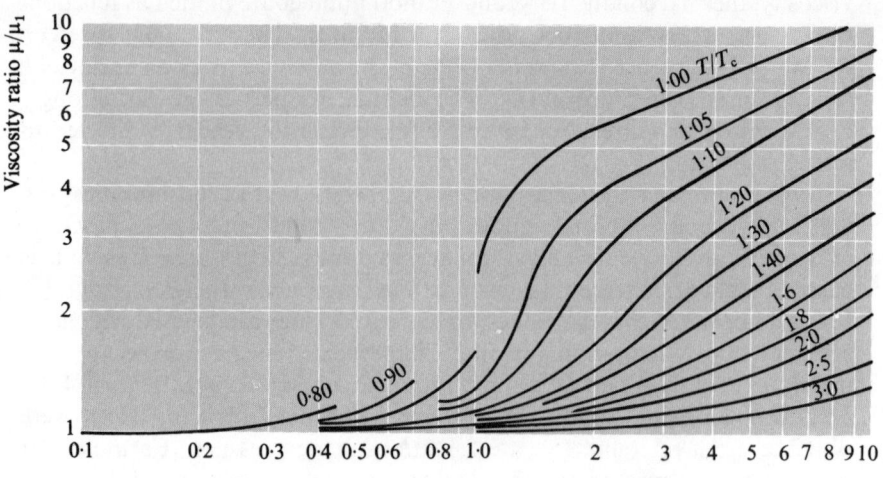

Fig. A1   The effect of high pressure on the viscosity of gases (after Comings et al.[8])

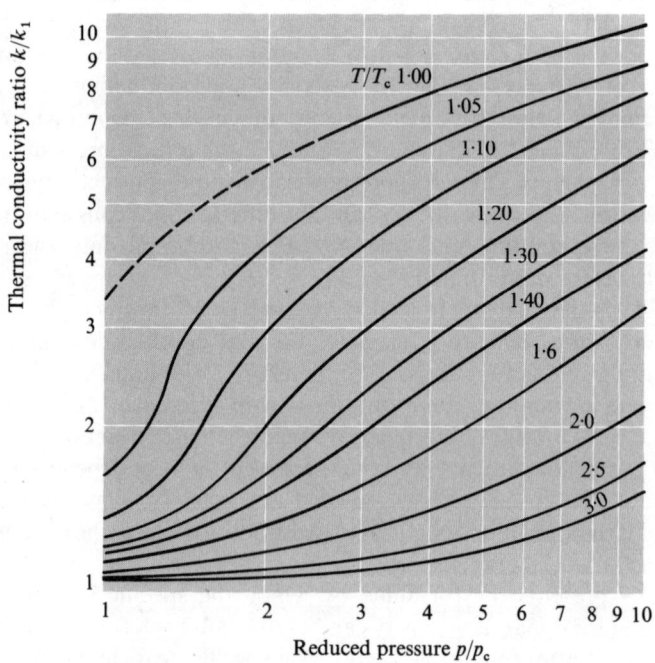

Fig. A2   The effect of high pressure on the thermal conductivity of gases (after Comings and Nathan[9])

and the viscosity can be evaluated by Sutherland's law where

$$\mu = S_1 T^{3/2}/(S_2 + T)$$

C_1, C_2, C_3, m, n, S_1, and S_2 are given in Table A8 for a number of gases. The values of the constants have been derived from data in Chapman's *Heat Transfer*[13].

In *Table A10* coefficients of diffusion and Schmidt numbers are given for a number of liquids and gases at 0°C and 1 atm. The data were obtained from reference (1), and additional material can be found in reference (5).

Table A11 presents emissivities of radiation in the direction normal to the surface, ε_n, and the total hemispherical radiation, ε, for a number of materials. The emissivities of metals increase with increasing temperature, but this is not necessarily true for metal oxides and non-metals. The data are derived from reference (1), and where values are not given $\varepsilon/\varepsilon_n \approx 1\cdot2$ for bright metal surfaces, $\varepsilon/\varepsilon_n \approx 0\cdot95$ for smooth non-metallic surfaces, and $\varepsilon/\varepsilon_n \approx 0\cdot98$ for rough surfaces. Additional values of emissivity have been compiled by Hottel in reference (14).

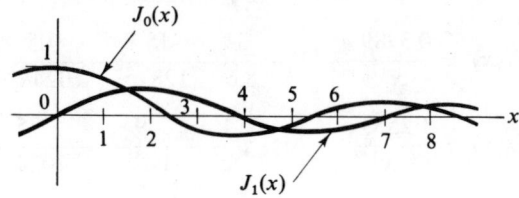

Fig. A3 Bessel functions of the first kind

In *Table A12* values of Bessel functions of the first and second kind, and orders zero and one, are given for a range of arguments between 0 and 15·8. For $x > 15\cdot8$:

$$J_0(x) \approx \sqrt{\left(\frac{2}{\pi x}\right)}\left\{\sin\left(x + \frac{1}{4}\pi\right) + \frac{1}{8x}\sin\left(x - \frac{1}{4}\pi\right)\right\}$$

$$J_1(x) \approx \sqrt{\left(\frac{2}{\pi x}\right)}\left\{\sin\left(x - \frac{1}{4}\pi\right) + \frac{3}{8x}\sin\left(x + \frac{1}{4}\pi\right)\right\}$$

$$Y_0(x) \approx \sqrt{\left(\frac{2}{\pi x}\right)}\left\{\sin\left(x - \frac{1}{4}\pi\right) - \frac{1}{8x}\sin\left(x + \frac{1}{4}\pi\right)\right\}$$

$$Y_1(x) \approx \sqrt{\left(\frac{2}{\pi x}\right)}\left\{\sin\left(x - \frac{3}{4}\pi\right) + \frac{3}{8x}\sin\left(x - \frac{1}{4}\pi\right)\right\}$$

The approximate behaviour of these functions is illustrated in Figs. A3 and A4. For more details the reader is referred to reference (15).

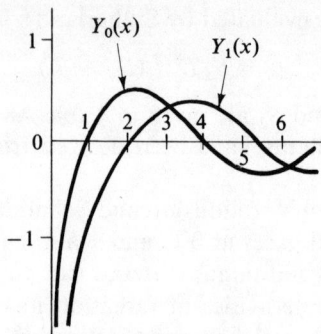

Fig. A4 Bessel functions of the second kind

Table A13 presents values of modified Bessel functions of the first and second kind, orders zero and one, for a range of arguments between 0 and 10. For $x > 10$:

$$I_0(x) \approx \frac{0\cdot3989\,e^x}{x^{1/2}}\left\{1 + \frac{1}{8x} + \frac{9}{128x^2} + \frac{75}{1024x^3}\right\}$$

$$I_1(x) \approx \frac{0\cdot3989\,e^x}{x^{1/2}}\left\{1 - \frac{3}{8x} - \frac{15}{128x^2} - \frac{105}{1024x^3}\right\}$$

$$K_0(x) \approx \frac{1\cdot2533\,e^{-x}}{x^{1/2}}\left\{1 - \frac{1}{8x} + \frac{9}{128x^2} - \frac{75}{1024x^3}\right\}$$

$$K_1(x) \approx \frac{1\cdot2533\,e^{-x}}{x^{1/2}}\left\{1 + \frac{3}{8x} - \frac{15}{128x^2} + \frac{105}{1024x^3}\right\}$$

The approximate behaviour of these functions is illustrated in Figs. A5 and A6.

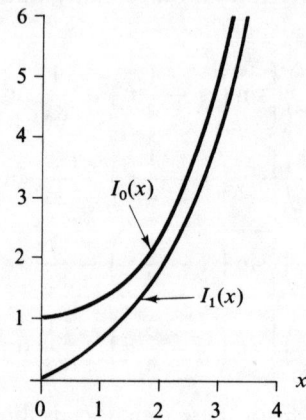

Fig. A5 Modified Bessel functions of the first kind

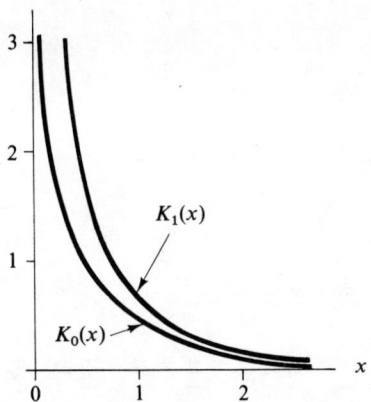

Fig. A6 Modified Bessel functions of the second kind

Table A1 SI–BS conversion factors

Length	1 m = 3·281 ft
	1 mm = 0·03937 in
	1 km = 0·6214 mile
Mass	1 kg = 2·205 lb
Force	1 N = 0·2248 lbf = 7·233 pdl
	1 kN = 0·1004 tonf
Pressure ⎫	1 bar = 10^5 N/m² = 14·50 lbf/in²
Stress ⎬	1 kN/m² = 7·5 mmHg = 0·2953 inHg
Volume ⎭	1 m³ = 35·32 ft³ = 220·0 U.K. gal = 264·2 U.S. gal
Density	1 kg/m³ = 0·06243 lb/ft³
Energy	1 kJ = 10^3 N m = 737·6 ft lbf = 0·9478 Btu
Power	1 kW = 1 kJ/s = 737·6 ft lbf/s = 1·341 hp
Dynamic viscosity	1 kg/ms = 10 poise = 2,419 lb/ft h = 0·6719 lb/ft s
Kinematic viscosity	1 m²/s = 10^4 stokes = 38,750 ft²/h = 10·76 ft²/s
Specific heat	1 kJ/kg K = 0·2388 Btu/lb °R
Thermal conductivity	1 kW/m K = 577·8 Btu/ft h °R
Heat transfer coefficient	1 kW/m² K = 176·1 Btu/ft² h °R
Gas constant	1 J/kg K = 0·1859 ft lbf/lb °R

Table A2 Thermal properties of solid metals

Metal	Properties at 20°C			k, W/m K						
	ρ, kg/m³	C_p, kJ/kg K	k, W/m K	−100°C	0°C	100°C	200°C	400°C	800°C	1,200°C
Aluminium										
Pure	2,710	0·896	204	215	202	228	228			
Al–Cu (Duralumin) 94–96 Al, 3–5 Cu, trace Mg	2,790	0·883	164	126	159	182	194			
Al–Mg (Hydronalium) 91–95 Al, 5–9 Mg	2,610	0·904	112	94	109	126	142			
Al–Si (Silumin) 87 Al, 13 Si	2,660	0·871	164	149	163	175	185			
Al–Si (Silumin, copper bearing) 86·5 Al; 12·5 Si; 1 Cu	2,660	0·867	137	119	137	144	152			
Al–Si (Alusil) 78–80 Al; 20–22 Si	2,630	0·854	161	144	157	168	175			
Al–Mg–Si 97 Al; 1 Mg; 1 Si; 1 Mn	2,710	0·892	177		175	189	204			
Lead	11,380	0·130	35	37	35	33	31			
Iron										
Pure	7,900	0·452	73	87	73	67	62	48	36	36
Wrought iron (C < 0·50%)	7,850	0·461	59		59	57	52	45	33	33
Cast iron (C ≈ 4%)	7,270	0·419	52							
Steel										
Carbon steel C ≈ 0·5%	7,840	0·465	54		55	52	48	42	29	31
1·0%	7,800	0·473	43		43	43	42	36	29	29
1·5%	7,760	0·486	36		36	36	36	33	28	29

Material	lb/ft³	Btu/lb °R	Btu/ft h °R						
Nickel steel									
Ni ≈ 10%	7,950	0·461	26						
20%	8,000	0·461	19						
40%	8,170	0·461	10						
60%	8,380	0·461	19						
80%	8,620	0·461	35						
Invar: Ni ≈ 36%	8,140	0·461	11						
Chrome Steel									
Cr ≈ 1%	7,870	0·461	61	62	55	52	42	33	29
2%	7,870	0·461	52	54	48	45	38	31	
5%	7,840	0·461	40	40	38	36	33	29	
10%	7,790	0·461	31	31	31	31	29	28	
20%	7,690	0·461	22	22	22	22	24	26	
30%	7,630	0·461	19						
Cr–Ni (chrome–nickel)									
15 Cr; 10 Ni	7,870	0·461	19						
18 Cr; 8 Ni	7,820	0·461	16	16	17	17	19	26	
20 Cr; 15 Ni	7,850	0·461	15						
25 Cr; 20 Ni	7,870	0·461	13						
Ni–Cr (nickel–chrome)									
80 Ni; 15 Cr	8,520	0·461	17						
60 Ni; 15 Cr	8,270	0·461	13						
40 Ni; 15 Cr	8,080	0·461	12						
20 Ni; 15 Cr	7,830	0·461	14	14	15	15	17	22	
Cr–Ni–Al: 6 Cr; 1·5 Al; 0·5 Si (Sicromal 8)	7,720	0·490	22						
24 Cr; 2·5 Al; 0·5 Si (Sicromal 12)	7,680	0·494	19						
To convert to	lb/ft³	Btu/lb °R	Btu/ft h °R						
Multiply by	0·06243	0·2388	0·5778						

Table A2—*contd.*

Metal	Properties at 20°C			k, W/m K						
	ρ, kg/m³	C_p, kJ/kg K	k, W/m K	−100°C	0°C	100°C	200°C	400°C	800°C	1,200°C
Manganese steel										
Ma = 0%	7,900	0·494	73							
1%	7,900	0·461	50							
2%	7,900	0·461	38		38	36	36	34		
5%	7,800	0·461	22							
10%	7,800	0·461	17							
Tungsten steel										
W = 0%	7,900	0·452	73							
1%	7,930	0·448	66							
2%	7,950	0·444	62		62	59	54	45		
5%	8,010	0·435	54							
10%	8,320	0·419	48							
20%	8,830	0·389	43							
Silicon steel										
Si = 0%	7,900	0·452	73							
1%	7,780	0·461	42							
2%	7,680	0·461	31							
5%	7,420	0·461	19							
Copper										
Pure	8,960	0·389	386	407	386	379	374	363		
Aluminium bronze 95 Cu; 5 Al	8,670	0·410	83							
Bronze 75 Cu; 25 Sn	8,660	0·343	26							
Red brass 85 Cu; 9 Sn; 6 Zn	8,720	0·385	61		59	71				
Brass 70 Cu; 30 Zn	8,520	0·385	111	88		128	144	147		
German silver 62 Cu; 15 Ni; 22 Zn	8,620	0·394	25	19		31	40	48		
Constantan 60 Cu; 40 Ni	8,930	0·410	23	21		22	26			

	lb/ft³	Btu/lb °R	Btu/ft h °R							
Magnesium										
Pure	1,750	1·013	171	178	171	168	163			
Mg–Al (electrolytic)	1,820	1·005	66		52	62	74			
6–8% Al; 1–2% Zn	1,780	1·005	114		111	125	130			
Mg–Mn 2% Mn				93						
Molybdenum	10,220	0·251	137	138	137	137				
Nickel										
Pure (99·9%)	8,910	0·446	90	104	93	83	73	59	62	69
Impure (99·2%)	8,910	0·444	69		69	64	59	52		
Ni–Cr:										
90 Ni; 10 Cr	8,670	0·444	17		17	19	21	25		
80 Ni; 20 Cr	8,320	0·444	13		12	14	16	19		
Silver										
Purest	10,530	2·324	419	419	417	415	412	360		
Pure (99·9%)	10,530	2·324	407	419	410	415	374			
Tungsten	19,360	0·134	163		166	151	142	126	76	
Zinc, pure	7,150	0·384	112	114	112	109	106	93		
Tin, pure	7,310	0·227	64		66	59	57			
To convert to	lb/ft³	Btu/lb °R	Btu/ft h °R							
Multiply by	0·06243	0·2388	0·5778							

Table A3 Thermal properties of solid non-metals

Material	T, °C	ρ, kg/m³	Cp, kJ/kg K	k, W/m K
Aerogel, silica	120	136		0·022
Asbestos	−200	469		0·074
Asbestos	0	469		0·155
Asbestos	0	577	0·816	0·151
Asbestos	100	577	0·816	0·192
Asbestos	200	577		0·208
Asbestos	400	577		0·223
Asbestos	−200	697		0·156
Asbestos	0	697		0·234
Brick, dry	20	1,760–1,810	0·837	0·38–0·52
Bakelite	20	1,258	1·59	0·232
Cardboard, corrugated				0·064
Clay	20	1,458	0·879	1·28
Concrete	20	1,910–2,310	0·879	0·81–1·40
Coal, anthracite	20	1,200–1,510	1·26	0·260
Coal, powdered	30	737	1·30	0·116
Cotton	20	80	1·30	0·059
Cork, board	30	160		0·043
Cork, expanded scrap	20	45–120	1·88	0·036
Cork, ground	30	151		0·043
Diatomaceous earth	38	320		0·062
Diatomaceous earth	850	320		0·142
Earth, coarse gravelly	20	2,050	1·84	0·520
Felt, wool	30	330		0·052
Fibre, insulating board	21	237		0·048
Fibre, red	20	1,290		0·467
Glass plate	20	2,710	0·837	0·762
Glass, borosilicate	30	2,230		1·090
Glass, wool	20	200	0·670	0·040
Granite				1·7–4·0
Ice	0	913	1·93	2·21
Marble	20	2,500–2,710	0·808	2·77
Rubber, hard	0	1,200		0·151
Sandstone	20	2,160–2,310	0·712	1·63–2·08
Silk	20	58	1·38	0·036
Wood, oak radial	20	610–800	2·39	0·17–0·21
Wood, fir (20% moisture) radial	20	417–421	2·72	0·138
To convert to		lb/ft³	Btu/lb °R	Btu/ft h °R
Multiply by		0·06243	0·2388	0·5778

Table A4 Thermal properties of liquid metals

Metal (melting point)	T, K	ρ, kg/m³	C_p, kJ/kg K	$\mu \times 10^5$, kg/m s	k, W/m K	$\beta \times 10^3$, K⁻¹	P
Lead–bismuth 44·5:55·5 eutectic (397 K)	400	10,570	0·146	336	10·9	0·111	0·0450
	500	10,450	0·146	234	12·0	0·114	0·0284
	600	10,330	0·146	184	12·9	0·117	0·0208
	800	10,090	0·146	133	15·0	0·122	0·0129
	1,000	9,840	0·146	110	17·0	0·127	0·0095
Mercury (234·3 K)	250	13,650	0·141	188	7·5	0·184	0·0353
	300	13,530	0·139	152	8·1	0·182	0·0261
	400	13,290	0·137	119	9·4	0·177	0·0173
	500	13,050	0·137	101	10·7	0·173	0·0129
	600	12,840	0·137	89	12·8	0·169	0·0095
	800	12,420	0·138	78	13·7	0·161	0·0079
Potassium (336·8 K)	400	812	0·805	41·7	46·5	0·275	0·0072
	500	789	0·786	31·9	45·4	0·287	0·0055
	600	766	0·772	25·8	42·5	0·299	0·0047
	800	721	0·768	17·9	33·7	0·318	0·0041
	1,000	675	0·775	13·3	27·8	0·332	0·0037
Sodium (370·5 K)	400	921	1·369	61	86	0·261	0·0097
	500	897	1·315	42	80	0·272	0·0069
	600	872	1·277	32	74	0·282	0·0055
	800	823	1·273	23	63	0·299	0·0047
	1,000	774	1·277	18	59	0·312	0·0039
Sodium–Potassium 22:78 eutectic (262° K)	300	869	0·977	78·0	22·2	0·280	0·0343
	400	845	0·929	46·7	23·6	0·284	0·0184
	500	821	0·904	34·8	24·9	0·291	0·0126
	600	797	0·886	27·7	26·2	0·300	0·0094
	800	749	0·871	19·3	28·7	0·324	0·0059
	1,000	700	0·882	14·6	31·2	0·353	0·0041
To convert to	°R	lb/ft³	Btu/lb °R	lb/ft s	Btu/ft h °R	°R⁻¹	—
Multiply by	1·8	0·06243	0·2388	0·6719	0·5778	0·5556	1

Table A5 Thermal properties of saturated liquids

$T,$ °C	$\rho,$ kg/m^3	$C_p,$ kJ/kg K	$v \times 10^6,$ m^2/s	$k,$ W/m K	$\beta,$ K^{-1}	P
Ammonia						
− 50	704	4·463	0·435	0·547		2·50
− 40	692	4·467	0·406	0·547		2·28
− 30	680	4·475	0·387	0·549		2·15
− 20	667	4·509	0·381	0·547		2·09
− 10	654	4·563	0·378	0·543		2·07
0	640	4·634	0·373	0·540		2·05
10	626	4·714	0·368	0·531		2·04
20	612	4·798	0·359	0·521	0·00245	2·02
30	597	4·890	0·349	0·508		2·01
40	581	4·999	0·340	0·493		2·00
50	564	5·116	0·330	0·476		1·99
Carbon dioxide						
− 50	1,157	1·842	0·119	0·0855		2·96
− 40	1,118	1·884	0·118	0·1011		2·46
− 30	1,077	1·967	0·117	0·1117		2·22
− 20	1,033	2·051	0·116	0·1151		2·12
− 10	984	2·177	0·113	0·1099		2·20
0	927	2·470	0·109	0·1045		2·38
10	860	3·140	0·101	0·0971		2·80
20	773	5·024	0·091	0·0872	0·0140	4·10
30	598	36·42	0·080	0·0703		28·7
Dichlorodifluoromethane (Freon)						
− 50	1,547	0·875	0·310	0·0675	0·00262	6·2
− 40	1,519	0·885	0·279	0·0692		5·4
− 30	1,490	0·896	0·253	0·0692		4·8
− 20	1,461	0·907	0·235	0·0710		4·4
− 10	1,430	0·920	0·221	0·0727		4·0
0	1,398	0·935	0·214	0·073		3·8
10	1,365	0·949	0·204	0·073		3·6
20	1,331	0·966	0·198	0·073		3·5
30	1,295	0·984	0·191	0·071		3·5
40	1,257	1·002	0·191	0·069		3·5
50	1,216	1·022	0·190	0·067		3·5
To convert to	lb/ft^3	Btu/lb °R	ft^2/s	Btu/ft h °R	°R^{-1}	—
Multiply by	0·06243	0·2388	10·76	0·5778	0·5556	1

Table A5—*contd.*

T, °C	ρ, kg/m³	C_p, kJ/kg K	$v \times 10^6$, m²/s	k, W/m K	β, K⁻¹	P
			Engine oil			
0	899	1·796	4,280	0·147		47,100
20	888	1·880	900	0·145	0·00070	10,400
40	876	1·964	240	0·144		2,870
60	864	2·047	84	0·140		1,050
80	852	2·131	37	0·138		490
100	840	2·219	20	0·137		276
120	829	2·307	12	0·135		175
140	817	2·395	8	0·133		116
160	806	2·483	6	0·132		84
			Ethylene glycol			
0	1,131	2·294	57·5	0·242		615
20	1,117	2·382	19·2	0·249	0·00065	204
40	1,102	2·474	8·69	0·256		93
60	1,088	2·562	4·75	0·260		51
80	1,078	2·650	2·98	0·261		32
100	1,059	2·742	2·03	0·263		22
		Eutectic calcium chloride solution (29·9% Ca)				
−50	1,320	2·608	36·3	0·402		312
−40	1,315	2·636	24·9	0·415		208
−30	1,311	2·661	17·1	0·429		139
−20	1,306	2·688	11·0	0·445		87·1
−10	1,301	2·713	6·96	0·459		53·6
0	1,296	2·738	4·40	0·472		33·0
10	1,292	2·763	3·35	0·485		24·6
20	1,287	2·788	2·72	0·498		19·6
30	1,282	2·814	2·27	0·511		16·0
40	1,278	2·839	1·92	0·523		13·3
50	1,273	2·868	1·65	0·535		11·3
To convert to	lb/ft³	Btu/lb °R	ft²/s	Btu/ft h °R	°R⁻¹	—
Multiply by	0·06243	0·2388	10·76	0·5778	0·5556	1

Table A5—*contd.*

T, °C	ρ, kg/m³	C_p, kJ/kg K	$v \times 10^6$, m²/s	k, W/m K	β, K⁻¹	P
			Glycerine			
0	1,276	2·261	8,310	0·282		84,700
10	1,270	2·320	3,000	0·284		31,000
20	1,264	2·387	1,180	0·286	0·00050	12,500
30	1,258	2·445	500	0·286		5,380
40	1,252	2·512	220	0·286		2,450
50	1,245	2·583	150	0·287		1,630
			Methylchloride			
−50	1,053	1·476	0·320	0·2146		2·31
−40	1,034	1·483	0·318	0·2094		2·32
−30	1,017	1·492	0·314	0·2025		2·35
−20	1,000	1·504	0·309	0·1956		2·38
−10	982	1·519	0·306	0·1870		2·43
0	963	1·538	0·302	0·1783		2·49
10	943	1·560	0·297	0·1713		2·55
20	924	1·587	0·293	0·1627		2·63
30	903	1·616	0·288	0·1540		2·72
40	883	1·650	0·281	0·1437		2·83
50	861	1·689	0·274	0·1333		2·97
			Sulphur dioxide			
−50	1,561	1·360	0·484	0·2423		4·24
−40	1,537	1·361	0·424	0·2354		3·74
−30	1,513	1·362	0·371	0·2302		3·31
−20	1,489	1·362	0·324	0·2250		2·93
−10	1,464	1·363	0·288	0·2181		2·62
0	1,439	1·364	0·257	0·2111		2·38
10	1,413	1·365	0·232	0·2042		2·18
20	1,387	1·365	0·210	0·1990	0·00194	2·00
30	1,360	1·366	0·199	0·1921		1·83
40	1,330	1·367	0·173	0·1852		1·70
50	1,299	1·368	0·162	0·1765		1·61
To convert to	lb/ft³	Btu/lb °R	ft²/s	Btu/ft h °R	°R⁻¹	—
Multiply by	0·06243	0·2388	10·76	0·5778	0·5556	1

Table A6 Thermal properties of saturated water and steam

(the volume expansion coefficient, β, for water is at atmospheric pressure, 101·3 kN/m²)

T, °C	Saturation pressure, kN/m²	ρ, kg/m³		C_p, kJ/kg K		$\mu \times 10^3$, kg/m s		k, W/m K		$\beta \times 10^3$, K⁻¹	P	
		Water	Steam	Water	Steam	Water	Steam	Water	Steam	Water	Water	Steam
0·01	0·611	1,000	0·00485	4·217	1·854	1·755	0·0088	0·569	0·0173	−0·06	13·02	0·942
10	1·227	1,000	0·00940	4·193	1·860	1·301	0·0091	0·587	0·0185	0·088	9·29	0·915
20	2·34	998	0·0173	4·182	1·866	1·002	0·0094	0·603	0·0191	0·207	6·95	0·918
30	4·24	996	0·0304	4·179	1·875	0·797	0·0097	0·618	0·0198	0·303	5·39	0·923
40	7·38	992	0·0513	4·179	1·885	0·651	0·0101	0·632	0·0204	0·385	4·31	0·930
50	12·34	988	0·0830	4·181	1·899	0·544	0·0104	0·643	0·0210	0·458	3·53	0·939
60	19·92	983	0·130	4·185	1·915	0·462	0·0107	0·653	0·0217	0·523	2·96	0·947
70	31·16	978	0·198	4·190	1·936	0·400	0·0111	0·662	0·0224	0·584	2·53	0·956
80	47·36	972	0·293	4·197	1·962	0·350	0·0114	0·670	0·0231	0·641	2·19	0·966
90	70·11	965	0·423	4·205	1·992	0·311	0·0117	0·676	0·0240	0·696	1·93	0·976
100	101·3	958	0·598	4·216	2·028	0·278	0·0121	0·681	0·0249	0·750	1·723	0·986
125	232·1	939	1·30	4·254	2·147	0·219	0·0133	0·687	0·0272		1·358	1·047
150	476·0	917	2·55	4·310	2·314	0·180	0·0144	0·687	0·0300		1·133	1·110
175	892·7	893	4·60	4·389	2·542	0·153	0·0156	0·679	0·0334		0·990	1·185
200	1,555	862	7·87	4·497	2·843	0·133	0·0167	0·665	0·0375		0·902	1·270
225	2,550	833	12·8	4·648	3·238	0·1182	0·0179	0·644	0·0427		0·853	1·36
250	3,978	800	20·0	4·867	3·772	0·1065	0·0191	0·616	0·0495		0·841	1·45
275	5,949	758	30·6	5·202	4·561	0·0972	0·0202	0·582	0·0587		0·869	1·56
300	8,592	712	46·3	5·762	5·863	0·0897	0·0214	0·541	0·0719		0·955	1·74
325	12,060	654	70·4	6·861	8·440	0·0790	0·0230	0·493	0·0929		1·100	2·09
350	16,530	575	114	10·10	17·15	0·0648	0·0258	0·437	0·1343		1·50	3·29
360	18,670	528	144	14·6	25·1	0·0582	0·0275	0·400	0·168		2·11	3·89
374·2 (critical point)	22,120	315	315	∞	∞	0·045	0·045	0·24	0·24		∞	∞
To convert to	lbf/in²	lb/ft³		Btu/lb °R		lb/ft s		Btu/ft h °R		°R⁻¹	—	
Multiply by	0·1450	0·06243		0·2388		0·6719		0·5778		0·5556	1	

Table A7   Thermal properties of gases at atmospheric pressure (0·1013 MN/m²)

T, K	ρ, kg/m³	C_{p}, kJ/kg K	$\mu \times 10^5$, kg/m s	k, W/m K	**P**
\multicolumn{6}{c}{_Ammonia_   ($p_{\mathrm{c}} = 11\cdot3$ MN/m², $T_{\mathrm{c}} = 405$ K)}
223	0·9304	2·198	0·7254	0·0171	0·93
273	0·7600	2·177	0·9352	0·0220	0·90
323	0·6423	2·177	1·100	0·0270	0·88
373	0·5562	2·235	1·288	0·0327	0·87
423	0·4905	2·315	1·467	0·0391	0·87
473	0·4386	2·394	1·648	0·0467	0·84
\multicolumn{6}{c}{_Carbon dioxide_ ($p_{\mathrm{c}} = 7\cdot40$ MN/m², $T_{\mathrm{c}} = 304$ K)}
250	2·166	0·8039	1·258	0·0129	0·793
300	1·797	0·8709	1·495	0·0166	0·770
350	1·536	0·9002	1·720	0·0205	0·755
400	1·342	0·9420	1·931	0·0246	0·738
450	1·192	0·9797	2·133	0·0290	0·721
500	1·073	1·0132	2·325	0·0335	0·702
550	0·974	1·0467	2·507	0·0382	0·685
600	0·894	1·0760	2·683	0·0431	0·668
\multicolumn{6}{c}{_Carbon monoxide_ ($p_{\mathrm{c}} = 3\cdot55$ MN/m², $T_{\mathrm{c}} = 134$ K)}
250	1·3649	1·042	1·541	0·0214	0·750
300	1·1374	1·042	1·784	0·0253	0·737
350	0·9745	1·043	2·008	0·0288	0·728
400	0·8539	1·048	2·218	0·0323	0·722
450	0·7587	1·055	2·418	0·0355	0·718
500	0·6824	1·063	2·605	0·0386	0·718
550	0·6204	1·075	2·788	0·0416	0·721
600	0·5687	1·087	2·959	0·0445	0·724
\multicolumn{6}{c}{_Helium_ ($p_{\mathrm{c}} = 0\cdot229$ MN/m², $T_{\mathrm{c}} = 526$ K)}
33	1·4772	5·200	0·5015	0·0353	0·74
144	0·3385	5·200	1·254	0·0928	0·70
200	0·2435	5·200	1·565	0·1177	0·69
255	0·1907	5·200	1·817	0·1357	0·70
366	0·1328	5·200	2·305	0·1691	0·71
477	0·1021	5·200	2·749	0·1973	0·72
589	0·0828	5·200	3·113	0·2250	0·72
700	0·0703	5·200	3·474	0·2510	0·72
811	0·0602	5·200	3·817	0·2752	0·72
922	0·0529	5·200	4·135	0·2977	0·72
To convert to	lb/ft³	Btu/lb °R	lb/ft s	Btu/ft h °R	—
Multiply by	0·06243	0·2388	0·6719	0·5778	1

Table A7—*contd.*

$T,$ K	$\rho,$ kg/m^3	$C_p,$ kJ/kg K	$\mu \times 10^5,$ kg/m s	$k,$ W/m K	P
Hydrogen ($p_c = 1\cdot30$ MN/m^2, $T_c = 33\cdot2$ K)					
30	0·8475	10·84	0·1606	0·0228	0·759
50	0·5097	10·50	0·2516	0·0362	0·721
100	0·2458	11·23	0·4211	0·0665	0·712
150	0·1638	12·60	0·5595	0·0981	0·718
200	0·1227	13·54	0·6812	0·1282	0·719
250	0·0982	14·06	0·7918	0·1561	0·713
300	0·0819	14·32	0·8962	0·1817	0·706
350	0·0702	14·44	0·9953	0·2060	0·697
400	0·0614	14·49	1·086	0·2285	0·690
450	0·0546	14·50	1·178	0·2510	0·682
500	0·0492	14·51	1·264	0·2717	0·675
550	0·0448	14·53	1·347	0·2925	0·668
600	0·0409	14·54	1·428	0·3150	0·664
700	0·0349	14·57	1·589	0·3513	0·659
800	0·0306	14·68	1·739	0·3842	0·664
900	0·0272	14·82	1·878	0·4119	0·676
1,000	0·0245	14·97	2·016	0·4396	0·686
1,100	0·0223	15·16	2·145	0·4638	0·703
1,200	0·0205	15·37	2·275	0·4881	0·715
1,300	0·0189	15·58	2·408	0·5123	0·733
1,350	0·0184	15·64	2·443	0·5192	0·736
Nitrogen ($p_c = 3\cdot40$ MN/m^2, $T_c = 126$ K)					
100	3·482	1·072	0·6861	0·0094	0·786
200	1·711	1·043	1·295	0·0182	0·747
300	1·142	1·041	1·784	0·0262	0·713
400	0·854	1·046	2·198	0·0334	0·691
500	0·682	1·055	2·569	0·0398	0·684
600	0·568	1·076	2·910	0·0458	0·686
700	0·493	1·097	3·213	0·0512	0·691
800	0·428	1·123	3·483	0·0561	0·700
900	0·379	1·146	3·748	0·0607	0·711
1,000	0·341	1·168	3·999	0·0647	0·724
1,100	0·311	1·185	4·227	0·0685	0·736
1,200	0·285	1·204	4·449	0·0718	0·748
To convert to	lb/ft^3	Btu/lb °R	lb/ft s	Btu/ft h °R	—
Multiply by	0·06243	0·2388	0·6719	0·5778	1

Table A7—*contd.*

$T,$ K	$\rho,$ kg/m^3	$C_{\mathrm{p}},$ kJ/kg K	$\mu \times 10^5,$ kg/m s	$k,$ W/m K	**P**
\multicolumn{6}{c}{*Oxygen* ($p_{\mathrm{c}} = 5\cdot04$ MN/m^2, $T_{\mathrm{c}} = 154$ K)}					
100	3·9930	0·9479	0·776	0·0090	0·815
150	2·6198	0·8801	1·148	0·0137	0·773
200	1·9564	0·9131	1·484	0·0182	0·745
250	1·5622	0·9157	1·787	0·0226	0·725
300	1·3011	0·9203	2·062	0·0268	0·709
350	1·1136	0·9291	2·315	0·0307	0·702
400	0·9758	0·9420	2·553	0·0346	0·695
450	0·8684	0·9559	2·776	0·0383	0·694
500	0·7803	0·9722	2·991	0·0417	0·697
550	0·7098	0·9881	3·196	0·0452	0·700
600	0·6505	1·004	3·391	0·0483	0·704
To convert to	lb/ft^3	Btu/lb °R	lb/ft s	Btu/ft h °R	—
Multiply by	0·06243	0·2388	0·6719	0·5778	1

Table A8 Thermal properties of dry air at atmospheric pressure
(0·1013 MN/m²; 14·7 lbf/in²)

$T,$ K	$\rho,$ kg/m³	$C_p,$ kJ/kg K	$\mu \times 10^5,$ kg/m s	$k \times 10^5,$ kW/m K	P
175	2·017	1·0023	1·182	1·593	0·744
200	1·765	1·0025	1·329	1·809	0·736
225	1·569	1·0027	1·467	2·020	0·728
250	1·412	1·0031	1·599	2·227	0·720
275	1·284	1·0038	1·725	2·428	0·713
300	1·177	1·0049	1·846	2·624	0·707
325	1·086	1·0063	1·962	2·816	0·701
350	1·009	1·0082	2·075	3·003	0·697
375	0·9413	1·0106	2·181	3·186	0·692
400	0·8824	1·0135	2·286	3·365	0·688
450	0·7844	1·0206	2·485	3·710	0·684
500	0·7060	1·0295	2·670	4·041	0·680
550	0·6418	1·0398	2·849	4·357	0·680
600	0·5883	1·0511	3·017	4·661	0·680
650	0·5430	1·0629	3·178	4·954	0·682
700	0·5043	1·0750	3·332	5·236	0·684
750	0·4706	1·0870	3·482	5·509	0·687
800	0·4412	1·0987	3·624	5·774	0·690
850	0·4153	1·1101	3·763	6·030	0·693
900	0·3922	1·1209	3·897	6·276	0·696
950	0·3716	1·1313	4·026	6·520	0·699
1,000	0·3530	1·1411	4·153	6·754	0·702
1,050	0·3362	1·1502	4·276	6·985	0·704
1,100	0·3209	1·1589	4·396	7·209	0·707
1,150	0·3069	1·1670	4·511	7·427	0·709
1,200	0·2941	1·1746	4·626	7·640	0·711
1,250	0·2824	1·1817	4·736	7·849	0·713
1,300	0·2715	1·1884	4·846	8·054	0·715
1,350	0·2615	1·1946	4·952	8·253	0·717
1,400	0·2521	1·2005	5·057	8·450	0·719
1,500	0·2353	1·2112	5·264	8·831	0·722
1,600	0·2206	1·2207	5·457	9·199	0·724
1,700	0·2076	1·2293	5·646	9·554	0·726
1,800	0·1961	1·2370	5·829	9·899	0·728
1,900	0·1858	1·2440	6·008	10·233	0·730
To convert to	°R	lb/ft³	Btu/lb °R	lb/ft s	Btu/ft h °R	—
Multiply by	1·8	0·06243	0·2388	0·6719	577·8	1

Table A9 Gas constant, specific heat equations, and Sutherland's law for a number of gases

Gas (temperature range)	Gas constant, kJ/kg K	$C_p = C_1 + C_2 T^m + C_3 T^n$, kJ/kg K					$\mu = S_1 T^{3/2}/(S_2 + T)$, kg/m s	
		C_1	C_2	C_3	m	n	$S_1 \times 10^6$	S_2
Air (280–1,500 K)	0.287	0.917	2.58×10^{-4}	-3.98×10^{-8}	1	2	1.46	110
Ammonia (300–1,000 K)	0.489	1.52	1.94×10^{-3}	-1.79×10^{-7}	1	2	1.54	472
Carbon dioxide (300–3,500 K)	0.189	1.54	-345	4.14×10^4	-1	-2	1.56	233
Carbon monoxide (300–5,000 K)	0.297	1.42	-27.3	4.94×10^4	-1	-2	1.40	109
Helium (all temperatures)	2.08	5.19	0	0	0	0	1.52	98
Hydrogen (300–2,200 K)	4.124	12.0	2.16×10^{-3}	31.0	1	$-\frac{1}{2}$	0.65	71
Nitrogen (300–5,000 K)	0.297	1.42	-288	5.35×10^4	-1	-2	1.39	102
Oxygen (300–2,800 K)	0.260	1.51	-16.8	111	$-\frac{1}{2}$	-1	1.65	110

Table A10 Coefficients of diffusion and Schmidt numbers for binary mixtures

Diffusing constituent		Medium of diffusion	T, °C	$D \times 10^5$, m^2/s	Sc
HCl	} liquid	H_2O	0	0·535	0·81
NH_3		H_2O	4	0·121	1·5
NH_3		Air	0	2·165	0·634
CO_2		Air	0	1·198	1·14
CO_2		H_2	18	6·048	0·158
Hg		N_2	19	325·1	0·00424
O_2		Air	0	1·533	0·895
O_2		N_2	12	2·025	0·681
H_2		Air	0	5·472	0·250
H_2		O_2	14	7·198	0·182
H_2		N_2	12·5	7·376	0·187
H_2O		Air	8	2·062	0·615
H_2O	gas	Air	16	2·815	0·488
C_6H_6		Air	0	0·750	1·83
C_6H_6		CO_2	0	0·527	1·37
C_6H_6		H_2	0	2·936	3·26
CS_2		Air	20	0·881	1·68
Ether		Air	20	0·769	1·93
Ethyl alcohol		Air	0	1·013	1·36
Ethyl alcohol		Air	40	1·180	1·45

Table A11 Normal emissivity, ε_n, and total emissivity, ε, for solid materials

Material	T, °C	ε_n	ε
Gold, polished	130	0·018	
Gold, polished	400	0·022	
Silver	20	0·020	
Copper, polished	20	0·030	
Copper, lightly oxidized	20	0·037	
Copper, scraped	20	0·070	
Copper, black oxidized	20	0·78	
Copper, oxidized	130	0·76	0·725
Aluminium, bright rolled	170	0·039	0·049
Aluminium, bright rolled	500	0·050	
Aluminium paint	100	0·20–0·40	
Silumin, cast, polished	150	0·186	
Nickel, bright matte	100	0·041	0·046
Nickel, polished	100	0·045	0·053
Manganin, bright rolled	118	0·048	0·057
Chrome, polished	150	0·058	0·071
Iron, bright etched	150	0·128	0·158
Iron, bright abrased	20	0·24	
Iron, red rusted	20	0·61	
Iron, hot rolled	20	0·77	
Iron, hot rolled	130	0·60	
Iron, hot cast	100	0·80	
Iron, heavily rusted	20	0·85	
Iron, heat-resistant oxidized	80	0·613	
Iron, heat-resistant oxidized	200	0·639	
Zinc, grey oxidized	20	0·23–0·28	
Lead, grey oxidized	20	0·28	
Bismuth, bright	80	0·340	0·366
Corundum, emery rough	80	0·855	0·84
Clay, fired	70	0·91	0·86
Lacquer, white	100	0·925	
Red lead	100	0·93	
Enamel, lacquer	20	0·85–0·95	
Lacquer, black matte	80	0·970	
Bakelite lacquer	80	0·935	
Brick, mortar, plaster	20	0·93	
Porcelain	20	0·92–0·94	
Glass	90	0·940	0·876
Ice, smooth, water	0	0·966	0·918
Ice, rough crystals	0	0·985	
Waterglass	20	0·96	
Paper	95	0·92	0·89
Wood, beech	70	0·935	0·91
Tar paper	20	0·93	

Table A12   Bessel functions of the first and second kinds

x	$J_0(x)$	$J_1(x)$	$Y_0(x)$	$Y_1(x)$
0·0	+1·0000	+0·0000	$-\infty$	$-\infty$
0·2	+0·9900	+0·0995	−1·0811	−3·324
0·4	+0·9604	+0·1960	−0·6060	−1·781
0·6	+0·9120	+0·2867	−0·3085	−1·260
0·8	+0·8463	+0·3688	−0·0868	−0·9781
1·0	+0·7652	+0·4401	+0·0883	−0·7812
1·2	+0·6711	+0·4983	+0·2281	−0·6211
1·4	+0·5669	+0·5420	+0·3379	−0·4792
1·6	+0·4554	+0·5699	+0·4204	−0·3476
1·8	+0·3400	+0·5815	+0·4774	−0·2237
2·0	+0·2239	+0·5767	+0·5104	−0·1070
2·2	+0·1104	+0·5560	+0·5208	+0·0015
2·4	+0·0025	+0·5202	+0·5104	+0·1005
2·6	−0·0968	+0·4708	+0·4813	+0·1884
2·8	−0·1850	+0·4097	+0·4359	+0·2636
3·0	−0·2601	+0·3391	+0·3769	+0·3247
3·2	−0·3202	+0·2613	+0·3071	+0·3707
3·4	−0·3643	+0·1792	+0·2296	+0·4010
3·6	−0·3918	+0·0955	+0·1477	+0·4154
3·8	−0·4026	+0·0128	+0·0645	+0·4141
4·0	−0·3972	−0·0660	−0·0169	+0·3979
4·2	−0·3766	−0·1386	−0·0938	+0·3680
4·4	−0·3423	−0·2028	−0·1633	+0·3260
4·6	−0·2961	−0·2566	−0·2235	+0·2737
4·8	−0·2404	−0·2985	−0·2723	+0·2136
5·0	−0·1776	−0·3276	−0·3085	+0·1479
5·2	−0·1103	−0·3432	−0·3313	+0·0792
5·4	−0·0412	−0·3453	−0·3402	+0·0101
5·6	+0·0270	−0·3343	−0·3354	−0·0568
5·8	+0·0917	−0·3110	−0·3178	−0·1192
6·0	+0·1507	−0·2767	−0·2882	−0·1750
6·2	+0·2018	−0·2329	−0·2483	−0·2223
6·4	+0·2433	−0·1816	−0·2000	−0·2596
6·6	+0·2740	−0·1250	−0·1452	−0·2858
6·8	+0·2931	−0·0625	−0·0864	−0·3002
7·0	+0·3001	−0·0047	−0·0260	−0·3027
7·2	+0·2951	+0·0543	+0·0339	−0·2934
7·4	+0·2786	+0·1096	+0·0907	−0·2732
7·6	+0·2516	+0·1592	+0·1424	−0·2428
7·8	+0·2554	+0·2014	+0·1872	−0·2039
8·0	+0·1717	+0·2346	+0·2235	−0·1581
8·2	+0·1222	+0·2580	+0·2501	−0·1072

Table A12—*contd.*

x	$J_0(x)$	$J_1(x)$	$Y_0(x)$	$Y_1(x)$
8·4	+0·0692	+0·2708	+0·2662	−0·0535
8·6	+0·0146	+0·2728	+0·2715	−0·0011
8·8	−0·0392	+0·2641	+0·2659	+0·0544
9·0	−0·0903	+0·2453	+0·2499	+0·1043
9·2	−0·1368	+0·2147	+0·2245	+0·1491
9·4	−0·1768	+0·1816	+0·1907	+0·1871
9·6	−0·2090	+0·1395	+0·1502	+0·2171
9·8	−0·2323	+0·0928	+0·1045	+0·2379
10·0	−0·2459	+0·0435	+0·0557	+0·2490
10·2	−0·2496	−0·0066	+0·0056	+0·2502
10·4	−0·2434	−0·0555	−0·0437	+0·2416
10·6	−0·2276	−0·1012	−0·0904	+0·2236
10·8	−0·2032	−0·1422	−0·1326	+0·1973
11·0	−0·1712	−0·1768	−0·1688	+0·1637
11·2	−0·1330	−0·2039	−0·1977	+0·1243
11·4	−0·0902	−0·2225	−0·2183	+0·0807
11·6	−0·0446	−0·2320	−0·2299	+0·0348
11·8	−0·0020	−0·2323	−0·2322	+0·0118
12·0	+0·0477	−0·2234	−0·2252	−0·0571
12·2	+0·0908	−0·2060	−0·2095	−0·0994
12·4	+0·1296	−0·1807	−0·1858	−0·1371
12·6	+0·1626	−0·1487	−0·1551	−0·1689
12·8	+0·1887	−0·1114	−0·1187	−0·1935
13·0	+0·2069	−0·0703	−0·0782	−0·2101
13·2	+0·2167	−0·0271	−0·0352	−0·2182
13·4	+0·2177	+0·0166	+0·0085	−0·2176
13·6	+0·2101	+0·0590	+0·0512	−0·2084
13·8	+0·1943	+0·0984	+0·0913	−0·1912
14·0	+0·1711	+0·1334	+0·1272	−0·1666
14·2	+0·1414	+0·1626	+0·1575	−0·1359
14·4	+0·1065	+0·1850	+0·1812	−0·1003
14·6	+0·0679	+0·1999	+0·1974	−0·0612
14·8	+0·0271	+0·2066	+0·2056	−0·0202
15·0	−0·0142	+0·2051	+0·2055	+0·0211
15·2	−0·0544	+0·1955	+0·1972	+0·0609
15·4	−0·0919	+0·1784	+0·1813	+0·0979
15·6	−0·1253	+0·1544	+0·1584	+0·1305
15·8	−0·1533	+0·1247	+0·1295	+0·1575

Table A13 Modified Bessel functions of the first and second kind

x	$I_0(x)$	$I_1(x)$	$(2/\pi)\,K_0(x)$	$(2/\pi)\,K_1(x)$
0·0	1·0000	0·0000	∞	∞
0·2	1·0100	0·1005	1·116	3·041
0·4	1·0404	0·2040	0·7095	1·391
0·6	1·0920	0·3137	0·4950	0·8294
0·8	1·1665	0·4329	0·3599	0·5486
1·0	1·2661	0·5652	0·2680	0·3832
1·2	1·3937	0·7147	0·2028	0·2768
1·4	1·5534	0·8861	0·1551	0·2043
1·6	1·7500	1·0848	0·1197	0·1532
1·8	1·9896	1·3172	$0·9290 \times 10^{-1}$	0·1163
2·0	2·2796	1·5906	0·7251	$0·8904 \times 10^{-1}$
2·2	2·6291	1·9141	0·5683	0·6869
2·4	3·0493	2·2981	0·4470	0·5330
2·6	3·5533	2·7554	0·3527	0·4156
2·8	4·1573	3·3011	0·2790	0·3254
3·0	4·8808	3·9534	0·2212	0·2556
3·2	5·7472	4·7343	0·1757	0·2014
3·4	6·7848	5·6701	0·1398	0·1592
3·6	8·0277	6·7028	0·1114	0·1261
3·8	9·5169	8·1404	$0·8891 \times 10^{-2}$	$0·9999 \times 10^{-2}$
4·0	11·302	9·7595	0·7105	0·7947
4·2	13·443	11·706	0·5684	0·6327
4·4	16·010	14·046	0·4551	0·5044
4·6	19·093	16·863	0·3648	0·4027
4·8	22·794	20·253	0·2927	0·3218
5·0	27·240	24·336	0·2350	0·2575
5·2	32·584	29·254	0·1888	0·2062
5·4	39·009	35·182	0·1518	0·1653
5·6	46·738	42·328	0·1221	0·1326
5·8	56·038	50·946	$0·9832 \times 10^{-3}$	0·1064
6·0	67·234	61·342	0·7920	$0·8556 \times 10^{-3}$
6·2	80·718	73·886	0·6382	0·6879
6·4	96·962	89·026	0·5146	0·5534
6·6	116·54	107·31	0·4151	0·4455
6·8	140·14	129·38	0·3350	0·3588
7·0	168·59	156·04	0·2704	0·2891
7·2	202·92	188·25	0·2184	0·2331
7·4	244·34	227·18	0·1764	0·1880
7·6	294·33	274·22	0·1426	0·1517
7·8	354·69	331·10	0·1153	0·1424
8·0	427·56	399·87	$0·9325 \times 10^{-4}$	$0·9891 \times 10^{-4}$
8·2	515·59	483·05	0·7543	0·7991
8·4	621·94	583·66	0·6104	0·6458
8·6	750·46	705·38	0·4941	0·5220
8·8	905·80	852·66	0·4000	0·4221
9·0	1,093·6	1,030·9	0·3239	0·3415
9·2	1,320·7	1,246·7	0·2624	0·2763
9·4	1,595·3	1,507·9	0·2126	0·2236
9·6	1,927·5	1,824·1	0·1722	0·1810
9·8	2,329·4	2,207·1	0·1396	0·1465
10·0	2,815·7	2,671·0	0·1131	0·1187

REFERENCES TO APPENDIX

1. Eckert, E. R. G. and R. M. Drake. *Heat and Mass Transfer*. McGraw-Hill, New York, 1959.
2. Mayhew, Y. R. and G. F. C. Rogers. *Thermodynamic and Transport Properties of Fluids, SI units* (2nd edn). Blackwell, Oxford, 1969.
3. Atomic Energy Commission. *Liquid Metals Handbook* (2nd edn). Dept of U.S. Navy, NAVEXOS P-733, 1954.
4. *International Critical Tables*. Published for U.S. National Research Council by McGraw-Hill, New York, 1933.
5. *Handbook of Chemistry and Physics* (38th edn). Chemical Rubber Publishing Company, Cleveland, Ohio, 1956.
6. Haywood, R. W. *Thermodynamic Tables in SI Units*. Cambridge University Press, 1968.
7. *The 1967 Steam Tables*. Published for the Electrical Research Association by Arnold, London, 1967.
8. Comings, E. W., B. J. Mayland and R. S. Egly. The viscosity of gases at high pressures. *Univ. Ill. Bull.* (Series No. 354), 1944.
9. Comings, E. W. and M. F. Nathan. Thermal conductivity of gases at high pressures. *Ind. Engng Chem.*, 1947, **39**, 964.
10. *Thermodynamic Functions of Gases* (Din, F., ed.). Butterworths, London, 1962.
11. Hilsenrath, J., *et al. Table of Thermal Properties of Gases*. U.S. N.B.S. Circular 564, Pergamon Press, Oxford, 1960.
12. *Handbook of Supersonic Aerodynamics*, vol. 5. Bureau of Ordnance, Dept of U.S. Navy, 1953.
13. Chapman, A. J. *Heat Transfer* (2nd edn). Macmillan Company, New York, 1967.
14. McAdams, W. H. *Heat Transmission* (3rd edn). McGraw-Hill, London, 1954.
15. McLachlan, N. W. *Bessel Functions for Engineers* (2nd edn). Oxford University Press, London, 1955.

Index